Modern Electric, Hybrid Electric, and Fuel Cell Vehicles

Third Edition

International Student Edition

Modern Electric, Hybrid Electric, and Fuel Cell Vehicles

Third Edition

International Student Edition

Mehrdad Ehsani
Yimin Gao
Stefano Longo
Kambiz M. Ebrahimi

CRC Press
Taylor & Francis Group
Boca Raton London New York

CRC Press is an imprint of the
Taylor & Francis Group, an **informa** business

CRC Press
Taylor & Francis Group
6000 Broken Sound Parkway NW, Suite 300
Boca Raton, FL 33487-2742

© 2019 by Taylor & Francis Group, LLC
CRC Press is an imprint of Taylor & Francis Group, an Informa business

No claim to original U.S. Government works

Printed on acid-free paper

International Standard Book Number-13: 978-1-138-33049-8 (International Student Edition Paperback)

To my Wonderful Wife, Zohreh

Mehrdad Ehsani

To my Wife Anni, and my Daughter, Yuan

Yimin Gao

To my Mum, Dad and Little Brother

Stefano Longo

To my Wife and Daughter

Kambiz Ebrahimi

Contents

Preface

Electric and Hybrid Electric vehicles are now well known products in the market and are accepted internationally. However, their full potential for penetrating the automobile market is not yet fulfilled, even with the ever expanding awareness of the global warming problem due to the fossil fuel use. This is, in part, due to the low cost and ever abundance of fossil fuels for the conventional internal combustion engine vehicles. The abundance of hydrocarbon fuels is not going the change for decades and perhaps for centuries. Therefore, the electric and hybrid electric vehicles will dominate the automobile market only if they provide better and more appropriate products for the present and future needs of the automobile user. For example, a very low cost electric vehicle can dominate the market in the developing countries with no fossil fuel resourses of their own, where conventional cars are too expensive for the middle class family to purchase and operate. Further, a conventional vehicle, with full size internal combustion engine, can be optimized for performance, fuel economy, emissions, and cost, by a small traction motor/generator hybridization. Such a vehicle can be a superior product, in comparison to the conventional cars, with a small incremental cost, for the developed world markets.

The present third edition of our book contributes to the above aims by introducing appropriate technologies and design methodologies. Further, the present edition benefits from the suggestions of many readers, students and academic adopters of the previous editions of this book, whose suggestions is gratefully acknowledged.

The development of the internal combustion engine automobiles is one of the greatest achievements of the modern technology. However, the highly developed automotive industry and the increasingly large number of automobiles in use around the world are causing serious problems for the environment and hydrocarbon resources. The deteriorating air quality and global warming issues are becoming serious threats to modern life. Progressively more rigorous emissions and fuel efficiency standards are stimulating aggressive development of safer, cleaner, and more efficient vehicles. It is now well recognized that electric, hybrid electric, and fuel cell powered drive train technologies are the most promising vehicle solutions for the foreseeable future.

To meet this challenge, an increasing number of engineering schools, in the US and around the world, have initiated academic programs in advanced energy and vehicle technologies at undergraduate and graduate levels. We stared our first graduate course, In 1998, on "Advanced Vehicle Technologies—Design Methodology of Electric and Hybrid Electric Vehicles" for students in mechanical and electrical engineering at Texas A&M University. While preparing the lectures for this course, we found that although there is a wealth of information in the form of technical papers and reports, there was no rigorous and comprehensive textbook for students and professors that may wish to offer such a course. Furthermore, practicing engineers also needed a systematic reference book to fully understand the essentials of this new technology. The first edition of this book was our attempt to fill this need. The present third edition introduces newer topics and deeper treatments to the previous editions.

The book deals with the fundamentals, theoretical bases, and design methodologies of conventional internal combustion engines (ICE) vehicles, electric vehicles (EV), hybrid electric vehicles (HEV), and fuel cell vehicles (FCV). It comprehensively covers vehicle performance characteristics, configurations, control strategies, design methodologies, modeling, and simulation for modern vehicles with mathematical rigor. It includes drive train architecture analysis, ICE based drive trains, EV and HEV configurations, electric propulsion systems, series/parallel/mild hybrid electric drive train design methodologies, energy storage systems, regenerative braking, fuel cells and their applications in vehicles, and fuel cell hybrid electric drive train design. The book perspective is from the overall drive train system and not just individual components. The design methodology is described in mathematical terms, step by step. Furthermore, in explaining the design methodology of each drive train, design examples are presented with simulation results.

More specifically, the third edition contains many corrections and updates of the material in the first and second editions. Four new chapters and one appendix have been added. They are Chapter 4: Vehicle Transmission; Chapter 18: Design of Full-Size Engine HEV with Optimal Hybridization Ratio; Chapter 19: A User Guide for the Multiobjective Optimization Toolbox; and Chapter 20: Power Train Optimization, plus the Appendix: Technical Overview of Toyota Prius. In addition, plenty of new materials have been added to the old chapters. All these new contributions to the third edition make it more complete and useful to the reader.

Overall, this book consists of twenty chapters and one appendix. In Chapter 1, the social and environmental importance of modern transportation is discussed. This includes the air pollution, global warming, and petroleum resource depletion issues associated with the development of the modern transportation. In this chapter, the impact of future vehicle technologies on the oil supplies is analyzed. The results are helpful for the development strategies of the next generation vehicles. In addition, the development history of EV, HEV, and FCV is briefly reviewed.

In Chapter 2, the basic understanding of vehicle performance, power plant characteristics, transmission characteristics, and the equations used to describe the vehicle performance are introduced. The main purpose of this chapter is to provide the basic knowledge that is necessary for vehicle drive train design. As an improvement to the first edition, the brake system and its design and performance material has been strengthened in order to provide a more solid base for the hybrid brake system design in EV, HEV and FCV.

In Chapter 3, major operating characteristics of different heat engines are introduced. As the primary power plant, the engine is the most important subsystem in conventional and hybrid drive train systems. Full understanding of the characteristics of engine is necessary in design and control of conventional as well as hybrid electric vehicles.

In Chapter 4, vehicle transmission for conventional and hybrid electric vehicle is introduced. Most kinds of conventional and advanced transmission systems are presented and analyzed.

In Chapter 5, electric vehicles are introduced. This chapter mainly includes design of the electric propulsion system and its energy storage device, design of the traction motor and its transmission, methodology of prediction of vehicle performance, and system simulation results.

In Chapter 6, the basic concept of hybrid traction is first established. Then, various configurations of hybrid electric vehicles are discussed. These include series hybrid, parallel hybrid, torque coupling and speed coupling hybrids, and other configurations. The main operating characteristics of these configurations are also presented.

In Chapter 7, several electric power plants are introduced. These include DC, AC, permanent magnet brushless DC, and switched reluctance motor drives. From traction system point of view, their basic structure, operating principles, control and operational characteristics are described.

In Chapter 8, the design methodology of series hybrid electric drive trains is presented. This chapter is focused on the system oriented design of the engine and energy storage, the traction motor, the transmission, the control strategy, and the power converters. A design example is also provided. As an improvement to the first edition, various power converter configurations have been added.

In Chapter 9, a design methodology of parallel hybrid electric drive trains is provided. This chapter includes driving patterns and driving mode analysis, control strategy, design of the major components, e.g., the engine, the energy storage, and the transmission, and vehicle performance simulation. In addition to the first edition, a constrained engine on and off control strategy, fuzzy logic control strategy and the concept of control optimization based on dynamic programming have been added.

In Chapter 10, the operating characteristics, design methodology, and control strategies of series-parallel hybrid drive train is presented. This was a new chapter in the second edition.

In Chapter 11, the design and control principles of the plug-in hybrid vehicle is introduced. This chapter mainly addresses the charge sustaining hybrid drive train with regard to the drivetrain control strategy, energy storage design, and electric motor design. This was also a new chapter.

In Chapter 12, a design methodology of mild hybrid drive trains is introduced with the two major configurations of parallel torque coupling and series/parallel, torque-speed coupling. This chapter is focused on the operational analysis, control system development, and system simulation.

In Chapter 13, different energy storage technologies are introduced, including batteries, ultra-capacitors, and flywheels. The discussion focuses on the power and energy capacities. The concept of hybrid energy storage is also introduced in this chapter.

In Chapter 14, the design and control principles of hybrid brake systems are introduced. Brake safety and the recoverable energy are the main concerns. The available braking energy characteristics, with regard to vehicle speed, and the braking power in typical driving cycles are investigated. The brake force distribution on the front and rear wheels is discussed for guaranteeing the vehicle braking performance for safety. Furthermore, this chapter discusses the important issue of distributing the total braking force between the mechanical and the electrical regenerative brakes. Two advanced hybrid brake systems, including their control strategies are introduced. This chapter has been rewritten based on our recent research.

In Chapter 15, different fuel cell systems are described, with the focus on their operating principles and characteristics, various technologies, and their fuels. Specifically, vehicle applications of fuel cells are explained.

In Chapter 16, a systematic design of fuel cell hybrid drive trains is introduced. First, the concept of fuel cell hybrid vehicles is established. Then, their operating principles and drivetrain control systems are analyzed. Lastly, a design methodology is provided, focusing on the system design of the fuel cell, the electric propulsion system, and the energy storage system. A design example and its corresponding performance simulation results are provided.

In Chapter 17, a design methodology of an off-road tracked series hybrid vehicle is developed. The discussion focuses on the motion resistance calculation on soft grounds, traction motor system design, the engine/generator system design and the peaking power source system design.

Chapters 18, 19, and 20 are new chapters for the third edition. They introduce the new full size engine optimal hybrid vehicle design concept, and general hybridization of vehicle power plant and its associated software.

A case study appendix is included in the present edition. This is an overview of the Toyota Prius hybrid system. The purpose is to give the reader a practical example of the architecture, operational modes, control system, among other things, of a commercial hybrid electric drive train.

This book is suitable for a graduate or senior-level undergraduate course in advanced vehicles. Software (©2018 by Ganesh Mohan, Francis Assadian, Marcin Stryszowski, and Stefano Longo) designed to accompany the material in this book will be hosted on the book's CRC Press website: www.crcpress.com/9781498761772. Depending on the background of the students in different disciplines such as mechanical or electrical engineering, course instructors have the flexibility of choosing the specialized material to suit their lectures. This text has been taught at Texas A&M University as a graduate level course for two decades. The manuscript of this text has been revised many times and over many years, based on the comments and feedback from the students in our course. We are grateful to our students for their help.

This book is also an in-depth resource and a comprehensive reference in modern automotive systems for engineers, students, researchers, and other professionals who are working in the automotive related industries, as well those in the government, and academia.

In addition to the work by others, many of the technologies and advances presented in this book are the collection of many years of research and development by the authors and other members of the Advanced Vehicle Systems Research Program at Texas A&M University. We are grateful to all the dedicated staff of the Advanced Vehicle Systems Research group and the Power Electronics and Motor Drives group at Texas A&M, who made great contributions to this book.

We would like to acknowledge the efforts and assistance of the staff of Taylor & Francis Group, especially Ms. Nora Konopka. Last but not least, we thank our families for their patience and support during the long effort in the writing of this book.

Mehrdad Ehsani
Yimin Gao
Stefano Longo
Kambiz Ebrahimi
January, 2018

MATLAB® and Simulink® are registered trademark of The MathWorks, Inc. For product information, please contact:
The MathWorks, Inc.
3 Apple Hill Drive
Natick, MA 01760-2098 USA
Tel: 508-647-7000
Fax: 508-647-7001
E-mail: info@mathworks.com
Web: www.mathworks.com

Acknowledgments

Many of our past and present students and research staff have contributed to the work that has resulted in this book, over the past two decades. In particular, we acknowledge the work of Lin Lai who contributed to Chapter 18 through his PhD dissertation at Texas A&M University. We also acknowledge the help of our PhD students Nima Ersahd, Ahmet Yeksan, Own Golden, and Yiqi Wang, all of Texas A&M University Advanced Vehicle Systems Research Program.

For the chapters "Optimization Techniques for Power Train Topology and Component Sizing" and "A User Guide for the Multiobjective Optimization Toolbox" and for the software provided with this book, the authors would like to acknowledge the contribution of Dr. Ganesh Mohan, Professor Francis Assadian, and Mr Marcin Staszowski. Dr. Mohan was a PhD student at Cranfield University, UK, when he developed the techniques for power train optimization and the MATLAB®-based software that is available with this book. He is now an engineer at Jaguar Land Rover. Professor Assadian, his primary supervisor, directed most of this work before accepting a professorship at the University of California Davis, USA. Mr Staszowski is a current PhD student at Cranfield University. He helped develop the material for the two aforementioned chapters and improved the software functionalities. We would like to express our gratitude to them for the exceptional contribution they have made with their work and for allowing us to publish it.

Acknowledgments

Many of our past and present students and research staff have contributed to the work that has resulted in this book. Over the past two decades, in particular, we acknowledge the work of Lee Loh, as described in Chapter 14 through his PhD dissertation at Texas A&M. Jasbir S. Arora also acknowledges the help of four PhD students, Nina Pradit, Aloni Palan, Qian Chen, and Tzu Wang, all of Texas A&M University Advanced Workstation Laboratory at Texas A&M.

For the chapter "Optimization Techniques," Dr. Naval Kumar and C. Luqman Aziz would like to thank for the Multidisciplinary Optimization Laboratory and for the suggestion and assistance with this book, the authors would like to acknowledge the contribution of Dr. Ganesh Mohan, Professor Hamza Assadian, and Mr. Marcus Stasewski. Dr. Mohan was a PhD student at Cranfield University, UK, when he developed the techniques for power train optimization and the MATLAB-based software that is available with this book. He is now an engineer at Jaguar Land Rover. Professor Assadian, his primary supervisor, directed most of this work before accepting a professorship at the University of California, Davis, USA. Mr. Stasewski is a current PhD student at Cranfield University. He helped develop the material for the various unconventional chapters and improved the software functionalities. We would like to express our gratitude to them for the exceptional contribution they have made with their work and for allowing us to publish it.

Authors

 **Mehrdad Ehsani** is the Robert M. Kennedy Professor of Electrical Engineering at Texas A&M University. From 1974 to 1981, he was a research engineer at the Fusion Research Center, University of Texas, and with Argonne National Laboratory, Argonne, Illinois, as a resident research associate. Since 1981 he has been at Texas A&M University, College Station, Texas, where he is now an endowed professor of electrical engineering and director of the Advanced Vehicle Systems Research Program and the Power Electronics and Motor Drives Laboratory. He is the author of over 400 publications in pulsed-power supplies, high-voltage engineering, power electronics, motor drives, advanced vehicle systems, and sustainable energy engineering. He is the recipient of several Prize Paper Awards from the IEEE-Industry Applications Society, as well as over 100 other international honors and recognitions, including the IEEE Vehicular Society 2001 Avant Garde Award for "contributions to the theory and design of hybrid electric vehicles." In 2003, he was selected for the IEEE Undergraduate Teaching Award "for outstanding contributions to advanced curriculum development and teaching of power electronics and drives." In 2005, he was elected Fellow of the Society of Automotive Engineers (SAE). He is the coauthor of 17 books on power electronics, motor drives, and advanced vehicle systems. He has over 30 granted or pending U.S. and EU patents. His current research work is in power electronics, motor drives, hybrid vehicles and their control systems, and sustainable energy engineering.

Dr. Ehsani has been a member of IEEE Power Electronics Society (PELS) AdCom, chairman of the PELS Educational Affairs Committee, chairman of the IEEE-IAS Industrial Power Converter Committee, and chairman of the IEEE Myron Zucker Student-Faculty Grant program. He was the general chair of the IEEE Power Electronics Specialist Conference in 1990. He is the founder of IEEE Power and Propulsion Conference, the founding chairman of IEEE VTS Vehicle Power and Propulsion, and chairman of Convergence Fellowship Committees. In 2002, he was elected to the board of governors of VTS. He has also served on the editorial board of several technical journals and was the associate editor of *IEEE Transactions on Industrial Electronics* and *IEEE Transactions on Vehicular Technology*. He is a life time fellow of IEEE, a past IEEE Industrial Electronics Society and Vehicular Technology Society distinguished speaker, and IEEE Industry Applications Society and Power Engineering Society distinguished lecturer. He is also a registered professional engineer in the state of Texas.

Yimin Gao received his BS, MS, and PhD degrees in mechanical engineering (major in the development, design, and manufacturing of automotive systems) in 1982, 1986, and 1991, respectively, all from Jilin University of Technology, Changchun, Jilin, China. From 1982 to 1983, he worked as a vehicle design engineer at Dongfeng Motor Company, Shiyan, Hubei, China. He finished a layout design of a 5-ton truck (EQ144) and participated in prototyping and testing. From 1983 to 1986, he was a graduate student in Automotive Engineering College of Jilin University of Technology, Changchun, Jilin, China. His working field was the improvement of vehicle fuel economy by optimal matching of engine and transmission.

From 1987 to 1992, he was a PhD student in the Automotive Engineering College of Jilin University of Technology, Changchun, Jilin, China. During this period, he worked on research and development of legged vehicles, which can potentially operate in harsh environments where mobility is difficult for wheeled vehicles. From 1991 to 1995, he was an associate professor and automotive design engineer in the Automotive Engineering College of Jilin University of Technology. In this period, he taught undergraduate students a course in automotive theory and design several times and graduate students a course in automotive experiment technique twice. Meanwhile, he also conducted vehicle performance, chassis, and components analysis, and he conducted automotive design, including chassis design, power train design, suspension design, steering system design, and brake design.

He joined the Advanced Vehicle Systems Research Program at Texas A&M University in 1995 as a research associate. Since then, he has been working in this program on research and development of electric and hybrid electric vehicles. His research areas are mainly in the fundamentals, architecture, control, modeling, design, and major components of electric and hybrid electric drivetrains. He is a member of SAE.

Stefano Longo, after graduating in electrical and electronic engineering, received his MSc in control systems from the University of Sheffield, UK, in 2007 and his PhD, also in control systems, from the University of Bristol, UK, in 2010. His PhD thesis was awarded the Institution of Engineering and Technology (IET) Control and Automation Prize for significant achievements in the area of control engineering. In 2010, he was appointed to the position of research associate at Imperial College London, UK, in the Control and Power Group within the Department of Electrical and Electronic Engineering, where he worked at the intersection of control systems design and hardware implementation. In 2012, he was appointed lecturer (assistant professor) in vehicle electrical and electronic systems at Cranfield University, UK, within the Automotive Engineering Department (now called the Advanced Vehicle Engineering Centre). From 2012 to 2016, he was also an honorary research associate at Imperial College London. In 2017, Dr. Longo was promoted to the position of senior lecturer (associate professor) in automotive control and optimization, and he has been the course director for the MSc in automotive mechatronics since 2014.

Dr. Longo has published over 70 peer-reviewed research articles and another book, *Optimal and Robust Scheduling for Networked Control Systems*, (CRC Press, 2017). He teaches various postgraduate courses in automotive mechatronics, optimization, and control, supervises PhD students, and conducts academic research and consultancy.

Dr. Longo is a senior member of the IEEE, an associate editor of the Elsevier journal *Mechatronics*, a technical editor and reviewer for many IEEE and IFAC journals, a chartered engineer and elected executive member of the IET Control & Automation Network, a member of the IFAC technical committee on Mechatronic Systems and Automotive Control, and a fellow of the Higher Education Academy.

Kambiz M. Ebrahimi, PhD, received his BSc degree in mechanical engineering from Plymouth Polytechnic, UK, his M.Eng degree in systems engineering from UWIST, University of Wales, and his PhD in dynamics and mathematical modeling from Cardiff University, UK.

Currently, he is professor of advanced propulsion in the Aeronautical and Automotive Engineering Department at Loughborough University, UK. Before joining Loughborough, he worked as a research assistant at the University of Wales working on model-based condition monitoring on an EU project and at the University of Bradford on distributed—lumped modeling and least effort control strategies. Subsequently, he became a lecturer, reader, and professor of mechanical engineering at the University of Bradford, UK.

His main research interests are in systems and control theory; multivariable and large-scale systems; modeling and characterization of mechatronic systems; energy management and control of hybrid power trains; system monitoring, fault diagnosis, and turbomachinery tip-timing; hybrid, electric, and L-category vehicles. He is the author or coauthor of more than 100 articles in national and international journals and conferences.

Dr. Ebrahimi is a chartered mechanical engineer and member of ASME and SAE and has served as the chair and organizer of the Powertrain Modelling and Control Conference since 2012; a member of the editorial board of *International Journal of Powertrains* since 2012; and the Organizer of Meeting the Challenges in Powertrain Testing, in 2009. He is also a member of the editorial board of the *Proceedings of the Institution of Mechanical Engineers, Part K, Journal of Multi-body Dynamics*, as well as the coeditor of the 1998 book *Application of Multi-Variable System Techniques*, Professional Engineering Publishing, and coeditor of the 2000 book *Multi-Body Dynamics*, Professional Engineering Publishing.

He is actively involved in research collaboration with industry through contacts such as AVL, Ford Motor Company, Cummins Turbocharger Technologies, Jaguar, and Land Rover.

1

Environmental Impact and History of Modern Transportation

The development of internal combustion (IC) engine vehicles, and especially automobiles, is one of the greatest achievements of modern technology. Automobiles have made great contributions to the growth of modern society by satisfying many of the needs for mobility in everyday life. The rapid development of the automotive industry, unlike that of any other industry, has prompted the progress of human beings from a primitive society to a highly developed industrial one. The automobile industry and the other industries that serve it constitute the backbone of the world's economy and employ the greatest share of the working population.

However, the large number of automobiles in use around the world has caused and continues to cause serious problems for the environment and human life. Air pollution, global warming, and the rapid depletion of the Earth's petroleum resources are now problems of paramount concern.

In recent decades, the research and development activities related to transportation have emphasized the development of high-efficiency, clean, and safe transportation. Electric vehicles (EVs), hybrid electric vehicles (HEVs), and fuel cell vehicles have been typically proposed to replace conventional vehicles in the near future.

This chapter reviews the problems of air pollution, gas emissions causing global warming, and petroleum resource depletion. It also briefly reviews the history of EVs, HEVs, and fuel cell technology.

1.1 Air Pollution

At present, all vehicles rely on the combustion of hydrocarbon (HC) fuels to derive the energy necessary for their propulsion. Combustion is a reaction between the fuel and the air that releases heat and combustion products. The heat is converted to mechanical power by an engine, and the combustion products are released into the atmosphere. An HC is a chemical compound with molecules made up of carbon and hydrogen atoms. Ideally, the combustion of an HC yields only carbon dioxide and water, which do not harm the environment. Indeed, green plants "digest" carbon dioxide by photosynthesis. Carbon dioxide is a necessary ingredient in vegetal life. Animals do not suffer by breathing carbon dioxide unless its concentration in air is such that oxygen is almost absent.

The combustion of HC fuel in combustion engines is never ideal. Besides carbon dioxide and water, the combustion products contain a certain amount of nitrogen oxides (NO_x), carbon monoxides (CO), and unburned HCs, all of which are toxic to human health.

1.1.1 Nitrogen Oxides

Nitrogen oxides (NO_x) result from the reaction between nitrogen in the air and oxygen. Theoretically, nitrogen is an inert gas. However, the high temperatures and pressures in engines create favorable conditions for the formation of nitrogen oxides. Temperature is by far the most important parameter in nitrogen oxide formation. The most commonly found nitrogen oxide is nitric oxide (NO), although small amounts of nitric dioxide (NO_2) and traces of nitrous oxide (N_2O) are present. Once released into the atmosphere, NO reacts with oxygen to form NO_2. This is later decomposed by the Sun's ultraviolet radiation back to NO and highly reactive oxygen atoms that attack the membranes of living cells. Nitrogen dioxide is partly responsible for smog; its brownish color makes smog visible. It also reacts with atmospheric water to form nitric acid (HNO_3), which dilutes in rain. This phenomenon is referred to as "acid rain" and is responsible for the destruction of forests in industrialized countries.[1] Acid rain also contributes to the degradation of historical monuments made of marble.[1]

1.1.2 Carbon Monoxide

Carbon monoxide results from the incomplete combustion of HCs due to a lack of oxygen.[1] It is a poison to human beings and animals that inhale/breathe it. Once carbon monoxide reaches blood cells, it attaches to the hemoglobin in place of oxygen, thereby diminishing the quantity of oxygen that reaches the organs and reducing the physical and mental abilities of the affected living beings.[1] Dizziness is the first symptom of carbon monoxide poisoning, which can rapidly lead to death. Carbon monoxide binds more strongly to hemoglobin than oxygen. The bonds are so strong that normal body functions cannot break them. People intoxicated by carbon monoxide must be treated in pressurized chambers, where the pressure makes it easier to break the carbon monoxide–hemoglobin bonds.

1.1.3 Unburned HCs

Unburned HCs are a result of the incomplete combustion of HCs.[1,2] Depending on their nature, unburned HCs may be harmful to living beings.[2] Some of these unburned HCs may be direct poisons or carcinogenic chemicals such as particulates, benzene, or others. Unburned HCs are also responsible for smog; the Sun's ultraviolet radiation interacts with the unburned HCs and NO in the atmosphere to form ozone and other products. Ozone is a molecule formed by three oxygen atoms. It is colorless but very dangerous and poisonous because it attacks the membranes of living cells, causing them to age prematurely or die. Toddlers, older people, and asthmatics suffer greatly from exposure to high ozone concentrations. Annually, deaths from high ozone peaks in polluted cities have been reported.[3]

1.1.4 Other Pollutants

Impurities in fuels result in the emission of pollutants. The major impurity is sulfur, mostly found in diesel and jet fuel but also in gasoline and natural gas.[1] The combustion of sulfur (or sulfur compounds such as hydrogen sulfide) with oxygen releases sulfur oxides (SO_x). Sulfur dioxide (SO_2) is the major product of this combustion. On contact with air, it forms sulfur trioxide, which later reacts with water to form sulfuric acid, a major component of acid rain. It should be noted that sulfur oxide emissions originate from transportation sources but also largely from the combustion of coal in power plants and steel factories. In addition, there is debate over the exact contribution of natural sources such as volcanoes.

Petroleum companies add chemical compounds to their fuels to improve the performance or lifetime of engines.[1] Tetraethyl lead, often referred to simply as "lead," was used to improve the knock resistance of gasoline and, thereby, produce better engine performance. However, the combustion of this chemical releases lead metal, which is responsible for a neurological disease called saturnism. Its use is now forbidden in most developed countries, and it has been replaced by other chemicals.[1]

1.2 Global Warming

Global warming is a result of the greenhouse effect induced by the presence of carbon dioxide and other gases, such as methane, in the atmosphere. These gases trap the Sun's infrared radiation reflected from the ground, thus retaining the energy in the atmosphere and increasing the temperature. An increased Earth temperature results in major ecological damage to ecosystems and in many natural disasters that affect human populations.[2]

Considering the ecological damage induced by global warming, the disappearance of some endangered species is a concern because this destabilizes the natural resources that feed some populations. There are also concerns about the migration of some species from warm seas to previously colder northern seas, where they can potentially destroy indigenous species and the economies that live off those species. This may be happening in the Mediterranean Sea, where barracudas from the Red Sea have been observed.

Natural disasters command our attention more than ecological disasters because of the magnitude of the damage they cause. Global warming is believed to have induced meteorological phenomena such as El Niño, which disturbs the South Pacific region and regularly causes tornadoes, floods, and droughts. The melting of the polar icecaps, another major result of global warming, raises the sea level and can cause the permanent inundation of coastal regions and sometimes of entire countries.

Carbon dioxide is the result of the combustion of HCs and coal. Transportation accounts for a large share (32% from 1980 to 1999) of carbon dioxide emissions. The distribution of carbon dioxide emissions is shown in Figure 1.1.[4]

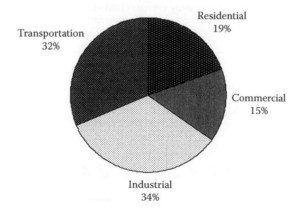

FIGURE 1.1
Carbon dioxide emission distribution from 1980 to 1999.

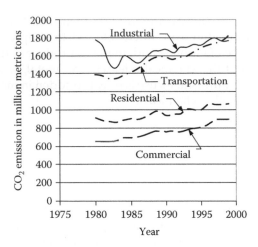

FIGURE 1.2
Evolution of CO_2 emission.

Figure 1.2 shows the trend in carbon dioxide emissions. The transportation sector is clearly now the major contributor to carbon dioxide emissions. It should be noted that developing countries are rapidly increasing their transportation sector, and these countries represent a very large share of the world's population. Further discussion of this issue is provided in the next subsection.

The large amounts of carbon dioxide released into the atmosphere by human activities are believed to be largely responsible for the increase in the global temperature on Earth observed in recent decades (Figure 1.3). It is important to note that carbon dioxide is indeed digested by plants and sequestrated by oceans in the form of carbonates. However, these natural assimilation processes are limited and cannot assimilate all emitted carbon dioxide, resulting in an accumulation of carbon dioxide in the atmosphere.

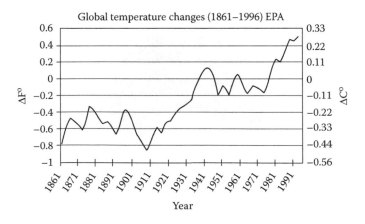

FIGURE 1.3
Global Earth atmospheric temperature. (IPCC (1995) updated.)

1.3 Petroleum Resources

The vast majority of fuels for transportation are liquid fuels originating from petroleum. Petroleum is a fossil fuel, resulting from the decomposition of living matter imprisoned millions of years ago (Ordovician, 600–400 million years ago) in geologically stable layers. The process is roughly as follows: living matter (mostly plants) dies and is slowly covered by sediments. Over time, these accumulating sediments form thick layers and transform in to rock. The living matter is trapped in a closed space, where it encounters high pressures and temperatures and slowly transforms into either HCs or coal, depending on its nature. This process takes millions of years to accomplish. This is what makes the Earth's fossil fuel resources finite.

Proved reserves are "those quantities that geological and engineering information indicates with reasonable certainty can be recovered in the future from known reservoirs under existing economic and operating conditions."[5] Therefore, they do not constitute an indicator of the Earth's total reserves. The proved reserves, as they are given in the British Petroleum 2001 estimate,[5] are given in billion tons in Table 1.1. The R/P ratio is the number of years that the proved reserves would last if production were to continue at its current level. This ratio is also given in Table 1.1 for each region.[5]

The oil extracted today is the easily extractable oil that lies close to the surface, in regions where the climate does not pose major problems. It is believed that far more oil lies underneath the Earth's crust in regions such as Siberia or the American and Canadian Arctic. In these regions, climate and ecological concerns are major obstacles to extracting or prospecting for oil. The estimation of the Earth's total reserves is a difficult task for political and technical reasons. A 2000 estimate of undiscovered oil resources by the U.S. Geological Survey is given in Table 1.2.[6]

Although the R/P ratio does not include future discoveries, it is significant. Indeed, it is based on proved reserves, which are easily accessible today. The amount of future oil discoveries is hypothetical, and newly discovered oil will not be easily accessible. The R/P ratio is also based on the hypothesis that production will remain constant. It is obvious, however, that consumption (and therefore production) is increasing yearly to keep up with the growth of developed and developing economies. Consumption is likely to increase in gigantic proportions with the rapid development of some highly populated countries,

TABLE 1.1

Proved Petroleum Reserves in 2000

Region	Proved Reserves in 2000 in Billion Tons	R/P Ratio
North America	8.5	13.8
South and Central America	13.6	39.0
Europe	2.5	7.7
Africa	10.0	26.8
Middle East	92.5	83.6
Former USSR	9.0	22.7
Asia Pacific	6.0	15.9
Total world	142.1	39.9

TABLE 1.2

U.S. Geological Survey Estimate of Undiscovered Oil in 2000

Region	Undiscovered Oil in 2000 in Billion Tons
North America	19.8
South and Central America	14.9
Europe	3.0
Sub-Saharan Africa and Antarctic	9.7
Middle East and North Africa	31.2
Former USSR	15.7
Asia Pacific	4.0
World (potential growth)	98.3 (91.5)

particularly in the Asia-Pacific region. Figure 1.4 shows the trend in oil consumption over the last 20 years.[7] Oil consumption is given in thousand barrels per day (one barrel is about 8 metric tons).

Despite the drop in oil consumption for Eastern Europe and the former USSR, the world trend is clearly increasing, as shown in Figure 1.5. The fastest-growing region is the Asia Pacific, where most of the world's population lives. An explosion in oil consumption is to be expected, with a proportional increase in pollutant emissions and CO_2 emissions.

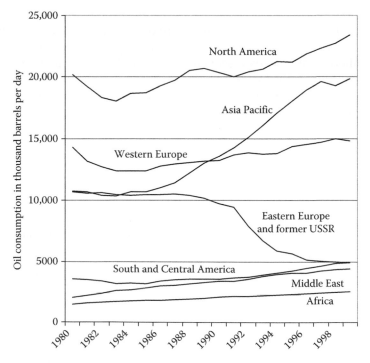

FIGURE 1.4
Oil consumption per region.

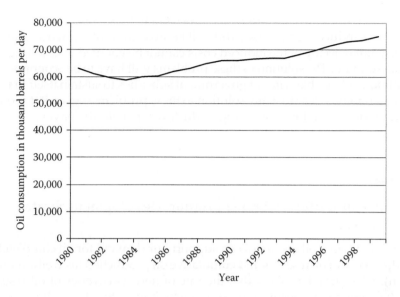

FIGURE 1.5
World oil consumption.

1.4 Induced Costs

The problems associated with the frenetic combustion of fossil fuels are many, including pollution, global warming, and foreseeable exhaustion of resources, among others. Although difficult to estimate, the costs associated with these problems are huge and indirect,[8] and they may be financial, human, or both.

Costs induced by pollution include, but are not limited to, health expenses, the cost of replanting forests devastated by acid rain, and the cost of cleaning and fixing monuments corroded by acid rain. Health expenses probably represent the largest share of these costs, especially in developed countries with socialized medicine or health-insured populations.

Costs associated with global warming are difficult to assess. They may include the cost of the damage caused by hurricanes, lost crops due to dryness, damaged properties due to floods, and international aid to relieve the affected populations. The amount is potentially huge.

Most of the petroleum-producing countries are not the largest petroleum-consuming countries. Most of the production is located in the Middle East, while most of the consumption is located in Europe, North America, and Asia Pacific. As a result, consumers must import their oil and depend on the producing countries. This issue is particularly sensitive in the Middle East, where political turmoil affected oil delivery to Western countries in 1973 and 1977. The Gulf War, the Iran–Iraq war, and the constant surveillance of the area by the United States and allied forces come at a cost that is both human and financial. The dependency of Western economies on a fluctuating oil supply is potentially expensive. Indeed, a shortage in oil supply causes a serious slowdown of the economy, resulting in damaged perishable goods, lost business opportunities, and the eventual impossibility of running businesses.

In searching for a solution to the problems associated with oil consumption, one must take into account those induced costs. This is difficult because the cost is not necessarily incurred where it is generated. Many of the induced costs cannot be counted in asserting the benefits of an eventual solution. The solution to these problems will have to be economically sustainable and commercially viable without government subsidies to sustain itself in the long run. Nevertheless, it remains clear that any solution to these problems—even if it is only a partial solution—will indeed result in cost savings, which will benefit the payers.

1.5 Importance of Different Transportation Development Strategies to Future Oil Supply

The number of years that oil resources can support our demand for oil completely depends on the new discovery of oil reserves and cumulative oil production (as well as cumulative oil consumption). Historical data show that the rate of new discoveries of oil reserves grows slowly. On the other hand, consumption shows a high growth rate, as shown in Figure 1.6. If oil discovery and consumption follow current trends, the world's oil resources will be used up by about 2038.[9,10]

It is becoming more and more difficult to discover new reserves of petroleum. Exploring new oil fields is becoming an increasingly expensive venture. It is believed that the scenario of oil supply will not change much if the consumption rate cannot be significantly reduced.

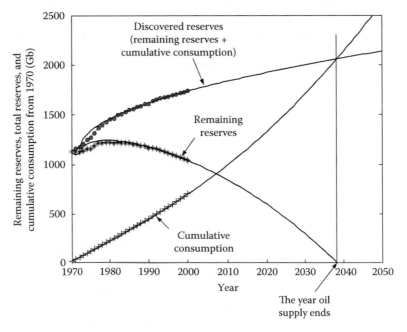

FIGURE 1.6
World oil discovery, remaining reserves, and cumulative consumption.

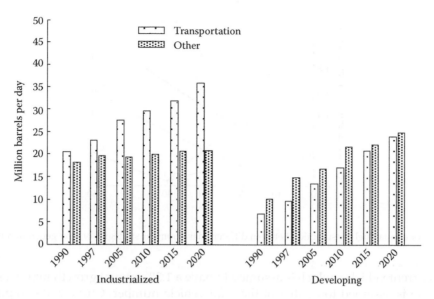

FIGURE 1.7
World oil consumption in transportation and other sectors.

As shown in Figure 1.7, the transportation sector is the primary user of petroleum, consuming 49% of the oil used in the world in 1997. The patterns of consumption of industrialized and developing countries are quite different, however. In the heat and power segments of the markets in industrialized countries, nonpetroleum energy sources were able to compete with and substitute for oil throughout the 1980s; by 1990, the oil consumption in other sectors was less than that in the transportation sector.

Most of the gains in worldwide oil use occur in the transportation sector. Of the total increase (11.4 million barrels per day) projected for industrialized countries from 1997 to 2020, 10.7 million barrels per day are attributed to the transportation sector, where few alternatives are economical until late in the forecast.

In developing countries, the transportation sector also shows the fastest projected growth in petroleum consumption, promising to rise nearly to the level of nontransportation energy use by 2020. In the developing world, however, unlike in industrialized countries, oil use for purposes other than transportation is projected to contribute 42% of the total increase in petroleum consumption. The growth in nontransportation petroleum consumption in developing countries is caused in part by the substitution of petroleum products for noncommercial fuels (such as wood burning for home heating and cooking).

Improving the fuel economy of vehicles has a crucial impact on oil supply. So far, the most promising technologies are HEVs and fuel cell vehicles. Hybrid vehicles, using current IC engines as their primary power source and batteries/electric motor as the peaking power source, have a much higher operational efficiency than those powered by an IC engine alone. The hardware and software of this technology are almost ready for industrial manufacturing. On the other hand, fuel cell vehicles, which are potentially more efficient and cleaner than HEVs, are still in the laboratory stage, and it will take a long time to overcome technical hurdles for commercialization.

Figure 1.8 shows the generalized annual fuel consumption of different development strategies of next-generation vehicles. Curve a–b–c represents the annual fuel consumption

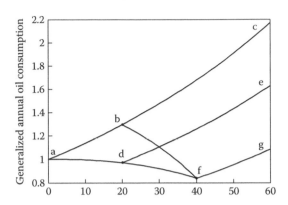

FIGURE 1.8
Comparison of annual fuel consumption between different development strategies of next-generation vehicles.

trend of current vehicles, which is assumed to have a 1.3% annual growth rate. This annual growth rate is assumed to be that of the total vehicle number. Curve a–d–e represents a development strategy in which conventional vehicles gradually become hybrid vehicles during the first 20 years, and after 20 years, all vehicles will be hybrids. In this strategy, it is assumed that the hybrid vehicle is 25% more efficient than a current conventional vehicle (25% less fuel consumption). Curve a–b–f–g represents a strategy in which, in the first 20 years, fuel cell vehicles are in a developing stage, while current conventional vehicles are still on the market. In the second 20 years, the fuel cell vehicles will gradually go to market, starting from point b and becoming totally fuel cell powered at point f. In this strategy, it is assumed that 50% less fuel will be consumed by fuel cell vehicles than by current conventional vehicles. Curve a–d–f–g represents a strategy whereby vehicles become hybrid in the first 20 years and fuel cell powered in the second 20 years.

Cumulative oil consumption is more meaningful because it involves annual consumption and the time effect, and it is directly associated with the reduction of oil reserves, as shown in Figure 1.6. Figure 1.9 shows the scenario of generalized cumulative oil consumption of the

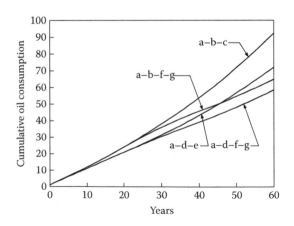

FIGURE 1.9
Comparison of cumulative fuel consumption between different development strategies of next-generation vehicles.

development strategies mentioned previously. Although fuel cell vehicles are more efficient than hybrid vehicles, the cumulative fuel consumption by strategy a–b–f–g (a fuel cell vehicle in the second 20 years) is higher than the strategy a–d–e (a hybrid vehicle in the first 20 years) within 45 years due to the time effect. From Figure 1.8 it is clear that strategy a–d–f–g (a hybrid vehicle in the first 20 years and a fuel cell vehicle in the second 20 years) is the best. Figures 1.6 and 1.9 reveal another important fact: fuel cell vehicles should not rely on oil products because of the difficulty of future oil supply 45 years later. Thus, the best development strategy of next-generation transportation would be to commercialize HEVs immediately and, at the same time, do the best to commercialize nonpetroleum fuel cell vehicles as soon as possible.

1.6 History of EVs

The first EV was built by Frenchman Gustave Trouvé in 1881. It was a tricycle powered by a 0.1-hp direct current (DC) motor fed by lead-acid batteries. The whole vehicle and its driver weighed approximately 160 kg. A vehicle similar to this was built in 1883 by two British professors.[11] These early realizations did not attract much attention from the public because the technology was not mature enough to compete with horse carriages. Speeds of 15 km/h and a range of 16 km were nothing exciting for potential customers. The 1864 Paris-to-Rouen race changed it all. The 1135-km race was run in 48 h and 53 min at an average speed of 23.3 km/h. This speed was by far superior to that possible with horse-drawn carriages. The public became interested in horseless carriages, or automobiles as these vehicles were now called.

The following 20 years were an era during which EVs competed with their gasoline counterparts. This was particularly true in the United States, where there were not many paved roads outside a few cities. The limited range of EVs was not a problem. However, in Europe, the rapidly increasing number of paved roads called for extended ranges, favoring gasoline vehicles.[11]

The first commercial EV was Morris and Salom's Electrobat. This vehicle was operated as a taxi in New York City by a company created by its inventors. The Electrobat proved to be more profitable than horse cabs despite a higher purchase price (around $3000 versus $1200). It could be used for three shifts of 4 h with 90-min recharging periods in between. It was powered by two 1.5-hp motors that allowed a maximum speed of 32 km/h and a 40-km range.[11]

The most significant technical advance of that era was the invention of regenerative braking by Frenchman M. A. Darracq on his 1897 coupe. This method makes it possible to recover the vehicle's kinetic energy while braking and recharging the batteries, which greatly enhances the driving range. It is one of the most significant contributions to electric and HEV technology as it contributes to energy efficiency more than anything else in urban driving.

In addition, among the most significant EVs of that era was the first vehicle ever to reach 100 km. It was "La Jamais Contente" built by Frenchman Camille Jenatzy. Note that Studebaker and Oldsmobile got started in business by building EVs.

As gasoline automobiles became more powerful, more flexible, and above all easier to handle, EVs started to disappear. Their high cost did not help, but it was their limited driving range and performance that really hurt them versus their gasoline counterparts. The last commercially significant EVs were released around 1905. For nearly 60 years, the only EVs sold were common golf carts and delivery vehicles.

In 1945, three researchers at Bell Laboratories invented a device that was meant to revolutionize the world of electronics and electricity: the transistor. It quickly replaced vacuum tubes for signal electronics, and soon the thyristor was invented, which made it possible to switch high currents at high voltages. This made it possible to regulate the power fed to an electric motor without the very inefficient rheostats, and it allowed the running of AC motors at variable frequency. In 1966, General Motors (GM) built the Electrovan, which was propelled by induction motors fed by inverters built with thyristors.

The most significant EV of that era was the Lunar Roving Vehicle, which the Apollo astronauts used on the Moon. The vehicle itself weighed 209 kg and could carry a payload of 490 kg. The range was around 65 km. The design of this extraterrestrial vehicle, however, has very little significance down on Earth. The absence of air and the lower gravity on the Moon, as well as the low speed, made it easier for engineers to reach an extended range with a limited technology.

During the 1960s and 1970s, concerns about the environment spurred research on EVs. However, despite advances in battery technology and power electronics, their range and performance remained obstacles.

The modern EV era culminated in the 1980s and early 1990s with the release of a few realistic vehicles by firms such as GM with the EV_1 and Peugeot Société Anonyme (PSA) with the 106 Electric. Although these vehicles represented a real achievement, especially compared with early realizations, it became clear during the early 1990s that electric automobiles could never compete with gasoline automobiles for range and performance. The reason is that in batteries the energy is stored in the metal of the electrodes, which weigh far more than gasoline for the same energy content. The automotive industry abandoned the EV to conduct research on HEVs. After a few years of development, these are far closer to the assembly line for mass production than EVs have ever been.

In the context of the development of EVs, battery technology is the weakest, preventing the EVs from making it to market. Great effort and investment have been put into battery research, with the intention of improving performance to meet EV requirements. Unfortunately, progress has been very limited. Performance lags far behind requirements, especially energy storage capacity per unit weight and volume. This poor energy storage capability of batteries limits EVs to specific applications, such as at airports, railroad stations, mail delivery routes, golf courses, and so on. In fact, basic study[12] shows that the EV will never be able to challenge the liquid-fueled vehicle even with the optimistic value of battery energy capacity. Thus, in recent years, advanced vehicle technology research has turned to HEVs as well as fuel cell vehicles.

1.7 History of HEVs

Surprisingly, the concept of an HEV is almost as old as the automobile itself. The primary purpose, however, was not so much to lower the fuel consumption but rather to assist the IC engine in providing an acceptable level of performance. Indeed, in the early days, IC engine engineering was less advanced than electric motor engineering.

The first hybrid vehicles reported were shown at the Paris Salon of 1899.[13] These were built by the Pieper establishments of Liège, Belgium and by the Vedovelli and Priestly

Electric Carriage Company, France. The Pieper vehicle was a parallel hybrid with a small air-cooled gasoline engine assisted by an electric motor and lead-acid batteries. It is reported that the batteries were charged by the engine when the vehicle coasted or was at a standstill. When the driving power required was greater than the engine rating, the electric motor provided additional power. In addition to being one of the first two hybrid vehicles and the first parallel hybrid vehicle, the Pieper was undoubtedly the first electric starter.

The other hybrid vehicle introduced at the Paris Salon of 1899 was the first series HEV and was derived from a pure EV commercially built by the French firm Vedovelli and Priestly.[13] This vehicle was a tricycle, with the two rear wheels powered by independent motors. An additional ¾-hp gasoline engine coupled to a 1.1-kW generator was mounted on a trailer and could be towed behind the vehicle to extend the range by recharging the batteries. In the French case, the hybrid design was used to extend its range by recharging the batteries. Also, the hybrid design was used to extend the range of an EV, not to supply additional power to a weak IC engine.

Frenchman Camille Jenatzy presented a parallel hybrid vehicle at the Paris Salon of 1903. This vehicle combined a 6-hp gasoline engine with a 14-hp electric machine that could either charge the batteries from the engine or assist them later. Another Frenchman, H. Krieger, built the second reported series hybrid vehicle in 1902. His design used two independent DC motors driving the front wheels. They drew their energy from 44 lead-acid cells that were recharged by a 4.5-hp alcohol spark-ignited engine coupled to a shunt DC generator.

Other hybrid vehicles, both parallel and series type, were built during the period 1899–1914. Although electric braking had been used in these early designs, there is no mention of regenerative braking. It is likely that most, possibly even all, designs used dynamic braking by short circuiting or by placing a resistance in the armature of the traction motors. The Lohner-Porsche vehicle of 1903 is a typical example of this approach.[13] The frequent use of magnetic clutches and magnetic couplings should be noted.

Early hybrid vehicles were built to assist the weak IC engines of that time or to improve the range of EVs. They made use of the basic electric technologies that were then available. Despite the great creativity that featured in their design, these early hybrid vehicles could no longer compete with the greatly improved gasoline engines that came into use after World War I. The gasoline engine made tremendous improvements in terms of power density, the engines became smaller and more efficient, and there was no longer a need to assist them with electric motors. The supplementary cost of having an electric motor and the hazards associated with the lead-acid batteries were key factors in the disappearance of hybrid vehicles from the market after World War I.

However, the greatest problem that these early designs had to cope with was the difficulty of controlling the electric machine. Power electronics did not become available until the mid-1960s, and early electric motors were controlled by mechanical switches and resistors. They had a limited operating range incompatible with efficient operation. Only with great difficulty could they be made compatible with the operation of a hybrid vehicle.

Dr. Victor Wouk is recognized as the modern investigator of the HEV movement.[13] In 1975, along with his colleagues, he built a parallel hybrid version of a Buick Skylark.[13] The engine was a Mazda rotary engine coupled to a manual transmission. It was assisted by a 15-hp, separately excited DC machine located in front of the transmission. Eight 12-V automotive batteries were used for energy storage. A top speed of 129 km/h (80 mph) was achieved, with acceleration from 0 to 60 mph in 16 s.

The series hybrid design was revived by Dr. Ernest H. Wakefield in 1967, when working for Linear Alpha Inc. A small engine coupled to an AC generator, with an output of 3 kW, was used to keep a battery pack charged. However, the experiments were quickly stopped because of technical problems. Other approaches used during the 1970s and early 1980s used range extenders, similar in concept to the French Vedovelli and Priestly 1899 design. These range extenders were intended to improve the range of EVs that never reached the market. Other prototypes of hybrid vehicles were built by the Electric Auto Corporation in 1982 and by the Briggs & Stratton Corporation in 1980. These were both parallel hybrid vehicles.

Despite the two oil crises of 1973 and 1977, and despite growing environmental concerns, no HEV made it to the market. Researchers' focus was on the EV, of which many prototypes were built during the 1980s. The lack of interest in HEVs during this period may be attributed to the lack of practical power electronics, modern electric motor, and battery technologies. The 1980s witnessed a reduction in conventional IC engine-powered vehicle sizes, the introduction of catalytic converters, and the generalization of fuel injection.

The HEV concept drew great interest during the 1990s when it became clear that EVs would never achieve the objective of saving energy. The Ford Motor Company initiated the Ford Hybrid Electric Vehicle Challenge, which drew efforts from universities to develop hybrid versions of production automobiles.

Automobile manufacturers around the world built prototypes that achieved tremendous improvements in fuel economy over their IC engine-powered counterparts. In the United States, Dodge built the Intrepid ESX 1, 2, and 3. The ESX-1 was a series hybrid vehicle, powered by a small turbocharged three-cylinder diesel engine and a battery pack. Two 100-hp electric motors were located in the rear wheels. The U.S. government launched the Partnership for a New Generation of Vehicles (PNGV), which included the goal of a midsize sedan that could achieve 80 mpg. The Ford Prodigy and GM Precept resulted from this effort. The Prodigy and Precept vehicles were parallel HEVs powered by small turbocharged diesel engines coupled to dry clutch manual transmissions. Both achieved the objective, but production did not follow.

Efforts in Europe are represented by the French Renault Next, a small parallel hybrid vehicle using a 750-cc spark-ignited engine and two electric motors. This prototype achieved 29.4 km/L (70 mpg) with maximum speed and acceleration performance comparable to conventional vehicles. Volkswagen also built a prototype, the Chico. The base was a small EV, with a nickel-metal hydride battery pack and a three-phase induction motor. A small two-cylinder gasoline engine was used to recharge the batteries and provide additional power for high-speed cruising.

The most significant effort in the development and commercialization of HEVs was made by Japanese manufacturers. In 1997, Toyota released the Prius sedan in Japan. Honda also released its Insight and Civic Hybrid. These vehicles are now available throughout the world. They achieve excellent figures of fuel consumption. Toyota's Prius and Honda's Insight vehicles have historical value in that they are the first hybrid vehicles commercialized in the modern era to respond to the problem of personal vehicle fuel consumption.

1.8 History of Fuel Cell Vehicles

As early as 1839, Sir William Grove (often referred to as the "Father of the Fuel Cell") discovered that it may be possible to generate electricity by reversing the electrolysis of water. It was not until 1889 that two researchers, Charles Langer and Ludwig Mond, coined the term "fuel cell" as they were trying to engineer the first practical fuel cell using air and coal gas. Although further attempts were made in the early 1900s to develop fuel cells that could convert coal or carbon into electricity, the advent of the IC engine temporarily quashed any hopes of further development of the fledgling technology.

Francis Bacon developed what was perhaps the first successful fuel cell device in 1932, with a hydrogen–oxygen cell using alkaline electrolytes and nickel electrodes—inexpensive alternatives to the catalysts used by Mond and Langer. Due to a substantial number of technical hurdles, it was not until 1959 that Bacon and company first demonstrated a practical 5-kW fuel cell system. Harry Karl Ihrig presented his now-famous 20-hp fuel-cell-powered tractor that same year.

The National Aeronautics and Space Administration (NASA) also began building compact electric generators for use on space missions in the late 1950s. NASA soon came to fund hundreds of research contracts involving fuel cell technology. Fuel cells now have a proven role in the space program, having supplied electricity for several space missions.

In more recent decades, several manufacturers—including major automakers—and various federal agencies have supported ongoing research into the development of fuel cell technology for use in fuel cell vehicles and other applications.[14] Hydrogen production, storage, and distribution are the biggest challenges. Truly, fuel-cell-powered vehicles still have a long way to go to enter the market.

Bibliography

1. C. R. Ferguson and A. T. Kirkpatrick, *Internal Combustion Engines—Applied Thermo-Sciences*, Second Edition, John Wiley & Sons, New York, 2001.
2. U.S. Environmental Protection Agency (EPA), Automobile emissions: An overview, *EPA 400-F-92-007*, Fact Sheet OMS-5, August 1994.
3. U.S. Environmental Protection Agency (EPA), Automobiles and ozone, *EPA 400-F-92-006*, Fact Sheet OMS-4, January 1993.
4. Energy Information Administration, U.S. Department of Energy, Carbon dioxide emissions from energy consumption by sector, 1980–1999, 2001, available at http://www.eia.doe.gov/emeu/aer/txt/tab1202.htm.
5. BP statistical review of world energy—oil, 2001, available at http://www.bp.com/downloads-/837/global_oil_section.pdf.
6. USGS World Energy Assessment Team, World undiscovered assessment results summary, *U.S. Geological Survey Digital Data Series 60*, available at http://greenwood.cr.usgs.gov/energy/WorldEnergy/DDS-60/sum1.html#TOP.
7. International Energy Database, Energy Information Administration, U.S. Department of Energy, World petroleum consumption, 1980–1999, 2000.
8. D. Doniger, D. Friedman, R. Hwang, D. Lashof, and J. Mark, Dangerous addiction: Ending America's oil dependence, *National Resources Defense Council and Union of Concerned Scientists*, 2002.
9. M. Ehsani, D. Hoelscher, N. Shidore, and P. Asadi, Impact of hybrid electric vehicles on the world's petroleum consumption and supply, *Society of Automotive Engineers (SAE) Future Transportation Technology Conference*, Paper No. 2003-01-2310, 2003.
10. J. E. Hake, International energy outlook—2000 with projection to 2020, available at http://tonto.eia.doe.gov/FTPROOT/presentations/ieo2000/sld008.htm.
11. E. H. Wakefield, *History of the Electric Automobile: Battery-only Powered Cars*, Society of Automotive Engineers (SAE), Warrendale, PA, 1994, ISBN: 1-56091-299-5.

12. Y. Gao and M. Ehsani, An investigation of battery technologies for the Army's hybrid vehicle application, *Proceedings of the IEEE 56th Vehicular Technology Conference*, Vancouver, British Columbia, Canada, September 2002.
13. E. H. Wakefield, *History of the Electric Automobile: Hybrid Electric Vehicles*, Society of Automotive Engineers (SAE), Warrendale, PA, 1998, ISBN: 0-7680-0125-0.
14. California Fuel Cell Partnership, available at http://www.fuelcellpartnership.org/
15. L. L. Christiansen, H. Frederick, K. Knechel, and E. L. Mussman, Meeting the Needs of Modern Transportation Researchers by Transforming the Iowa Department of Transportation Library-Early Efforts & Results. *Transportation Research Board 95th Annual Meeting*, no. 16-3827. 2016.
16. C. Ergas, M. Clement, and J. McGee, Urban density and the metabolic reach of metropolitan areas: A panel analysis of per capita transportation emissions at the county-level. *Social Science Research* 58, 2016: 243–253.
17. E. Weiner, *Urban transportation planning in the United States: History, policy, and practice*. Springer, 2016.
18. M. Alam, J. Ferreira, and J. Fonseca, Introduction to intelligent transportation systems. *Intelligent Transportation Systems*, pp. 1–17. Springer International Publishing, 2016.
19. R. Jedwab and A. Moradi, The permanent effects of transportation revolutions in poor countries: Evidence from Africa. *Review of Economics and Statistics* 98(2), 2016: 268–284.
20. S. A. Bagloee, M. Tavana, M. Asadi, and T. Oliver, Autonomous vehicles: Challenges, opportunities, and future implications for transportation policies. *Journal of Modern Transportation* 24 (4), 2016: 284–303.
21. M. Lawry, A. Mirza, Y. W. Wang, and D. Sundaram, Efficient Transportation-Does the Future Lie in Vehicle Technologies or in Transportation Systems? *International Conference on Future Network Systems and Security*, pp. 126–138. Springer, Cham, 2017.
22. M. Chowdhury and K. Dey, Intelligent transportation systems-a frontier for breaking boundaries of traditional academic engineering disciplines [Education]. *IEEE Intelligent Transportation Systems Magazine* 8(1), 2016: 4–8.
23. J. Hogerwaard, I. Dincer, and C. Zamfirescu, Thermodynamic and environmental impact assessment of NH3 diesel–fueled locomotive configurations for clean rail transportation. *Journal of Energy Engineering* 143(5), 2017: 04017018.
24. J. D. K. Bishop, N. Molden, and A. M. Boies, Real-world environmental impacts from modern passenger vehicles operating in urban settings. *International Journal of Transport Development and Integration* 1(2), 2017: 203–211.
25. L. A. W. Ellingsen, B. Singh, and A. H. Strømman, The size and range effect: Lifecycle greenhouse gas emissions of electric vehicles. *Environmental Research Letters* 11(5), 2016: 054010.
26. H. I. Abdel-Shafy and M. S. M. Mansour, A review on polycyclic aromatic hydrocarbons: source, environmental impact, effect on human health and remediation. *Egyptian Journal of Petroleum* 25(1), 2016: 107–123.
27. W. Ren, B. Xue, Y. Geng, C. Lu, C. Y. Zhang, L. Zhang, T. Fujita, and H. Hao, Inter-city passenger transport in larger urban agglomeration area: Emissions and health impacts. *Journal of Cleaner Production* 114, 2016: 412–419.
28. W. Ke, S. Zhang, X. He, Y. Wu, and J. Hao, Well-to-wheels energy consumption and emissions of electric vehicles: Mid-term implications from real-world features and air pollution control progress. *Applied Energy* 188, 2017: 367–377.
29. S. S. Sosale, Performance Analysis of Various 4-Wheelers with IC Engines for Hybridization, 2017.
30. D. Karner and J. Francfort, Hybrid and plug-in hybrid electric vehicle performance testing by the US Department of Energy Advanced Vehicle Testing Activity. *Journal of Power Sources* 174, 2007: 69–75. http://dx.doi.org/10.1016/j.jpowsour.2007.06.069.
31. X. Hu, N. Murgovski, L. M. Johannesson, and B. Egardt, Comparison of three electrochemical energy buffers applied to a hybrid bus powertrain with simultaneous optimal sizing and energy management. *IEEE Transactions on Intelligent Transportation Systems* 15, 2014: 1193–1205. http://dx.doi.org/10.1109/TITS.2013.2294675.

2

Fundamentals of Vehicle Propulsion and Braking

Vehicle operation fundamentals mathematically describe vehicle behavior based on the general principles of mechanics. A vehicle, consisting of thousands of components, is a complex system. To describe its behavior fully, sophisticated mechanical and mathematical knowledge is needed. A large body of literature in this field already exists. Since this book proposes to discuss electric, hybrid electric, and fuel cell power trains, the discussion of vehicle fundamentals will be restricted to one-dimensional movement. This chapter will therefore focus on aspects of vehicle performance, such as speed, gradeability, acceleration, fuel consumption, and braking performance.

2.1 General Description of Vehicle Movement

The movement behavior of a vehicle along its moving direction is completely determined by all the forces acting on it in this direction. Figure 2.1 shows the forces acting on a vehicle moving up a grade. The tractive effort, F_t, in the contact area between the tires of the drive wheels and the road surface propels the vehicle forward. It is produced by the power plant torque and transferred through transmission and final drive to the drive wheels. While the vehicle is moving, there is resistance that tries to stop its movement. The resistance usually includes tire rolling resistance, aerodynamic drag, and uphill resistance. According to Newton's second law, vehicle acceleration can be written

$$\frac{dV}{dt} = \frac{\sum F_t - \sum F_r}{\delta M},$$ (2.1)

where V is the speed of the vehicle, ΣF_t the total tractive effort of the vehicle, ΣF_r the total resistance, M the total mass of the vehicle, and δ the mass factor that equivalently converts the rotational inertias of rotating components into translational mass.

2.2 Vehicle Resistance

As shown in Figure 2.1, vehicle resistances opposing its movement include rolling resistance of the tires, appearing in Figure 2.1 as rolling resistance torques T_{rf} and T_{rr}, aerodynamic drag, F_w, and hill climbing resistance (the term $Mg \sin \alpha$ in Figure 2.1). All resistances will be discussed in detail in the following sections.

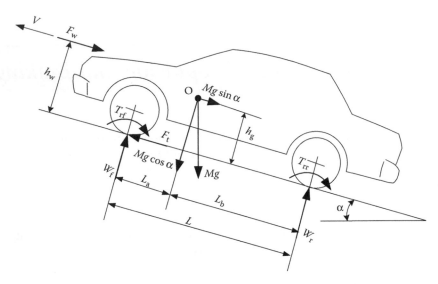

FIGURE 2.1
Forces acting on a vehicle moving uphill.

2.2.1 Rolling Resistance

The rolling resistance of tires on hard surfaces is primarily caused by hysteresis in the tire materials. Figure 2.2 shows a tire at standstill, on which a force, P, is acting at its center. The pressure in the contact area between the tire and the ground is distributed symmetrically to the central line, and the resultant reaction force, P_z, is aligned to P. The deformation, z, versus the load, P, in the loading and unloading process is shown in Figure 2.3. Due to hysteresis in the deformation of rubber material, the load at loading is larger than that at unloading at the same deformation, z, as shown in Figure 2.3. When the tire is rolling, as shown in Figure 2.4a, the leading half of the contact area is loading, and the trailing half is unloading. Consequently, the hysteresis causes an asymmetric distribution of the ground

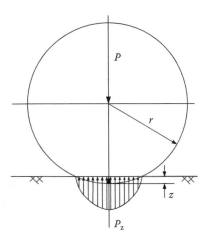

FIGURE 2.2
Pressure distribution in contact area.

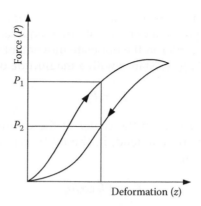

FIGURE 2.3
Force acting on a tire versus tire deformation in loading and unloading.

reaction forces. The pressure in the leading half of the contact area is larger than that in the trailing half, as shown in Figure 2.4a. This phenomenon results in the ground reaction force shifting forward somewhat. This forward shifted ground reaction force, with the normal load acting on the wheel center, creates a moment that opposes rolling of the wheel. On soft surfaces, the rolling resistance is primarily caused by deformation of the ground surface, as shown in Figure 2.4b. The ground reaction force almost completely shifts to the leading half.

The moment produced by the forward shift of the resultant ground reaction force is called rolling resistance moment, as shown in Figure 2.4a, and can be expressed as

$$T_r = Pa. \tag{2.2}$$

To keep the wheel rolling, a force, F, acting on the center of the wheel is required to balance this rolling resistant moment. This force is expressed as

$$F = \frac{T_r}{r_d} = \frac{Pa}{r_d} = Pf_r, \tag{2.3}$$

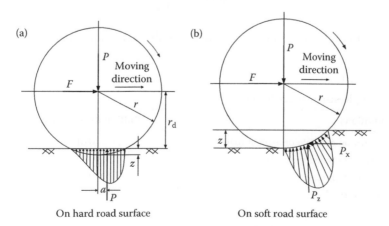

On hard road surface   On soft road surface

FIGURE 2.4
Tire deflection and rolling resistance on a (a) hard and (b) soft road surface.

where r_d is the effective radius of the tire, and $f_r = a/r_d$ is called the rolling resistance coefficient. In this way, the rolling resistance moment can be equivalently replaced by a horizontal force acting on the wheel center in the opposite movement direction of the wheel. This equivalent force is called rolling resistance with a magnitude of

$$F_r = Pf_r, \tag{2.4}$$

where P is the normal load acting on the center of the rolling wheel. When a vehicle is operated on a slope road, the normal load, P, should be replaced by the component that is perpendicular to the road surface:

$$F_r = Pf_r \cos \alpha, \tag{2.5}$$

where α is the road angle (refer to Figure 2.1).

The rolling resistance coefficient, f_r, is a function of the tire material, tire structure, tire temperature, tire inflation pressure, tread geometry, road roughness, road material, and presence or absence of liquids on the road. Typical values of rolling resistance coefficients on various roads are given in Table 2.1.[1] For fuel saving, low-resistance tires for passenger cars have been developed in recent years. Their rolling resistance coefficient is less than 0.01.

The values given in Table 2.1 do not consider coefficient variations with speed. Based on experimental results, many empirical formulas have been proposed for calculating rolling resistance on a hard surface. For example, the rolling resistance coefficient of passenger cars on a concrete road may be calculated from the equation

$$f_r = f_0 + f_s \left(\frac{V}{100} \right)^{2.5}, \tag{2.6}$$

where V is vehicle speed in km/h, and f_0 and f_s depend on the inflation pressure of the tire.[2]

In vehicle performance calculations, it is sufficient to consider the rolling resistance coefficient as a linear function of speed. For the most common range of inflation pressure, the following equation can be used for a passenger car on a concrete road[2]:

$$f_r = 0.01 \left(1 + \frac{V}{160} \right). \tag{2.7}$$

This equation predicts the values of f_r with acceptable accuracy for speeds up to 128 km/h.

TABLE 2.1

Rolling Resistance Coefficients

Conditions	Rolling Resistance Coefficient
Car tires on concrete or asphalt road	0.013
Car tires on rolled gravel road	0.02
Tar macadam road	0.025
Unpaved road	0.05
Field	0.1–0.35
Truck tire on concrete or asphalt road	0.006–0.01
Wheel on iron rail	0.001–0.002

2.2.2 Aerodynamic Drag

A vehicle traveling at a particular speed in air encounters a force resisting its motion. This force is referred to as aerodynamic drag. It mainly results from two components: shape drag and skin friction.

Shape drag: The forward motion of the vehicle pushes the air in front of it. However, the air cannot instantaneously move out of the way, and its pressure is thus increased, resulting in high air pressure. In addition, the air behind the vehicle cannot instantaneously fill the space left by the forward motion of the vehicle. This creates a zone of low air pressure. The motion of the vehicle, therefore, creates two zones of pressure that oppose the motion by pushing (high pressure in front) and pulling it backward (low pressure at the back), as shown in Figure 2.5. The resulting force on the vehicle is the shape drag. The term "shape drag" comes from the fact that this drag is completely determined by the shape of the vehicle body.

Skin friction: Air close to the skin of the vehicle moves almost at the speed of the vehicle, while air away from the vehicle remains still. In between, air molecules move at a wide range of speeds. The difference in speed between two air molecules produces a friction that results in the second component of aerodynamic drag.

Aerodynamic drag is a function of vehicle speed, V, vehicle frontal area, A_f, shape of the vehicle body, and air density, ρ:

$$F_W = \frac{1}{2} \rho A_f C_D (V - V_w)^2, \tag{2.8}$$

where C_D is the aerodynamic drag coefficient that characterizes the shape of the vehicle body, and V_w is a component of the wind speed in the vehicle moving direction, which has a positive sign when this component is in the same direction of the moving vehicle and a negative sign when it is opposite to the vehicle speed. The aerodynamic drag coefficients for typical vehicle body shapes are shown in Figure 2.6.

2.2.3 Grading Resistance

When a vehicle goes up or down a slope, its weight produces a component that is always directed in the downward direction, as shown in Figure 2.7. This component either opposes

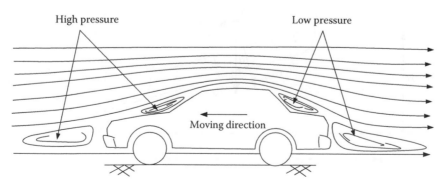

High pressure   Low pressure

Moving direction

FIGURE 2.5
Shape drag.

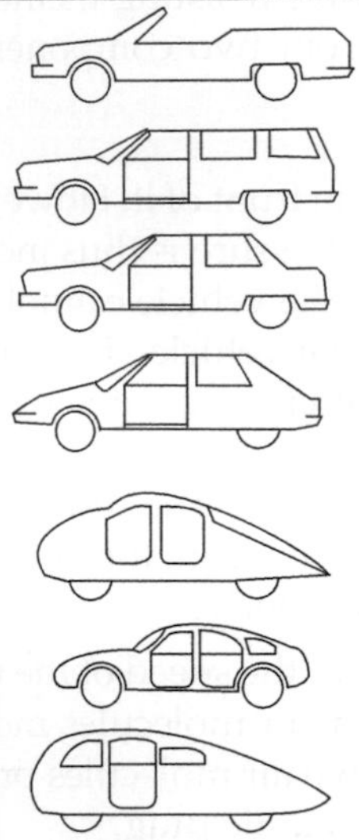

Vehicle type		Coefficient of aerodynamic resistance
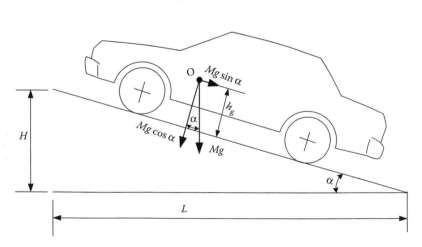	Open convertible	0.5...0.7
	Van body	0.5...0.7
	Ponton body	0.4...0.55
	Wedged-shaped body; headlamps and bumpers are integrated into the body, covered underbody, optimized cooling air flow	0.3...0.4
	Headlamp and all wheels in body, covered underbody	0.2...0.25
	K-shaped (small breakaway section)	0.23
	Optimum streamlined design	0.15...0.20
Trucks, road trains		0.8...1.5
Buses		0.6...0.7
Streamlined buses		0.3...0.4
Motorcycles		0.6...0.7

FIGURE 2.6
Indicative drag coefficients for different body shapes.

the forward motion (grade climbing) or helps the forward motion (grade descending). In vehicle performance analysis, only uphill operation is considered. This grading force is usually called the grading resistance.

Grading resistance, referring to Figure 2.7, can be expressed as

$$F_g = Mg \sin \alpha. \tag{2.9}$$

FIGURE 2.7
Vehicle climbing a grade.

To simplify the calculation, the road angle, α, is usually replaced by the grade value when the road angle is small. As shown in Figure 2.7, grade is defined as

$$i = \frac{H}{L} = \tan \alpha \approx \sin \alpha. \tag{2.10}$$

In some literature, the tire rolling resistance and grading resistance together are called road resistance, which is expressed as

$$F_{\text{rd}} = F_f + F_g = Mg(f_r \cos \alpha + \sin \alpha). \tag{2.11}$$

When the road angle is small, the road resistance can be simplified to

$$F_{\text{rd}} = F_f + F_g = Mg(f_r + i). \tag{2.12}$$

2.3 Dynamic Equation

In the longitudinal direction, the major external forces acting on a two-axle vehicle, as shown in Figure 2.1, include the rolling resistance of the front and rear tires, F_{rf} and F_{rr}, which are represented by the rolling resistance moment, T_{rf} and T_{rr}, aerodynamic drag, F_w, climbing resistance, F_g ($Mg \sin \alpha$), and tractive effort of the front and rear tires, F_{tf} and F_{tr}. F_{tf} is zero for a rear-wheel-drive vehicle, whereas F_{tr} is zero for a front-wheel-drive vehicle.

The dynamic equation of vehicle motion in the longitudinal direction is expressed by

$$M\frac{dV}{dt} = (F_{\text{tf}} + F_{\text{tr}}) - (F_{\text{rf}} + F_{\text{rr}} + F_w + F_g), \tag{2.13}$$

where dV/dt is the linear acceleration of the vehicle in the longitudinal direction, and M is the vehicle mass. The first term on the right-hand side of Equation 2.13 is the total tractive effort, and the second term is the resistance.

To predict the maximum tractive effort that the tire–ground contact can support, the normal loads on the front and rear axles must be determined. By summing the moments of all the forces about point R (center of the tire–ground area), the normal load on the front axle W_f can be determined as

$$W_f = \frac{MgL_b \cos \alpha - (T_{\text{rf}} + T_{\text{rr}} + F_w h_w + Mgh_g \sin \alpha + Mh_g \, dV/dt)}{L}. \tag{2.14}$$

Similarly, the normal load acting on the rear axle can be expressed as

$$W_r = \frac{MgL_a \cos \alpha + (T_{\text{rf}} + T_{\text{rr}} + F_w h_w + Mgh_g \sin \alpha + Mh_g \, dV/dt)}{L}. \tag{2.15}$$

For passenger cars, the height of the center of application of aerodynamic resistance, h_w, is assumed to be near the height of the gravity center of the vehicle, h_g. Equations 2.14

and 2.15 can be simplified as

$$W_f = \frac{L_b}{L} Mg \cos \alpha - \frac{h_g}{L} \left(F_w + F_g + Mg f_r \frac{r_d}{h_g} \cos \alpha + M \frac{dV}{dt} \right) \qquad (2.16)$$

and

$$W_r = \frac{L_a}{L} Mg \cos \alpha + \frac{h_g}{L} \left(F_w + F_g + Mg f_r \frac{r_d}{h_g} \cos \alpha + M \frac{dV}{dt} \right), \qquad (2.17)$$

where r_d is the effective radius of the wheel. Referring to Equations 2.5 and 2.13, Equations 2.16 and 2.17 can be rewritten as

$$W_f = \frac{L_b}{L} Mg \cos \alpha - \frac{h_g}{L} \left(F_t - F_r \left(1 - \frac{r_d}{h_g} \right) \right) \qquad (2.18)$$

and

$$W_r = \frac{L_a}{L} Mg \cos \alpha + \frac{h_g}{L} \left(F_t - F_r \left(1 - \frac{r_d}{h_g} \right) \right), \qquad (2.19)$$

where $F_t = F_{tf} + F_{tr}$ is the total tractive effort of the vehicle, and F_r is the total rolling resistance of the vehicle. The first term on the right-hand side of Equations 2.18 and 2.19 is the static load on the front and rear axles when the vehicle is at rest on level ground. The second term is the dynamic component of the normal load.

The maximum tractive effort that the tire–ground contact can support (any small amount over this maximum tractive effort will cause the tire to spin on the ground) is usually described by the product of the normal load and the coefficient of road adhesion, μ, or referred to as frictional coefficient in some of the literature (more details are given in Section 2.4). For a front-wheel-drive vehicle:

$$F_{t\max} = \mu W_f = \mu \left[\frac{L_b}{L} Mg \cos \alpha - \frac{h_g}{L} \left(F_{t\max} - F_r \left(1 - \frac{r_d}{h_g} \right) \right) \right] \qquad (2.20)$$

and

$$F_{t\max} = \frac{\mu Mg \cos \alpha [L_b + f_r(h_g - r_d)]/L}{1 + \mu h_g/L}, \qquad (2.21)$$

where f_r is the coefficient of the rolling resistance. Similarly, for a rear-wheel-drive vehicle:

$$F_{t\max} = \mu W_r = \mu \left[\frac{L_a}{L} Mg \cos \alpha + \frac{h_g}{L} \left(F_{t\max} - F_r \left(1 - \frac{r_d}{h_g} \right) \right) \right] \qquad (2.22)$$

and

$$F_{t\max} = \frac{\mu Mg \cos \alpha [L_a - f_r(h_g - r_d)]/L}{1 - \mu h_g/L}. \qquad (2.23)$$

In vehicle operation, the maximum tractive effort on the drive wheels, transferred from the power plant through transmission, should not exceed the maximum values determined by the tire–ground cohesion in Equations 2.21 and 2.23. Otherwise, the drive wheels will spin on the ground, leading to vehicle instability.

2.4 Tire–Ground Adhesion and Maximum Tractive Effort

When the tractive effort of a vehicle exceeds the limitation of the maximum tractive effort due to the adhesive capability between the tire and the ground, the drive wheels will spin on the ground. The adhesive capability between the tire and the ground sometimes is the main limitation of vehicle performance. This is especially true when the vehicle drives on wet, icy, snow-covered, or soft soil roads. In this case, a tractive torque on the drive wheel would cause the wheel to slip significantly on the ground. The maximum tractive effort on the drive wheel depends on the longitudinal force that the adhesive capability between the tire and the ground can supply, rather than the maximum torque that an engine can supply.

Experimental results show that, on various types of ground, the maximum tractive effort of the drive wheel closely relates to the slippage of the running wheel. This is also true on a good paved, dry road where the slippage is very small due to the elasticity of the tire. The slip, s, of a tire is defined in traction as

$$s = \left(1 - \frac{V}{r\omega}\right) \times 100\%, \tag{2.24}$$

where V is the translational speed of the tire center, ω is the angular speed of the tire, and r is the rolling radius of the free rolling tire. In traction, the speed, V, is less than $r\omega$; therefore, the slip of a tire has a positive value between 0 and 1.0. During braking, however, the slip of a tire can be defined as

$$s = \left(1 - \frac{r\omega}{V}\right) \times 100\%, \tag{2.25}$$

which has a positive value between 0 and 1.0, similar to traction. The maximum traction effort of a tire corresponding to a certain tire slippage is usually expressed as

$$F_x = P\mu(s), \tag{2.26}$$

where P is the vertical load of the tire, and μ is the tractive effort coefficient, which is a function of tire slippage. The tractive effort coefficient and the tire slippage have the relationship shown in Figure 2.8.

In the small-slip range (section OA in Figure 2.8), the tractive effort is almost linearly proportional to the slip value. This small slip is caused by the elasticity of the tire rather than the relative slipping between the tire and the ground at the contact patch, as shown in Figure 2.9. When a tractive torque is applied to the tire, a tractive force is developed at the tire–ground contact patch. At the same time, the tire tread in front and within the contact patch is subject

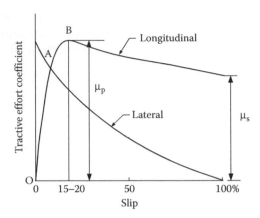

FIGURE 2.8
Variation of tractive effort coefficient with longitudinal slip of a tire.

to compression. A corresponding shear deformation of the side wall of the tire is also developed. As tread elements are compressed before entering the contact region, the distance that the tire travels will be less than the distance in a free rolling tire. Because of the nearly linear elastic property of the tire, the tractive effort–slip curve is almost linear. Further increase of wheel torque and tractive force results in having part of the tire tread slide on the ground. Under these circumstances, the relationship between tractive force and slip is nonlinear. This corresponds to section AB of the curve, as shown in Figure 2.8. The peak tractive effort is reached at a slip of 15%–20%. Further increase of the slip beyond that results in an unstable condition. The tractive effort coefficient falls rapidly from the peak value to the purely sliding value, as shown in Figure 2.8. For normal driving, the slip of the tire must be limited in

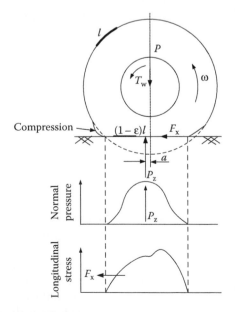

FIGURE 2.9
Behavior of tire under action of driving torque.

TABLE 2.2

Average Values of Tractive Effort Coefficient on Various Roads

Surface	Peaking Values (μ_p)	Slippage Values (μ_s)
Asphalt and concrete (dry)	0.8–0.9	0.75
Concrete (wet)	0.8	0.7
Asphalt (wet)	0.5–0.7	0.45–0.6
Gravel	0.6	0.55
Earth road (dry)	0.68	0.65
Earth road (wet)	0.55	0.4–0.5
Snow (hard packed)	0.2	0.15
Ice	0.1	0.07

a range of less than 15%–20%. Table 2.2 shows the average values of tractive effort coefficients on various roads.[2]

2.5 Power Train Tractive Effort and Vehicle Speed

An automotive power train, as shown in Figure 2.10, consists of a power plant (engine or electric motor), a clutch in a manual transmission or a torque converter in an automatic transmission, a gearbox (transmission), final drive, differential, drive shaft, and drive wheels. The torque and rotating speed from the output shaft of the power plant are transmitted to the drive wheels through the clutch or torque converter, gearbox, final drive, differential, and drive shaft. The clutch is used in a manual transmission to couple or decouple the gearbox to the power plant. The torque converter in an automatic transmission is a hydrodynamic device functioning as the clutch in a manual transmission with a continuously variable gear ratio (for more details, see Section 2.6). The gearbox supplies a few gear ratios from its input shaft to its output shaft for the power plant torque–speed profile to match the requirements of the load. The final drive is usually a pair of gears that supply a further speed reduction and distribute the torque to each wheel through the differential.

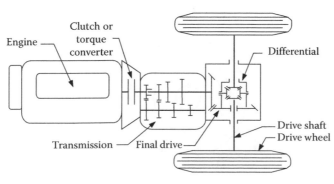

FIGURE 2.10
Conceptual illustration of an automobile power train.

The torque on the drive wheels, transmitted from the power plant, is expressed as

$$T_w = i_g i_0 \eta_t T_p, \tag{2.27}$$

where i_g is the gear ratio of the transmission defined as $i_g = N_{in}/N_{out}$ (N_{in}—input rotating speed, N_{out}—output rotating speed), i_0 is the gear ratio of the final drive, η_t is the efficiency of the driveline from the power plant to the drive wheels, and T_p is the torque output from the power plant.

The tractive effort on the drive wheels, as shown in Figure 2.11, can be expressed as

$$F_t = \frac{T_w}{r_d}. \tag{2.28}$$

Substituting Equation 2.27 into Equation 2.28 yields the following result:

$$F_t = \frac{T_p i_g i_0 \eta_t}{r_d}. \tag{2.29}$$

The friction in the gear teeth and bearings creates losses in the mechanical gear transmission. The following are representative values of the mechanical efficiency of various components:

Clutch: 99%.

Each pair of gears: 95%–97%.

Bearing and joint: 98%–99%.

The total mechanical efficiency of the transmission between the engine output shaft and drive wheels is the product of the efficiencies of all the components in the driveline. As a first approximation, the following average values of the overall mechanical efficiency of a manual gear-shift transmission may be used:

Direct gear: 90%.

Other gear: 85%.

Transmission with very high reduction ratio: 75%–80%.

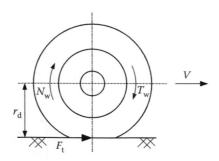

FIGURE 2.11
Tractive effort and torque on a drive wheel.

The rotating speed, in revolutions per minute (rpm), of the drive wheel can be expressed as

$$N_w = \frac{N_P}{i_g i_0},$$ (2.30)

where N_p is the transmission rotating speed (rpm), which is equal to the engine speed in a vehicle with a manual transmission and the turbine speed of a torque converter in a vehicle with an automatic transmission (for more details, see Section 2.6). The translational speed of the wheel center (vehicle speed) can be expressed as

$$V = \frac{\pi N_w r_d}{30} \, (\text{m/s}).$$ (2.31)

Substituting Equation 2.30 into Equation 2.31 yields

$$V = \frac{\pi N_p r_d}{30 i_g i_0} \, (\text{m/s}).$$ (2.32)

2.6 Vehicle Performance

The performance of a vehicle is usually described by its maximum cruising speed, gradeability, and acceleration. The prediction of vehicle performance is based on the relationship between tractive effort and vehicle speed, discussed in Section 2.5. For on-road vehicles, it is assumed that the maximum tractive effort is limited by the maximum torque of the power plant rather than the road adhesion capability. Tractive effort (Equation 2.29) and resistance $(F_r + F_w + F_g)$ as depicted in a diagram are used for vehicle performance analysis, as shown in Figures 2.12 and 2.13, for a gasoline-engine-powered, four-gear, manual-transmission vehicle and an electric-motor-powered, single-gear transmission vehicle, respectively.

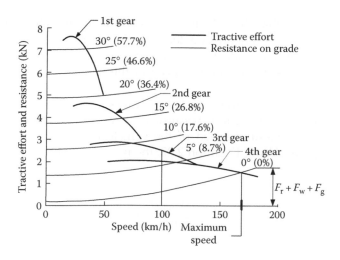

FIGURE 2.12
Tractive effort of gasoline-engine-powered vehicle with manual multispeed transmission and its resistance.

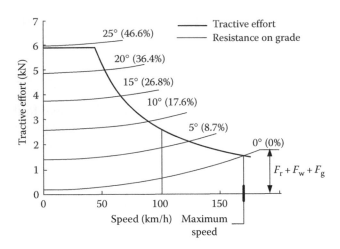

FIGURE 2.13
Tractive effort of electric-motor-powered vehicle with single-speed transmission and its resistance.

2.6.1 Maximum Speed of a Vehicle

The maximum speed of a vehicle is defined as the constant cruising speed that the vehicle can attain with full power plant load (full throttle of engine or full power of motor) on a flat road. The maximum speed of a vehicle is determined by the equilibrium between the tractive effort of the vehicle and the resistance or the maximum speed of the power plant and gear ratios of the transmission. The tractive effort and resistance equilibrium can be expressed as

$$\frac{T_p i_g i_0 \eta_t}{r_d} = M g f_r \cos\alpha + \frac{1}{2}\rho_a C_D A_f V^2. \tag{2.33}$$

This equation indicates that the vehicle reaches its maximum speed when the tractive effort, represented by the left-hand-side term in Equation 2.33, equals the resistance, represented by the right-hand-side terms. The intersection of the tractive effort curve and the resistance curve is the maximum speed of the vehicle, as shown in Figures 2.12 and 2.13.

It should be noted that for some vehicles no intersection exists between the tractive effort curve and the resistance curve because of a large power plant or a large gear ratio. In this case, the maximum speed of the vehicle is determined by the maximum speed of the power plant. Using Equation 2.32, the maximum speed of the vehicle can be obtained using

$$V_{max} = \frac{\pi n_{p\,max} r_d}{30 i_0 i_{g\,min}}\,(\text{m/s}), \tag{2.34}$$

where $n_{p\,max}$ and $i_{g\,min}$ are the maximum speed of the engine (or electric motor) and the minimum gear ratio of the transmission, respectively.

2.6.2 Gradeability

Gradeability is usually defined as the grade (or grade angle) that a vehicle can overcome at a certain constant speed, for instance the grade at a speed of 100 km/h (60 mph). For heavy commercial vehicles or off-road vehicles, gradeability is usually defined as the maximum grade or grade angle that a vehicle can overcome in the whole speed range.

When a vehicle drives on a road with a relatively small grade and at a constant speed, the tractive effort and resistance equilibrium can be written

$$\frac{T_p i_0 i_g \eta_t}{r_d} = Mg f_r + \frac{1}{2} \rho a C_D A_f V^2 + Mgi. \tag{2.35}$$

Thus,

$$i = \frac{T_p i_0 i_g \eta_t / r_d - Mg f_r - 1/2 \rho_a C_D A_f V^2}{Mg} = d - f_r, \tag{2.36}$$

where

$$d = \frac{F_t - F_w}{Mg} = \frac{T_p i_0 i_g \eta_t / r_d - 1/2 \rho_a C_D A_f V^2}{Mg} \tag{2.37}$$

is called the performance factor. When the vehicle drives on a road with a large grade, the gradeability of the vehicle can be calculated as

$$\sin \alpha = \frac{d - f_r \sqrt{1 - d^2 + f_r^2}}{1 + f_r^2}. \tag{2.38}$$

The gradeability of the vehicle can also be obtained from the diagram in Figure 2.12 or Figure 2.13, in which the tractive effort and resistance are plotted.

2.6.3 Acceleration Performance

The acceleration performance of a vehicle is usually described by its acceleration time and distance covered from zero speed to a certain high speed (0–96 km/h or 60 mph, for example) on level ground. Using Newton's second law (Equation 2.13), the acceleration of the vehicle can be written

$$a = \frac{dV}{dt} = \frac{F_t - F_f - F_w}{M\delta}$$
$$= \frac{T_p i_0 i_g \eta_t / r_d - Mg f_r - 1/2 \rho_a C_D A_f V^2}{M\delta} = \frac{g}{\delta}(d - f_r), \tag{2.39}$$

where δ is called the rotational inertia factor, considering the equivalent mass increase due to the angular moments of the rotating components. The mass factor can be written

$$\delta = 1 + \frac{I_w}{M r_d^2} + \frac{i_0^2 i_g^2 I_p}{M r^2}, \tag{2.40}$$

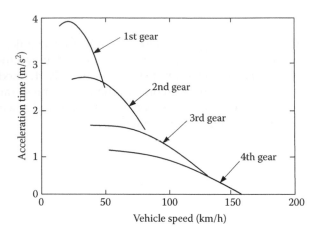

FIGURE 2.14
Acceleration rate of gasoline-engine-powered vehicle with four-gear transmission.

where I_w is the total angular inertial moment of the wheels, and I_p is the total angular inertial moment of the rotating components associated with the power plant. Calculation of the mass factor, δ, requires knowing the values of the mass moments of inertia of all the rotating parts. In cases where these values are not known, the rotational inertia factor, δ, for a passenger car would be estimated using the following empirical relation:

$$\delta = 1 + \delta_1 + \delta_2 i_g^2 i_0^2, \tag{2.41}$$

where δ_1 represents the second term on the right-hand side of Equation 2.56, with a reasonable estimate value of 0.04, and δ_2 represents the effect of the power-plant-associated rotating parts, with a reasonable estimate value of 0.0025.

Figures 2.14 and 2.15 show the acceleration rate along with vehicle speed for a gasoline-engine-powered vehicle with a four-gear transmission and an electric-motor-powered vehicle with a single-gear transmission.

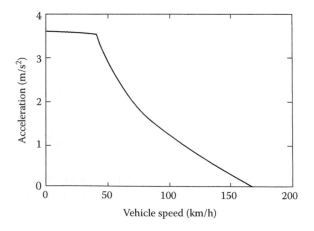

FIGURE 2.15
Acceleration rate of an electric-machine-powered vehicle with a single-gear transmission.

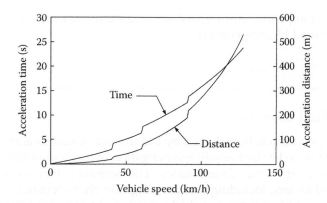

FIGURE 2.16
Acceleration time and distance along with vehicle speed for a gasoline-engine-powered passenger car with four-gear transmission.

From Equation 2.39, the acceleration time, t_a, and distance, S_a, from low-speed V_1 to high-speed V_2 can be written, respectively, as

$$t_a = \int_{V_1}^{V_2} \frac{M\delta}{T_p i_g i_0 \eta_t / r_d - Mgf_r - 1/2\rho_a C_D A_f V^2} dV \tag{2.42}$$

and

$$S_a = \int_{V_1}^{V_2} \frac{M\delta V}{T_p i_g i_0 \eta_t / r_0 - Mgf_r - 1/2\rho_a C_D A_f V^2} dV. \tag{2.43}$$

In Equations 2.42 and 2.43, the torque of the power plant T_p is a function of speed (Figures 2.13 and 2.14), which in turn is a function of vehicle speed (Equation 2.23) and gear ratio of the transmission. This makes it difficult to solve Equations 2.42 and 2.43 analytically. Numerical methods are usually used. Figures 2.16 and 2.17 show the acceleration

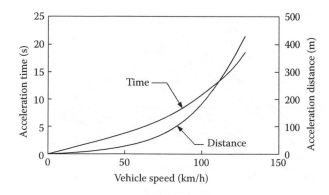

FIGURE 2.17
Acceleration time and distance along with vehicle speed for an electric-motor-powered passenger car with single-gear transmission.

time and distance along with vehicle speed for a gasoline-engine-powered vehicle and an electric-motor-powered EV, respectively.

2.7 Operating Fuel Economy

The fuel economy of a vehicle is evaluated by the amount of fuel consumption per 100-km traveling distance (liters/100 km) or mileage per gallon of fuel consumption (miles/gallon), which is currently used in the United States. The operating fuel economy of a vehicle depends on several factors, including fuel consumption characteristics of the engine, gear number and ratios, vehicle resistance, vehicle speed, and traffic conditions.

2.7.1 Fuel Economy Characteristics of IC Engines

The fuel economy characteristic of an internal combustion (IC) engine is evaluated by the amount of fuel per kilowatt-hour of energy output, which is referred to as the specific fuel consumption (g/kWh). The typical fuel economy characteristic of a gasoline engine is shown in Figure 2.18. The fuel consumption is quite different from one operating point to another. The optimum operating points are close to the points of full load (wide-open throttle). The speed of the engine also has a significant influence on the fuel economy. With a given power output, the fuel consumption is usually lower at low speed than at high speed. For instance, when the engine shown in Figure 2.18 has a power output of 40 kW, its minimum specific fuel consumption would be 270 g/kWh at a speed of 2080 rpm.

For a given power output at a given vehicle speed, the engine operating point is determined by the gear ratio of the transmission (refer to Equations 2.32). Ideally, a continuous variable transmission can choose the gear ratio, in a given driving condition, to operate the engine at its optimum operating point. This advantage has stimulated the development of a variety of continuous variable transmissions, including frictional drive, hydrodynamic drive, hydrostatic drive, and hydromechanical variable drive.

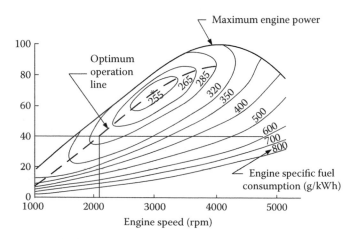

FIGURE 2.18
Fuel economy characteristics of a typical gasoline engine.

2.7.2 Computation of Vehicle Fuel Economy

Vehicle fuel economy can be calculated by finding the load power and speed and, thus, the specific fuel consumption of the engine. The engine power output is always equal to the resistance power plus the dynamic power for acceleration of the vehicle:

$$P_e \frac{V}{\eta_t} \left(F_f + F_w + F_g + M_v \delta \frac{dV}{dt} \right). \tag{2.44}$$

Equation 2.44 can be written

$$P_e \frac{V}{1000\eta_t} \left(Mg f_r \cos\alpha + \frac{1}{2}\rho_a C_D A_f V^2 + Mg \sin\alpha + M\delta \frac{dV}{dt} \right) (kW). \tag{2.45}$$

The engine speed, related to vehicle speed and gear ratio, can be expressed as

$$N_e = \frac{30 V i_g i_0}{\pi r_d}. \tag{2.46}$$

After determination of the engine power and speed by Equations 2.44 and 2.45, the value of the specific fuel consumption, g_e, can be found in the graph of the engine fuel economy characteristics as shown in Figure 2.18. The time rate of fuel consumption can be calculated using

$$Q_{fr} = \frac{P_e g_e}{1000\gamma_f} (1/h), \tag{2.47}$$

where g_e is the specific fuel consumption of the engine in g/kWh, and γ_f is the mass density of the fuel in kg/L. The total fuel consumption within the total distance, S, at a constant cruising speed, V, is obtained by

$$Q_s = \frac{P_e g_e}{1000\gamma_f} \frac{S}{V}. \tag{2.48}$$

Figure 2.19 shows an example of the fuel economy characteristics of a gasoline vehicle at a constant cruising speed on level ground. This figure indicates that at high speed the fuel consumption increases because the aerodynamic resistance power increases with the speed cubed. This figure also indicates that with a high-speed gear (small gear ratio), the fuel economy of the vehicle can be enhanced due to the reduced engine speed at a given vehicle speed and decreased gear ratio.

Figure 2.20 shows the operating points of the engine at a constant vehicle speed, with the highest gear and the second highest gear. It indicates that the engine has a much lower operating efficiency in low gear than in high gear. This is why the fuel economy of a vehicle can be improved with more gear transmission and continuous variable transmission.

It should be noted that because of the complexity of vehicle operation in the real world, fuel consumption at a constant speed (as shown in Figure 2.12) cannot accurately represent

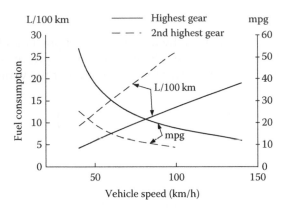

FIGURE 2.19
Fuel economy characteristics of a typical vehicle at constant speed.

fuel consumption for a vehicle under real driving conditions. Thus, various drive cycles have been developed to simulate real driving conditions, such as EPA FTP75 urban and highway, LA92, ECE-15, Japan1015, and so on. The drive cycles are usually represented by the speed of the vehicle along with the driving time. Figure 2.21 shows the urban and highway drive cycles of EPA FTP75 used in the United States.

To calculate fuel consumption in a drive cycle, the total fuel consumption can be obtained by the summation of fuel consumption in each time interval, Δt_i:

$$Q_{tc} = \sum_i \frac{P_{ei}g_{ei}}{1000\gamma_f}\Delta t_i, \qquad (2.49)$$

where P_{ei} is the average power of the engine in the ith time interval in kW, g_{ei} is the average specific fuel consumption of the engine in the ith time interval in g/kWh, and Δt_i is the ith time interval in h. This calculation can be performed with a numerical method using a

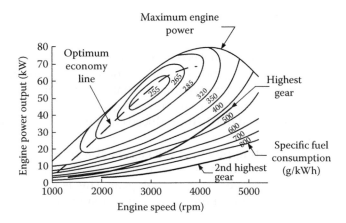

FIGURE 2.20
Operating point of the engine at a constant speed with highest gear and second highest gear.

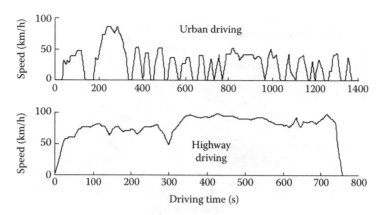

FIGURE 2.21
EPA FTP75 urban and highway drive cycles.

computer program. Figures 2.22 and 2.23 show examples of the fuel economy and engine operating points in EPA FTP75 urban and highway drive cycles, respectively.

2.7.3 Basic Techniques to Improve Vehicle Fuel Economy

The effort to improve the fuel economy of vehicles has always been an ongoing process in the automobile industry. Fundamentally, the techniques used mainly include the following aspects:

1. *Reducing vehicle resistance:* Using light materials and advanced manufacturing technologies can reduce the weight of vehicles, in turn reducing the rolling resistance and inertial resistance in acceleration and, therefore, reducing the power demanded on the engine. The use of advanced technologies in tire production is another important method in reducing the rolling resistance of vehicles. For instance, steel

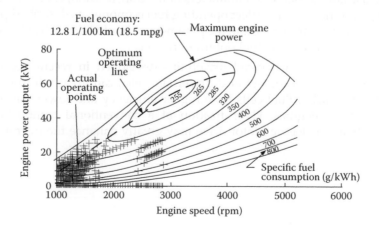

FIGURE 2.22
Fuel economy and engine operating points in EPA FTP75 urban drive cycle superimposed on engine fuel consumption characteristics map.

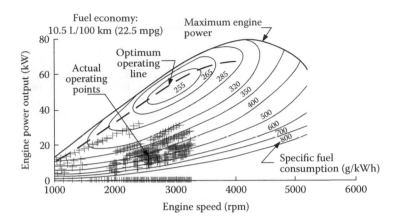

FIGURE 2.23
Fuel economy and engine operating points in EPA FTP75 highway drive cycle superimposed on engine fuel consumption characteristics map.

wire plied radial tires have a much lower rolling resistance coefficient than conventional bias ply tires. Reducing aerodynamic resistance is also quite important at high speeds. This can be achieved by using a flow-shaped body style, a smooth body surface, and other techniques. Furthermore, improving transmission efficiency can reduce energy losses in the transmission. Proper transmission construction, good lubrication, proper adjustment and tightening of moving parts in the transmission, and so on will achieve this purpose.

2. *Improving engine operation efficiency:* Improving engine operation efficiency has great potential to contribute to the improvement of vehicle fuel economy. There are many effective advanced techniques, such as accurate air/fuel ratio control with computer-controlled fuel injection, high thermal isolated materials for reducing thermal loss, varying ignition-timing techniques, active controlled valve and port, and so on.

3. *Properly matched transmission:* Parameters of the transmission, especially gear number and gear ratios, greatly affect operating fuel economy, as described previously. In the design of the transmission, the parameters should be constructed so that the engine operates close to its fuel optimum region.

4. *Advanced drivetrains:* Advanced drivetrains developed in recent years, such as new power plants, various hybrid drivetrains, etc., can greatly improve the fuel economy of vehicles. Fuel cells have higher efficiency and lower emissions than conventional IC engines. Hybridization of a conventional combustion engine with an advanced electric motor drive may greatly enhance the overall efficiency of vehicles.

2.8 Brake Performance

The braking performance of a vehicle is undoubtedly one of the most important concerns in vehicle safety. In urban area driving, a significant amount of energy is consumed in braking. In recent years, more and more electric drives have been involved in vehicle traction, such as EVs, HEVs, and fuel-cell-powered vehicles; the electrification of the vehicle drivetrain makes it feasible to recover some of the energy lost in braking. Nevertheless, braking performance is still the first concern in the design of a vehicle brake system. When electric braking is introduced for braking energy recovery, mechanical braking using friction is still required to ensure that the vehicle stops quickly. Consequently, a hybrid braking system was developed. The design and control objectives of such a hybrid braking system are (1) sufficient braking force to quickly reduce vehicle speed, (2) proper braking force distribution on the front and rear wheels to ensure vehicle stability during braking, and (3) recovery of as much braking energy as possible. This chapter discusses only the design principle of a vehicle brake system from a braking performance point of view. Regenerative braking is discussed in Chapter 13.

2.8.1 Braking Force

The function of the vehicle brake system is to reduce the vehicle speed quickly while keeping the vehicle traveling direction stable and controllable under various road conditions. These requirements are satisfied by applying sufficient braking force on the wheels and properly allocating the total braking force on the front and rear wheels.

Figure 2.24a shows a wheel during braking. The brake pad is pressed against the brake plate hydraulically or pneumatically, thereby developing a frictional torque on the brake plate. This braking torque results in a braking force in the tire–ground contact area. It is just this braking force that tries to stop the vehicle. The braking force can be expressed as

$$F_b = \frac{T_b}{r_d}. \tag{2.50}$$

The braking force increases with an increase in the braking torque. However, when the braking force reaches the maximum braking force that the tire–ground adhesion can support, it will not increase further, although the braking torque may still increase, as shown

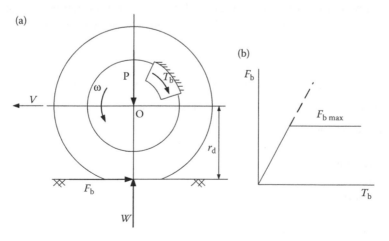

FIGURE 2.24
(a) Braking torque and braking force; (b) relationship between braking torque and braking force.

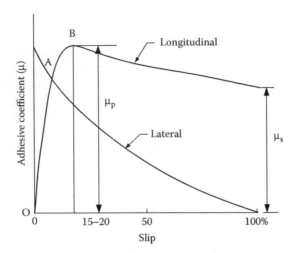

FIGURE 2.25
Variation of tractive effort coefficient with longitudinal slip of a tire.

in Figure 2.24b. This maximum braking force limited by the adhesive capability can be expressed as

$$F_{b\,max} = \mu W, \tag{2.51}$$

where μ is the adhesive coefficient of the tire–ground contact. Similar to the traction case, the adhesive coefficient varies with the slipping of the tire, as shown in Figure 2.25. However, the slip is defined in braking as

$$s = \left(1 - \frac{r\omega}{V}\right) \times 100\%, \tag{2.52}$$

where V is the vehicle translational speed, ω is the wheel rotation speed, and r is the wheel radius. In this definition, when $\omega = 0$, that is, the wheel is completely locked, $s = 100\%$. Figure 2.25 shows the typical relationship between adhesive coefficient and wheel slip. There exists a maximum value in the slip range of 15%–20% and somewhat declining at 100% slip. Table 2.3 shows the average values of tractive effort coefficients on various roads.[2]

TABLE 2.3

Average Values of Tractive Effort Coefficient on Various Roads

Surface	Peaking Values (μ_p)	Slipping Values (μ_s)
Asphalt and concrete (dry)	0.8–0.9	0.75
Concrete (wet)	0.8	0.7
Asphalt (wet)	0.5–0.7	0.45–0.6
Gravel	0.6	0.55
Earth road (dry)	0.68	0.65
Earth road (wet)	0.55	0.4–0.5
Snow (hard packed)	0.2	0.15
Ice	0.1	0.07

2.8.2 Braking Distribution on Front and Rear Axles

Figure 2.26 shows the forces acting on a vehicle during braking on a flat road. Rolling resistance and aerodynamic drag are ignored in this figure because they are quite small compared to the braking forces. j is the deceleration of the vehicle during braking, which can be easily expressed as

$$j = \frac{F_{bf} + F_{br}}{M},$$

(2.53)

where F_{bf} and F_{br} are the braking forces acting on the front and rear wheels, respectively.

The maximum braking force is limited by the tire–ground adhesion and is proportional to the normal load acting on the tire. The actual braking force developed by the brake torque should also be proportional to the normal load so that both the front and rear wheels obtain their maximum braking force at the same time. During braking, there is load transfer from the rear axle to the front axle. By considering the equilibrium of moments about the front and rear tire–ground contact points A and B, as shown in Figure 2.26, the normal loads on the front and rear axles W_f and W_r, with a vehicle deceleration rate, j, can be expressed as

$$W_f = \frac{Mg}{L}\left(L_b + h_g \frac{j}{g}\right)$$

(2.54)

and

$$W_r = \frac{Mg}{L}\left(L_a - h_g \frac{j}{g}\right).$$

(2.55)

The braking forces applied on the front axle and the rear axle should be proportional to their normal load; thus, one obtains

$$\frac{F_{bf}}{F_{br}} = \frac{W_f}{W_r} = \frac{L_b + h_g j/g}{L_a - h_g j/g}.$$

(2.56)

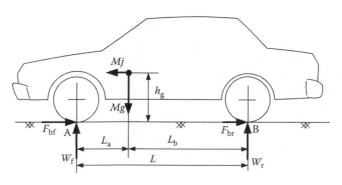

FIGURE 2.26
Force acting on a vehicle during braking on a flat road.

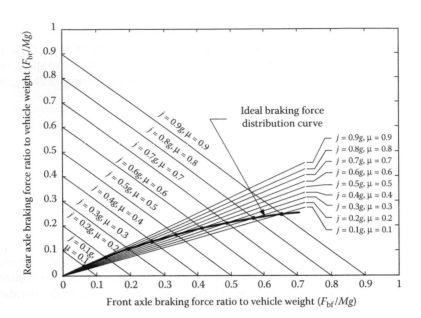

FIGURE 2.27
Ideal braking force distribution curve on front and rear axles.

Combining Equations 2.53 and 2.56, the ideal braking force distribution on the front and rear axles can be obtained as shown in Figure 2.27. When the braking is strong, both the front and rear wheels obtain their maximum ground braking force, which is limited by the capability of the tire–ground adhesion (wheel lock for non-antilock brake system (ABS) or by the activation of the ABS, which prevents the braking force from rising so the wheels don't become locked). In this case, the vehicle achieves its maximum deceleration rate as

$$|j_{max}|_\mu = \frac{F_{bf\text{-}max} + F_{br\text{-}max}}{M} = \frac{(W_f + W_r)\mu}{M} = g\mu. \tag{2.57}$$

The ideal braking force distribution curve (I curve), as shown in Figure 2.27, is a nonlinear hyperbolic curve. If it is desired for the front and rear wheels to lock or the ABS to function at the same time on any road, the braking force on the front and rear axles must follow this curve exactly.

Completely following the I curve for the braking force distribution makes the system very complex in terms of structure and control. However, with rapid advances in electronics and microcontrol technologies, electric braking systems (EBSs) are being developed, which can greatly improve the braking performance compared with the traditional design currently used in most vehicles. This technology is briefly described in the chapter on regenerative braking (Chapter 13).

Traditionally, the actual braking forces applied to the front and rear axles by the brake system are usually designed to have a fixed linear proportion. This proportion is represented by the ratio of the front axle braking force to the total braking force of the vehicle:

$$\beta = \frac{F_{bf}}{F_b}, \tag{2.58}$$

where F_b is the total braking force of the vehicle ($F_b = F_{bf} + F_{br}$). β depends only on the braking system design, such as the diameters of the wheel cylinders in the front and rear wheels, and has nothing to do with the vehicle parameters. With a value of β, the actual braking forces on the front and rear axles produced by the brake system can be expressed as

$$F_{bf} = \beta F_b \tag{2.59}$$

and

$$F_{br} = (1 - \beta)F_b. \tag{2.60}$$

Thus, one obtains

$$\frac{F_{bf}}{F_{br}} = \frac{\beta}{1 - \beta}. \tag{2.61}$$

Figure 2.27 shows the ideal and actual braking force distribution curves (labeled I and β curve). It is obvious that only one intersection point exists, at which the front and rear axles lock at the same time. This point represents one specific road adhesive coefficient, μ_0. Referring to Equation 2.56 in which j/g is replaced by μ_0 and Equation 2.61, one obtains (Figure 2.28)

$$\frac{\beta}{1 - \beta} = \frac{L_b + \mu_0 h_g}{L_a - \mu_0 h_g}. \tag{2.62}$$

From Equation 2.62 one can obtain μ_0 and β using

$$\mu_0 = \frac{L\beta - L_b}{h_g} \tag{2.63}$$

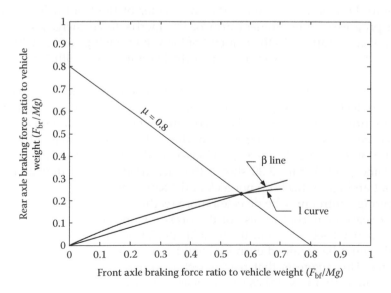

FIGURE 2.28
Ideal and actual braking force distribution curves.

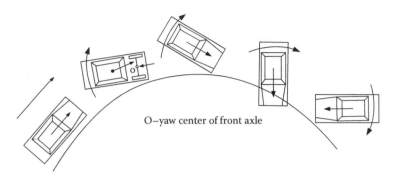

FIGURE 2.29
Loss of directional stability due to the lockup of rear wheels.

and

$$\beta = \frac{\mu_0 h_g + L_b}{L}.$$ (2.64)

During braking on roads with an adhesive coefficient less than μ_0 (the region in which the β curve is below the I curve), the front wheels lock first, whereas when the road adhesive coefficient is larger than μ_0 (the region in which the β curve is above the I curve), the rear wheels lock first.

When the rear wheels lock first, the vehicle will lose directional stability, as shown in Figure 2.29, which shows the top view of a two-axle vehicle acted upon by the braking force and the inertia force. When the rear wheels lock, the ability of the rear tires to resist lateral forces is reduced to zero (refer to Figure 2.25). If some slight lateral movement of the rear wheels is initiated by side wind, road camber, or centrifugal force, a yawing moment due to the inertia force about the yaw center of the front axle develops. As the yaw motion progresses, the moment arm of the inertia force increases, resulting in an increase in yaw acceleration. As the rear end of the vehicle swings around 90°, the moment arm gradually decreases, and eventually the vehicle rotates 180° with the rear end leading the front end.

The lockup of front wheels causes a loss of directional control, and the driver is no longer able to exercise effective steering. It should be pointed out, however, that front wheel lockup does not cause directional instability. This is because whenever the lateral movement of the front wheels occurs, a self-correcting moment due to the inertial force of the vehicle about the yaw center of the rear axle develops. Consequently, it tends to bring the vehicle back to a straight-line path. Figure 2.30 shows the measured angular deviation of a vehicle when the front and rear wheels do not lock at the same instant.[2]

Loss of steering control may be detected more readily by the driver, and control may be regained by release or partial release of the brakes. Contrary to the case of front wheel lockup, when rear wheels lock and the angular deviation of the vehicle exceeds a certain level, control cannot be regained even by complete release of the brakes and by the most skillful driving. This suggests that rear wheel lockup is a more critical situation, particularly on a road with a low adhesive coefficient. Because the value of the braking force is low on slippery surfaces, the kinetic energy of the vehicle dissipates at a low rate, and the vehicle will experience a serious loss of directional stability over a considerable distance.

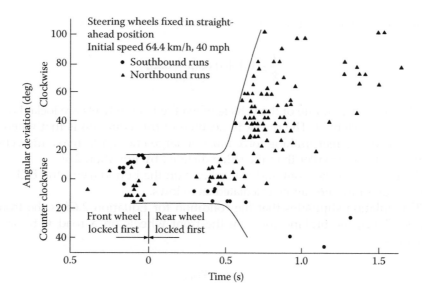

FIGURE 2.30
Angular deviation of a car when all four wheels do not lock at the same instant.

Therefore, designers of vehicle brake systems must ensure that the rear wheels do not lock first.

The ABS, developed in recent years, can effectively prevent wheels from locking up. This system employs speed sensors to detect the wheel rotation speed. When a wheel lockup is detected, the braking pressure control system reduces the pressure and brings the wheel back to its rotation.[3]

2.8.3 Braking Regulation and Braking Performance Analysis

2.8.3.1 Braking Regulation

As described previously, if the real braking force distribution line β is below the ideal braking force distribution curve I, as shown in Figure 2.28, the front wheels will be locked earlier than the rear wheels. This situation leads to stable behavior of the vehicle. This usually is the design, especially for passenger cars which run at high speed. However, when the β line is much below the I curve, most of the braking force is applied to the front wheel and a very small force to the rear wheels. This design causes the problem of reduced utilization of road adhesive capability. That is, when the front wheels are locked and the rear wheels are not locked, the maximum braking force on the rear wheels will never be used. To avoid this situation, some brake design regulations have been developed. A typical one is the ECE brake regulation.

The ECE brake design regulation for passenger cars is expressed by

$$\frac{F_{bf}}{W_f} \geq \frac{F_{br}}{W_r}. \tag{2.65}$$

Equation 2.65 shows that the rear wheels are never locked before the front wheels. In other words, the real braking force distribution curve β is always below the I curve. ECE also dictates the minimum braking force on rear wheels, as expressed by

$$\frac{j}{g} \geq 0.1 + 0.85(\mu - 0.2), \tag{2.66}$$

where j is the deceleration rate of the vehicle when the front wheels are locked on the road with adhesive coefficient, μ. The physical meaning of this equation is that when the front wheels are locked, the rear braking force must be large enough to make the vehicle yield a deceleration rate not smaller than the value dictated by Equation 2.66.

The braking forces of the front and rear wheels on the boundary of the ECE regulation described by Equation 2.66 can be calculated as follows.

The ECE regulation stipulates that the condition for Equation 2.65 is the front wheels being locked. Thus, the braking force on the front wheels on a road with an adhesive coefficient, μ, is

$$F_{bf} = W_{f\mu}, \tag{2.67}$$

where W_f is the vertical loading on the front wheels, which is expressed at the vehicle deceleration rate by Equation 2.54, and the total braking force of the vehicle at the deceleration rate, j, is expressed by Equations 2.53 and 2.66 related to μ. Using all the equations mentioned above, the front and rear braking force can be calculated as shown in Figure 2.30. It must be noted that in Figure 2.30, the front and rear wheel braking forces on the ECE regulation curve at a deceleration rate j (point A with $j = 0.6g$, for example) do not mean that the road adhesive coefficient is $\mu = 0.6$, but larger than that, due to the unlocked rear wheels (Figure 2.31).

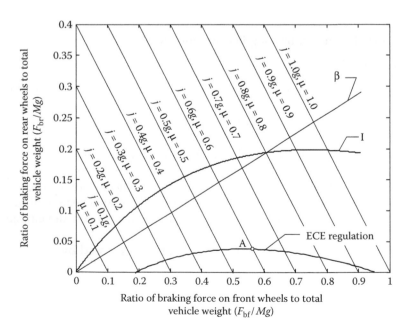

FIGURE 2.31
Minimum braking force on rear wheels stipulated by ECE regulation.

Obviously, the real braking force distribution achieved by the brake system design must fall into the area between the I curve and the ECE regulation curve.

2.8.3.2 Braking Performance Analysis

As mentioned previously, a vehicle with a traditional brake system design of a straight line real braking force distribution locks up its front and rear wheels simultaneously only on one kind of road with adhesive coefficient μ_0. On other roads, the front or rear wheels will lock up first. To fully understand the braking force scenarios on the front and rear wheels after the front or rear wheels lock, further analysis is introduced. This is helpful for the design of advanced braking systems for electric, hybrid electric, and fuel cell vehicles, which not only need to meet the braking performance requirements but also are capable of recapturing braking energy as much as possible.

1. *The case of locked front wheels and unlocked rear wheels:* When the front wheels are locked, the braking force on them is expressed as Equation 2.67. With the vehicle deceleration rate, j, the vertical load on the front wheels is expressed as Equation 2.54. Thus, the braking force on the front wheels can be expressed as

$$F_{bf} = \frac{Mg\mu}{L} \geq \left(L_b + \frac{j}{g}h_g\right). \tag{2.68}$$

Since

$$F_{bf} + F_{br} = Mj, \tag{2.69}$$

then

$$F_{bf} = \frac{Mg\mu}{L}\left(L_b + \frac{F_{bf} + F_{br}}{Mg}h_g\right). \tag{2.70}$$

Finally, we obtain

$$F_{br} = \frac{L - \mu h_g}{\mu h_g}F_{bf} - \frac{M_g L_b}{h_g}. \tag{2.71}$$

With a different road adhesive coefficient, μ, Equation 2.71 generates a group of lines (referred to as f lines) to represent the relationship between the braking forces and the front and rear wheels when the front wheels are locked and the rear wheels are not locked, as shown in Figure 2.32.

2. *The case of locked rear wheels and unlocked front wheels:* Similarly, when the rear wheels are locked and the front wheels are not, the braking force on the rear wheels against the braking force on the front wheels on roads with different adhesive coefficients can be expressed as

$$F_{br} = \frac{-\mu h_g}{L + \mu h_g}F_{bf} + \frac{\mu M_g L_a}{L + \mu h_g}. \tag{2.72}$$

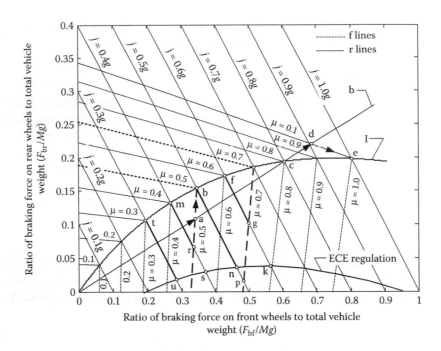

FIGURE 2.32
Braking process analysis on roads with different adhesive coefficients.

The lines generated by Equation 2.72 on roads with different adhesive coefficients are referred to as r lines, as shown in Figure 2.32.

3. *Braking process analysis:* Using the diagram in which the I curve, the β line, the ECE regulation curve, the f lines, and the r lines are plotted, as shown in Figure 2.32, the detailed braking process can be analyzed as follows:

a. *On a road with μ < μ₀ (μ = 0.5 and μ₀ = 0.8 in Figure 2.32 as the analyzing case)* On a road with $\mu < \mu_0$, the braking forces on the front and rear wheels are increased, as the brake pedal is depressed, along the real braking distribution line β, until point a: the intersection point of the β line and the f line with $\mu = 0.5$, where the front wheels are locked, but the rear wheels are not. Further depressing the brake pedal will cause a faster increase in the braking force on the rear wheels and a slow increase in the braking force on the front wheels, along the f line of $\mu = 0.5$, until we reach point b, where the rear wheels are also locked, and the vehicle achieves its maximum deceleration of $j = \mu g = 0.5g$. This case will not cause vehicle instability.

b. *On a road with μ > μ₀ (μ = 1.0 and μ₀ = 0.8 in Figure 2.32 as the analyzing case)* Similarly, when the brake pedal is depressed, the braking forces on the front and rear wheels rise along the β line until point d: the intersection point of the β line and the r line with $\mu = 1.0$, where the rear wheels are locked but the front wheels are not. Further depressing the brake pedal will cause the braking forces to develop along the r line with $\mu = 1.0$, finally reaching point e, where the front wheels are locked, and the vehicle achieves its maximum deceleration rate of $j = \mu g = 1.0g$. In this process, the braking force on the rear wheels has slightly decreased due to the load shifting from the rear to the front wheels. This case will cause vehicle instability.

c. *On a road with $\mu = \mu_0$ ($\mu = 0.8$ and $\mu_0 = 0.8$ in Figure 2.32 as the analyzing case)* Obviously, the front and rear wheels will be locked simultaneously at point c, where the vehicle achieves its maximum deceleration rate of $j = \mu g = 0.8g$. This is the ideal case.

4. *Maximum available braking force on front wheels*: In electric, hybrid electric, and fuel cell passenger cars, electric motors are mostly employed to drive the front wheels. This means that regenerative braking is only available for the front wheels. In the braking system design and control (mechanical and electrical), more braking energy should be allocated to the front wheels to increase the braking energy that is available for recovery, under the conditions of meeting brake regulation.

As shown in Figure 2.32, when the commanded braking deceleration rate, j, is smaller than μg, the braking forces on the front and rear wheels can be varied in a range rather than a specified point. For example, when $\mu = 0.7$ and $j = 0.6g$, this range is between point f and point g specified by the heavy solid line in Figure 2.32. Obviously, the maximum braking force on the front wheels is dictated by point g. However, if a lower deceleration rate is commanded on the same road, the variation range of the braking force will be larger. For instance, when $j = 0.5g$ and $\mu = 0.7$, the range is from point b to point p. However, this violates the ECE regulation, and therefore the maximum braking force on the front wheels is dictated by point n. Similarly, when $\mu = 0.5$ and $j = 0.4g$, the maximum braking on the front wheels is specified by point r, rather than s, and when $j = 0.3$ by point u. It is obvious that, with a small difference between the deceleration rate (g) and road adhesive coefficient, the maximum braking force is usually dictated by the f lines and, naturally, meeting the ECE regulation. However, when the deceleration rate (g) is much smaller than the road adhesive coefficient (e.g., slight braking on a good road), the ECE regulation will dictate the maximum braking force on the front wheels.

The preceding analysis provides the basis for the design and control of the hybrid brake system (mechanical plus electrical) of electric, hybrid electric, and fuel cell vehicles. More details will be discussed in the chapter on regenerative braking.

Bibliography

1. J. Y. Wong, *Theory of Ground Vehicles*, John Wiley & Sons, New York, 1978.
2. R. Bosch, *Automotive Handbook*, Robert Bosch GmbH, Karlsruhe, Germany, 2000.
3. S. Mizutani, *Car Electronics*, Sankaido Co., Minato-Ku, Tokyo, Japan, 1992.
4. I. Bolvashenkov, J. Kammermann, and H.-G. Herzog, Methodology for selecting electric traction motors and its application to vehicle propulsion systems, *Power Electronics, Electrical Drives, Automation and Motion (SPEEDAM), 2016 International Symposium* on, pp. 1214–1219, IEEE, 2016.
5. D. Manoharan, S. Chandramohan, S. Chakkath, and S. Maurya, Design, Development & Testing of Test Rig Setup for Unmanned Aerial Vehicle Propulsion Systems. No. 2017-01-2064. SAE Technical Paper, 2017.
6. W. Kriegler and S. Winter, A3PS: Austrian association for advanced propulsion systems, In *Automated Driving*, pp. 617–620, Springer International Publishing, 2017.

7. C. Capasso, M. Hammadi, S. Patalano, R. Renaud, and O. Veneri, A multi-domain modelling and verification procedure within MBSE approach to design propulsion systems for road electric vehicles. *Mechanics & Industry* 18(1), 2017: 107.

8. N. P. D. Martin, J. D. K. Bishop, and A. M. Boies, How well do we know the future of CO2 emissions? Projecting fleet emissions from light duty vehicle technology drivers. *Environmental Science & Technology* 51(5), 2017: 3093–3101.

9. H. Mirzaeinejad, M. Mirzaei, and R. Kazemi, Enhancement of vehicle braking performance on split-μ roads using optimal integrated control of steering and braking systems. *Proceedings of the Institution of Mechanical Engineers, Part K: Journal of Multi-Body Dynamics* 230(4), 2016: 401–415.

10. J. Lu, H. Hammoud, T. Clark, O. Hofmann, M. Lakehal-Ayat, S. Farmer, J. Shomsky, and R. Schaefer, A System for Autonomous Braking of a Vehicle Following Collision. No. 2017-01-1581. SAE Technical Paper, 2017.

11. L. Martinotto, F. Merlo, and D. Donzelli, Vehicle braking systems and methods. U.S. Patent Application 15/184,806, filed June 16, 2016.

12. B. Anthonysamy, A. K. Prasad, and B. Shinde, Tuning of Brake Force Distribution for Pickup Truck Vehicle LSPV Brake System During Cornering Maneuver. No. 2017-01-2491. SAE Technical Paper, 2017.

13. M. Rosenberger, M. Plöchl, K. Six, and J. Edelmann, eds, The dynamics of vehicles on roads and tracks. *Proceedings of the 24th Symposium of the International Association for Vehicle System Dynamics (IAVSD 2015)*, Graz, Austria, August 17–21, 2015, Crc Press, 2016.

14. X.-T. Nguyen, V.-D. Tran, and N.-D. Hoang, An investigation on the dynamic response of cable stayed bridge with consideration of three-axle vehicle braking effects. *Journal of Computational Engineering* 2017, 2017: 4584657:1–4584657:13.

15. H. Zhang and J. Wang, Vehicle lateral dynamics control through AFS/DYC and robust gain-scheduling approach. *IEEE Transactions on Vehicular Technology* 65(1), 2016: 489–494.

16. M. Corno, F. Roselli, L. Onesto, S. Savaresi, F. Molinaro, E. Graves, and A. Doubek, Longitudinal and Lateral Dynamics Evaluation of an Anti-Lock Braking System for Trail Snowmobiles. No. 2017-01-2512. SAE Technical Paper, 2017.

3

Internal Combustion Engines

The internal combustion (IC) engine is, and will be in the foreseeable future, the most popular power plant for motor vehicles. In hybrid electric vehicles (HEVs), the IC engine is also the first selection as the primary power source. However, its operation in HEVs differs from that in a conventional motor vehicle. The engine in a HEV runs for a longer time at high power and does not require changing its power rapidly. A specifically designed and controlled engine for HEV applications has not been fully developed. This chapter briefly reviews the key characteristics and performance of the commonly used spark ignited (SI) or gasoline IC engines, which are more related to HEV development. This chapter also reviews other types of engines that are possible for use in HEVs, such as four-stroke compression ignition (CI) engines (mostly fueled with diesel) and alternative-fuel engines.

3.1 Spark Ignition Engine

3.1.1 Basic Structure and Operation Principle with Otto Cycle

Most spark ignition (SI) engines are fueled with gasoline. In recent years, technologies involving the burning of alternative fuels, such as natural gas and ethanol, have been developed without major changes in engine structure. A conceptual structure of SI engines fueled with gasoline and operating on the Otto cycle is illustrated in Figure 3.1. It consists of a powering system (crankshaft, connection rod, piston, and cylinders), intake and exhaust system (air filter, throttle, inlet and exhaust manifolds, inlet and exhaust valves, and valve control cams), fuel supply (fuel tank [not shown], fuel pump [not shown], and fuel injectors), ignition system (battery [not shown], ignition coils [not shown], and spark plugs), cooling system (coolant, water pump radiator [not shown]), and lubricating system (not shown).

In most cases, the combustion of the air–fuel mixture formed within the inlet manifold and trapped in the cylinder produces heat, and then the temperature and pressure in the cylinder increase quickly. Thus, the piston is forced to move down. The connection rod transfers the linear movement of the piston into rotation of the crankshaft.

A four-stroke SI engine has four instinctive processes corresponding to the four strokes of each piston,[1,2] as shown in Figure 3.2.

1. Induction-stroke (cylinder filling process): While the piston travels down the cylinder from its top dead center (TDC), the valve cam opens the inlet valves and closes the exhaust valves. The air–fuel mixture formed in the inlet manifold is drawn into the cylinder, as shown in Figure 3.2a, until the piston arrives at its bottom dead center (BDC), where the valve cam closes the inlet valves (both inlet and exhaust valves are in a closed state).

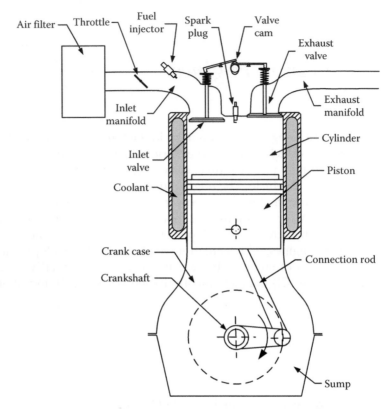

FIGURE 3.1
Spark-ignited gasoline engine.

2. Compression-stroke: While both the inlet and exhaust valves are closed, and the piston travels up the cylinder from its BDC, the in-charged air–fuel mixture in the cylinder is compressed, as shown in Figure 3.2b. As the piston approaches its TDC, the spark plug produces a spark, igniting the compressed air–fuel mixture in the cylinder, as shown in Figure 3.2c, and then starting a very quick combustion

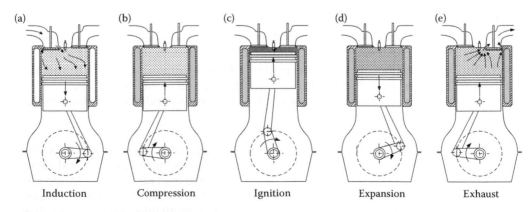

(a)	(b)	(c)	(d)	(e)
Induction	Compression	Ignition	Expansion	Exhaust

FIGURE 3.2
Four strokes of a SP engine.

of the compressed air–fuel mixture, which in turn causes a very sharp rise in the pressure and temperature in the cylinder.

3. Expansion stroke (power producing or working process): After the piston goes across the TDC, the high pressure in the cycler pushes the piston downward, as shown in Figure 3.2d. The piston transmits the pressure in the cylinder, through the connection rod, into force, which then turns the crankshaft. At the end the expansion stroke, the exhaust valve opens, and irreversible expansion of the exhaust gases blows out the exhaust valve.

4. Exhaust stroke: After the piston goes across the BDC, the exhaust valve remains open, and as the piston travels up the cylinder, the remaining combustion residual gases in the cylinder are expelled, as shown in Figure 3.2e. At end of the exhaust stroke, the exhaust valve closes. However, some exhaust gas residuals will be left. This exhaust dilutes the next charge. Following this stroke, the induction stroke of the next cycle starts.

One cycle is completed every two revolutions of the crankshaft (720 degrees of crank shaft movement). The power-producing stroke takes only one-fourth (180 degrees of crankshaft) of the complete cycle. The gear driver camshaft (for opening and closing the valves) must be driven by the mechanism operating at half crankshaft speed (engine speed). Some of the power from the expansion stroke is stored in the flywheel to provide the energy for another three strokes.

3.1.2 Operation Parameters

3.1.2.1 Rating Values

The common parameters that engine manufacturers advertise in their product technical specifications to indicate the key performance of the engine must include the following:

Maximum Rated Power: The highest power that an engine is allowed to develop for a short period of operation.

Normal Rated Power: The highest power that an engine is allowed to develop in continuous operation.

Rated Speed: The rotational speed of the crankshaft, at which the rated power is developed. For vehicle application, engine performance is more precisely defined by

1. The maximum power (or maximum torque): available at each speed within the useful engine operating speed range,
2. The range of speed and the power over which engine operation is satisfactory.

3.1.2.2 Indicated Torque and Indicated Mean Effective Pressure

The engine torque can be determined by the pressure variation in the cylinders along the crankshaft rotation. The term "indicated" means that the torque or power of the engine is evaluated in the scope of thermodynamics (pressure and volume of cylinder), not including any mechanical losses in the whole power development and transmission process. Figure 3.3 conceptually illustrates pressure variation in a cylinder along with crankshaft rotation angle. A more comprehensive diagram for demonstrating the working process of pressure in a cylinder is a pressure volume (PV) diagram, as shown in Figure 3.4.

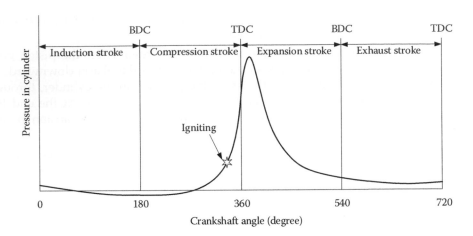

FIGURE 3.3
Pressure in cylinder along crankshaft rotation angle.

Referring to Figure 3.4, the pressure within the cylinder in an induction stroke (g→h→a) is usually lower than the atmospheric pressure into the cylinder due to the resistance of the air flow. In the compression stroke (a→b→c), the pressure increases with the upward movement of the piston. When the piston approaches the TDC, the spark plug produces a spark, igniting the air–fuel mixture trapped in the cylinder. With combustion of the air–fuel

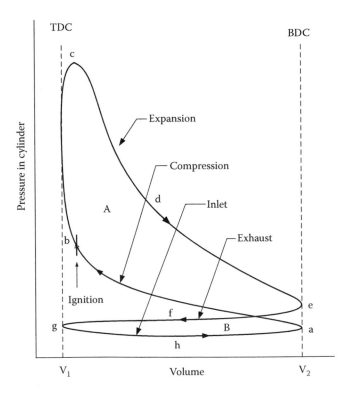

FIGURE 3.4
Pressure-volume diagram of a SP engine.

mixture, the pressure in the cylinder increases, quickly reaching its maximum at a certain crankshaft angle after the TDC (point c) and entering the expansion stroke. In the expansion stroke (c→d→e), the gases with high pressure within the cylinder push the piston downward, producing torque on the crankshaft through the connection rod. When the piston approaches the BDC, the exhaust value opens, the gases within the cylinder blow out the exhaust valve, and the process enters an exhaust stroke. In the exhaust stroke (e→f→g), the residual gases in the cylinder are propelled out of the cylinder at a higher pressure than the induction stroke.

The indicated torque is usually evaluated by the gross work done in one cycle, called the "gross indicated work," $W_{c,in}$, which is expressed as

$$W_{c,in} = \int_{\text{area } A} p \, dV - \int_{\text{area } B} p \, dV, \tag{3.1}$$

where p is the pressure in the cylinder, and V is the volume. The work done in area B is negative because the pressure in the induction stroke is lower than that in the exhaust stroke. To achieve more work in one cycle, area A should be made as large as possible by increasing the pressure in the expansion stroke, and area B should be made as small as possible by increasing the pressure in the induction stroke and decreasing it in the exhaust stroke. When the pressure in the induction stroke is greater than that in the exhaust stroke, the work in this area becomes positive. This is the case in a supercharged or turbocharged engine.

The indicated torque of an engine depends on the engine size (engine displacement, defined as the total geometric volume of all cylinders, usually measured in liters). A more useful relative performance measure associated with the indicated torque is the indicated mean effective pressure (imep), which is defined as the work per cycle per displacement:

$$\text{imep} = \frac{\text{work per cycle}}{\text{cylinder displacement}}. \tag{3.2}$$

The physical meaning of the imep is the work (Nm or joule or J) done by a unit cylinder displacement in one cycle, which is an important indicator for evaluating the performance of an engine. Engine indicator instruments are usually used to measure the pressure in cylinders, which is associated with variation of the cylinder volume (refer to Figure 3.4), and to calculate the work per cycle using Equation 3.1.

The imep is related to indicated torque as

$$\text{imep (kPa)} = \frac{2\pi n_R T_i \text{ (Nm)}}{V_d \text{ (Liter)}}, \tag{3.3}$$

where n_R is the number of revolutions of the crankshaft for each power stroke per cylinder ($n_R = 2$ for four-stroke engines and $n_R = 1$ for two-stroke engines), T_i is the indicated torque in Nm, and V_d is the displacement of the engine in liters. The indicated torque of an engine depends only on the imep in the cylinder and the engine displacement V_d. For a given engine size, increasing the imep is the only method for increasing the engine indicated torque.

With the imep measurement, the indicated torque can be obtained for Equation 3.3. It should be noted that imep or indicated torque is only an indicator of utilization efficiency of the cylinder volume, which is associated with volumetric efficiency (measurement of intake system), air–fuel ratio, valve timing, spark timing, and others. Indicated torque is only the equivalent crankshaft torque, mathematically transferred from the pressure inside the cylinders, not actually measured torque at the crankshaft.

The terms "indicated thermal efficiency" and "indicated specific fuel consumption" are usually used to evaluate the fuel utilization efficiency of the power-producing system (cylinder and piston). Indicated thermal efficiency is defined as indicated work done by the per-unit heat input, which is expressed as

$$\eta_i = \frac{P_i}{f h_u},$$

(3.4)

where P_i is the indicated power in W, which is expressed as $P_i = T_i \omega_e$, where T_i is the indicated torque in Nm, ω_e is the angular velocity of the engine in rad/s, f is the flow rate of the fuel into the cylinders in g/s, and h_u is the low heat value of the fuel in J/g. Correspondingly, the indicated specific fuel consumption (isfc) is expressed as

$$\text{isfc} = \frac{q}{P_i/1000},$$

(3.5)

where isfc is the indicated specific fuel consumption rate of grams per kilowatt-hour (g/kWh), and q is the fuel consumption in grams per hour (g/h).

3.1.2.3 Brake Mean Effective Pressure (bmep) and Brake Torque

Not all of the power and torque produced in the cylinder (indicated power and torque) are available on the crankshaft. Some is used to drive engine accessories and overcome the frictions inside the engine. The basic engine accessories include a water pump, oil pump, fuel pump, and valve shaft. Other accessories may depend on applications, such as cooling fan, alternator, power steering pump, and air conditioner compressor. In an engine test, the accessories that are driven by the engine crankshaft should be clearly specified. All power and torque requirements for driving accessories and overcoming frictions are grouped together and called the friction power P_f and friction torque T_f. The available power and torque on the crankshaft are the difference between the indicated power and torque and friction power and torque, respectively, expressed as

$$P_b = P_i - P_f$$

(3.6)

and

$$T_b = T_i - T_f,$$

(3.7)

where P_b and T_b are the brake power and brake torque measured on the crankshaft, P_i and T_i are the indicated power and indicated torque, as discussed in the previous section, and P_f and T_f are the friction power and torque measured on the crankshaft.

The ratio of brake power (useful power on the crankshaft) to indicated power is called the mechanical efficiency, η_m:

$$\eta_m = \frac{P_b}{P_i} = 1 - \frac{P_f}{P_i}. \tag{3.8}$$

The mechanical efficiency of an engine depends on the throttle position as well as on the design and engine speed. Typical values for modern automotive engines with wide-open throttles are 90% at speeds below about 1800–2400 rpm, decreasing to 75% at maximum rated speed. As the engine is throttled, the mechanical efficiency decreases, eventually to zero at idle operation. By removing the engine mechanical loss from the indicated work or imep, one can obtain the net work or bmep that is measured on the crankshaft. The maximum bmep of good engine designs is well established and is essentially constant over a large range of engine sizes. Typical values for bmep are as follows.[1] For naturally aspirated SI engines, maximum values are in the range 850–1059 kPa (125–150 psi) at the engine speed where maximum torque is produced. At the maximum rated power, bmep values are 10%–15% lower. For turbocharged automotive SI engines, the maximum bmep is in the range 1250–1700 kPa (180–250 psi). At the maximum rated power, bmep is in the range 900–1400 kPa (130–200 psi).

Brake specific fuel consumption (bsfc) and brake efficiency are usually used for evaluating the fuel utilization efficiency. In engine tests, fuel consumption is measured as a flow rate, *f*—mass flow per unit time, usually in grams per second (g/s). The physical meaning of the bsfc is the fuel consumption (grams) per unit work output (kWh) from the crankshaft, which is expressed as

$$\text{bsfc} = \frac{3600f}{P_b}, \tag{3.9}$$

where f is the fuel flow rate in g/s, and P_b is the brake power measured in the crankshaft in kW. Under the International System of Units, the bsfc is measured in g/kWh. Low bsfc values are obviously desirable. For SI engines, typical best values of bsfc are about 250–270 g/kWh.

A dimensionless term, "engine efficiency," is also used to measure engine fuel utilization efficiency, which is defined as the brake work (kWh) produced by unit heat (energy) contained in the fuel consumed as

$$\eta_f = \frac{1000\,P_b}{f\,h_u} = \frac{3.6 \times 10^6}{\text{bsfc} \times h_u}, \tag{3.10}$$

where P_b is brake power on the crankshaft in kW, f is the fuel flow rate in g/s, and h_u is the low heat value of the fuel in J/g.

3.1.2.4 Emission Measurement

The toxic emissions of an engine are mostly composed of carbon monoxide (CO), nitrogen oxides (NO_x), unburned hydrocarbons (HCs), and particulate matter (PM). The level of each emission element is usually measured by the volumetric concentration of the emission element to the exhaust gases in milligrams per cubic meter (mg/m^3) or micrograms per cubic meter ($\mu g/m^3$) (or ppm [parts per million]) of exhaust gas. For comparison, at the same base, the volume of the exhaust gas should be adjusted to standard conditions at a total pressure of 101.3 kPa, water steam pressure of 1 kPa, and temperature of 273 K. PM can be characterized by particle numbers (PN), and the units are $\#/m^3$ or $\#/km$.

Specific emission, similar to specific fuel consumption, is also commonly used to measure the level of emissions. Specific emission is defined as the emission mass (gram) per kWh shaft work output (g/kWh).

3.1.2.5 Engine Operation Characteristics

One of the most serious concerns in terms of engine operation characteristics is torque and power variations with engine rotation speed at wide-open throttle, which is usually referred to as the torque or power performance of the engine and strongly dictates vehicle performance with a given design. The typical wide-open throttle operating characteristics of an SI engine are shown in Figure 3.5. Indicated power is the average rate of work transfer from gases in the cylinders to the piston during the compression and expansion strokes. Brake power is obtained by subtracting friction power from indicated power. The brake power shows a maximum value at about a speed slightly less than the maximum speed of the engine. Indicated torque shows a maximum value in the mid-speed range, which approximately corresponds to the speed at which volumetric efficiency has the maximum value. Brake torque decreases more than indicated torque at high speed owing to greater friction loss. At partial load and fixed throttle position, these parameters behave similarly; however, at high speeds, torque decreases more rapidly than at full load, as shown in Figure 3.6. A partially opened throttle causes more resistance to flowing air at higher speed,

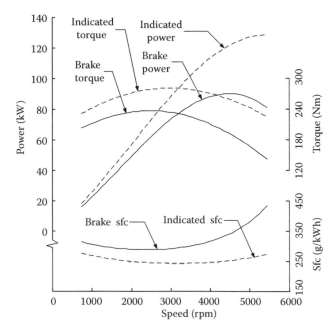

FIGURE 3.5
Indicated and brake powers, torques, and specific fuel consumptions varying with engine speed.

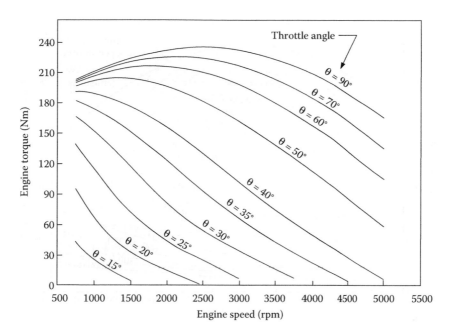

FIGURE 3.6
Torque characteristics with engine throttle opening and engine speed.

and volumetric efficiency decreases. The pumping components of total friction also increase as the engine is throttled.

Another major concern in connection with engine operation are the fuel consumption characteristics. The fuel consumption characteristics (brake-specific fuel consumption) of an engine vary widely with engine speed and load, as shown in Figure 3.7. Generally, an engine has its optimal operating region when the specific fuel consumption is minimized. This region is usually located in the middle of the speed range, corresponding to the maximum torque, where the losses in the induction and exhaust strokes are minimized. On the other hand, this region is close to full-load operation (wide-open throttle), where the percentage of losses to the total indicated power is small. In vehicle design, the operating points of the engine should be close to this region to achieve high operating fuel economy.

3.1.3 Basic Techniques for Improving Engine Performance, Efficiency, and Emissions

3.1.3.1 Forced Induction

The amount of torque produced in an IC engine depends on the amount of air inducted into its cylinders. An easy way of increasing the amount of air inducted is to increase the pressure in the intake manifold. This can be done by three means: variable intake manifold, supercharging, or turbocharging.

The intake manifold is like a wind instrument: it has resonant frequencies. A variable intake manifold tunes itself according to engine speed to exploit those resonant frequencies. If the tuning is done properly, the amount of air inducted into the cylinders can be optimized because the pressure in the intake manifold is increased. This technique improves the "breathing" of the engine but does not result in a very large increase of torque output.

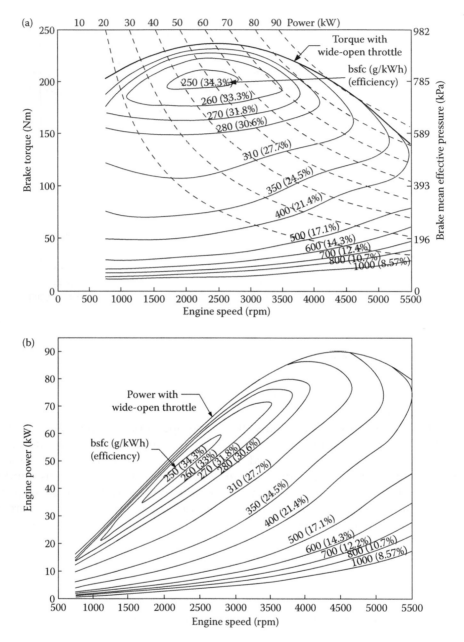

FIGURE 3.7
Fuel consumption characteristics of a typical SI engine.

Electrification of these components brings new boosting possibilities for decoupling the turbocharger, such as e-boosting (electric compressor) or electric turbocharger (e-turbo), for torque assist, or a turbo generator (turbine connected to an e-machine) for energy recovery.

A supercharger is an air compressor turned by the engine crankshaft. The compressed air is fed to the intake manifold. The advantage of a supercharger is that it can significantly increase the pressure in the intake manifold, even at low speed. The most significant disadvantage is that the supercharging power is taken from the engine crankshaft. This reduces the engine output and harms fuel consumption.

A turbocharger consists of a turbine driven by exhaust gases and of a compressor turned by the turbine. A turbocharger has the great advantage of taking its energy from the exhaust gases, which are normally wasted. Therefore, the efficiency of the engine does not suffer from the addition of the turbocharger. Turbocharging can tremendously increase the power output of the engine, especially if coupled to a charge cooling system. It also significantly improves efficiency because the higher intake pressure reduces the negative work associated with the induction stroke. The disadvantages of turbocharging include slow response, little or no effectiveness at low engine speed, and high rotational speed for the turbocharger, which increases the cost of forced induction.

Supercharging and turbocharging both suffer from two disadvantages: knock and emissions. Compressing the intake air also increases its temperature. An increased temperature means a greater risk of auto-ignition and knocks for the mixture and increased nitric oxide emissions. The solution to this problem consists in cooling down the intake air after compression by means of an intercooler or heat exchanger. The compressed air is passed through a radiator, while the ambient air or water is passed on the exterior of the radiator, removing the heat from the charge. The temperature of inducted air can be reduced sufficiently to avoid auto-ignition and knock. Nitric oxide emissions are also reduced. It should be noted that an engine designed for forced induction has a lower compression ratio than an engine designed for normal induction. Cooling the inducted air is also beneficial for torque production because cooler air is denser air. Therefore, more air can be inducted into the cylinder if it is cooler.

3.1.3.2 Gasoline Direct Injection and Lean-Burn Engines

HC and CO emissions can be reduced if the engine burns a lean mixture. If an SI engine could run on extremely lean mixtures, then the emissions would be very significantly reduced. However, ultralean mixtures pose problems because the flame has trouble propagating, and NO_x emission increases.

Gasoline direct injection is one means of achieving a very efficient mixing. Because the injector is located in the cylinder, it must inject the fuel at high pressure, which reduces the size of the fuel droplets. The fuel can be injected close to the spark plug, thereby enriching locally the mixture and allowing better ignition of the intake air. Additional advantages of gasoline direct injection include the cooling of intake air, which reduces knock and allows for operation at a higher compression ratio. This further improves the engine efficiency. Besides the cost increases, gasoline direct injection results in higher NO_x emissions.

3.1.3.3 Multivalve and Variable Valve Timing

While many engines use only two valves, high-performance engines use three, four, or five valves to increase the intake flow area. Multiple valves provide a significant increase of torque at high engine speed but sacrifice low-speed torque because the larger intake flow area results in slower flows at low speed. Multiple valves imply multiple camshafts, which increases the cost and complexity of the engine. Modern high-performance engines can feature variable valve lift and timing.

3.1.3.4 Variable Compression Ratio

A variable compression ratio allows for operating a forced induction engine at an optimal compression ratio at any intake pressure. If the charging mechanism does not provide maximum intake pressure, the compression ratio of the engine can be increased without risking auto-ignition or knock. The increased compression ratio results in improved fuel economy at partial torque output.

Variable valve systems that completely switch off cylinders, so-called cylinder deactivation, modify the compression ratio. Fully variable control of the valves can control the engine torque in a fully variable way without the use of the intake throttle, as in the skip fire concept.

3.1.3.5 Exhaust Gas Recirculation

Exhaust gas recirculation (EGR) consists in readmitting some of the exhaust gases into the combustion chamber to reduce the effective displacement of the engine. This technique is used in conventional vehicles to decrease the fuel consumption at a partial torque output, while preserving the acceleration capabilities of the engine. The greatest benefit of EGR is in emission reduction because it reduces the amount of fuel burned in the chamber and, therefore, the temperature of the exhaust gases. The nitric oxide emissions are greatly reduced.

Further classification distinguishes between *internal* (modifying the exhaust valve opening to retain combusted gases) and *external* (which can be tapped off before or after the turbine).

3.1.3.6 Intelligent Ignition

Intelligent ignition systems can set the spark advances at their optimum valve at any operating speed and load for optimum performance, efficiency, and exhaust emissions. High-power ignition systems can prevent the loss of fire in any cylinder, especially for engines with a lean mixture combination or external EGR.

3.1.3.7 New Engine Materials

New materials developed for engine components will contribute to better fuel economy in two ways. First, ceramic materials can be expected to offer better thermal insulation than metallic ones, with corresponding lower heat transfer (and therefore lower heat loss) and, hence, higher thermal efficiency. Second, lightweight materials such as fiber-reinforced plastics with high tensile strength can save a lot of weight.

3.1.4 Brief Review of SI Engine Control System

A modern engine is equipped with an integrated electronic-based control system to control engine operation. The control activities are performed in connection with fuel injection control, controlling the fuel injection system to inject the correct amount of fuel into the cylinders based on the current engine operation, such as throttle opening, air flow rate, and temperature. The control system also performs ignition timing control, knock control, idle speed control, and engine diagnosis.

Figure 3.8 conceptually illustrates the basic engine control structure (not all signals are shown). A microprocessor electronic control unit (ECU) receives the engine operation signals from various sensors and generates control signals for control actuators based on the control algorithm code, which is preinstalled in the ECU. The sensors basically include an engine speed sensor, coolant temperature sensor, throttle opening sensor, air intake flow rate sensor, knock sensor, and manifold pressure sensor.

The fuel injection control is basically to correctly control the fuel injection amount. Lean combustion is usually desired for better fuel economy. However, too lean an air–fuel

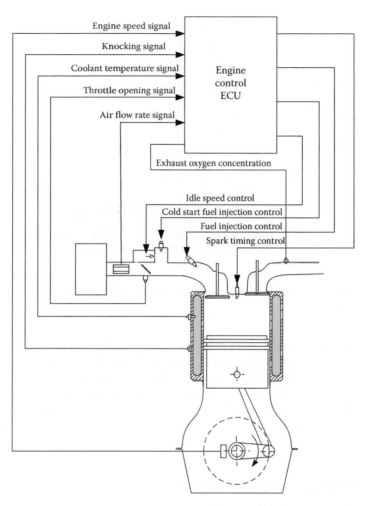

Engine speed signal

Knocking signal

Coolant temperature signal

Throttle opening signal

Air flow rate signal

Engine
control
ECU

Exhaust oxygen concentration

Idle speed control

Cold start fuel injection control

Fuel injection control

Spark timing control

FIGURE 3.8
SI engine control system.

mixture would cause misfire, deteriorating the engine performance (e.g., low torque output, high pollutants emission, bad fuel economy, high vibration). In addition, all modern engines are equipped with three-way catalytic converters for purification of the exhaust emission gases. Three-way catalytic converters are only effective in a narrow air–fuel ratio window, as shown in Figure 3.9,[3] which is very close to the stoichiometric air–fuel ratio. In the case of an engine started in cold conditions, the air–fuel mixture tends to be lean, since there is less evaporation of the fuel. An enrichment compensation is performed through cold start fuel injection control, as shown in Figure 3.8, for a certain time after the engine is started in the cold. After the time of enrichment for cold start, the enrichment may continue for a certain time while the engine is warming up, aiming to improve the engine performance, especially for heavy acceleration. The engine control system also regulates the engine idle speed by controlling the air flow to stabilize the idle speed.

The engine control system also performs spark timing control for optimizing the advance of the spark timing. In conditions of low temperature, high speed, and large throttle opening, spark timing is more advanced.

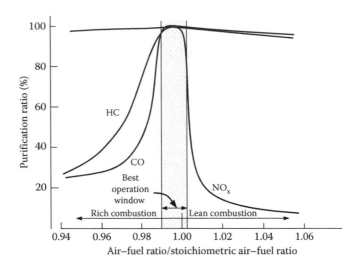

FIGURE 3.9
Purification effect of three-way catalytic converter versus air–fuel ratio. (Mizutani, S., 1992, *Car Electronics*, Sankaido Co., LTD, Warrendale, PA.)

The engine control system also performs "knocking" control. A knock detection sensor is used to detected engine "knocking." Once "knocking" is detected, the spark timing immediately retards the sparking advance. For more details on engine control, readers may consult the relevant literature. External EGR can be used to further improve knock relief.

Additional systems may include PM devices, such as a gasoline particle filter, either as an integrated four-way catalyst or as a separate device. Further active means may be required to purge the filter.

3.1.5 Operation Principle with Atkinson Cycle

3.1.5.1 Original Engine with Atkinson Cycle

The original Atkinson cycle engine has a structure that allows the intake, compression, expansion (powering), and exhaust stokes to occur in a single revolution of the crankshaft, utilization of cylinder volume in expansion (power), and exhaust strokes that are larger than that in the intake and compression strokes. Figure 3.10 shows the conceptual structure and operation principle of an Atkinson cycle engine.

The power-producing system of the original Atkinson cycle engine is composed of a crankshaft, link rod, swing arm, connection rod, piston, and cylinder, as shown in Figure 3.10. When all components are in position 1, the piston is in its TDC position. As the crankshaft rotates from position 1 to position 2, the piston goes down its first dead center. This stroke is the intake stroke, inhaling air–fuel mixture into the cylinder with opened intake valve and closed exhaust valve (intake stroke). As the crankshaft further rotates to position 3, the piston goes back up to its TDC, compressing the air–fuel mixture trapped in the cylinder, with both closed intake and exhaust valves (compression stroke). Meanwhile, as the cylinder approaches its TDC, the spark plug produces a spark, igniting the compressed air–fuel mixture in the cylinder. It is obvious that the seeping volume of the piston in the intake and compression strokes are from its TDC to the first BDC, as shown in Figure 3.10. After going across the TDC, the piston is forced to move down the cylinder

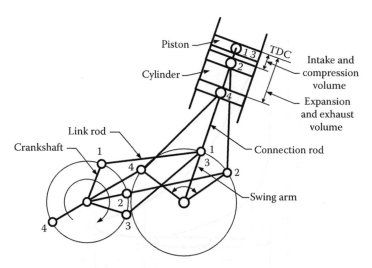

FIGURE 3.10
Conceptual structure and operation principle of Atkinson cycle engine.

to its second BDC; correspondingly, the crankshaft rotates to position 4 (expansion or powering stroke). At this position (position 4), the exhaust valve is open, and the crankshaft continuously rotates from position 4 to position 1, forcing the piston to move upward from the second dead center to the TDC, propelling the exhaust gases out of the cylinder (exhaust stroke). In this stroke, the sweeping volume of the piston is from the second dead center to the TDC, as shown in Figure 3.10.

Figure 3.11 conceptually illustrates the P-V diagram of an Atkinson cycle engine. Compared with an Otto cycle engine with the same cylinder displacement (V_1 to V_2), the Atkinson cycle engine produces more work in one cycle (the area labeled C in Figure 3.11), which could improve fuel utilization efficiency.

Another distinctive feature of the Atkinson cycle engine is its one power stroke per revolution of the crankshaft. This feature could enhance the engine power density.

3.1.5.2 Modern Engine with Atkinson Cycle

The original engine with an Atkinson cycle has not been widely manufactured and used in vehicles, perhaps due to its more complicated structure than modern, popularly used engines. However, in recent years, some engine manufacturers have employed the basic feature of the original Atkinson engine in modern engines without much change in the modern engine structure. The basic feature of the original Atkinson engine that appears in modern engines is the use of different cylinder volumes in the compression stroke and expansion stroke. This has been accomplished by changing only slightly the closed timing of the intake value—delaying the close timing of the intake valve to a point where the cylinder is moving up the cylinder.

Figure 3.12 conceptually illustrates the operation process of a four-stroke engine with an Atkinson cycle. While the piston moves down the cylinder from TDC with an opened intake valve, the air–fuel mixture in the intake manifold is inducted into the cylinder, as shown in Figure 3.12a. Rather than closing the intake valve as in an Otto cycle, the intake valve stays open as the piston moves up the cylinder, as shown in Figure 3.12b, and part of the induced air–fuel mixture is forced to flow back into the intake manifold until the piston reaches the

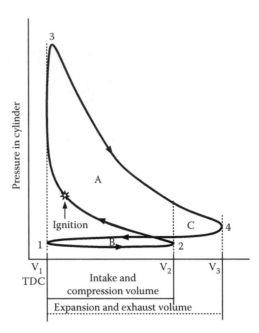

FIGURE 3.11
P-V diagram of an Atkinson cycle engine.

compression start center (CSC) position, where the intake valve is closed, and a compression stroke starts. Continuous upward movement of the piston compresses the air–fuel mixture in the cylinder, as shown in Figure 3.12c. As the piston approaches the TDC, the spark plug produces a spark, igniting the compressed air–fuel mixture in the cylinder. After the piston crosses the TDC, the high pressure inducted by the combustion of the air–fuel mixture in the cylinder pushes the piston down the cylinder, as shown in Figure 3.12d. As the piston approaches the BDC, the exhaust valve is opened, and the exhaust gases in the cylinder are pushed out. After the piston crosses the BDC, the upward movement of the piston propels the residual exhaust gases out of the exhaust valve, as shown in Figure 3.12e. The valve timing along the crankshaft rotation angle of an Atkinson engine is shown in Figure 3.13, where the valve timing of an Otto cycle is also shown.

It is obvious that the cylinder volume swept by the piston in intake and compression strokes is smaller than that in expansion and exhaust strokes. Figure 3.14 shows the *P-V*

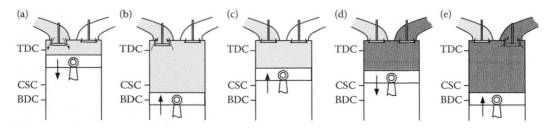

FIGURE 3.12
Four strokes with Atkinson cycle.

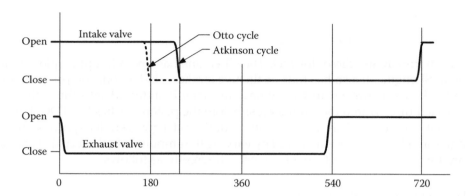

FIGURE 3.13
Valve timing diagram of Atkinson cycle engine. Labels in figure: intake, compression, combustion, exhaust.

diagram of an Atkinson engine. Compared with an Otto cycle engine that has the same compression volume as shown in Figure 3.14, the Atkinson engine produces more work, indicated by area C minus area D, which could improve the engine fuel utilization efficiency.

One disadvantage of the Atkinson cycle engine is the reduced utilization efficiency of cylinder volume (part of the geometric volume of the cylinder is used in intake and compressing strokes), thereby reducing the power density of the engine (power capacity per unit cylinder geometric volume). This disadvantage may not be serious for an hybrid electric vehicle (HEV) thanks to its reduced engine power requirement.

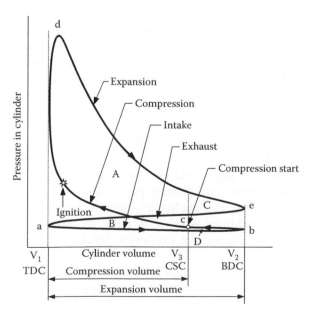

FIGURE 3.14
P-V diagram of Atkinson cycle engine.

3.2 Compression Ignition Engine

CI engines normally use diesel for fuel. The CI engine has operation principles similar to those of an SI engine. It also has four distinctive strokes—induction, compression, expansion, and exhaust. However, in CI engines, air alone is inducted into the cylinder. The fuel is injected directly into the cylinder just before the piston reaches the TDC. High temperature in the compressed air ignites the fuel. Air flow at a given engine speed is essentially unchanged, and load control is achieved by varying the amount of fuel injected at each cycle. Compared with SI engines, a CI engine works differently as follows:

1. The compression ratio is higher.
2. During the initial part of compression, only air is present.
3. The fuel–air mixture is always stoichiometrically weak.

These operation characteristics result in better fuel efficiency than in an SI engine. Furthermore, the CI engine is, in general, designed to operate at lower speeds, and consequently, friction losses are smaller.

Since the fuel–air ratio in a CI engine is always lean, CO emission is much lower than in SI engines and can be negligible. Unburned HCs in a properly regulated diesel engine come from two sources. First, around the perimeters of the reaction zone there is a mixture that is too lean to burn, and the longer the delay period, the greater the amount of HC emissions from this source. However, there is a delay period below which no further reductions in HC emissions are obtained. Under these conditions, the HC emissions mostly originate from a second source: the fuel retained in the nozzle sac (the space between the nozzle seat and the spray holes) and the spray holes. Fuel from these sources can enter the combustion chamber late in the cycle, thereby producing HC emissions.

The formation of NOx is strongly dependent on temperature, the local concentration of oxygen, and the duration of combustion. Thus, in diesel engines, NOx is formed during the diffusion combustion phase, on the weak side of the reaction zone. Reducing the diffusion-controlled combustion duration by increasing the rate of injection leads to a reduction in NOx emissions. Retarding the injection timing also reduces the NOx emissions since the later injection leads to lower temperature. However, injection retardation reduces the fuel efficiency of the engine and increases the exhaust temperature.

The black smoke from diesel engines originates from the fuel-rich side of the reaction zone in the diffusion-controlled combustion phase. After the rapid combustion at the end of the delay period, the subsequent combustion of the fuel is controlled by the rates of diffusion of air into the fuel vapor and vice versa and the diffusion of the combustion products away from the reaction zone. Carbon particles are formed by the thermal decomposition (cracking) of the large HC molecules, and soot particles form by agglomeration. The soot particles are oxidized when they enter the lean side of the reaction zone, and further oxidation occurs during the expansion stroke after the end of the diffusion combustion phase.

Smoke generation is increased by the high temperature in the fuel-rich zone during diffusion combustion. Smoke emission can be reduced by shortening the diffusion combustion phase because this gives less time for soot formation and more time for soot oxidation. The diffusion phase can be shortened by increased swirl, more rapid rejection, and a finer fuel spray. Advancing the injection timing can also reduce smoke.

3.3 Alternative Fuels and Alternative Fuel Engines

To date, fuels consumed in transportation have been dominated by petroleum products, gasoline and diesel. With concerns over Petroleum supply and the environment, interest is growing in switching a portion of traditional transportation fuels to alternative, nonpetroleum-based fuels. The development of electrical vehicles is a good practice. Due to the limitations of battery technologies, however, utilization of electric vehicles is still constrained in a relatively small scope. Charge-sustainable hybrid vehicles discussed in this book are still categorized in the vehicle section of traditional-fueled vehicles but with a significant improvement since their primary power sources, IC engines, are still fueled by gasoline or diesel. Plug-in hybrid vehicles can displace a portion of gasoline or diesel to electrical energy, but they still need an IC engine as the primary power source. This section will briefly review the technologies of alternative-fuel engines that can be used to replace the single power sources of conventional vehicles and the primary power sources of hybrid vehicles.

3.3.1 Alternative Fuels

The alternative fuels considered to be the most promising at present for fueling IC engines are ethanol, compressed natural gas (CNG), propane, and biodiesel. Table 3.1 shows comparisons of the afore mentioned fuels with gasoline and diesel.

3.3.1.1 Ethanol and Ethanol Engine

Ethanol is produced by fermentation from various plant materials collectively known as "biomass," such as sugar cane stalk, corn grain, sweet potato roots, cassava roots, and wood. Ethanol is a renewable energy source. Compared with gasoline, ethanol as SI engine fuel has the following advantages over gasoline:

1. High-octane value and high auto-ignition temperature allow an SI engine to use a high compression ratio, potentially enhancing the fuel utilization efficiency.
2. High latent heat value of vaporization effectively reduces the temperature of the air–fuel mixture in the manifold, potentially enhancing the volumetric efficiency of the engine (more air–fuel mixture is drawn into the cylinders).

Ethanol has some disadvantages, as compared with gasoline:

1. High latent heat value of vaporization may cause low vaporization degree at compression stroke, perhaps making cold starts difficult.
2. Its low energy density requires a larger fuel tank for similar traveling distances per tank of fuel.

Burning ethanol in a gasoline engine does not require changing the engine structure and control system hardware, but the control software should be changed to allow greater fuel injection, compared to burning gasoline, to maintain the air–fuel ratio at around its stoichiometric value (14.7 for gasoline, 9 for ethanol).

TABLE 3.1

Property Comparison of Alternative Fuels with Gasoline and Diesel

Property	Gasoline	No. 2 Diesel	Ethanol	Compressed Natural Gas	Propane	Hydrogen	Biodiesel
Chemical formula	C_4 to C_{12} hydrocarbon	C_8 to C_{25} hydrocarbon	C_2H_5OH	CH_4 (83%–89%) C_2H_6 (1%–13%)	C_3H_8	H_2	C_{12}-C_{22} FAME
Molecular weight	100–105	~200	46.07	16.04	44.1	2.02	~292
Carbon mass %	85–88	87	52.2	75	82	0	77
Hydrogen mass %	12–15	13	13.1	25	18	100	12
Oxygen mass %	0	0	49.9	–	–	0	11
Low heating value, mJ/kg	43.4	42.8	26.9	47.1	46.3	121.5	37.5
Octane No.							
Research octane No.	88–98	–	108.6	127+	112	130+	–
Motor octane No.	80–88	–	89.7	122	97	–	–
Autoignition temperature, °C	257	210	365	482–632	450	500	–
Flammability limits, volume %							
Lower	1.4	1	4.3	5.3	2.2	4.1	–
Higher	7.6	6	19	15	9.5	74	–
Stoichiometric air–fuel ratio, weight	14.7	14.7	9	17.2	15.7	34.3	13.8
Latent heat of evaporation, kJ/kg, at 15°C	348	232	921	509	449	447	–

An ethanol-fueled engine may experience difficulties with cold starts in a cold environment due to a higher latent heat value of vaporization. A common method for overcoming this difficulty is to blend in a certain portion of gasoline, such as E85, which is widely used in the United States. In addition, a heavy-duty ignition system may be helpful by increasing the ignition energy. The spark timing may be more advanced than burning gasoline.

3.3.1.2 Compressed Natural Gas and Natural Gas Engine

Natural gas is a gaseous fuel at room temperature with a much lower energy density than liquid fuels. A general way of storing natural gas onboard a vehicle is to compress it into a high-strength cylinder with a pressure of 20–30 MPa; a pressure regulator is used to reduce the high pressure to feed the engine.

Natural gas, compared with gasoline, has the advantage of having a very high octane number and auto-ignition temperature. This unique characteristic allows engines to be designed with a much higher compression ratio than burning gasoline, without the "knocking" problem. It is well known that a high compression ratio can significantly enhance the thermal efficiency of an engine. Its disadvantage is also well known: it requires a bulky and heavy gas storage tank (cylinders).

There are three types of natural-gas-fueled engines:

1. Dedicated

   A dedicated CNG engine is designed to run only on CNG. The engine is designed to have a higher compression ratio and a dedicated fuel supply and control system. Due to its dedicated characteristic, it generally has better performance than the two other types of engine.

2. Bi-fuel

   Bi-fuel engines have two separate fueling systems that enable them to run on either natural gas or gasoline. The bi-fuel systems are controlled by an engine management system that enables the engine to switch between CNG and gasoline mode.

   In CNG mode, the compressed gas in the fuel tank is fed via the fuel rail to a pressure regulator that reduces the gas pressure. Natural gas injectors inject precisely the required amount of gas into the inlet manifold. The air–gas mixture is then ignited by a spark plug. Currently, there are many conversion kit packages available on the market that can be used to easily convert a gasoline engine into a bi-fuel (gasoline and CNG) engine easily.

   The main advantage of a bi-fuel engine is that it can fully utilize cheaper CNG and, at the same time, retain the flexibility of burning gasoline. Obviously, the engine cannot use a high compression ratio as a dedicated CNG engine; thus, the high-octane value of CNG cannot be fully used.

3. Dual-fuel

   A typical dual-fuel engine is a CI engine that is fueled with natural gas and diesel. Burning both diesel and natural gas does not require changing the basic configuration of a CI engine. What is needed is to add a gas fueling system that operates in parallel with the diesel fueling system. The operating principle is shown in Figure 3.15.

In the intake stroke, as shown in Figure 3.15a, the natural gas, which is injected into the manifold by the gas injector and mixed with air, is drawn into the cylinder as the piston moves downward. After the piston goes across the BDC and continuously moves upward, the mixture of air/natural gas is compressed, as illustrated in Figure 3.15b. As the piston approaches the TDC, the diesel injector injects a certain amount of diesel into the cylinder, as illustrated in Figure 3.15c. The hot mixture of air/natural gas immediately ignites the

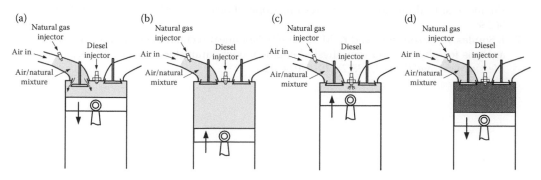

FIGURE 3.15
Natural gas–diesel dual-fuel engine.

diesel. The burning diesel further ignites the mixture of air/natural gas. The engine then goes into its expansion stroke, as shown in Figure 3.15d.

A CI engine usually has a higher compression ratio than a gasoline engine. The high compression ratio would not cause auto-ignition of the mixture of air/natural gas due to the high auto-ignition temperature of the natural gas. A high compression ratio can significantly enhance the thermal efficiency of an engine.

This duel-fuel engine can also be operated only with diesel, which makes the engine more fuel flexible.

3.3.1.3 Enhanced Hydrogen (H2 Combustion)

The addition of H2, which has a much higher lower heating value (LHV) compared to gasoline (Table 3.1), to the combustion process increases the laminar velocity of combustion, which improves combustion stability and permits the engine to run closer to the lean limit, perhaps with increased levels of external EGR, thereby improving emissions and performance. Further, knock resistance can be improved due to the reduced burn duration and slower auto-ignition chemistry, which facilitates an increase in compression ratio ϵ, improving the overall engine efficiency and fuel consumption.

The addition of H2 can be done externally, either by direct storage or onboard generation using methanol reformers or electrolysis of water.

Bibliography

1. J. B. Heywood, *Internal Combustion Engine Fundamentals*, McGraw-Hill Inc., New York, 1988.
2. R. Stone, *Introduction to Combustion Engines*, Second Edition, Society of Automotive Engineers (SAE), New York, NY, 1992.
3. S. Mizutani, *Car Electronics*, Sankaido Co., LTD, Warrendale, PA, 1992.
4. S. Petrovich, K. Ebrahimi, and A. Pezouvanis, MIMO (Multiple-Input-Multiple-Output) Control for Optimising the Future Gasoline Powertrain-A Survey (No. 2017-01-0600). 2017, SAE Technical Paper.
5. T. Q. Dinh, J. Marco, D. Greenwood, L. Harper, and D. Corrochano, Powertrain modelling for engine stop–start dynamics and control of micro/mild hybrid construction machines. *Proceedings of the Institution of Mechanical Engineers, Part K: Journal of Multi-Body Dynamics*, 2017, https://doi.org/10.1177/1464419317709894.
6. I. Souflas, A. Pezouvanis, B. Mason, and K. M. Ebrahimi, Dynamic Modeling of a Transient Engine Test Cell for Cold Engine Testing Applications. 2014, ASME Paper No. IMECE2014-36286.
7. I. Souflas, A. Pezouvanis, and K. M. Ebrahimi, Nonlinear recursive estimation with estimability analysis for physical and semiphysical engine model parameters. *Journal of Dynamic Systems, Measurement, and Control*, 138(2), 2016: 024502.

4

Vehicle Transmission

The vehicle transmission regulates the transfer of power (torque and speed) from the power plant (prime mover) to the driveline and the wheels. In the case of hybrid vehicles, the transmission becomes even more complex than in conventional or electric vehicles with two or more prime movers (inputs) and an output to the driveline/wheels. Figure 4.1 shows a typical automotive powertrain, which consists of a power plant (IC engine), drivetrain (transmission, final drive, differential, driveshaft), and drive wheels. The generated torque in the power plants is transmitted to the drive wheels through the drivetrain. There are different drivetrain configurations, and it is an expanding area of technology development in the automotive industry.

The transmission system plays a central role in determining the tractive force and the fuel consumption and energy regulation in the overall system. The torque and rotating speed from the output shaft of the power plant are transmitted to the drive wheels through the clutch or torque converter, gearbox, final drive, differential, and driveshaft. The clutch is used in a manual transmission to couple or decouple the gearbox to the power plant. The torque converter in an automatic transmission is a hydrodynamic device, functioning as the clutch in manual transmissions with a continuously variable gear ratio. The gearbox supplies a few gear ratios from its input shaft to its output shaft for the power plant torque–speed profile to match the requirements of the load. The final drive is usually a pair of gears that supply a further speed reduction and distribute the torque to each wheel through the differential.

In the design and analysis of the transmission systems, two major factors should be considered, first the degree of freedom (DOF). For example, a typical manual gearbox can have five DOFs (or five different gear ratios). Second, the transmission topology defines the geometrical properties, spatial relations, and the ways that the gears are arranged. This topology is further defined by the modes of operation of the transmission, for example the inputs/output in hybrid transmissions, interchanges during operation (e.g., in the case of regeneration or using brakes for stopping the spinning of a part of the transmission). In the case of electric vehicles, as mentioned in Section 4.2, it is not always necessary to have a gearbox; however, there are cases where a two-speed gearbox allows a smaller motor or battery system to provide the same power at optimum efficiencies.

4.1 Power Plant Characteristics

For vehicular applications, the ideal performance characteristic of a power plant is a constant power output over the full speed range. However, the torque varies with speed hyperbolically, as shown in Figure 4.2. With this ideal profile, the maximum power of the power plant will be available at any vehicle speed, yielding optimal vehicle performance. However, in practice, the torque is constrained to be constant at low speeds, so as not to be

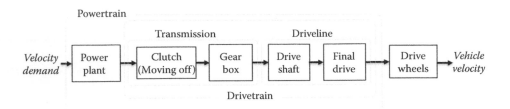

FIGURE 4.1
Transmission of an automobile power train.

over the maxima limited by the adhesion in the tire–ground contact area. This constant power characteristic provides the vehicle with high tractive effort at low speeds where demands for acceleration, drawbar pull, or grade climbing capability are high.

Internal combustion (IC) engines are the most commonly used power plants for land vehicles to date. Representative characteristics of a gasoline engine in wide-open throttle are shown in Figure 4.3, which has torque–speed characteristics far from the ideal performance characteristic required by traction. It starts operating smoothly at the idle speed. Good combustion quality and maximum torque are reached at an intermediate engine speed. As the speed further increases, torque decreases as a result of less air being inducted into the cylinders, caused by the growing losses in the air-induction manifold and grossing power losses caused by mechanical friction and hydraulic viscosity. Power output, however, increases to its maximum at a certain higher speed. Beyond this speed, the engine power starts declining. In vehicular applications, the maximum permissible speed of the engine is usually set just slightly above the speed of the maximum power output. An IC engine has a relatively flat torque–speed profile (as compared with an ideal power plant), as shown in Figure 4.3. Consequently, a multigear transmission is usually employed to modify it, as shown in Figure 4.4.

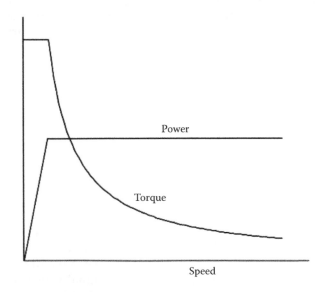

FIGURE 4.2
Ideal performance characteristics for a vehicle traction power plant.

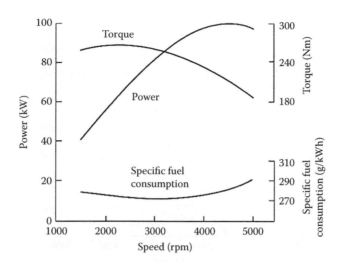

FIGURE 4.3
Typical performance characteristics of gasoline engines.

The electric motor is another candidate as a vehicle power plant and is becoming extremely important with the rapid development of electric, hybrid electric, and fuel cell vehicles. Electric motors with good speed adjustment control usually have a speed–torque characteristic that is much closer to the ideal, as shown in Figure 4.5.

Generally, the electric motor starts from zero speed. As it increases to its base speed, the voltage increases to its rated value while the flux remains constant. In this speed range of zero to base speed, the electric motor produces a constant torque. Beyond the base speed, the voltage remains constant, and the flux is weakened. This results in a constant output power while the torque declines hyperbolically with speed. Since the speed–torque profile

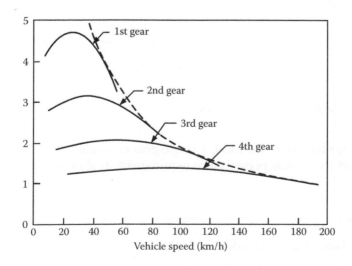

FIGURE 4.4
Tractive effort of an IC engine and a multigear transmission vehicle versus vehicle speed.

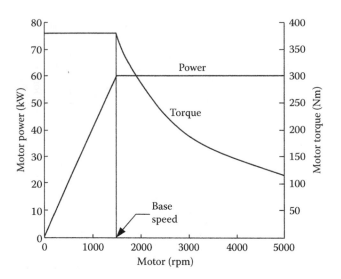

FIGURE 4.5
Typical performance characteristics of electric motors for traction.

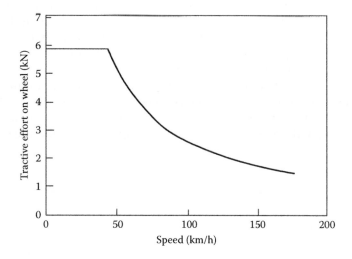

FIGURE 4.6
Tractive effort of a single-gear EV versus vehicle speed.

of an electric motor is close to the ideal, a single-gear or double-gear transmission may be employed to meet the vehicle performance requirement, as shown in Figure 4.6.

4.2 Transmission Characteristics

The transmission basically consists of a gearbox (set of gear trains with different gear ratios) and a power interruption device (clutch mechanism) that can be actuated mechanically, electrically, or hydraulically. The transmission requirements of a vehicle depend on the characteristics of the power plant and the performance requirements of the vehicle.

As mentioned previously, a well-controlled electric machine, such as the power plant of an electric vehicle (EV), would not need a multigear transmission, and currently, most EVs use single-speed transmissions; however, the application of multispeed transmissions in the future cannot be ruled out. Therefore, a review of conventional transmission technologies is given in this section that is also applicable to stop-start vehicles and mild hybrids.

Dedicated hybrid transmission (DHT) refers to purpose-built, full-hybrid transmissions that are introduced as the transmission system for electric/ IC engine hybrid vehicles. For example, the DHT introduced with the Toyota Prius can operate using at least two sources of propulsion: an internal combustion engine (ICE) and one or more electric motors or generators, in series, parallel, split power, or purely IC engine modes.

However, an IC engine must use a multigear or continuously varying transmission to multiply its torque at low speed. The term transmission here includes all those systems employed for transmitting the engine power to the drive wheels. For conventional IC engine automobile applications, there are usually two basic types of transmission: manual gear transmission and automatic transmission. Figure 4.7 shows the classification and types of the automotive transmission system. Transmission systems are divided into two main categories based on the discrete (fixed gear ratio) or continuously varying gear ratios (stepless gear ratio).

Furthermore, the attributes and characteristics of different transmission systems, such as the type of power transmission, mode of operation, and actuation system, are presented in Table 4.1.

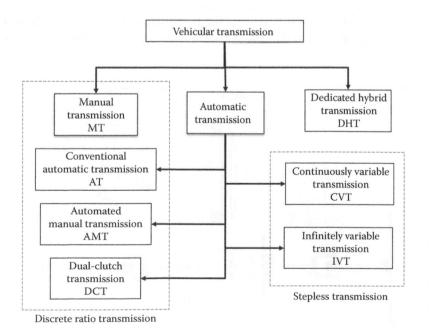

FIGURE 4.7
Classification and types of conventional automotive transmission systems.

TABLE 4.1

Attributes and Characteristics of Transmission Systems

Type	Clutch Type	Actuation System	Gear System	Mode of Operation	Gear Ratio
MT	Friction clutch	Foot pedal	Standard gearbox	Manual gear selection	Fixed gear ratios
AT	Torque converter	Hydrodynamic	Planetary gearbox	Automatic gear shift	Fixed gear ratios
AMT	Friction clutch	Electric actuator	Standard gearbox	Automatic gear shift	Fixed gear ratios
DCT	Friction clutch	Electric actuator	Standard gearbox	Automatic gear shift	Fixed gear ratios
CVT	Torque converter	Hydrodynamic	N/A	Automatic gear shift	Variable ratios
IVT	N/A	N/A	Planetary gearbox	Automatic gear shift	Variable ratios

4.3 Manual Gear Transmission (MT)

A manual gear transmission (MT) consists of a clutch, a gearbox, a final drive, and a driveshaft, as shown in Figure 4.8. The final drive has a constant gear ratio. The common practice of requiring direct drive (nonreducing) in the gearbox to be in the highest gear determines this ratio. The gearbox provides several gear ratios ranging from three to five for passenger cars and more for heavy commercial vehicles that are powered by gasoline or diesel engines.

Manual and automated manual and dual-clutch transmissions generally use a layshaft (or countershaft) design gearbox transmission, as shown in Figure 4.9, schematically for a typical five-speed MT gearbox, where all the gears, including reverse, use synchronizers. The friction clutch in an MT enables the gradual establishment and interruption of power flow from the engine to the gearbox, and it controls the engagement of the gearbox to the engine flywheel. The power enters the gearbox and transfers to the layshaft through the selected meshed gears to the output shaft of the transmission. The synchronizers are placed between every two gears, and in the case shown in Figure 4.9, the separation of synchronizer

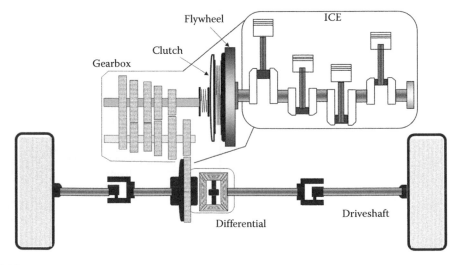

FIGURE 4.8

A typical conventional front wheel powertrain with a manual gearbox.

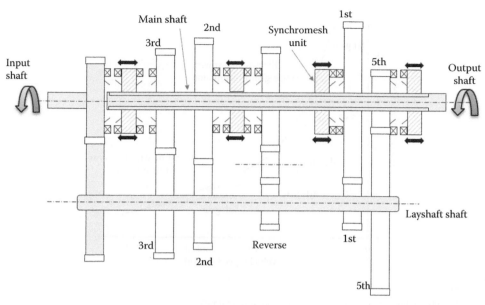

FIGURE 4.9
Speed layshaft manual gearbox.

functions and their operation are achieved by dog clutch "lock" for the first gear. The fourth gear has a ratio of 1:1 and is referred to as direct drive, and the top gear (fifth gear) is an overdrive where the output shaft runs faster than the input shaft (power plant speed).

The maximum speed of the vehicle determines the gear ratio of the highest gear (i.e., the smallest ratio). On the other hand, the gear ratio of the lowest gear (i.e., the maximum ratio) is determined by the requirement of the maximum tractive effort or the gradeability. Ratios between them should be spaced in such a way that they provide tractive effort–speed characteristics as close to the ideal as possible, as shown in Figure 4.10. In the first iteration of transmission design, gear ratios between the highest and lowest gears may be selected in such a way that the engine can operate in the same speed range for all the gears. This approach benefits the fuel economy and the performance of the vehicle. For instance, in normal driving, the proper gear can be selected according to vehicle speed to operate the engine in its optimum speed range for fuel-saving purposes. In fast acceleration, the engine can be operated in its speed range with high power output. This approach is depicted in Figure 4.11.

For a five-speed gearbox, the following relationship can be established:

$$\frac{i_{g1}}{i_{g2}} = \frac{i_{g2}}{i_{g3}} = \frac{i_{g3}}{i_{g4}} = \frac{i_{g4}}{i_{g5}} = K_g \tag{4.1}$$

and

$$K_g = \sqrt[3]{\frac{i_{g1}}{i_{g5}}}, \tag{4.2}$$

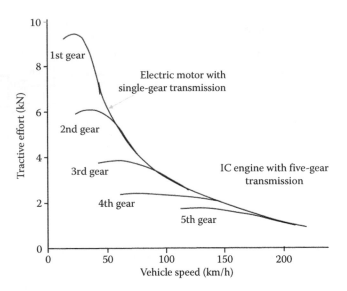

FIGURE 4.10
Tractive effort characteristics of a gasoline-engine-powered vehicle.

where i_{g1}, i_{g2}, i_{g3}, i_{g4}, and i_{g5} are the gear ratios for the first, second, third, fourth, and fifth gears, respectively. In the more general case, if the ratio of the highest gear, i_{gn} (smallest gear ratio), and the ratio of the lowest gear, i_{g1} (largest gear ratio), have been determined and the number of the gear n_g is known, the factor K_g can be determined as

$$K_g = \sqrt[n_{g-1}]{\frac{i_{g1}}{i_{gn}}} \tag{4.3}$$

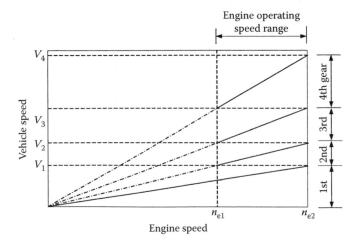

FIGURE 4.11
Demonstration of vehicle and speed ranges for each gear.

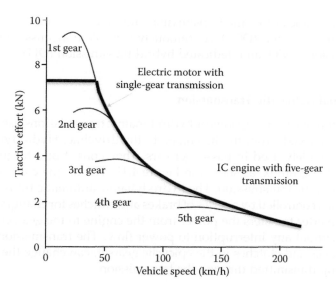

FIGURE 4.12
Tractive efforts of gasoline-engine vehicle with a five-gear transmission and EV with single-gear transmission.

and each gear ratio can be obtained by

$$i_{gn-1} = K_g i_{gn},$$

$$i_{gn-2} = K_g^2 i_{gn},$$

$$\vdots$$

$$i_{g2} = K_g^{n_g-1} i_{gn}. \tag{4.4}$$

For passenger cars that usually use a high gear in normal driving, the step between the ratios of the upper two gears is often slightly closer than that calculated from Equation 4.4:

$$\frac{i_{g1}}{i_{g2}} > \frac{i_{g2}}{i_{g3}} > \frac{i_{g3}}{i_{g4}} > \frac{i_{g4}}{i_{g5}}, \tag{4.5}$$

This in turn affects the selection of the ratios of the lower gears. For commercial vehicles, however, the gear ratios in the gearbox are often arranged based on Equation 4.5.

Figure 4.12 shows the tractive effort of a gasoline engine vehicle with a five-gear transmission and that of an EV with a single-gear transmission. Electric machines with favorable torque–speed characteristics can satisfy the tractive effort with a simple single-gear transmission.

4.4 Automatic Transmission

This type of transmission can automatically change gear ratios without any manual input from the driver. It can be broadly divided into a conventional automatic transmission

(CAT) or hydrodynamic automatic transmission, automated manual transmission (AMT) and dual-clutch transmission (DCT), continuously variable transmission (CVT), infinitely variable transmissions (IVT), and dedicated hybrid transmission (DHT).

4.4.1 Conventional Automatic Transmission

Conventional automatic transmissions or hydrodynamic transmissions use fluid to transmit power (torque and speed) from the IC engine to the driveline. Hydrodynamic automatic transmissions are widely used in passenger cars. They consist of a torque converter and planetary (epicyclic) gearbox, as shown in Figure 4.13. The torque converter is connected to the input shaft of the transmission. The hydrodynamic automatic transmission employs epicyclic gear trains controlled by a series of brakes and clutches for changing the gear ratios.

The torque converter transfers the power from the engine to the gearset in the transmission smoothly, without any interruption to power flow. The transmission's controller, by activating the brakes and clutches in the epicyclic gearset, can change the gear ratio while power is still being transmitted through the transmission.

4.4.1.1 Torque Converter Operation

The torque converter is a fluid coupling consisting of at least three rotary elements known as the impeller (pump), the turbine, and the reactor, as shown in Figure 4.14. The impeller is connected to the engine shaft, and the turbine is connected to the output shaft of the converter, which in turn is coupled to the input shaft of the multispeed gearbox. The reactor is coupled to the external housing to provide a reaction on the fluid circulating in the converter. The function of the reactor is to enable the turbine to develop output torque higher than the input torque of the converter, thereby obtaining torque multiplication. The reactor is usually mounted on a free wheel (one-way clutch) so that when the starting period is completed and the turbine speed is approaching that of the pump, the reactor is in free rotation. At this point, the converter operates as a fluid coupler with a 1:1 ratio of output torque to input torque.

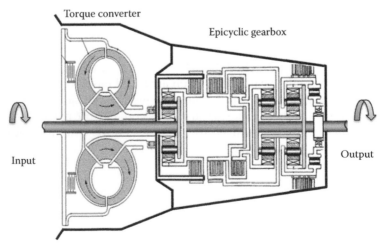

FIGURE 4.13
Conventional automatic transmission or hydrodynamic transmissions.

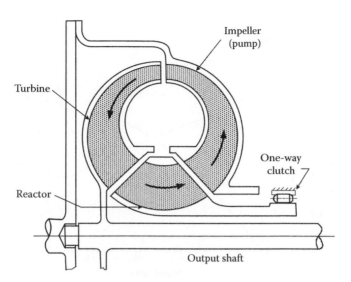

FIGURE 4.14
Schematic view of torque converter.

The major advantages of a hydrodynamic transmission may be summarized as follows:

- When properly matched, the engine will not stall.
- It provides flexible coupling between the engine and the drive wheels.
- Together with a suitably selected multispeed gearbox, it provides torque–speed characteristics that approach the ideal.

The major disadvantages of a hydrodynamic transmission are its low efficiency in a stop–go driving pattern and complex structure.

The performance characteristics of a torque converter are described in terms of the following four parameters:

$$\text{Speed ratio } C_{sr} = \frac{\text{output\_speed}}{\text{input\_speed}}, \tag{4.6}$$

which is the reciprocal of the gear ratio mentioned previously:

$$\text{Torque ratio } C_{tr} = \frac{\text{output\_torque}}{\text{input\_torque}}, \tag{4.7}$$

$$\text{Efficiency } \eta_c = \frac{\text{output\_speed} \times \text{output\_torque}}{\text{input\_speed} \times \text{input\_torque}} = C_{sr}C_{tr}, \tag{4.8}$$

$$\text{Capacity factor(size factor) } K_{tc} = \frac{\text{speed}}{\sqrt{\text{torque}}}. \tag{4.9}$$

The capacity factor, K_c, is an indicator of the ability of the converter to absorb or transmit torque, which is closely related to the size and geometric shape of the blades.

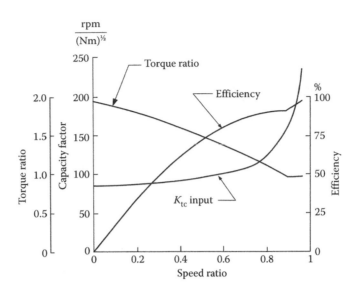

FIGURE 4.15
Performance characteristics of torque converter.

The typical performance characteristics of a torque converter are shown in Figure 4.15, in which the torque ratio, efficiency, and input capacity factor (the ratio of input speed to the square root of input torque) are plotted against the speed ratio. The torque ratio has the maximum value in a stalled state, where the output speed is zero. The torque ratio decreases as the speed ratio increases (gear ratio decreases), and the converter eventually acts as a hydraulic coupling with a torque ratio of 1.0. At this point, a small difference between the input and output speed exists because of the slip between the impeller (pump) and the turbine. The efficiency of the torque converter is zero in a stalled state (zero speed ratio) and increases with an increase in the speed ratio. It reaches its maximum when the converter acts as a fluid coupling (torque ratio equal to 1.0).

To determine the actual operating condition of the torque converter, the engine operating point must be specified since the engine directly drives the torque converter. To characterize the engine operating condition for the purpose of determining the combined performance of the engine and the converter, an engine capacity factor, K_e, is introduced and defined as

$$K_e = \frac{n_e}{\sqrt{T_e}}, \qquad (4.10)$$

where n_e and T_e are engine speed and torque, respectively. The variation of the capacity factor with speed for a typical engine is shown in Figure 4.16. To achieve proper matching, the engine and the torque converter should have a similar range in the capacity factor.

As mentioned previously, the engine shaft is usually directly connected to the input shaft of the torque converter:

$$K_e = K_{tc}. \qquad (4.11)$$

The matching procedure begins by specifying the engine speed and engine torque. Knowing the engine operating point, one can determine the engine capacity factor, K_e, using Equation 4.9 (Figure 4.15). Since $K_e = K_{tc}$, the input capacity factor of the torque converter corresponding to the specific engine operating point is then known. As shown in

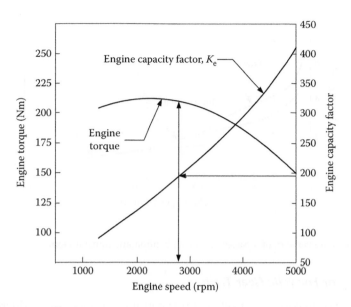

FIGURE 4.16
Capacity factor of typical engine.

Figure 4.15, for a particular value of the input capacity factor of the torque converter, K_{tc}, the converter speed ratio, C_{sr}, and the torque ratio, C_{tr}, can be determined from the torque converter performance characteristics, as shown in Figure 4.15. The output torque and the output speed of the converter are then given by

$$T_{tc} = T_e C_{tr} \tag{4.12}$$

and

$$n_{tc} = n_e C_{sr}, \tag{4.13}$$

where T_{tc} and n_{tc} are the output torque and the output speed of the converter, respectively.

Since the torque converter has a limited torque ratio range (usually less than 2), a multi-speed gearbox is usually connected to it. The gearbox comprises several planetary gear sets and is automatically shifted. With the gear ratios of the gearbox, the tractive effort and the speed of the vehicle can be calculated by

$$F_t = \frac{T_e C_{tr} i_g i_0 \eta_t}{r} \tag{4.14}$$

and

$$V = \frac{\pi n_e C_{sr} r}{30 i_g i_0}\,(\text{m/s}) = 0.377\frac{n_e C_{sr} r}{i_t}\,(\text{km/h}). \tag{4.15}$$

Figure 4.17 shows the variation of the tractive effort with speed for a passenger car equipped with a torque converter and a three-speed gearbox.

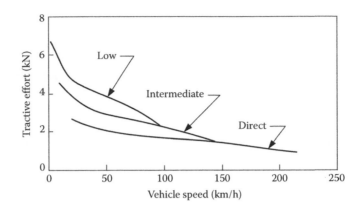

FIGURE 4.17
Tractive effort–speed characteristics of a passenger car with automatic transmission.

4.4.1.2 Planetary or Epicyclic Gear Train

The epicyclic gear train could be used as direct coupling to increase or decrease the power or speed as well as reverse the direction of rotation. Figure 4.18 shows an epicyclic train with an annulus gear (or ring gear) A and planet gears P orbiting around the sun gear S. The planet gears are in constant mesh with the sun gear S and the teeth on the internal circumference of the annulus gear. Each planet gear P rotates freely on a pin connected to a cage arm, referred to as the planet carrier (or arm) C. A brake could be applied to either the sun or the annulus gear or the carrier arm, creating zero speed to that particular part of the gear train.

Standard gear trains used in manual gearboxes have one DOF (one input and one output), while the epicyclic gear train has two DOFs (i.e., two inputs are needed to obtain an output).

To find the speed relationship between the three different shafts, the gear ratios can be determined from the numbers of teeth on the sun and annulus gears. The planet gears act as idlers and do not affect the gear ratio. The basic ratio R_{SA} between the sun and the annulus

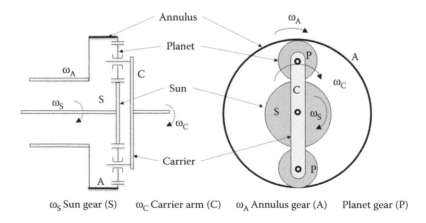

ω_S Sun gear (S) ω_C Carrier arm (C) ω_A Annulus gear (A)   Planet gear (P)

FIGURE 4.18
Planetary gear train arrangement.

is given by the number of teeth on the sun wheel with respect to the number of teeth on the annulus gear:

$$R_{SA} = \frac{-t_A}{t_S}.$$ (4.16)

The minus sign indicates that the two wheels are rotating in opposite directions. To derive the speed ratio between the sun and the annulus wheel, consider a case where the carrier shaft is kept stationary ($\omega_c = 0$). If ω_A^O and ω_S^O are the speeds of the annulus and the sun gear when the carrier arm is fixed, then the speed ratio between the sun and the annulus is

$$R_{SA} = \frac{\omega_A^O}{\omega_S^O}.$$ (4.17)

If the whole epicyclic gear train starts to rotate at a speed of ω_c, the speed of the annulus becomes

$$\omega_A = \omega_A^o + \omega_C,$$ (4.18)

and for the sun wheel it becomes

$$\omega_S = \omega_S^o + \omega_C.$$ (4.19)

Replacing Equations 4.18 and 4.19 in Equation 4.17 gives the basic ratio

$$R_{SA} = \frac{\omega_A - \omega_C}{\omega_S - \omega_C},$$ (4.20)

or it could be written

$$\omega_A = R_{SA}\omega_S + \omega_C(1 - R_{SA}).$$ (4.21)

Therefore, the speed ratio for any two shafts of the epicyclic gear train can be derived using the previous relationships. This is shown pictorially in Figure 4.19, based on the method presented by G.G. Lucas. The diagram shows the signal flow in the epicyclic gear train.

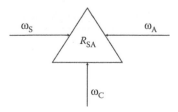

FIGURE 4.19
Block representation of signal flow in epicyclic gear train.

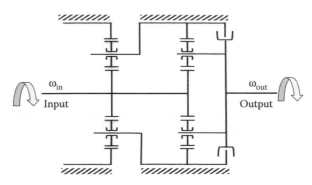

FIGURE 4.20
Simpson compound epicyclic gear train system.

4.4.1.3 Compound Epicyclic Gear

By connecting two or more epicyclic gear trains together, a compound gear train can be designed whereby using a number of clutches and brakes, different speed ratios are realizable. This technique is the basis of many automatic or dedicated hybrid transmission systems. A Simpson gear train is a simple example of a compounded gearbox containing two epicyclic gear trains, as shown in Figure 4.20.

The pictorial representation of the signal flow for a Simpson compound gear train is represented in Figure 4.21.

Another example of a compounded epicyclic gear train is the Wilson gearbox, which can provide four forward gear ratios and one reverse.

The other example is a Lepelletier gear train design, which is used in a ZF 6HP six-speed automatic transmission where a simple epicyclic gear train on the input side is followed by a Ravigneaux gear train on the output side using three clutches and two brakes (five shift elements enable six forward and one reverse speeds). Figure 4.22 shows schematically a Lepelletier transmission.

Planetary gear trains are particularly flexible devices, and in addition to being employed in automatic gearboxes, they are also commonly used in applications such as full hybrids, CVTs, IVTs, differentials, transfer boxes, and overdrive units.

The gear ratio of these devices can be varied by changing the speed of a third shaft, while the torque distribution is fixed by the number of teeth on the sun and annulus gears.

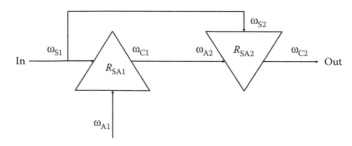

FIGURE 4.21
Block representation of signal flow in Simpson compounded epicyclic gear train.

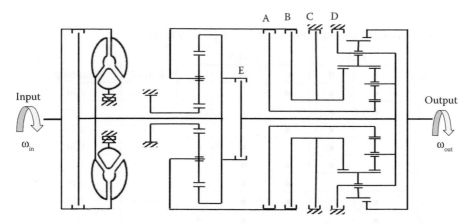

FIGURE 4.22
Lepelletier transmission.

The layout of a typical compound gear train used in conventional automatic transmissions such as a ZF eight-speed automatic (8HP) employing four simple epicyclic gearsets used in the Lexus is shown in Figure 4.23.

The gearset uses a total of five shift elements (three clutches and two brakes). The steps and the ratios for all gears are shown in Table 4.2.

4.4.2 Automated Manual and Dual-Clutch Transmission

The AMT or sequential manual transmission and the DCT employ computer-controlled servo systems to change the gear ratios of synchromesh gearboxes automatically. To change gears, clutches are needed to interrupt the power flow between the engine and the gearbox. Therefore, an AMT or DCT consists of friction clutches, gearbox (gear sets and synchromesh system similar to the manual transmissions), gear change servo mechanism, and transmission control unit (TCU). By employing hydraulic or electric motor servos to operate the clutch and gear shift mechanisms, the desirable speed and torque based on the throttle

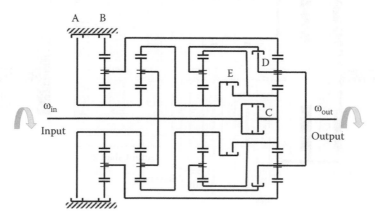

FIGURE 4.23
Schematic showing layout for ZF 8HP eight-speed transmission.

TABLE 4.2

Shift Control Element Status and Gear Steps for ZF 8HP

Gear	Brake A	Brake B	Clutch C	Clutch D	Clutch E	Ratio i	Gear Step
1	■	■	■			4.696	1.50
2	■	■			■	3.130	1.49
3		■	■		■	2.104	1.26
4		■		■	■	1.667	1.30
5		■	■	■		1.285	1.29
6			■	■	■	1.000	1.19
7	■		■	■		0.839	1.25
8	■			■	■	0.667	Total 7.05
R	■	■		■		−3.297	

position can be achieved. AMTs and DCTs usually operate in fully automatic, economy, or sporty modes. They usually have better gearshift performance and fuel economy than manual gearboxes with the convenience of automatic transmissions.

Figure 4.24 shows a DCT, also known as a direct shift gearbox (DSG) or twin-clutch transmission, which is an automatic transmission system based on two layout synchromesh gearboxes with twin clutches. It is an automated manual transmission with two clutch and gearbox systems. It operates by preselecting the required gears, thereby reducing the gear change delay.

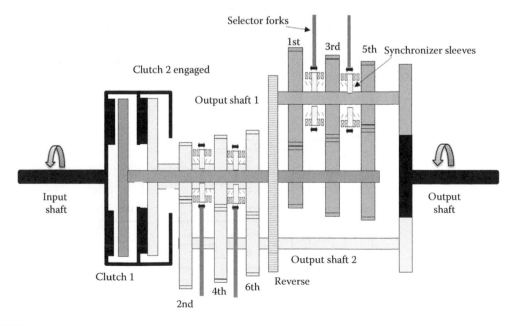

FIGURE 4.24
Gear diagram representation of DCT.

4.5 Continuously Variable Transmission

A CVT has a gear ratio that can be varied continuously within a certain range, thus providing an infinite number of gears. The continuous variation makes it possible to match virtually any engine speed and torque to any wheel speed and torque. It is, therefore, possible to achieve an ideal torque–speed profile (constant power profile).

The commonly used CVT in automobiles uses a pulley-and-belt assembly. One pulley is connected to the engine shaft, while the other is connected to the output shaft. The belt links the two pulleys. The distance between the two half pulleys can be varied, thereby varying the effective diameter on which the belt grips. The transmission ratio is a function of two effective diameters:

$$i_g = \frac{D_2}{D_1},\tag{4.22}$$

where D_1 and D_2 are the effective diameters of the output pulley and the input pulley, respectively.

Until recently, this implementation was affected by the limited belt–pulley adhesive contact. The design has been improved using metallic belts that provide better solidity and improved contact. Furthermore, an interesting concept has been developed and is being used by Nissan. This concept uses three friction gears: one is connected to the engine shaft, another to the output shaft, and the third grips on the particular profile of the other two gears. It can be rotated to grip on different effective diameters, making it possible to achieve a variable gear ratio.

4.6 Infinitely Variable Transmissions

The IVT provides the full range of forward and reverse speeds, as well as neutral gearing continuously without the need for a clutch or torque converter. It is a split-path design CVT that can provide unlimited transmission gear ratio span without any need for clutch or torque converter using a device called a "variator." The IVT has two power transfer paths in parallel between the input and the output, and it transfers the minimum energy through one of the paths that contains the variator. The main path controls most of the energy transfer from input to output. The second input to the gear train is the output of the variator, and it effectively controls the gear ratio or the output speed. The variator could be an electric motor or mechanical or hydraulic actuator. IVTs are also referred to as split torque, split power (shunt and speed split) transmission system. Figure 4.25 shows a split-path shunt (SPT) IVT system. The power from the IC engine is split into two paths, and the lower path shows the flow of the epicyclic gear train, which is controlled by an input from the variator.

Figures 4.26 and 4.27 show different types and configurations of infinitely variable transmission (IVT) systems (G.G. Locus).

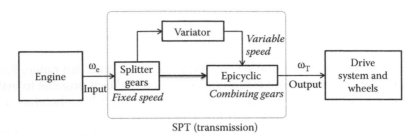

FIGURE 4.25
Split-path shunt infinitely variable transmission.

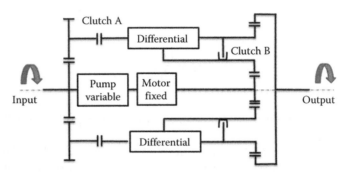

FIGURE 4.26
Sundstrand IVT system.

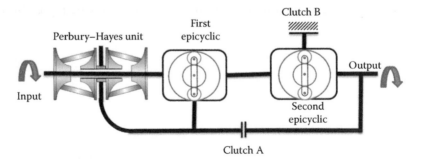

FIGURE 4.27
Perbury–Hayes friction roller-type IVT system.

4.7 Dedicated Hybrid Transmission (DHT)

By integrating an electric motor and a generator in the transmission system and using planetary gear sets, the power from the electric motor and an IC engine could be combined to provide a hybrid transmission system capable of operating in different drive modes. For example, the DHT enables the IC engine to drive the vehicle in combination with an electric motor or drive the generator and charge the batteries.

The Toyota Prius transmission is one of the earliest developed DHT concepts. The following chapters will review the DHT in more detail.

Bibliography

1. H. Naunheimer, B. Bertsche, J. Ryborz, and W. Novak, *Automotive Transmissions, Fundamentals, Selection, Design and Application*. Springer International Publishing, 2009.
2. R. Fischer, F. Küçükay, G. Jürgens, R. Najork, and B. Pollak, *The Automotive Transmission*. Springer International Publishing, 2015.
3. G. G. Lucas, Road vehicle performance, methods of measurement and calculation. Gordon and Breach, 1986, ISBN 0-677-21400-6.
4. Robert Bosch GmbH, *Bosch Automotive Handbook*. Eighth Edition, July 16, 2011.
5. L. Guzzella and A. Sciarretta, *Vehicle Propulsion Systems: Introduction to Modeling and Optimization*. Springer, January 29, 2015, ISBN-13: 978-3642438479.
6. H. Scherer, ZF 6-speed automatic transmission for passenger cars. SAE Technical Paper 2003-01-0596, 2003, doi: 10.4271/2003-01-0596.
7. H. Naunheimer, B. Bertsche, J. Ryborz, and W. Novak, *Automotive Transmissions Fundamentals, Selection, Design and Application*. 2011.
8. A. Emadi, (Ed.) *Advanced Electric Drive Vehicles*. CRC Press, 2014.
9. K. Rahman et al. Design and performance of electrical propulsion system of extended range electric vehicle (EREV) Chevrolet Voltec. *IEEE Energy Conversion Congress and Exposition (ECCE)*, 2012.
10. J. De Santiago et al. Electrical motor drivelines in commercial all-electric vehicles: A review. *IEEE Transactions on Vehicular Technology*, 61(2), 2012: 475–484.
11. J. O. Estima and A. J. Marques Cardoso, Efficiency analysis of drive train topologies applied to electric/hybrid vehicles. *IEEE Transactions on Vehicular Technology*, 61(3), 2012: 1021–1031.
12. T. Imamura et al. Concept and approach of multi stage hybrid transmission. SAE Technical Paper 2017-01-1098, 2017, doi: 10.4271/2017-01-1098.
13. M. Awadallah, P. Tawadros, P. Walker, and N. Zhang, Dynamic modelling and simulation of a manual transmission based mild hybrid vehicle. *Mechanism and Machine Theory*, 112, 2017: 218–239.
14. L. Zhenzhen, S. Dongye, L. Yonggang, Q. Datong, Z. Yi, Y. Yang, and C. Liang, Analysis and coordinated control of mode transition and shifting for a full hybrid electric vehicle based on dual clutch transmissions. *Mechanism and Machine Theory*, 114, 2017: 125–140.

5

Electric Vehicles

5.1 Configurations of Electric Vehicles

Previously, the electric vehicles (EV) was mainly converted from the exiting internal combustion engine vehicle (ICEV) by replacing the internal combustion engine and fuel tank with an electric motor drive and battery pack while retaining all the other components, as shown in Figure 5.1. Drawbacks such as its heavy weight, lower flexibility, and performance degradation have caused the use of this type of EV to fade out. In its place, the modern EV is purposely built, based on original body and frame designs. This satisfies the structure requirements unique to EVs and makes use of the greater flexibility of electric propulsion.[1]

A modern electric drive train is conceptually illustrated in Figure 5.2.[1] The drive train consists of three major subsystems: electric motor propulsion, energy source, and auxiliary. The electric propulsion subsystem comprises the vehicle controller, power electronic converter, electric motor, mechanical transmission, and driving wheels. The energy source subsystem involves the energy source, the energy management unit, and the energy refueling unit. The auxiliary subsystem consists of the power steering unit, the hotel climate control unit, and the auxiliary supply unit.

Based on the control inputs from the accelerator and brake pedals, the vehicle controller provides proper control signals to the electronic power converter, which functions to regulate the power flow between the electric motor and energy source. The backward power flow is due to the regenerative braking of the EV and this regenerated energy can be restored into the energy source, provided the energy source is receptive. Most EV batteries as well as ultracapacitors and flywheels readily posses the ability to accept regenerative energy. The energy management unit cooperates with the vehicle controller to control the regenerative braking and its energy recovery. It also works with the energy refueling unit to control the refueling unit and to monitor the usability of the energy source. The auxiliary power supply provides the necessary power with different voltage levels for all the EV auxiliaries, especially the hotel climate control and power steering units.

There are a variety of possible EV configurations due to the variations in electric propulsion characteristics and energy sources, as shown in Figure 5.3.[1]

a. Figure 5.3a shows the configuration of the first alternative, in which an electric propulsion replaces the IC engine of a conventional vehicle drive train. It consists of an electric motor, a clutch, a gearbox, and a differential. The clutch and gearbox may be replaced by an automatic transmission. The clutch is used to connect or disconnect the power of the electric motor from the driven wheels. The gearbox provides a set of gear ratios to modify the speed-power (torque) profile to match the load requirement (refer to Chapter 2). The differential is a mechanical device

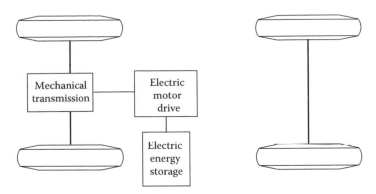

FIGURE 5.1
Primary electric vehicle power train.

(usually a set of planetary gears), which enables the wheels of both sides to be driven at different speeds when the vehicle runs along a curved path.

b. With an electric motor that has constant power in a long speed range (refer to Chapter 2), a fixed gearing can replace the multispeed gearbox and reduce the need for a clutch. This configuration not only reduces the size and weight of the mechanical transmission, it also simplifies the drive train control because gear shifting is not needed.

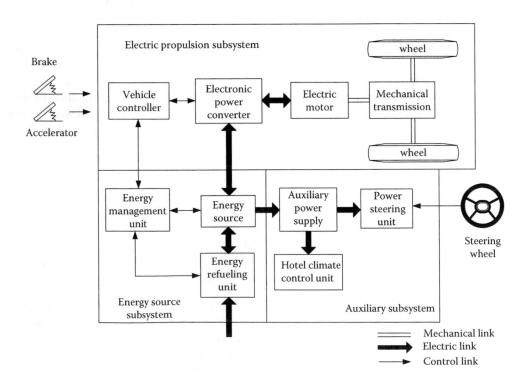

FIGURE 5.2
Conceptual illustration of general EV configuration. (From C. C. Chan and K. T. Chau, *Modern Electric Vehicle Technology*, Oxford University Press, New York, 2001.)

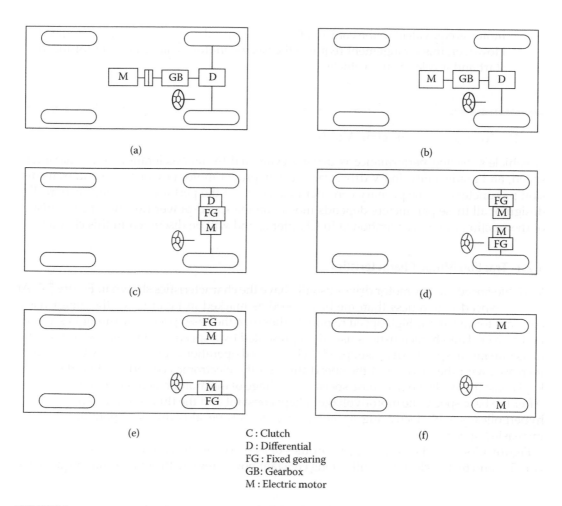

C : Clutch
D : Differential
FG : Fixed gearing
GB: Gearbox
M : Electric motor

FIGURE 5.3
Possible EV configurations. (From C. C. Chan and K. T. Chau, *Modern Electric Vehicle Technology*, Oxford University Press, New York, 2001.)

c. Similar to the drive train in (b), the electric motor, the fixed gearing, and the differential can be further integrated into a single assembly while both axles point at both driving wheels. The whole drive train is further simplified and compacted.

d. In Figure 5.3d, the mechanical differential is replaced by using two traction motors. Each of them drives one side wheel and operates at a different speed when the vehicle is running along a curved path.

e. In order to further simplify the drive train, the traction motor can be placed inside a wheel. This arrangement is the so-called in-wheel drive. A thin planetary gear set may be employed to reduce the motor speed and enhance the motor torque. The thin planetary gear set offers the advantage of a high-speed reduction ratio as well as an inline arrangement of the input and output shaft.

f. By fully abandoning any mechanical gearing between the electric motor and the driving wheel, the out-rotor of a low-speed electric motor in the in-wheel drive can be directly connected to the driving wheel. The speed control of the electric motor is equivalent to the control of the wheel speed and hence the vehicle speed. However, this arrangement requires the electric motor to have a higher torque, to start and accelerate the vehicle.

5.2 Performance of Electric Vehicles

A vehicle's driving performance is usually evaluated by its acceleration time, maximum speed, and gradeability. In EV drive train design, proper motor power rating and transmission parameters are the primary considerations to meet the performance specification. The design of all these parameters depends mostly on the speed-power (torque) characteristics of the traction motor, as mentioned in Chapter 2, and will be discussed in this chapter.

5.2.1 Traction Motor Characteristics

Variable-speed electric motor drives usually have the characteristics shown in Figure 5.4. At the low-speed region (less than the base speed as marked in Figure 5.4), the motor has a constant torque. In the high-speed region (higher than the base speed), the motor has a constant power. This characteristic is usually represented by a speed ratio x, defined as the ratio of its maximum speed to its base speed. In low-speed operation, voltage supply to the motor increases with the increase of the speed through the electronic converter while the flux is kept constant. At the point of base speed, the voltage of the motor reaches the source voltage. After the base speed, the motor voltage is kept constant and the flux is weakened, dropping hyperbolically with increasing speed. Hence, its torque also drops hyperbolically with increasing speed.[2,3,4]

Figure 5.5 shows the torque-speed profiles of a 60 kW motor with different speed ratios x ($x = 2$, 4 and 6). It is clear that with a long constant power region, the maximum torque of the

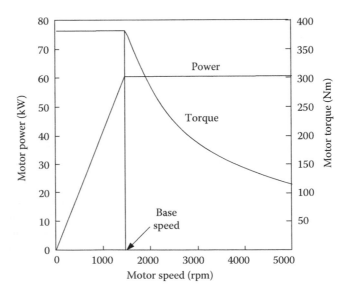

FIGURE 5.4
Typical variable-speed electric motor characteristics.

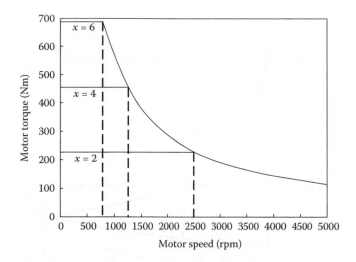

FIGURE 5.5
Speed-torque profile of a 60 kW electric motor with $x = 2$, 4, and 6.

motor can be significantly increased, and hence, vehicle acceleration and gradeability performance can be improved and the transmission can be simplified. However, each type of motor inherently has its limited maximum speed ratio. For example, a permanent magnet motor has a small x (<2) because of the difficulty of field weakening due to the presence of the permanent magnet. Switched reluctance motors may achieve $x > 6$ and induction motors about $x = 4$.[2,5]

5.2.2 Tractive Effort and Transmission Requirement

The tractive effort developed by a traction motor on driven wheels and the vehicle speed are expressed as:

$$F_t = \frac{T_m i_g i_0 \eta_t}{r_d} \tag{5.1}$$

and

$$V = \frac{\pi N_m r_d}{30 i_g i_0} \quad (\text{m/s}), \tag{5.2}$$

where T_m and N_m are the motor torque output in Nm and speed in rpm, respectively, i_g is gear ratio of transmission, i_0 is the gear ratio of final drive, η_t is the efficiency of the whole driveline from the motor to the driven wheels, and r_d is the radius of the driven wheels.

The use of a multigear or single-gear transmission depends mostly on the motor speed-torque characteristic. That is, at a given rated motor power, if the motor has a long constant power region, a single-gear transmission would be sufficient for a high tractive effort at low speeds. Otherwise, a multigear (more than two gears) transmission has to be used. Figure 5.6 shows the tractive effort of an EV, along with the vehicle speed with a traction motor of $x = 2$ and a three-gear transmission. The first gear covers the speed region of $a–b–c$, the second

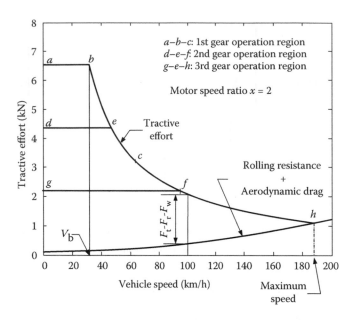

FIGURE 5.6
Tractive effort versus vehicle speed with a traction motor of $x = 2$ and 3-gear transmission.

gear covers d–e–f and the third gear covers g–f–h. Figure 5.7 shows the tractive effort with a traction motor of $x = 4$ and a two-gear transmission. The first gear covers the speed region of a–b–c and the second gear d–e–f. Figure 5.8 shows the tractive effort with a traction motor of $x = 6$ and a single-gear transmission. These three designs have the same tractive effort versus vehicle speed profiles. Therefore the vehicles will have the same acceleration and gradeability performance.

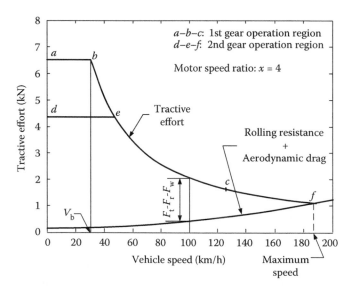

FIGURE 5.7
Tractive effort versus vehicle speed with a traction motor of $x = 4$ and 2-gear transmission.

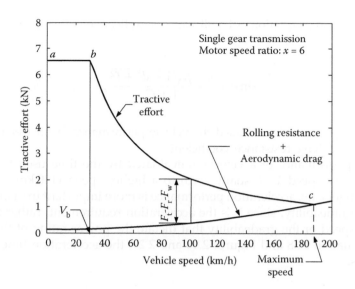

FIUGRE 5.8
Tractive effort versus vehicle speed with a traction motor of $x = 6$ and single-gear transmission.

5.2.3 Vehicle Performance

Basic vehicle performance includes maximum cruising speed, gradeability, and accelera-tion. The maximum speed of a vehicle can be easily found by the intersection point of the tractive effort curve with the resistance curve (rolling resistance plus aerodynamic drag), in the tractive effort versus vehicle speed diagram shown in Figures 5.6 through 5.8. It should be noted that such an intersection point does not exist in some designs, which usually use a larger traction motor or a large gear ratio. In this case, the maximum vehicle speed is determined by the maximum speed of the traction motor as:

$$V_{max} = \frac{\pi N_{m\ max} r_d}{30 i_{g\ min} i_0} \quad (m/s),$$ (5.3)

where $N_{m\ max}$ is the allowed maximum rpm of the traction motor and $i_{g\ min}$ is the minimum gear ratio of the transmission (highest gear).

Gradeability is determined by the net tractive effort of the vehicle, $F_{t\text{-net}}$ ($F_{t\text{-net}} = F_t - F_r - F_w$), as shown in Figures 5.6 through 5.8. At mid- and high speeds, the gradeability is smaller than the gradeability at low speeds. The maximum grade that the vehicle can overcome at the given speed can be calculated by

$$i = \frac{F_{t\text{-net}}}{Mg} = \frac{F_t - (F_r + F_w)}{Mg},$$ (5.4)

where F_t is the tractive effort on the driven wheels, F_r is the tire rolling resistance, and F_w is the aerodynamic drag. However, at low speeds, the gradeability is much larger.

Calculations based on Equation 5.4 will cause significance error; instead, Equation 5.5 should be used

$$\sin\alpha = \frac{d - f_r\sqrt{1 - d^2 + f_r^2}}{1 + f_r^2},$$ (5.5)

where $d = (F_t - F_w)/Mg$ which is called the vehicle performance factor (refer to Chapter 2), and f_r is the tire rolling resistance coefficient.

Acceleration performance of a vehicle is evaluated by the time used to accelerate the vehicle from a low speed V_1 (usually zero) to a higher speed (100 km/h for passenger cars). For passenger cars, acceleration performance is more important than maximum cruising speed and gradeability, since it is the acceleration requirement, rather than the maximum cruising speed or the gradeability that dictate the power rating of the motor drive. Referring to Equation 2.58 and Figures 2.28 and 2.29, the acceleration time for an EV can be expressed as:

$$t_a = \int_0^{V_b} \frac{M\delta}{(P_t/V_b) - Mgf_r - (1/2)\rho_a C_D A_f V^2} dV$$

$$+ \int_{V_b}^{V_f} \frac{M\delta}{(P_t/V) - Mgf_r - (1/2)\rho_a C_D A_f V^2} dV,$$ (5.6)

where V_b and V_f are the vehicle base speed as shown in Figures 5.6 through 5.8, and the final acceleration speed, respectively, and P_t is the tractive power on the driven wheels transmitted from the traction motor corresponding the vehicle base speed. The first term on the right side of Equation 5.6 is in correspondence with the speed region less than the vehicle base speed; the second term is in correspondence with the speed region beyond the vehicle base speed.

It is difficult to obtain the analytical solution from Equation 5.6. For initial evaluation of the acceleration time versus the tractive power, one can ignore the rolling resistance and the aerodynamic drag and obtain

$$t_a = \frac{\delta M}{2P_t}(V_f^2 + V_b^2),$$ (5.7)

where the vehicle rotational inertial factor, δ, is a constant. The tractive power, P_t, can then be expressed as:

$$P_t = \frac{\delta M}{2t_a}(V_f^2 + V_b^2).$$ (5.8)

It should be noted that the power rating obtained from Equation 5.8 is only the power consumed for vehicle acceleration. To accurately determine the tractive power rating, the power consumed in overcoming the rolling resistance and dynamic drag should be considered. The average drag power during acceleration can be expressed as:

$$\bar{P}_{drag} = \frac{1}{t_a} \int_0^{t_a} \left(Mgf_r V + \frac{1}{2}\rho_a C_D A_f V^3 \right) dt. \tag{5.9}$$

Referring to Figures 2.28 and 2.29, the vehicle speed V may be expressed using time t, as:

$$V = V_f \sqrt{\frac{t}{t_a}}. \tag{5.10}$$

Substituting Equations 5.10 into 5.9 and integrating, one obtains

$$\bar{P}_{drag} = \frac{2}{3}Mgf_r V_f + \frac{1}{5}\rho_a C_D A_f V_f^3. \tag{5.11}$$

The total tractive power for accelerating the vehicle from zero to speed, V_f in t_a seconds can be finally obtained as:

$$P_t = \frac{\delta M}{2t_a}(V_f^2 + V_b^2) + \frac{2}{3}Mgf_r V_f + \frac{1}{5}\rho_a C_D A_f V_f^3. \tag{5.12}$$

Equation 5.12 indicates that for a given acceleration performance, low vehicle base speed will result in a small motor power rating. However, the power rating decline rate to the vehicle base speed reduction is not identical. Differentiating Equation 5.12 with respect to the vehicle speed V_b, one can obtain

$$\frac{dP_t}{dV_b} = \frac{\delta M_v}{t_a}V_b. \tag{5.13}$$

Figure 5.9 shows an example of the tractive power rating and the power rating decline rate to the vehicle speed reduction (dP_t/dV_b) versus the speed fact x. In this example, acceleration time is 10 seconds, vehicle mass is 1200 kg, the rolling resistance coefficient is 0.01, the aerodynamic drag coefficient is 0.3, and front area is 2 m^2. This figure clearly indicates that a low x (high V_b) reduction in V_b will result in significant decline in the power rating requirement. But with a high x (low V_b), $x > 5$ for example, it is not so effective. Figure 5.10 gives an example of the acceleration time and the distance versus the vehicle speed, using Equation 5.6 and numerical methods.

5.3 Tractive Effort in Normal Driving

The vehicle performance described in the previous section dictates the vehicle capabilities with respect to speed, gradeability, and acceleration, thus dictating the power capacity of the power train. However, in normal driving conditions these maximum capabilities are rarely used. During most of the operation time, the power train operates with partial load. Actual tractive effort (power) and vehicle speed vary widely with operating conditions, such as acceleration, deceleration, uphill, downhill motion and so on. These variations

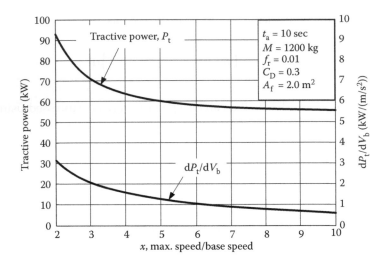

FIGURE 5.9
Power rating versus speed factor.

are associated with the traffic environment as well as the type of vehicles. City and highway traffic conditions vary greatly, as do the different missions of the vehicles, such as passenger cars and vehicles with regular operation routes and schedules.

It is difficult to describe the tractive effort and vehicle speed variations in all actual traffic environments accurately and quantitatively. However, some representative driving cycles (driving schedules) have been developed to emulate typical traffic environments. These driving cycles are represented by the vehicle speeds versus the operating time while driving on a flat road. Some typical drive cycles are illustrated in Figure 5.12, which include: (a) FTP75 urban cycle, (b) FTP75 highway cycle, (c) US06 cycle, which is a high-speed and high-acceleration drive cycle, (d) J227a schedule B, (e) J227a schedule C, and (f) J227a

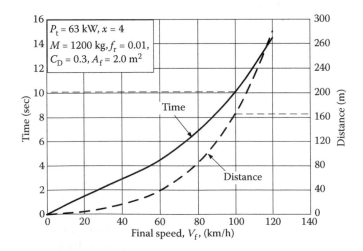

FIGURE 5.10
Acceleration time and distance versus final speed.

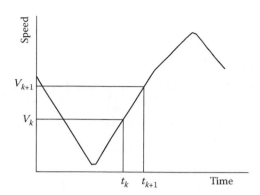

FIGURE 5.11
Acceleration being consumed constant with a short time period.

schedule D. The J227a series are recommended by the Society of Automotive Engineers in the United States.[6] and is applied in the evaluation of EVs and batteries.

In a specific drive cycle, the tractive effort of a vehicle can be expressed as:

$$F_t = Mgf_r \cos\alpha + \frac{1}{2}\rho_a C_D A_f V^2 + M\delta\frac{dV}{dt}. \tag{5.14}$$

In a short time period, the speed is assumed to be linear with time, and acceleration is constant as shown in Figure 5.11, The acceleration, dV/dt in a driving cycle, can be obtained by:

$$\frac{dV}{dt} = \frac{V_{k+1} - V_k}{t_{k+1} - t_k} \quad (k = 1, 2, \ldots n, \quad n\text{--total member of points}). \tag{5.15}$$

By using Equation (5.14), the tractive efforts in any instant in a driving cycle can be calculated, as shown in Figure 5.12. The operating points of the tractive effort versus the vehicle speed scatter over the plane, and they clearly show the operating area in which the power train operates most of the time. Furthermore, the time distribution of the vehicle speed and tractive effort can be generated as shown in Figure 5.13. This time distribution information is very helpful for power train design, in which the most efficient region of the power train is designed to overlap the greatest operation time area.

5.4 Energy Consumption

In transportation, the unit of energy is usually kilowatt-hour (kWh) rather than Joule or kilojoule (J or kJ). The energy consumption per unit distance in kWh/km is generally used to evaluate the vehicle energy consumption. However, for ICE vehicles the commonly used unit is a physical unit of fuel volume per unit distance, such as liters per 100 km (l/100 km). In the United States, the distance per unit volume of fuel is usually used; this

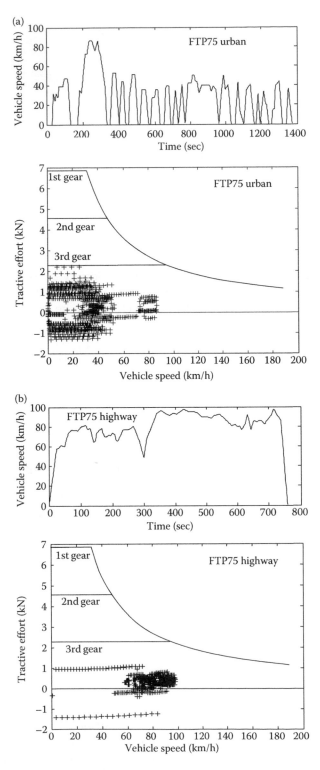

FIGURE 5.12
Speed profile and tractive effort in different representative drive cycles, operating points are marked by "+,"
(a) FTP75 urban, (b) FTP75 highway. *(Continued)*

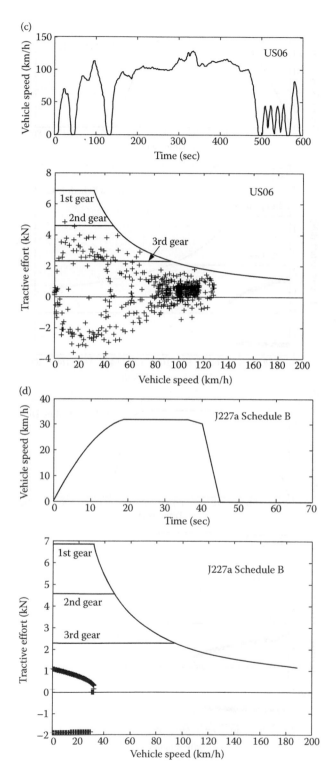

FIGURE 5.12
Speed profile and tractive effort in different representative drive cycles, operating points are marked by "+," (c) US06, (d) J227a schedule B. (*Continued*)

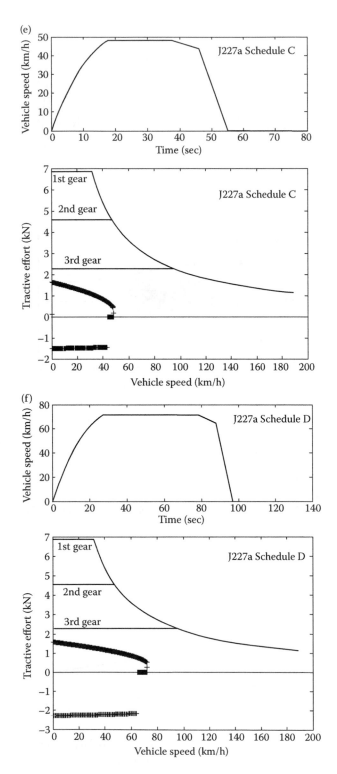

FIGURE 5.12
Speed profile and tractive effort in different representative drive cycles, operating points are marked by "+," (e) J227a Schedule C, and (f) J227a schedule D.

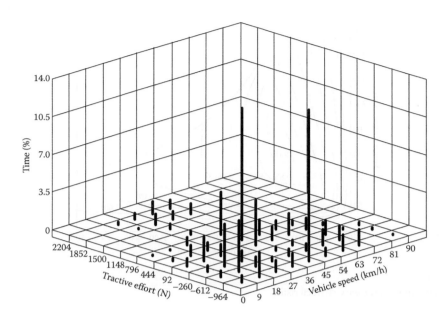

FIGURE 5.13
Time distribution on vehicle speed and tractive effort in FTP75 urban cycle.

is expressed as miles per gallon (mpg). On the other hand, for battery powered EVs, the original energy consumption unit in kWh, measured at the battery terminals, is more suitable. The battery energy capacity is usually measured in kWh and the driving range per battery charge can be easily calculated. Similar to ICE vehicles, l/100 km (for liquid fuels) or kg/100 km (for gas fuels, such as hydrogen) or mpg or miles per kilogram is a more suitable unit of measurement for vehicles that use gaseous fuels.

Energy consumption is an integration of the power output at the battery terminals. For propelling, the battery power output is equal to the resistance power and power losses in the transmission and the motor drive, including power losses in the electronics. The power losses in transmission and motor drive are represented by their efficiencies η_t and η_m, respectively. Thus, the battery power output can be expressed as:

$$P_{\text{b-out}} = \frac{V}{\eta_t \eta_m}\left(Mg(f_r + i) + \frac{1}{2}\rho_a C_D A_f V^2 + M\delta\frac{dV}{dt}\right). \tag{5.16}$$

Here, the nontraction load (auxiliary load) is not included. In some cases, the auxiliary loads may be too significant to be ignored and should be added to the traction load. When regenerative braking is effective on an EV, a part of the braking energy—wasted in conventional vehicles—can be recovered by operating the motor drive as a generator and restoring it into the batteries. The regenerative braking power at the battery terminals can also be expressed as:

$$P_{\text{b-in}} = \frac{\alpha V}{\eta_t \eta_m}\left(Mg(f_r + i) + \frac{1}{2}\rho_a C_D A_f V^2 + M\delta\frac{dV}{dt}\right), \tag{5.17}$$

where road grade i or acceleration dV/dt or both of them are negative, and α ($0 < \alpha < 1$) is the percentage of the total braking energy that can be regenerated by the electric motor,

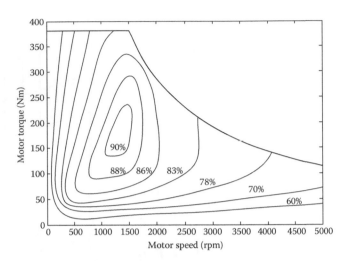

FIGURE 5.14
Typical electric motor efficiency characteristics.

called the regenerative braking factor. The regenerative braking factor α is a function of the applied braking strength and the design and control of the braking system, which will be discussed in detail in the later chapters. The net energy consumption from the batteries is

$$E_{out} = \underbrace{\int P_{b\text{-out}} \, dt}_{\text{traction}} + \underbrace{\int P_{b\text{-in}} \, dt}_{\text{braking}}. \tag{5.18}$$

It should be noted that the braking power in Equation 5.17 has a negative sign. When the net battery energy consumption reaches the total energy in the batteries, measured at their terminal, the batteries are empty and need to be charged. The traveling distance between two charges (usually called effective travel range) is determined by the total energy carried by the batteries, the resistance power and the effectiveness of the regenerative braking (α).

Efficiency of a traction motor varies with its operating points on the speed-torque (speed-power) plane as shown in Figure 5.14, where the most efficient operating area exists. In power train design, this area should overlap or at least be as close as possible to the area of the greatest operation as mentioned in the previous section.

Bibliography

1. C. C. Chan and K. T. Chau, *Modern Electric Vehicle Technology*, Oxford University Press, New York, 2001.
2. Y. Gao, H. Maghbelli, M. Ehsani et al., "Investigation of proper motor drive characteristics for military vehicle propulsion," In *Society of Automotive Engineers (SAE) Journal*, Warrendale, PA, 2003, Paper No. 2003-01-2296.
3. Z. Rahman. M. Ehsani, and K. Butler, "An investigation of electric motor drive characteristics for EV and HEV propulsion systems," In *Society of Automotive Engineers (SAE) Journal*, Warrendale, PA, 2003, Paper No. 2000-01-3062.

4. Z. Rahman, M. Ehsani, and K. Butler, "Effect of extended-speed, constant-power operation of electric drives on the design and performance of EV-HEV propulsion system," In *Society of Automotive Engineers (SAE) Journal*, Warrendale, PA, 2003, Paper No. 2000-01-1557.

5. K. M. Rahman and M. Ehsani, "Performance analysis of electric motor drives for electric and hybrid electric vehicle application." *IEEE Power Electronic in Transportation* 1996: 49–56.

6. D. A. J. Rand, R. Woods, and R. M. Dell, *Batteries for Electric Vehicles*, Research Studies Press, Ltd., Austin, TX, 1998.

7. A. Bouscayrol, L. Boulon, T. Hofman, and C. C. Chan, "Special section on advanced power-trains for more electric vehicles." *IEEE Transactions on Vehicular Technology* 65(3), 2016: 995–997.

8. C. C. Chan and M. Cheng, Vehicle traction motors, In *Encyclopedia of Sustainability Science and Technology*, 2015, pp. 1–34.

9. F. Lin, K. T. Chau, C. Liu, C. C. Chan, and T. W. Ching, "Comparison of hybrid-excitation fault-tolerant in-wheel motor drives for electric vehicles," In *International Electric Vehicle Symposium and Exhibition, EVS28. EVS28 International Electric Vehicle Symposium and Exhibition*, 2015.

10. C. C. Chan, Overview of electric, hybrid, and fuel cell vehicles, In *Encyclopedia of Automotive Engineering*, 2015.

11. S. Cui, S. Han, and C. C. Chan, "Overview of multi-machine drive systems for electric and hybrid electric vehicles," In *Transportation Electrification Asia-Pacific (ITEC Asia-Pacific), 2014 IEEE Conference and Expo. IEEE*, pp. 1–6, 2014.

12. C. Mi and M. Abul Masrur, *Hybrid Electric Vehicles: Principles and Applications with Practical Perspectives*, John Wiley & Sons, New York, 2017.

13. J. Du and D. Ouyang, "Progress of Chinese electric vehicles industrialization in 2015: A review." *Applied Energy* 188, 2017: 529–546.

14. M. Quraan, P. Tricoli, S. D'Arco, and L. Piegari, "Efficiency assessment of modular multilevel converters for battery electric vehicles." *IEEE Transactions on Power Electronics* 32(3), 2017: 2041–2051.

15. V. Ivanov, D. Savitski, and B. Shyrokau, "A survey of traction control and antilock braking systems of full electric vehicles with individually controlled electric motors." *IEEE Transactions on Vehicular Technology* 64(9), 2015: 3878–3896.

16. J.-R. Riba, C. López-Torres, L. Romeral, and A. Garcia, "Rare-earth-free propulsion motors for electric vehicles: A technology review." *Renewable and Sustainable Energy Reviews* 57, 2016: 367–379.

17. M. Hernandez, M. Messagie, O. Hegazy, L. Marengo, O. Winter, and J. Van Mierlo, "Environmental impact of traction electric motors for electric vehicles applications." *The International Journal of Life Cycle Assessment* 22(1), 2017: 54–65.

18. V. Ivanov, D. Savitski, K. Augsburg, and P. Barber, "Electric vehicles with individually controlled on-board motors: Revisiting the ABS design," In *Mechatronics (ICM), 2015 IEEE International Conference on. IEEE*, pp. 323–328, 2015.

19. L. De Novellis, A. Sorniotti, and P. Gruber, "Driving modes for designing the cornering response of fully electric vehicles with multiple motors." *Mechanical Systems and Signal Processing* 64, 2015: 1–15.

20. Y. Chen and J. Wang, "Design and experimental evaluations on energy efficient control allocation methods for overactuated electric vehicles: Longitudinal motion case." *IEEE/ASME Transactions on Mechatronics* 19(2), 2014: 538–548.

21. Y. Miyama, M. Hazeyama, S. Hanioka, N. Watanabe, A. Daikoku, and M. Inoue, "PWM carrier harmonic iron loss reduction technique of permanent-magnet motors for electric vehicles." *IEEE Transactions on Industry Applications* 52(4), 2016: 2865–2871.

22. R. de Castro, M. Tanelli, R. E. Araújo, and S. M. Savaresi, "Minimum-time manoeuvring in electric vehicles with four wheel-individual-motors." *Vehicle System Dynamics* 52(6), 2014: 824–846.

23. S. Dang, A. Odonde, T. Mirza, C. Dissanayake, and R. Burns, "Sustainable energy management: An analysis report of the impacts of electric vehicles," In *Environment and Electrical Engineering (EEEIC), 2014 14th International Conference on. IEEE*, pp. 318–322, 2014.

24. A. V. Sant, V. Khadkikar, W. Xiao, and H. H. Zeineldin, "Four-axis vector-controlled dual-rotor PMSM for plug-in electric vehicles." *IEEE Transactions on Industrial Electronics* 62(5), 2015: 3202–3212.
25. M. Yildirim, M. Polat, and H. Kürüm, "A survey on comparison of electric motor types and drives used for electric vehicles," In *Power Electronics and Motion Control Conference and Exposition (PEMC), 2014 16th International. IEEE*, pp. 218–223, 2014.

6

Hybrid Electric Vehicles

Conventional vehicles with internal combustion (IC) engines provide good performance and a long operating range by utilizing the high-energy-density advantages of petroleum fuels. However, conventional IC engine vehicles have the disadvantages of poor fuel economy and environmental pollution. The main reasons for their poor fuel economy are (1) the mismatch of engine fuel efficiency characteristics with real operation requirements (Figures 2.22 and 2.23); (2) the dissipation of vehicle kinetic energy during braking, especially while operating in urban areas; and (3) the low efficiency of hydraulic transmission in current automobiles in stop-and-go driving patterns (Figure 2.21). Battery-powered electric vehicles (EVs), on the other hand, possess some advantages over conventional IC engine vehicles, such as high energy efficiency and zero environmental pollution. However, the performance, especially the operation range per battery charge, is far less competitive than IC engine vehicles due to the much lower energy density of the batteries than that of gasoline. Hybrid electric vehicles (HEVs), which use two power sources (a primary power source and a secondary power source), have the advantages of both IC engine vehicles and EVs and overcome their disadvantages.[1,2] In this chapter, the basic concept and operating principles of HEV power trains are discussed.

6.1 Concept of Hybrid Electric Drivetrains

Basically, any vehicle power train is required to (1) develop sufficient power to meet the demands of vehicle performance, (2) carry sufficient energy onboard to support the vehicle while driving a sufficient range, (3) demonstrate high efficiency, and (4) emit few environmental pollutants. Broadly speaking, a vehicle may have more than one power train. Here, the power train is defined as the combination of the energy source and the energy converter or power source, such as the gasoline (or diesel)–heat engine system, the hydrogen–fuel cell–electric motor system, the chemical battery–electric motor system, and so on. A vehicle that has two or more power trains is called a hybrid vehicle. A hybrid vehicle with an electrical power train is called an HEV. The drivetrain of a vehicle is defined as the aggregation of all the power trains.

A hybrid vehicle drivetrain usually consists of no more than two power trains. More than two power trains makes the drivetrain very complicated. To recapture the braking energy that is dissipated in the form of heat in conventional IC engine vehicles, a hybrid drivetrain usually has a power train that allows energy to flow bidirectionally. The other one is either bidirectional or unidirectional. Figure 6.1 shows the concept of a hybrid drivetrain and the possible different power flow routes.

A hybrid drivetrain can supply its power to the load by a selective power train. There are many available patterns of operating two power trains to meet the load requirement:

1. Power train 1 alone delivers its power to the load.
2. Power train 2 alone delivers its power to the load.

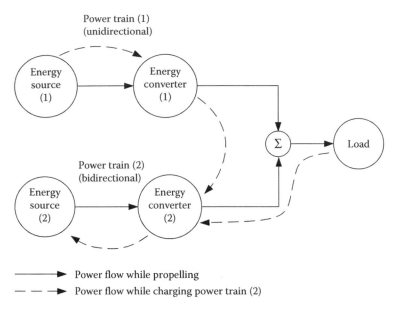

Power train (1)
(unidirectional)

Energy
source
(1)

Energy
converter
(1)

Σ

Load

Power train (2)
(bidirectional)

Energy
source
(2)

Energy
converter
(2)

⟶ Power flow while propelling

⟶ Power flow while charging power train (2)

FIGURE 6.1
Conceptual illustration of hybrid electric drivetrain.

3. Both power train 1 and power train 2 deliver their power to the load simultaneously.

4. Power train 2 obtains power from the load (regenerative braking).

5. Power train 2 obtains power from power train 1.

6. Power train 2 obtains power from power train 1 and the load simultaneously.

7. Power train 1 delivers power to the load and to power train 2 simultaneously.

8. Power train 1 delivers its power to power train 2, and power train 2 delivers its power to the load.

9. Power train 1 delivers its power to the load, and the load delivers the power to power train 2.

In the case of hybridization with a gasoline (diesel)–IC engine (power train 1) and a battery–electric motor (power train 2), pattern (1) is the engine-alone propelling mode. This may be used when the batteries are almost completely depleted, and the engine has no remaining power to charge the batteries, or when the batteries have been fully charged, and the engine is able to supply sufficient power to meet the power demands of the vehicle. Pattern (2) is the pure electric propelling mode, in which the engine is shut off. This pattern may be used for situations where the engine cannot operate effectively, such as at very low speed or in areas where emissions are strictly prohibited. Pattern (3) is the hybrid traction mode and may be used when high power is needed, such as during sharp acceleration or steep hill climbing. Pattern (4) is the regenerative braking mode, by which the kinetic or potential energy of the vehicle is recovered through the electric motor functioning as a generator. The recovered energy is then stored in the batteries and reused later. Pattern (5) is the mode in which the engine charges the batteries while the vehicle is at a standstill, coasting, or descending on a slight grade, in which no power goes into or comes from the load. Pattern (6) is the mode in which both regenerating braking and the IC engine charge the batteries simultaneously. Pattern (7) is the mode in which the engine propels the vehicle and charges the batteries simultaneously. Pattern (8) is the mode in which the engine charges the batteries, and the batteries supply power to the load. Pattern (9) is the mode in which the power flows into the batteries from the heat engine through the vehicle mass. The typical configuration of this mode is that the two power trains are separately mounted on the front and rear axles of the vehicle, which is discussed in the following sections.

The abundant operation modes in a hybrid vehicle create much more flexibility over a single power train vehicle. With proper configuration and control, applying a specific mode for a special operating condition can potentially optimize the overall performance, efficiency, and emissions. However, in a practical design, deciding which mode should be implemented depends on many factors, such as the physical configuration of the drivetrain, power train efficiency characteristics, load characteristics, and so on.

Operating each power train in its optimal efficiency region is essential for the overall efficiency of the vehicle. An IC engine generally has the highest efficiency operating region with a wide throttle opening. Operating away from this region causes low operating efficiency (refer to Figures 2.18, 2.20, 2.22, 2.23, and 3.6). On the other hand, the efficiency drop in an electric motor is not as detrimental when compared to an IC engine that operates away from its optimal region (Figure 4.14).

The load power of a vehicle varies randomly in real operation due to frequent acceleration, deceleration, and climbing up and down grades, as shown in Figure 6.2. The load power is composed of two components: steady (average) power, which has a constant value, and dynamic power, which has a zero average. In designing the control strategy of a hybrid vehicle, one power train that favors steady-state operation, such as an IC engine and fuel cell, may be used to supply the average power. On the other hand, another power train, such as an electric motor, may be used to supply the dynamic power. The total energy output from the dynamic power train will be zero in an entire driving cycle. This implies that the energy source of the dynamic power train does not lose its energy capacity at the end of the driving cycle. It functions only as a power damper.

In a hybrid vehicle, steady power may be provided by an IC engine, a Stirling engine, a fuel cell, and so on. The IC engine or the fuel cell can be much smaller than that in a single power train design because the dynamic power is taken by the dynamic power source and then the engine can operate steadily in its most efficient region. The dynamic power may be provided by an electric motor powered by batteries, ultracapacitors, flywheels (mechanical batteries), and their combinations.[1,3]

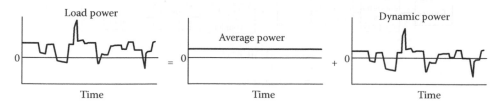

FIGURE 6.2
A load power is decomposed into steady and dynamic components.

6.2 Architectures of Hybrid Electric Drivetrains

The architecture of a hybrid vehicle is loosely defined as the connection between the components that define the energy flow routes and control ports. Traditionally, HEVs were classified into two basic types: series and parallel. It is interesting to note that in 2000, some newly introduced HEVs could not be classified into these kinds.[4] Hence, HEVs are presently classified into four kinds—series hybrid, parallel hybrid, series–parallel hybrid, and complex hybrid—that are functionally shown in Figure 6.3.[5] Scientifically, the preceding classifications are not very clear and may cause confusion. In an HEV, there are two kinds of energy flowing in the drivetrain: mechanical energy and electrical energy. Adding two powers together or splitting one power into two at the power-merging point always occurs with the same power type, that is, electrical or mechanical, not electrical and mechanical. Perhaps a more accurate definition for HEV architecture may be to take the power coupling or decoupling features such as an electrical coupling drivetrain, a mechanical coupling drivetrain, and a mechanical–electrical coupling drivetrain.

Figure 6.3a functionally shows the architecture that is traditionally called a series hybrid drivetrain. The key feature of this configuration is that two electric powers are added together in the power converter, which functions as an electric power coupler to control the power flows from the batteries and generator to the electric motor, or in the reverse direction from the electric motor to the batteries. The fuel tank, the IC engine, and the generator constitute the primary energy supply, and the batteries function as the energy bumper.

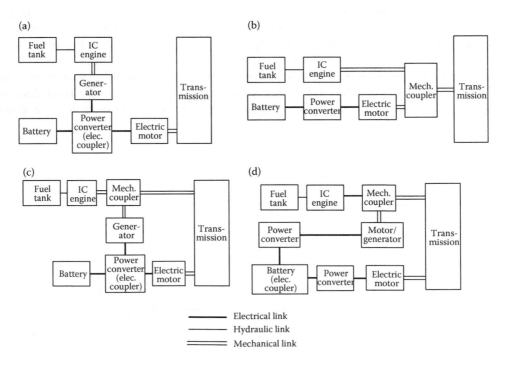

FIGURE 6.3
Classifications of HEVs. (a) Series (electrical coupling), (b) parallel (mechanical coupling), (c) series–parallel (mechanical and electrical coupling), and (d) complex (mechanical and electrical coupling).

Figure 6.3b shows the configuration that is traditionally called a parallel hybrid drivetrain. The key of this configuration is that two mechanical powers are added together in a mechanical coupler. The IC engine is the primary power plant, and the batteries and electric motor drive constitute the energy bumper. The power flows can be controlled only by the power plants—the engine and the electric motor.

Figure 6.3c shows the configuration that is traditionally called a series–parallel hybrid drivetrain. The distinguishing feature of this configuration is the employment of two power couplers—mechanical and electrical. This configuration is the combination of series and parallel structures, possessing the major features of both and more plentiful operation modes than those of the series or parallel structure alone. On the other hand, it is relatively more complicated and may be more expensive.

Figure 6.3d shows a configuration of the so-called complex hybrid, which has a similar structure to the series–parallel one. The only difference is that the electrical coupling function is moved from the power converter to the batteries, and one more power converter is added between the motor/generator and the batteries.

We will concentrate more on the first three configurations—series, parallel, and series–parallel.

6.2.1 Series Hybrid Electric Drivetrains (Electrical Coupling)

A series hybrid drivetrain is one in which two electric power sources feed a single electrical power plant (electric motor) that propels the vehicle. The configuration that is most often used is the one shown in Figure 6.4. The unidirectional energy source is a fuel tank, and the unidirectional energy converter (power plant) is an IC engine coupled to an electric generator. The output of the electric generator is connected to a power DC bus through a controllable electronic converter (rectifier). The bidirectional energy source is a battery pack connected to the power DC bus by means of a controllable, bidirectional power electronic converter (DC/DC converter). The power bus is also connected to the controller of the electric motor. The traction motor can be controlled as either a motor or a generator and in

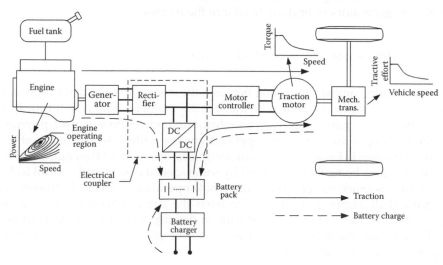

FIGURE 6.4
Configuration of a series hybrid electric drivetrain.

forward or reverse motion. This drivetrain may need a battery charger to charge the batteries by wall plug-in from a power grid. The series hybrid drivetrain originally came from an EV on which an additional engine–generator was added to extend the operating range that is limited by the poor energy density of the batteries.

The drivetrain needs a vehicle controller to control the operation and power flows based on the driver's operating command through accelerator and brake pedals and other feedback information from the components (not shown in Figure 6.4, but for details see Figure 8.1). The vehicle controller controls the IC engine through its throttle, electrical coupler (controllable rectifier and DC/DC converter), and traction motor to produce the demanded propelling torque or regenerative braking torque with one of the following operation modes:

1. *Pure electric traction mode:* The engine is turned off, and the vehicle is propelled only from the batteries.

2. *Pure engine traction mode:* The vehicle traction power comes only from the engine–generator, while the batteries neither supply nor accept any power from the drivetrain. The electric machines serve as an electric transmission from the engine to the drive wheels.

3. *Hybrid traction mode:* The traction powers are drawn from both the engine–generator and the batteries, merging together in the electrical coupler.

4. *Engine traction with battery charging mode:* The engine–generator supplies power to charge the batteries and propel the vehicle simultaneously. The engine–generator power is split in the electrical coupler.

5. *Regenerative braking mode:* The engine–generator is turned off, and the traction motor is operated as a generator powered by the vehicle kinetic or potential energy. The power generated is charged to the batteries and reused in later propelling.

6. *Battery charging mode:* The traction motor receives no power, and the engine–generator is operated only to charge the batteries.

7. *Hybrid battery charging mode:* Both the engine–generator and the traction motor operate as generators in braking to charge the batteries.

Series hybrid drivetrains offer several advantages:

1. There is no mechanical connection between the engine and the drive wheels. Consequently, the engine could be potentially at any point on its speed–torque (power) map. This distinct advantage, with a sophisticated power flow control, provides the engine with opportunities to be operated always within its maximum efficiency region, as shown in Figure 6.4. The efficiency and emissions of the engine in this narrow region may be further improved by some special design and control technologies, which is much easier than in the whole operating domain. Furthermore, the mechanical decoupling of the engine from the drive wheels allows the use of high-speed engines, where it is difficult to directly propel the wheels through a mechanical link, such as gas turbines or power plants that have slow dynamic responses (e.g., Stirling engine).

2. Because electric motors have a torque–speed profile that is very close to the ideal for traction, as shown in Figures 2.12, 2.14, and 4.4, the drivetrain may not need a

multigear transmission, as discussed in Chapter 3. Therefore, the structure of the drivetrain can be greatly simplified and cost less. Furthermore, two motors may be used, each powering a single wheel, and the mechanical differential can be removed. Such an arrangement also has the advantages of decoupling the speeds of two wheels, a function similar to that of a mechanical differential, and an additional function of antislip similar to conventional traction control. Furthermore, four in-the-wheel motors may be used, each one driving a wheel. In such a configuration, the speed and torque of each wheel can be independently controlled. Consequently, the drivability of the vehicle is significantly enhanced. This is very important for off-road vehicles, which usually operate on difficult terrain, such as ice, snow, and soft ground.

3. The control strategy of the drivetrain may be simple compared to other configurations because of its fully mechanical decoupling between the engine and wheels.

However, series hybrid electric drivetrains have some disadvantages, such as the following:

1. The energy from the engine changes its form twice to reach its destination—drive wheels (mechanical to electrical in the generator and electrical to mechanical in the traction motor). The inefficiencies of the generator and the traction motor may cause significant losses.
2. The generator adds additional weight and cost.
3. Because the traction motor is the only power plant propelling the vehicle, it must be sized to produce enough power for optimal vehicle performance in terms of acceleration and gradeability.

The design and control principle of a series HEV is discussed in Chapter 8.

6.2.2 Parallel Hybrid Electric Drivetrains (Mechanical Coupling)

A parallel hybrid drivetrain is one in which the engine supplies its mechanical power directly to the drive wheels in a manner similar to a conventional IC engine vehicle. The engine is assisted by an electric motor that is mechanically coupled to the driveline. The powers of the engine and the electric motor are coupled together by mechanical coupling, as shown in Figure 6.5. The distinguishing feature of this architecture is that two mechanical powers from the engine and the electric motor are added together by a mechanical coupler.

All the possible operating modes mentioned in the series hybrid drivetrain are still effective. The major advantages of the parallel hybrid drivetrain over the series one are: (1) both the engine and the electric motor directly supply torques to the drive wheels, and no energy form conversion occurs, so the energy loss may be less; and (2) it is compact because there is no need for an additional generator, and the traction motor is smaller than in series. Its major disadvantage is the mechanical coupling between the engine and the drive wheels, since then the engine operating points cannot be fixed in a narrow speed and torque region. Another disadvantage may be the complex structure and control.

Generally, mechanical coupling consists of torque coupling and speed coupling. In torque coupling, the mechanical coupler adds the torques of the engine and the motor together and delivers the total torque to the drive wheels. The engine and motor torque can be independently controlled, but the speeds of the engine, the motor, and the vehicle are linked together

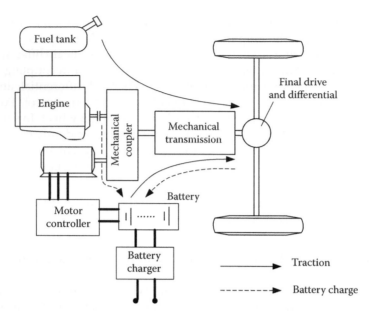

FIGURE 6.5
Configuration of a parallel hybrid electric drivetrain.

in a fixed relationship and cannot be independently controlled because of the power conservation constraint. Similarly, in speed coupling, the speeds of the engine and the motor can be added together, and all the torques are linked together and cannot be independently controlled. The details of these two kinds of mechanical coupler are described hereafter.

6.2.2.1 Parallel Hybrid Drivetrain with Torque Coupling

6.2.2.1.1 Torque-Coupling Devices

Figure 6.6 conceptually shows a mechanical torque coupling, which is a three-port, two-degree-of-freedom mechanical device. Port 1 is a unidirectional input, and ports 2 and 3 are bidirectional input or output, but both are not input at the same time. Here, input means the energy flow into the device, and output means the energy flow out of the device. In a hybrid vehicle application, port 1 is connected to an IC engine directly or through a mechanical transmission. Port 2 is connected to the shaft of an electric motor directly or through a mechanical transmission. Port 3 is connected to the drive wheels through a mechanical linkage.

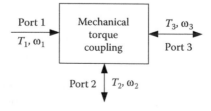

FIGURE 6.6
Torque-coupling device.

If the losses are ignored and in steady state, the power input to the torque coupler is always equal to the power output from it. Suppose here port 2 (electric motor) is in propelling, that is, input mode. The power output to the vehicle wheels is

$$T_3\omega_3 = T_1\omega_1 + T_2\omega_2. \tag{6.1}$$

The torque coupler can be expressed as

$$T_3 = k_1T_1 + k_2T_2, \tag{6.2}$$

where k_1 and k_2 are the structural parameters of the torque coupler, which are described by the gear ratios and usually are constant when the device design is fixed. For the torque coupler, T_3 is load torque, and T_1 and T_2 are propelling torques that are independent of each other and can be independently controlled. However, due to the constraint of Equation 6.1, the angular velocities ω_1, ω_2, and ω_3 are linked together and cannot be independently controlled, as expressed by

$$\omega_3 = \frac{\omega_1}{k_1} = \frac{\omega_2}{k_2}. \tag{6.3}$$

Figure 6.7 shows some common mechanical torque-coupling devices.

6.2.2.1.2 Drivetrain Configurations with Torque Coupling

Torque couplers can be used to constitute hybrid drivetrains with many different configurations. Based on the torque coupler used, a two- or one-shaft configuration may be constituted. In each, the transmission may be placed in different positions with different gears, resulting in various tractive characteristics. A good design depends mostly on the tractive requirements, engine size, motor size, and motor speed–torque characteristics.

Figure 6.8 shows a two-shaft configuration, in which two transmissions are used. One is placed between the engine and the torque coupler and the other between the motor and the torque coupler. Both transmissions may be single-gear or multigear. Figure 6.9 shows the general tractive effort–speed profiles of a vehicle with different transmission gears. It is evident that two multigear transmissions produce many tractive effort profiles. The performance and overall efficiency of the drivetrain may be superior to those of other designs because two multigear transmissions provide more opportunities for both the engine and the electric traction system (electric machine and batteries) to operate in their optimum region. This design also provides great flexibility in the design of the engine and electric motor characteristics. However, two multigear transmissions will significantly complicate the drivetrain and increase the burden of the control system for selecting the proper gear in each transmission.[6,7]

As shown in Figure 6.8, a multigear transmission, 1, and a single-gear transmission, 2, may be used. Referring to the relative positions of transmissions and the electric motor, this configuration is referred to as a pretransmission configuration (the electric motor is in front of the transmission). The tractive effort–speed profiles are shown in Figure 6.9b. In the design of a hybrid drivetrain, the maximum tractive effort with this transmission arrangement may be sufficient for hill climbing performance; greater tractive effort would not be needed because of the limitation of tire–ground contact adhesion. Utilizing a

Gear Box

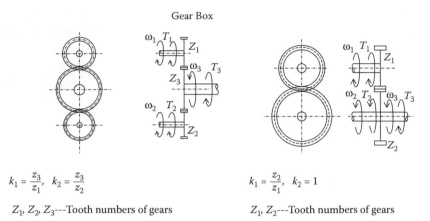

$$k_1 = \frac{z_3}{z_1}, \quad k_2 = \frac{z_3}{z_2}$$

Z_1, Z_2, Z_3---Tooth numbers of gears

$$k_1 = \frac{z_2}{z_1}, \quad k_2 = 1$$

Z_1, Z_2---Tooth numbers of gears

Pulley or chain assembly

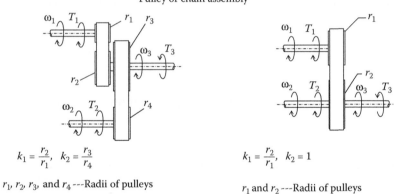

$$k_1 = \frac{r_2}{r_1}, \quad k_2 = \frac{r_3}{r_4}$$

$r_1, r_2, r_3,$ and r_4 ---Radii of pulleys

$$k_1 = \frac{r_2}{r_1}, \quad k_2 = 1$$

r_1 and r_2 ---Radii of pulleys

Shaft

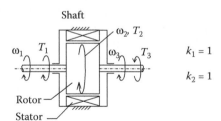

FIGURE 6.7
Commonly used mechanical torque-coupling devices.

single-gear transmission, 2, takes advantage of the high torque of an electric machine at low speed. The multigear transmission, 1, is used to overcome the disadvantages of the IC engine speed–torque characteristics (flat torque output in its entire speed range). The multigear transmission, 1, also tends to improve the operating efficiency of the engine and reduces the speed range of the vehicle, in which the electric machine must propel the vehicle alone, thus preventing the batteries from quickly discharging.

In contrast to the previous design, Figure 6.9c shows the tractive effort–speed profile of a drivetrain that has a single-gear transmission, 1, for the engine and a multigear transmission, 2, for the electric motor. This configuration is an unfavorable design because it does not use the advantages of the two power plants.

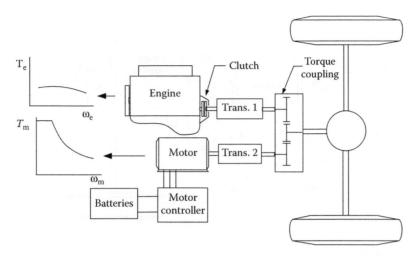

FIGURE 6.8
Two-shaft configuration.

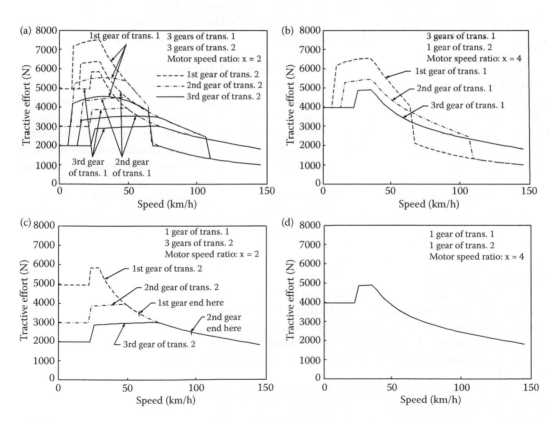

FIGURE 6.9
Tractive effort along with vehicle speed with different transmission schemes: (a) two multigear transmissions, (b) multigear engine transmission and single-gear motor transmission, (c) single-gear engine transmission and multigear motor transmission, and (d) both single-gear transmissions.

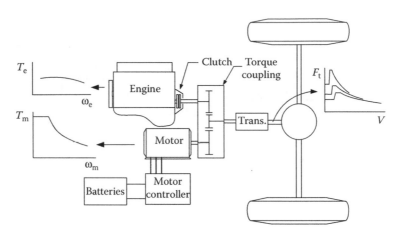

FIGURE 6.10
Two-shaft configuration.

Figure 6.9d shows the tractive effort–speed profile of a drivetrain that has two single-gear transmissions. This arrangement results in a simple configuration and control. The limitation to the application of this drivetrain is the maximum tractive effort of the drivetrain. When the power ratings of the engine, electric motor, batteries, and transmission parameters are properly designed, this drivetrain gives the vehicle satisfactory performance and efficiency.

Another configuration of a two-shaft parallel hybrid drivetrain is shown in Figure 6.10, in which the transmission is located between the torque coupler and the drive shaft and may be categorized as a pretransmission. The transmission amplifies the torques of both the engine and the electric motor with the same scale. The design of the gear ratios k_1 and k_2 in the torque coupler (Equation 6.3) allows the electric motor and the engine to reach their maximum speeds at the same time. This configuration would be suitable when a relatively small engine and electric motor are used, where a multigear transmission is needed to enhance the tractive effort at low speeds.

The simplest and most compact architecture of the torque-coupling parallel hybrid is the single-shaft configuration, where the rotor of the electric motor functions as the torque coupler ($k_1 = 1$ and $k_2 = 1$ in Equations 6.2 and 6.3). The electric motor may be located either between the engine and the transmission, as shown in Figure 6.11, referred to as pretransmission, or between the transmission and the final drive, as shown in Figure 6.12, referred to as post-transmission.

In the pretransmission configuration, as shown in Figure 6.11, torques of both the engine and the motor are modified by the transmission. However, the engine and the motor are required to have the same speed range. This configuration is usually used in the case of a small motor, referred to as a mild hybrid drivetrain, in which the electric motor functions as an engine starter, an electrical generator, an engine power assistant, and for regenerative braking.

In the post-transmission configuration, as shown in Figure 6.12, the transmission only modifies the engine torque, while the motor directly delivers its torque to the final drive without modification. This configuration may be used in a drivetrain where a large electric motor with a long constant power region is employed. The transmission is only used to change the engine operating points for improving the vehicle performance and engine operating efficiency. It should be noted that the batteries cannot be charged from the engine by

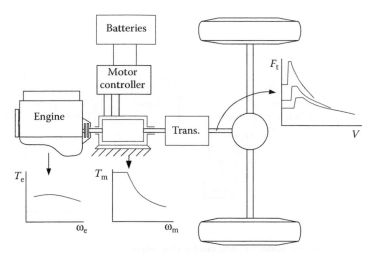

FIGURE 6.11
Pretransmission single-shaft torque combination parallel hybrid electric drivetrain.

running the electric motor as a generator when the vehicle is at a standstill since the motor is rigidly connected to the drive wheels.

Another torque-coupling parallel hybrid drivetrain is the separated axle architecture, in which one axle is powered by the engine and the other by the electric motor, as shown in Figure 6.13. The tractive efforts produced by the two power trains are added together through the vehicle chassis and road. The operating principle is similar to that of the two-shaft configuration shown in Figure 6.8. Both transmissions for the engine and the electric motor may be single-gear or multigear. This configuration has similar tractive effort characteristics, as shown in Figure 6.9.

The separated axle architecture offers some of the advantages of a conventional vehicle. It keeps the original engine and transmission unaltered and adds an electrical traction system

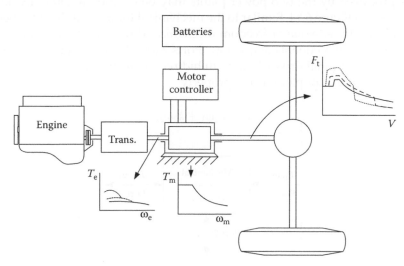

FIGURE 6.12
Post-transmission single-shaft torque combination parallel hybrid electric drivetrain.

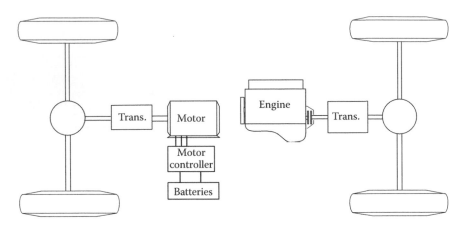

FIGURE 6.13
Separated axle torque combination parallel hybrid electric drivetrain.

on the other axle. It is also a four-wheel drive, which improves the traction on slippery roads and reduces the tractive effort on a single tire.

However, electric machines and the eventual differential gear system occupy a lot of space and may reduce the available passenger space and luggage space. This problem may be solved if the transmission behind the electric motor is single-gear, and the single electric motor is replaced by two small-sized electric motors that are placed within the two drive wheels. It should be noted that the batteries cannot be charged from the engine when the vehicle is at a standstill.

6.2.2.2 Parallel Hybrid Drivetrain with Speed Coupling

6.2.2.2.1 Speed-Coupling Devices

The powers produced by the two power plants may be coupled together by adding their speeds, as shown in Figure 6.14. Similar to the mechanical torque coupler, the speed coupler is also a three-port, two-degree-of-freedom mechanical device. Port 1 may be connected to an IC engine with unidirectional energy flow. Ports 2 and 3 may be connected to an electric motor and to the load (final drive), both with bidirectional energy flow.

The mechanical speed coupler has the property

$$\omega_3 = k_1\omega_1 + k_2\omega_2, \tag{6.4}$$

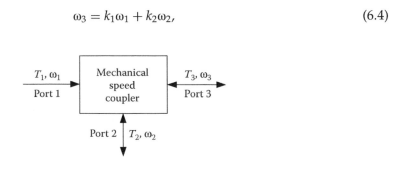

FIGURE 6.14
Speed coupling.

where k_1 and k_2 are constants associated with the structural and geometric design. Among the three speeds, ω_1, ω_2, and ω_3, two are independent of each other and can be controlled independently. Due to the constraint of power conservation, the torques are linked together by

$$T_3 = \frac{T_1}{k_1} = \frac{T_2}{k_2},\qquad(6.5)$$

in which the minimum torque determines the other two.

A typical speed-coupling device is the planetary gear unit, as shown in Figure 6.15. The planetary gear unit is a three-port unit consisting of a sun gear, ring gear, and yoke, labeled 1, 2, and 3, respectively. The speed relationship in the sun gear, ring gear, and yoke can be obtained as follows.

First, let the yoke be attached to a stationary frame, that is, $\omega_3 = 0$; the gear ratio from the sun gear to ring gear is

$$i_{1-2}^3 = \frac{\omega_2^3}{\omega_1^3} = -\frac{R_2}{R_1} = -\frac{Z_2}{Z_1},\qquad(6.6)$$

where ω_1^3 and ω_2^3 are the angular velocities of the sun gear and the ring gear with respect to the yoke (when the yoke is at a standstill); R_1 and R_2 are the radii of the sun gear and the ring gear, respectively; and Z_1 and Z_2 are the tooth numbers of the sun gear and the ring gear, respectively, which are proportional to the radii of the sun gear and the ring gear. Here, rotating in the counterclockwise direction is defined as positive angular velocity, whereas rotating in the clockwise direction is defined as negative angular velocity, as shown in Figure 6.15. Equation 6.6 indicates that ω_1^3 and ω_2^3 have different rotating directions, and thus the gear ratio i_{1-2}^3 is negative. When the yoke is free from the stationary frame, the absolute angular velocities of the sun gear, the ring gear, and the yoke can be expressed by

$$\frac{\omega_1 - \omega_3}{\omega_2 - \omega_3} = i_{1-2}^3.\qquad(6.7)$$

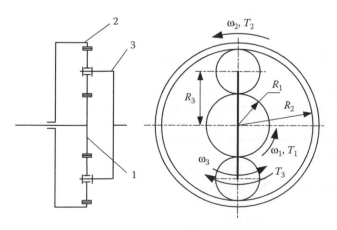

FIGURE 6.15
Planetary gear unit used as a speed-coupling device.

Then we obtain

$$\omega_1 - i_{1-2}^3 \omega_2 - (1 - i_{1-2}^3)\omega_3 = 0. \tag{6.8}$$

Conventionally, we are not accustomed to a negative gear ratio. If we define the gear ratio as a positive number, as

$$i_g = i_{1-2}^3 = \frac{R_2}{R_1} = \frac{Z_2}{Z_1}, \tag{6.9}$$

then Equation 6.8 can be rewritten as

$$\omega_1 + i_g \omega_2 - (1 + i_g)\omega_3 = 0 \tag{6.10}$$

or

$$\omega_3 = \frac{1}{1 + i_g}\omega_1 + \frac{i_g}{1 + i_g}\omega_2. \tag{6.11}$$

Comparing Equation 6.11 with Equation 6.4, $k_1 = 1/(1 + i_g)$ and $k_2 = i_g/(1 + i_g)$ are obtained.

Similar to the definition of speed, when the torque acting on each element of the planetary gear unit is defined to be positive in the counterclockwise direction and negative in the clockwise direction, the total power into the unit should be zero (output power is negative) when the loss inside the unit is ignored, that is,

$$T_1\omega_1 + T_2\omega_2 + T_3\omega_3 = 0. \tag{6.12}$$

Combining Equations 6.11 and 6.12 yields

$$T_3 = -(1 + i_g)T_1 = -\frac{1 + i_g}{i_g}T_2. \tag{6.13}$$

Equation 6.13 indicates that the torques acting on the sun gear, T_1, and the ring gear, T_2, always have the same sign (both positive or negative), and the torque acting on the yoke, T_3, always has the direction opposite to T_1 and T_2, as shown in Figure 6.15.

When one element of the sun gear, the ring gear, or the yoke is locked to the stationary frame, that is, one degree of freedom is constrained, the unit will become a single-gear transmission (one input and one output). The speed and torque relationships, while different elements are fixed, are shown in Figure 6.16.

Another interesting device used as a speed coupler is an electric motor with a floating stator (called transmotor in this book), in which the stator, generally fixed to a stationary frame in a traditional motor, is released to form a double-rotor machine—outer and inner rotor. The outer rotor, inner rotor, and air gap are the three ports. Electric power is converted into mechanical power through the air gap, as shown in Figure 6.17. The motor speed, in conventional terms, is the relative speed of the inner rotor with respect to the outer

Element fixed	Speed	Torque
Sun gear	$\omega_3 = \dfrac{i_g}{1 + i_g}\, \omega_2$	$T_3 = -\dfrac{1 + i_g}{i_g}\, T_2$
Ring gear	$\omega_3 = \dfrac{1}{1 + i_g}\, \omega_1$	$T_3 = -(1 + i_g)\, T_1$
Yoke	$\omega_1 = -i_g\, \omega_2$	$T_1 = \dfrac{1}{i_g}\, T_2$

FIGURE 6.16
Speed and torque relationships while one element is fixed.

rotor. Because of the action and reaction effect, the torques acting on both rotors are always the same and result in constants $k_1 = 1$ and $k_2 = 1$. The speed relationship can be expressed as

$$\omega_{or} = \omega_{ir} + \omega_{oi}, \tag{6.14}$$

where ω_{io} is the inner rotor speed relative to the outer rotor stator. The torque relationship can be expressed as

$$T_{ir} = T_{os} = T_e. \tag{6.15}$$

6.2.2.2.2 Drivetrain Configurations with Speed Coupling

Similar to the torque-coupling device, speed-coupling units can be used to constitute various hybrid drivetrains. Figures 6.18 and 6.19 show two examples of hybrid drivetrains with speed coupling using a planetary gear unit and an electric transmotor. In Figure 6.18, the engine supplies its power to the sun gear through a clutch and transmission. The transmission is used to modify the speed–torque profile of the engine to match the traction requirements. The transmission may be multigear or single-gear based on the engine speed–torque profile. The electric motor supplies its power to the ring gear through a pair of gears. Lock 1 and lock 2 are used to lock the sun gear and the ring gear to the

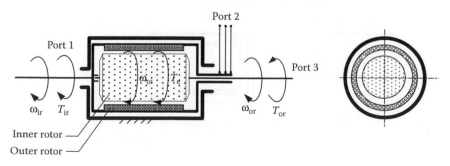

FIGURE 6.17
Transmotor used as a speed coupler.

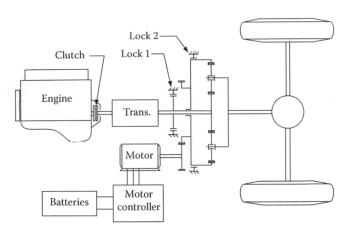

FIGURE 6.18
Hybrid electric drivetrain with speed coupling of a planetary gear unit.

stationary frame of the vehicle in order to implement different operation modes. The following operation modes can be carried out:

1. *Hybrid traction:* When lock 1 and lock 2 are released (the sun gear and the ring gear can rotate), both the engine and the electric machine supply positive speed and torque (positive power) to the drive wheels. The output speed and torque from the yoke of the planetary unit are described by Equations 6.11 and 6.13. That is, the rotational speed of the yoke is the summation of the sun gear speed (engine speed, or proportional to engine speed) and the ring gear speed (electric motor speed, or proportional to motor speed). However, the output torque from the yoke is proportional to the engine torque and the motor torque. Torque control will be studied in Chapter 9.

2. *Engine-alone traction:* When lock 2 locks the ring gear to the vehicle frame and lock 1 is released, the engine alone supplies power to the drive wheels. From Equations 6.11 and 6.13, the speed of the yoke is proportional to the speed of the sun gear as $\omega_3 = \omega_1/(1 + i_g)$, and the torque output from the yoke is proportional to the torque applied on the sun gear from the engine as $T_3 = (1 + i_g)T_1$.

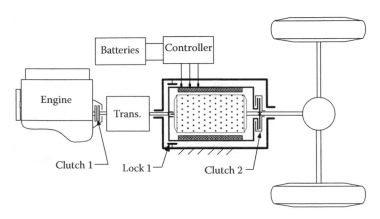

FIGURE 6.19
Hybrid electric drivetrain with speed coupling of an electric transmotor.

3. *Motor-alone traction:* When lock 1 locks the sun gear to the vehicle frame (engine is shut off and engine clutch is disengaged) and lock 2 is released, only the electric motor supplies its power to the drive wheels. From Equations 6.11 and 6.13, the speed of the yoke is proportional to the speed of the ring gear as $\omega_3 = (\omega_2 i_g)/(1 + i_g)$, and the torque output from the yoke is proportional to the torque applied on the ring gear by the electric motor as $T_3 = (1 + i_g)/(i_g T_1)$.

4. *Regenerative braking:* The states of lock 1 and lock 2 are the same as in motor-alone traction, the engine is also shut off, the engine clutch is disengaged, and the electric machine is controlled in regenerating mode (negative torque). The kinetic or potential energy of the vehicle can be absorbed by the electrical system.

5. *Battery charging from the engine:* The engine clutch and lock 1 and lock 2 are in the same state as in the hybrid traction mode. However, the electric motor is controlled to rotate in the opposite direction, that is, negative speed. Thus, the electric motor operates with positive torque and negative speed (negative power) and absorbs energy from the engine and delivers it to the batteries. In this case, the engine power is split into two parts by decomposing its speed.

The drivetrain, consisting of the transmotor as shown in Figure 6.19, has a structure similar to that in Figure 6.18. Lock 1 and clutch 2 are used to lock the outer rotor to the vehicle frame and the outer rotor to the inner rotor, respectively. This drivetrain can fulfill all the operation modes mentioned above. The operating mode analysis is left to the reader.

Figure 6.20 shows an implementation of speed coupling with a transmotor. Clutch 1 is the substitution for clutch 1, as shown in Figure 6.19; clutch 2 has the same function as clutch 2 in Figure 6.19; and clutch 3 has the same function as lock 1 in Figure 6.19.

The main advantage of the hybrid drivetrain with speed coupling is that the speed of two power plants is decoupled from the vehicle speed. Therefore, the speed of both power plants can be chosen freely. This advantage is important to power plants, such as the Stirling engine

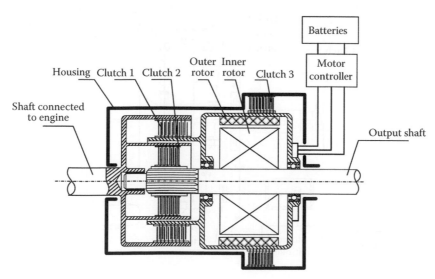

FIGURE 6.20
Implementation of speed coupling with a transmotor.

and the gas turbine engine, in which the operating efficiencies are sensitive to speed and less sensitive to torque.

6.2.2.3 Hybrid Drivetrains with Both Torque and Speed Coupling

6.2.2.3.1 With Optional Coupling Mode

By combining torque and speed coupling, one may establish a hybrid drivetrain in which torque- and speed-coupling states can be alternately chosen. Figure 6.21 shows such a drivetrain.[8] When the torque-coupling operation mode is chosen, lock 2 locks the ring gear of the planetary unit to the vehicle frame while clutches 1 and 3 are engaged, and clutch 2 is disengaged. The power of the engine and the electric motor are added together by adding their torques together through gear Z_a, Z_b and clutch 3 to the sun gear shaft. In this case, the planetary gear unit functions only as a speed reducer. The gear ratio from the sun gear to the yoke, defined as ω_1/ω_3, equals $(1 + i_g)$. This is a typical parallel hybrid drivetrain with torque coupling.

When the speed-coupling mode is chosen as the current operating mode, clutches 1 and 2 are engaged, whereas clutch 3 is disengaged, and locks 1 and 2 release the sun gear and the ring gear. The speed of the yoke, connected to the drive wheels, is a combination of engine speed and motor speed (Equation 6.11), but the engine torque, the electric motor torque, and the torque on the drive wheels are kept in a fixed relationship, as described by Equation 6.13.

With the option to choose the power-coupling mode (torque or speed coupling), the power plant has more opportunities to choose its operation manner and operation region to optimize its performance. For instance, at low vehicle speeds, the torque combination operation mode may be suitable for high acceleration or hill climbing. On the other hand, at high vehicle speeds, the speed combination mode would be used to keep the engine speed in its optimal region.

The planetary gear unit and the traction motor in Figure 6.21 can be replaced by a transmotor to constitute a similar drivetrain, as shown in Figure 6.22. When clutch 1 is engaged to

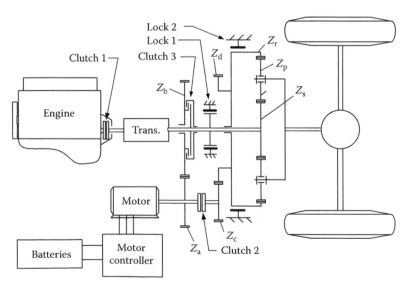

FIGURE 6.21
Alternative torque and speed hybrid electric drivetrain with a planetary gear unit.

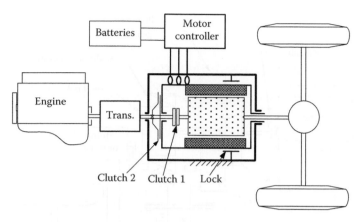

FIGURE 6.22
Alternative torque- and speed-coupling hybrid electric drivetrain with a transmotor.

couple the output shaft of the transmission to the inner rotor shaft of the transmotor, clutch 2 is disengaged to release the engine shaft from the inner rotor of the transmotor, and the lock is activated to fix the outer rotor of the transmotor to the vehicle frame. The drivetrain then works in torque-coupling mode. On the other hand, when clutch 1 is disengaged, clutch 2 is engaged, and the lock is released, the drivetrain works in speed-coupling mode.

The distinguishing characteristic of the previous hybrid drivetrains is that the drivetrain can optionally choose the best coupling mode in different driving situations to achieve the best vehicle performance and efficiency. However, they cannot run in both coupling modes at the same time, since only two power plants are available.

6.2.2.3.2 With Both Coupling Modes

By adding another power plant, a hybrid drivetrain with both speed- and torque-coupling modes at the same time can be realized. A good example is the one developed and implemented in the Toyota Prius by Toyota Motor Company.[9] This drivetrain is schematically illustrated in Figure 6.23. The drivetrain uses a planetary gear unit as the speed-coupling device and a set of fixed-axle gears as the torque-coupling device. An IC engine is connected to the yoke of the planetary gear unit, and a small motor/generator (a few kilowatts) is connected to the sun gear of a planetary gear unit to constitute the speed-coupling configuration. The ring gear is connected to the drive wheels through the axle-fixed gear unit (torque coupler). Meanwhile, a traction motor is also connected to the fixed-axle gear unit to constitute the torque-coupling configuration.

From Equation 6.11, the rotational speed of the ring gear or gear Z_a, which is proportional to vehicle speed, is related to the rotational speed of the engine (yoke) and of the motor/generator (sun gear) and is expressed as

$$\omega_r = \frac{1+i_g}{i_g}\omega_{ice} - i_g\omega_{m/g},$$

(6.16)

where i_g is the gear ratio defined by Equation 6.9, and ω_{ice} and $\omega_{m/g}$ are the rotational speeds of the engine and motor/generator, respectively. The load torque, acting on the ring gear of

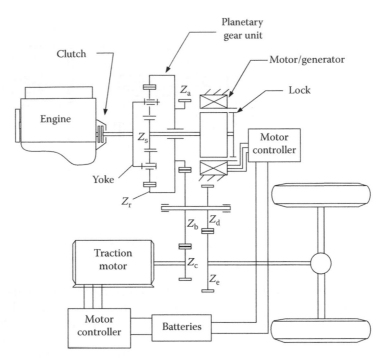

FIGURE 6.23
Integrated speed- and torque-coupling hybrid electric drivetrain.

the planetary gear unit by gear Z_4, is related to the engine torque and the motor/generator torque by

$$- T_r = \frac{i_g}{1 + i_g} T_{ice} = -i_g T_{m/g}. \tag{6.17}$$

Equation 6.17 indicates that the torque, acting on the sun gear, supplied by the motor/generator has a direction opposite to that of the engine torque and the same direction as the load torque on the ring gear. At low vehicle speed (small ω_r) and a not very low engine speed (larger than its idle speed), the motor/generator must rotate in the positive direction (same direction as engine speed). In this condition, the motor/generator operates with a negative power, that is, generating. The power of the engine is split into two parts: one part goes to the motor/generator and the other to the vehicle load through the ring gear. This is how the drivetrain gets its name of power-split hybrid drivetrain. However, at high vehicle speed, while trying to maintain the engine speed below a given speed, for high engine operating efficiency, the motor/generator may be operated at negative speed, that is, rotating in the opposite direction to engine speed. In this case, the motor/generator delivers positive power to the planetary gear unit, that is, motoring. It becomes clear through the previous analysis that the major function of a motor/generator is to control the engine speed, that is, decouple the engine speed from the wheel speed.

The traction motor adds additional torque to the torque output from the ring gear of the planetary gear unit in torque-coupling mode through gears Z_c, Z_b, Z_d, and Z_e, by which the engine torque is decoupled from the vehicle load.

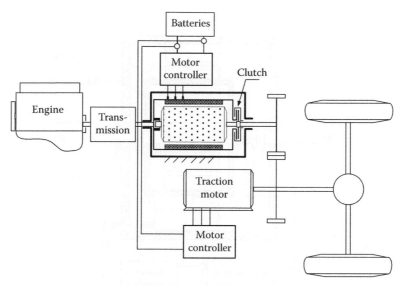

FIGURE 6.24
Hybrid electric drivetrain with speed and torque coupling of a transmotor and double shaft.

The small motor and the planetary gear unit in Figure 6.23 can be replaced by an individual transmotor, as shown in Figure 6.24.[10] This drivetrain has characteristics very similar to those of the drivetrain shown in Figure 6.23. Another variation of the drivetrain in Figure 6.24 is the single-shaft design, as shown in Figure 6.25. A more compact design of the drivetrain in Figure 6.25 would integrate the transmotor and the traction motor together, as shown in Figure 6.26. The design and control may be more complicated than the separated structure due to a correlated magnetic field in the double air gaps.

In the literature, the integrated or separated transmotor and traction motor in Figures 6.25 and 6.26 is called electrical variable transmission (EVS).[11,12] This name is derived from the fact that engine speed is electrically decoupled from vehicle speed by the speed-coupling

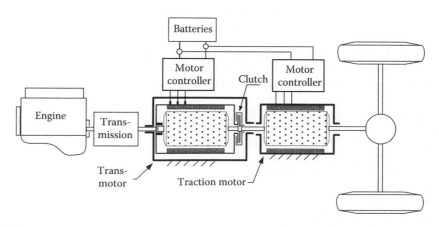

FIGURE 6.25
Hybrid electric drivetrain with speed and torque coupling of transmotor and single shaft.

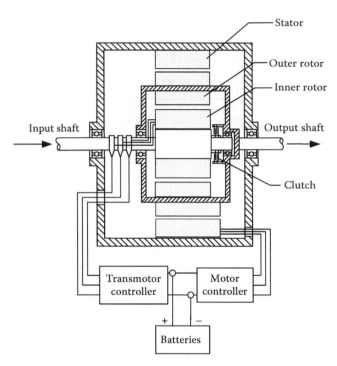

FIGURE 6.26
Integrated transmotor and traction motor.

device: the transmotor. The operating characteristics and drivetrain control will be discussed later.

Bibliography

1. M. Ehsani, Y. Gao, and J. M. Miller, Hybrid electric vehicles: Architecture and motor drives, *Proceedings of the IEEE, Special issue on Electric, Hybrid Electric and Fuel Cells Vehicle*, 95(4), April 2007.
2. M. Ehsani, K. L. Butler, Y. Gao, and K. M. Rahman, Next generation passenger cars with better range, performance, and emissions: The ELPH car concept, In *Horizon in Engineering Symposium*, Texas A&M University Engineering Program Office, College Station, Texas, September 1998.
3. M. Ehsani, *The Electrically Peaking Hybrid System and Method*, U.S. Patent No. 5,586,613, December 1996.
4. K. Yamaguchi, S. Moroto, K. Kobayashi, M. Kawamoto, and Y. Miyaishi, Development of a new hybrid system-dual system, *Society of Automotive Engineers (SAE) Journal*, Paper No. 960231, Warrendale, PA, 1997.
5. C. C. Chan and K. T. Chau, *Modern Electric Vehicle Technology*, Oxford University Press, New York, 2001.
6. Y. Gao, K. M. Rahman, and M. Ehsani, The energy flow management and battery energy capacity determination for the drive train of electrically peaking hybrid, *Society of Automotive Engineers (SAE) Journal*, Paper No. 972647, Warrendale, PA, 1997.

7. Y. Gao, K. M. Rahman, and M. Ehsani, Parametric design of the drive train of an electrically peaking hybrid (ELPH) vehicle, *Society of Automotive Engineers (SAE) Journal*, Paper No. 970294, Warrendale, PA, 1997.

8. Y. Gao and M. Ehsani, *New Type of Transmission for Hybrid Vehicle with Speed and Torque Summation*, U.S. Patent pending.

9. Available at http://www.toyota.com, Toyota Motor Company, visited in September 2003.

10. Y. Gao and M. Ehsani, *Series–Parallel Hybrid Drive Train with an Electric Motor of Floating Stator and Rotor*, U.S. Patent pending.

11. M. J. Hoeijimakes and J. A. Ferreira, The electrical variable transmission, *IEEE on Industry Application*, 42(4), July–August 2006: 1092–1100.

12. S. Cui, Y. Cheng, and C. C. Chan, A basic study of electrical variable transmission and its application in hybrid electric vehicle, In *IEEE on Vehicle Power and Propulsion Conference, (VPPC)*, 2006.

13. K. Kimura et al. Development of new IGBT to reduce electrical power losses and size of power control unit for hybrid vehicles, *SAE International Journal of Alternative Powertrains*, 6, 2017-01-1244, 2017: 303–308.

14. S. Shili, A. Sari, A. Hijazi, and P. Venet, Online lithium-ion batteries health monitoring using balancing circuits, In *Industrial Technology (ICIT) 2017 IEEE International Conference* on, pp. 484–488, 2017.

15. A. Lievre, A. Sari, P. Venet, A. Hijazi, M. Ouattara-Brigaudet, and S. Pelissier, Practical online estimation of lithium-ion battery apparent series resistance for mild hybrid vehicles, *IEEE Transactions on Vehicular Technology*, 65(6), June 2016: 4505–4511, doi: 10.1109/TVT.2015.2446333.

16. V. Saharan and K. Nakai, High power cell for mild and strong hybrid applications including Chevrolet Malibu, SAE Technical Paper 2017-01-1200, 2017, doi: 10.4271/2017-01-1200.

17. IEEE 2009: 1286–92. http://dx.doi.org/10.1109/VPPC.2009.5289703.

18. A. Poullikkas, Sustainable options for electric vehicle technologies, *Renew Sustain Energy Reviews*, 41, 2015: 1277–1287. http://dx.doi.org/10.1016/j.rser.2014.09.016.

19. S. F. Tie and C. W. Tan, A review of energy sources and energy management system in electric vehicles, *Renew Sustain Energy Rev* 20, 2013: 82–102. http://dx.doi.org/10.1016/j.rser.2012.11.077.

20. M. A. Delucchi et al. An assessment of electric vehicles: technology, infrastructure requirements, greenhouse-gas emissions, petroleum use, material use, lifetime cost, consumer acceptance and policy initiatives, *Philosophical Transaction Series A Mathematicla Physical and Engineering Science*, 372, 2014: 20120325. http://dx.doi.org/10.1098/rsta.2012.0325.

7

Electric Propulsion Systems

Electric propulsion systems are at the heart of EVs and HEVs. They consist of electric motors, power converters, and electronic controllers. The electric motor converts the electric energy into mechanical energy to propel the vehicle, or vice versa, to enable regenerative braking and/or to generate electricity for charging the on-board energy storage. The power converter is used to supply the electric motor with proper voltage and current. The electronic controller commands the power converter by providing control signals to it, and then it controls the operation of the electric motor to produce proper torque and speed, according to the command from the driver. The electronic controller can be further divided into three functional units—sensor, interface circuitry, and processor. The sensor is used to translate the measurable quantities, such as current, voltage, temperature, speed, torque, and flux, into electric signals through the interface circuitry. These signals are conditioned to the appropriate level before being fed into the processor. The processor output signals are usually amplified via the interface circuitry to drive power semiconductor devices of the power converter. The functional block diagram of an electric propulsion system is illustrated in Figure 7.1.

The choice of electric propulsion systems for EVs and HEVs mainly depends on several factors, including the driver's expectation, vehicle constraints, and energy source. The driver's expectation is defined by a driving profile, which includes the acceleration, maximum speed, climbing capability, braking, and range. The vehicle constraints, including volume and weight, depend on the vehicle type, vehicle weight, and payload. The energy source relates to batteries, fuel cells, ultracapacitors, flywheels, and various hybrid sources. Thus, the process of identifying the preferred feature and package options for electric propulsion must be carried out at the system level. The interaction of subsystems and the likely impacts of system trade-offs must be examined.

Unlike the industrial applications of motors, the motors used in EVs and HEVs usually require frequent starts and stops, high rates of acceleration/deceleration, high torque and low-speed hill climbing, low torque and high-speed cruising, and a very wide speed range of operation. The motor drives for EVs and HEVs can be classified into two main groups, commutator motors and commutatorless motors, as illustrated in Figure 7.2. Commutator motors mainly are the traditional DC motors, which include series-excited, shunt-excited, compound-excited, separately excited, and permanent magnet (PM)-excited motors. DC motors need commutators and brushes to feed current into the armature, making them less reliable and unsuitable for maintenance-free operation and high speed. In addition, winding-excited DC motors have low specific power density. Nevertheless, because of their mature technology and simple control, DC motor drives have been prominent in electric propulsion systems.

Technological developments have recently pushed commutatorless electric motors into a new era. Advantages include higher efficiency, higher power density, and lower operating cost. They are also more reliable and maintenance-free compared to commutator DC motors; thus, commutatorless electric motors have now become more attractive.

Induction motors are widely accepted as a commutatorless motor type for EV and HEV propulsion. This is because of their low cost, high reliability, and maintenance-free

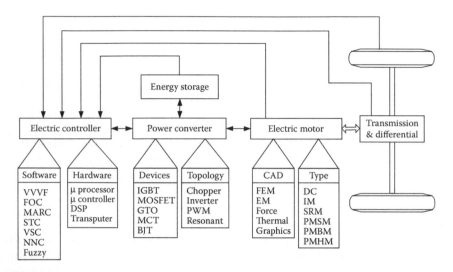

FIGURE 7.1
Functional block diagram of a typical electric propulsion system. (Adapted from C. C. Chan and K. T. Chau, *Modern Electric Vehicle Technology*, Oxford University Press, Oxford, 2001.)

operation. However, conventional control of induction motors such as variable voltage, variable frequency cannot provide the desired performance. With the advent of the power electronics and microcomputer era, the principle of field-oriented control (FOC) or vector control of induction motors has been accepted to overcome their control complexity due to their nonlinearity.[2] However, these EV and HEV motors using FOC still suffer from low efficiency at light loads and a limited constant-power operating range.

By replacing the field winding of conventional synchronous motors with PMs, PM synchronous motors can eliminate conventional brushes, slip rings, and field copper losses.[3] These PM synchronous motors are also called PM brushless AC motors, or sinusoidal-fed PM brushless motors, because of their sinusoidal AC current and brushless configuration. Since these motors are essentially synchronous motors, they can run from a sinusoidal or pulse width modulation (PWM) supply without electronic commutation. When PMs are mounted on the rotor surface, they behave as nonsalient synchronous motors because the permeability of PMs is similar to that of air. With those PMs buried inside the magnetic circuit of the rotor, the saliency causes an additional reluctance torque, which leads to facilitating a wider speed range at constant power operation.

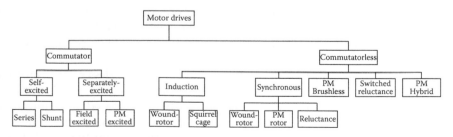

FIGURE 7.2
Classification of electric motor drives for EV and HEV applications.

On the other hand, by abandoning field winding or PMs while purposely making use of the rotor saliency, synchronous reluctance motors are generated. These motors are generally simple and inexpensive but with relatively low output power. Similar to induction motors, these PM synchronous motors usually use FOC for high-performance applications.[3] Because of their inherently high power density and high efficiency, they have been accepted as having great potential to compete with induction motors for EV and HEV applications.

By virtually inverting the stator and rotor of PM DC motors (commutator), PM brushless DC (BLDC) motors are generated. It should be noted that the term "DC" may be misleading, since it does not refer to a DC current motor. These motors are fed by rectangular AC current and hence are also rectangular-fed PM brushless motors.[4] The most obvious advantage of these motors is the removal of brushes. Another advantage is the ability to produce a large torque because of the rectangular interaction between current and flux. Moreover, the brushless configuration allows more cross-sectional area for the armature windings. Since the conduction of heat through the frame is improved, an increase in electric loading causes higher power density. Unlike PM synchronous motors, these PM BLDC motors generally operate with shaft position sensors. Recently, sensorless control technologies have been developed in the Power Electronics and Motor Drive Laboratory at Texas A&M University.

Switched reluctance motors (SRMs) have been recognized to have considerable potential for EV and HEV applications. Basically, they are direct derivatives of single-stack, variable-reluctance stepping motors. SRMs have definite advantages of simple construction, low manufacturing cost, and outstanding torque–speed characteristics for EV and HEV applications. Although they possess simplicity in construction, this does not imply any simplicity of their design and control. Because of the heavy saturation of pole tips and the fringe effect of pole and slots, their design and control are difficult and subtle. Traditionally, SRMs operate with shaft sensors to detect the relative position of the rotor to the stator. These sensors are usually vulnerable to mechanical shock and sensitive to temperature and dust. Therefore, the presence of the position sensor reduces the reliability of SRMs and constrains some applications. Recently, sensorless technologies have been developed in the Power Electronics and Motor Drive Laboratory—again, at Texas A&M University. These technologies can ensure smooth operation from zero speed to maximum speed.[5] This will be discussed in detail in the following sections.

7.1 DC Motor Drives

DC motor drives have been widely used in applications requiring adjustable speed, good speed regulation, and frequent starting, braking, and reversing. Various DC motor drives have been widely applied to different electric traction applications because of their technological maturity and control simplicity.

7.1.1 Principle of Operation and Performance

The operation principle of a DC motor is straightforward. When a wire carrying electric current is placed into a magnetic field, a magnetic force acting on the wire is produced. The force is perpendicular to the wire and the magnetic field, as shown in Figure 7.3.

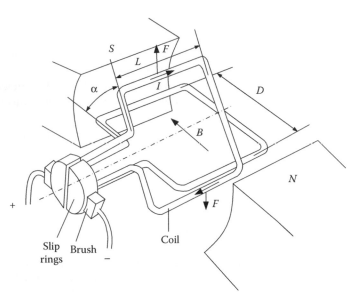

FIGURE 7.3
Operation principle of a DC motor.

The magnetic force is proportional to the wire length, magnitude of the electric current, and the density of the magnetic field, that is,

$$F = BIL. \tag{7.1}$$

When the wire is shaped into a coil, as shown in Figure 7.3, the magnetic forces acting on both sides produce a torque, which is expressed as,

$$T = BIL \cos \alpha, \tag{7.2}$$

where α is the angle between the coil plane and the magnetic field, as shown in Figure 7.3. The magnetic field may be produced by a set of windings or PMs. The former is called a wound-field DC motor, and the latter is called a PM DC motor. The coil carrying electric current is called the armature. In practice, the armature consists of several coils. To obtain continuous and maximum torque, slip rings and brushes are used to conduct each coil at a position of $\alpha = 0$.

Practically, the performance of DC motors can be described by the armature voltage, back electromotive force (EMF), and field flux.

Typically, there are four types of wound-field DC motors, depending on the mutual interconnection between the field and armature windings. They are separately excited, shunt excited, series excited, and compound excited, as shown in Figure 7.4. In the case of a separately excited motor, the field and the armature voltage can be controlled independently of one another. In a shunt motor, the field and the armature are connected in parallel to a common source. Therefore, an independent control of field or armature currents can only be achieved by inserting a resistance in the appropriate circuit. This is an inefficient method of control. The efficient method is to use power-electronics-based DC–DC converters in the appropriate circuit to replace the resistance. The DC–DC converters can be actively controlled to produce proper armature and field voltage. In the case of a series motor, the field current is the same as the armature current; therefore, field flux is a function of

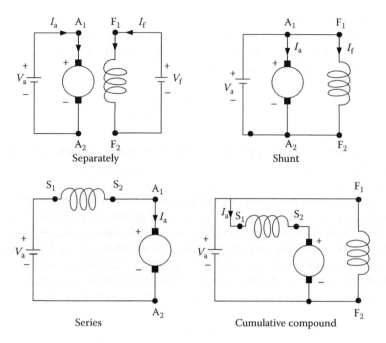

FIGURE 7.4
Wound-field DC motors.

armature current. In a cumulative compound motor, the magnetomotive force (mmf) of a series field is a function of the armature current and is in the same direction as the mmf of the shunt field.[6]

The steady-state equivalent circuit of the armature of a DC motor is shown in Figure 7.5. The resistor R_a is the resistance of the armature circuit. For separately excited and shunt DC motors, it is equal to the resistance of the armature windings; for series and compound motors, it is the sum of armature and series field winding resistances. The basic equations of a DC motor are

$$V_a = E + R_a I_a, \quad E = K_e \phi \omega_m, \tag{7.3}$$

$$T = K_e \phi I_a, \tag{7.4}$$

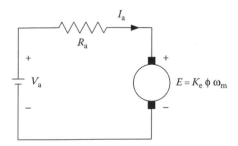

FIGURE 7.5
Steady-state equivalent circuit of armature circuit of a DC motor.

where ϕ is the flux per pole in webers, I_a is the armature current in A, V_a is the armature voltage in volts, R_a is the resistance of the armature circuit in ohms, ω_m is the speed of the armature in rad/s, T is the torque developed by the motor in N-m, and K_e is a constant.

From Equations 7.3 and 7.4 one can obtain

$$T = \frac{K_e\phi}{R_a}V - \frac{(K_e\phi)^2}{R_a}\omega_m. \tag{7.5}$$

Equations 7.3 through 7.5 are applicable to all DC motors, namely separately (or shunt) excited, series, and compound motors. In the case of separately excited motors, if the field voltage is maintained as constant, one can assume the flux to be practically constant as the torque changes. In this case, the speed–torque characteristic of a separately excited motor is a straight line, as shown in Figure 7.6. The nonload speed ω_m is determined by the armature voltage and the field excitation. Speed decreases as torque increases, and speed regulation depends on the armature circuit resistance. Separately excited motors are used in applications requiring good speed regulation and proper adjustable speed.

In the case of series motors, the flux is a function of armature current. In an unsaturated region of the magnetization characteristic, ϕ can be assumed to be proportional to I_a. Thus,

$$\phi = K_f I_a. \tag{7.6}$$

Using Equations 7.4 through 7.6, the torque for series excited DC motors can obtained as

$$T = \frac{K_e K_f V_a^2}{(R_a + K_e K_f \omega_m)^2}, \tag{7.7}$$

where the armature circuit resistance R_a is now the sum of armature and field winding resistance.

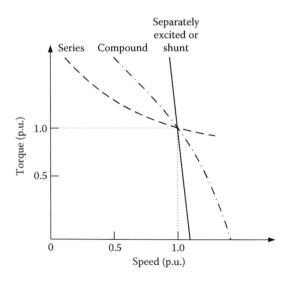

FIGURE 7.6
Speed characteristics of DC motors.

A speed–torque characteristic of a series DC motor is shown in Figure 7.6. In the case of a series, any increase in torque is accompanied by an increase in the armature current and, therefore, an increase in magnetic flux. Because flux increases with the torque, the speed drops to maintain a balance between the induced voltage and the supply voltage. The characteristic, therefore, shows a dramatic drop. A motor of standard design works at the knee point of the magnetization curve at the rated torque. At heavy torque (large current) overload, the magnetic circuit saturates, and the speed–torque curve approaches a straight line.

Series DC motors are suitable for applications requiring high starting torque and heavy torque overload, such as traction. This was just the case for electric traction before the power electronics and microcontrol era. However, series DC motors for traction application have some disadvantages. They are not allowed to operate without the load torque with full supply voltage. Otherwise, their speed quickly increases up to a very high value (Equation 7.7). Another disadvantage is the difficulty of regenerative braking.

Performance equations for cumulative compound DC motors can be derived from Equations 7.3 and 7.4. The speed–torque characteristics are between series-excited and separately excited (shunt) motors, as shown in Figure 7.6.

7.1.2 Combined Armature Voltage and Field Control

The independence of armature voltage and field provides more flexible control of the speed and torque than other types of DC motors. In EV and HEV applications, the most desirable speed–torque characteristic is to have a constant torque below a certain speed (base speed) and a constant power in a speed range above the base speed, as shown in Figure 7.7. In a speed range lower than the base speed, the armature current and field are set at their rated values, producing the rated torque. From Equations 7.3 and 7.4, it is clear that the armature voltage must be increased proportionally to the increase in speed. At the base speed, the armature voltage reaches its rated value (equal to the source voltage) and cannot be increased further. To further increase the speed, the field must be weakened with the increase in speed, then the back EMF E and armature current constant maintained. The

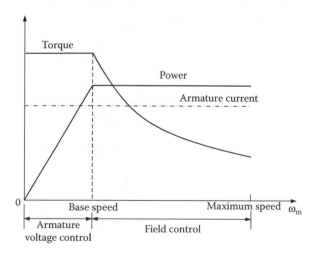

FIGURE 7.7
Torque and power limitations in combined armature voltage and field control.

torque produces drops parabolically with the increase in speed, and the output power remains constant, as shown in Figure 7.7.

7.1.3 Chopper Control of DC Motors

Choppers are used for the control of DC motors because of their several advantages such as high efficiency, flexibility in control, light weight, small size, quick response, and regeneration down to very low speeds. At present, the separately excited DC motors are usually used in traction due to the control flexibility of armature voltage and field.

For a DC motor control in open-loop and closed-loop configurations, the chopper offers many advantages due to the high operation frequency. High operation frequency results in high-frequency output voltage ripple and, therefore, fewer ripples in the motor armature current and a smaller region of discontinuous conduction in the speed–torque plane. A reduction in the armature current ripple reduces the armature losses. A reduction or elimination of the discontinuous conduction region improves speed regulation and transient response of the drive.

The power electronic circuit and the steady-state waveform of a DC chopper drive are shown in Figure 7.8. A DC voltage source, V, supplies an inductive load through a self-commutated semiconductor switch S. The symbol of a self-commutated semiconductor switch has been used because a chopper can be built using any devices among thyristors with a forced commutation circuit: GTO, power transistor, MOSFET, and IGBT. The diode shows the direction in which the device can carry current. A diode D_F is connected in parallel with the load. The semiconductor switch S is operated periodically over a period T and remains closed for a time $t_{on} = \delta T$ with $0 < \delta < 1$. The variable $\delta = t_{on}/T$ is called the duty

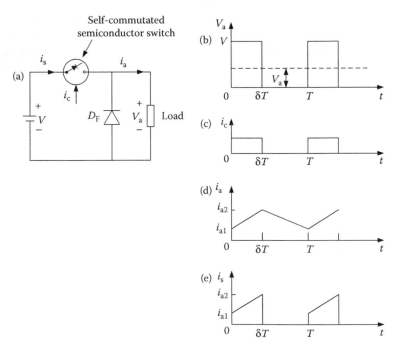

FIGURE 7.8
Principle of operation of a step-down (or class A) chopper: (a) basic chopper circuit and (b)–(e) waveforms.

ratio or duty cycle of a chopper. Figure 7.8 also shows the waveform of control signal i_c. Control signal i_c will be the base current for a transistor chopper and a gate current for the GTO of a GTO chopper or the main thyristor of a thyristor chopper. If a power MOSFET is used, it will be a gate to the source voltage. When a control signal is present, the semiconductor switch S will conduct, if it is forward biased. It is assumed that the circuit operation has been arranged such that the removal of i_c will turn off the switch.

During the on interval of the switch ($0 \leq t \leq \delta T$), the load is subjected to a voltage V, and the load current increases from i_{a1} to i_{a2}. The switch is opened at $t = \delta T$. During the off period of the switch ($\delta T \leq t \leq 1$), the load inductance maintains the flow of current through diode D_F. The load terminal voltage stays zero (if the voltage drop on the diode is ignored in comparison to V), and the current decreases from i_{a2} to i_{a1}. The internal $0 \leq t \leq \delta T$ is called the duty interval, and the interval $\delta T \leq t \leq T$ is known as the freewheeling interval. Diode D_F provides a path for the load current to flow when switch S is off and thus improves the load current waveform. Furthermore, by maintaining the continuity of the load current at turn-off, it prevents transient voltage from appearing across switch S due to the sudden change of the load current. The source current waveform is also shown in Figure 7.8e. The source current flows only during the duty interval and is equal to the load current.

The direct component or average value of the load voltage V_a is given by

$$V_a = \frac{1}{T} \int_0^T v_a \, dt = \frac{1}{T} \int_0^{\delta T} V \, dt = \delta V. \tag{7.8}$$

By controlling δ between 0 and 1, the load voltage can be varied from 0 to V; thus, a chopper allows a variable DC voltage to be obtained from a fixed voltage DC source.

The switch S can be controlled in various ways for varying the duty ratio δ. The control technologies can be divided into two categories:

1. Time ratio control (TRC).
2. Current limit control (CLC).

In TRC, also known as pulse width control, the ratio of on time to chopper period is controlled. The TRC can be further divided as follows:

1. *Constant-frequency TRC*: The chopper period T is kept fixed, and the on period of the switch is varied to control the duty ratio δ.
2. *Variable-frequency TRC*: Here, δ is varied either by keeping t_{on} constant and varying T or by varying both t_{on} and T.

In variable-frequency control with constant on time, low-output voltage is obtained at very low chopper frequencies. The operation of a chopper at low frequencies adversely affects the motor performance. Furthermore, the operation of a chopper with variable frequencies makes the design of an input filter very difficult. Thus, variable-frequency control is rarely used.

In CLC, also known as point-by-point control, δ is controlled indirectly by controlling the load current between certain specified maximum and minimum values. When the load current reaches a specified maximum value, the switch disconnects the load from the source and reconnects it when the current reaches a specified minimum value. For a DC motor load, this type of control is, in effect, a variable-frequency variable on time control.

The following important points can be noted from the waveform of Figure 7.8:

1. The source current is not continuous but flows in pulses. The pulsed current makes the peak input power demand high and may cause fluctuations in the source voltage. The source current waveform can be resolved into DC and AC harmonics. The fundamental AC harmonic frequency is the same as the chopper frequency. The AC harmonics are undesirable because they interfere with other loads connected to the DC source and cause radio frequency interference through conduction and electromagnetic radiation. Therefore, an L-C filter is usually incorporated between the chopper and the DC source. At higher chopper frequencies, harmonics can be reduced to a tolerable level by a cheaper filter. From this point, a chopper should be operated at the highest possible frequency.

2. The load terminal voltage is not a perfect direct voltage. In addition to a direct component, it has the harmonics of the chopping frequency and its multiples. The load current also has an AC ripple.

The chopper of Figure 7.8 is called a class A chopper. It is one of the number of chopper circuits that are used for the control of DC motors. This chopper can provide only a positive voltage and a positive current. It is called a single-quadrant chopper, only providing separately excited DC motor control in the first quadrant, that is, positive speed and positive torque. Since it can vary the output voltage from V to 0, it is also a step-down chopper or a DC-to-DC buck converter. The basic principle involved can also be used to realize a step-up chopper or DC-to-DC boost converter.

The circuit diagram and steady-state waveforms of a step-up chopper are shown in Figure 7.9. This chopper is known as a class B chopper. The presence of control signal i_c indicates the duration for which the switch can conduct if forward-biased. During a chopping period T, it remains closed for an interval $0 \leq t \leq \delta T$ and remains open for an

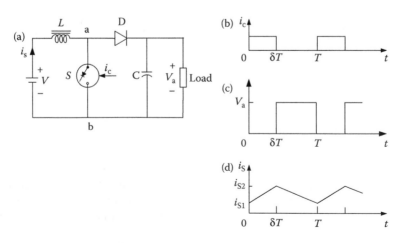

FIGURE 7.9
Principle of operation of a step-up (or class B) chopper: (a) basic chopper circuit; (b)–(d) waveforms.

interval $\delta T \le t \le T$. During the on period, i_S increases from i_{S1} to i_{S2}, thus increasing the magnitude of energy stored in inductance L. When the switch is opened, current flows through the parallel combination of the load and capacitor C. Since the current is forced against the higher voltage, the rate of change of the current is negative. It decreases from i_{S2} to i_{S1} in the switch's off period. The energy stored in the inductance L and the energy supplied by the low-voltage source are given to the load. Capacitor C serves two purposes. At the instant of opening of switch S, the source current, i_S, and load current, i_a, are not the same. In the absence of C, the turn-off of S forces the two currents to have the same values. This causes high induced voltage in the inductance L and the load inductance. Another reason for using capacitor C is to reduce the load voltage ripple. The purpose of diode D is to prevent any flow of current from the load into switch S or source V.

For understanding of the step-up operation, capacitor C is assumed to be large enough to maintain a constant voltage V_a across the load. The average voltage across terminals a and b is given as

$$V_{ab} = \frac{1}{T} \int_0^T v_{ab} \, dt = V_a(1 - \delta). \tag{7.9}$$

The average voltage across the inductance L is

$$V_L = \frac{1}{T} \int_0^T \left(L \frac{di}{dt} \right) dt = \frac{1}{T} \int_{i_{S1}}^{i_{S2}} L \, di = 0. \tag{7.10}$$

The source voltage is

$$V = V_L + V_{ab}. \tag{7.11}$$

Substituting from Equations 7.9 and 7.10 into Equation 7.11 gives

$$V = V_a(1 - \delta) \quad \text{or} \quad V_a = \frac{V}{1 - \delta}. \tag{7.12}$$

According to Equations 7.12, theoretically, the output voltage V_a can be changed from V to ∞ by controlling δ from 0 to 1. In practice, V_a can be controlled from V to a higher voltage, which depends on capacitor C and the parameters of the load and chopper.

The main advantage of a step-up chopper is the low ripple in the source current. While most applications require a step-down chopper, the step-up chopper finds application in low-power, battery-driven vehicles. The principle of the step-up chopper is also used in the regenerative braking of DC motor drives.

7.1.4 Multiquadrant Control of Chopper-Fed DC Motor Drives

The application of DC motors on EVs and HEVs requires the motors to operate in multiquadrants, including forward motoring, forward braking, backward motoring, and backward braking, as shown in Figure 7.10. For vehicles with reverse mechanical gears, two-quadrant operation (forward motoring and forward braking, or quadrant I and quadrant IV) is

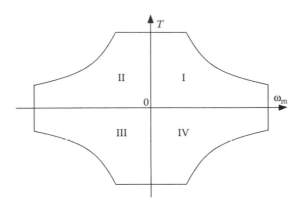

FIGURE 7.10
Speed–torque profiles of a multiquadrant operation.

required. However, for vehicles without reverse mechanical gears, four-quadrant operation is needed. Multiquadrant operation of a separately excited DC motor is implemented by controlling the voltage poles and magnitude through power-electronics-based choppers.

7.1.4.1 Two-Quadrant Control of Forward Motoring and Regenerative Braking

A two-quadrant operation consisting of forward motoring and forward regenerative braking requires a chopper capable of giving a positive voltage and current in either direction. This two-quadrant operation can be realized in the following two schemes.[6]

7.1.4.1.1 Single Chopper with a Reverse Switch

The chopper circuit used for forward motoring and forward regenerative braking is shown in Figure 7.11, where S is a self-commutated semiconductor switch, operated periodically such that it remains closed for a duration of δT and remains open for the duration of $(1 - \delta)T$. C is the manual switch. When C is closed and S is in operation, the circuit is similar to that of Figure 7.6, permitting forward motoring operation. Under this condition, terminal a is positive, and terminal b is negative.

The regenerative braking in the forward direction is obtained when C is opened, and the armature connection is reversed with the help of the reversing switch RS, making terminal b positive and terminal a negative. During the on period of the switch S, the motor current

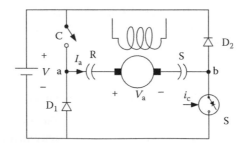

FIGURE 7.11
Forward motoring and regenerative braking control with a single chopper.

flows through a path consisting of the motor armature, switch S, and diode D_1 and increases the energy stored in the armature circuit inductance. When S is opened, the current flows through the armature diode D_2, source V, diode D_1, and back to the armature, thus feeding energy into the source.

During motoring, the changeover to regeneration is done in the following steps. Switch S is deactivated, and switch C is opened. This forces the armature current to flow through diode D_2, source V, and diode D_1. The energy stored in the armature circuit is fed back to the source, and the armature current falls to zero. After an adequate delay to ensure that the current has indeed become zero, the armature connection is reversed, and switch S is reactivated with a suitable value of δ to start regeneration.

7.1.4.1.2 Class C Two-Quadrant Chopper

In some applications, a smooth transition from motoring to braking and vice versa is required. For such applications, a class C chopper is used, as shown in Figure 7.12. The self-commutated semiconductor switch S_1 and diode D_1 constitute one chopper, and the self-commutator switch S_2, and diode D_2 form another chopper. Both the choppers are controlled simultaneously, both for motoring and regenerative braking. The switches S_1 and S_2 are closed alternately. In the chopping period T, S_1 is kept on for a duration δT, and S_2 is kept on from δT to T. To avoid a direct short-circuit across the source, care is taken to ensure that S_1 and S_2 do not conduct at the same time. This is generally achieved by providing some delay between the turn-off of one switch and the turn-on of another switch.

The waveforms of the control signals, v_a, i_a, and i_s and the devices under conducting during different intervals of a chopping period are shown in Figure 7.11b. In drawing these waveforms, the delay between the turn-off of one switch and turn-on of another switch was ignored because it is usually very small. The control signals for switches S_1 and S_2 are denoted by i_{c1} and i_{c2}, respectively. It is assumed that a switch conducts only when a control signal is present, and the switch is forward biased.

The following points clarify the operation of this two-quadrant circuit:

1. In this circuit, discontinuous conduction does not occur, irrespective of its frequency of operation. Discontinuous conduction occurs when the armature current falls to zero and remains zero for a finite interval of time. The current may become zero either during the freewheeling interval or in the energy transfer interval. In this circuit, freewheeling occurs when S_1 is off and the current is flowing through D_1. This happens in interval $\delta T \leq t \leq T$, which is also the interval for which S_2 receives the control signal. If i_a falls to zero in the freewheeling interval, the back EMF immediately drives a current through S_2 in the reverse direction, preventing the armature current from remaining zero for a finite interval of time. Similarly, the energy transfer is present when S_2 is off and D_2 is conducting—that is, during the interval $0 \leq t \leq \delta T$. If the current falls to zero during this interval, S_1 conducts immediately because i_c is present and $V > E$. The armature current flows, preventing discontinuous conduction.

2. Since discontinuous conditions are absent, the motor current is flowing all the time. Thus, during the interval $0 \leq t \leq \delta T$, the motor armature is connected through either S_1 or D_2. Consequently, the motor terminal voltage is V, and the rate of change of i_a is positive because $V > E$. Similarly, during the interval $\delta T \leq t \leq T$, the motor armature is shorted through either D_1 or S_2. Consequently, the motor voltage is zero, and the rate of change of i_a is negative.

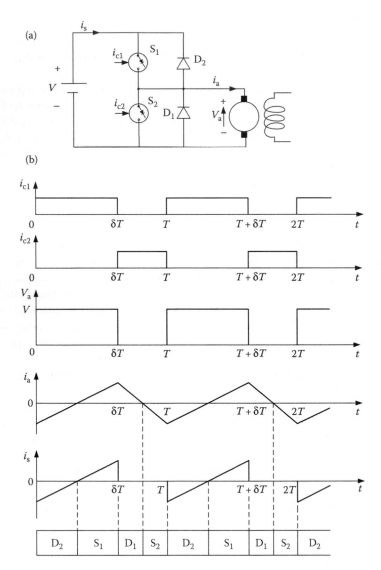

FIGURE 7.12
Forward motoring and regenerative braking control using class C two-quadrant chopper: (a) chopper circuit and (b) waveforms.

3. During the interval $0 \leq t \leq \delta T$, the positive armature current is carried by S_1 and the negative armature current is carried by D_2. The source current flows only during this interval, and it is equal to i_a. During the interval $\delta T \leq t \leq T$, the positive current is carried by D_1 and the negative current by S_2.

4. From the motor terminal voltage waveform of Figure 7.12b, $V_a = \delta V$. Hence:

$$I_a = \frac{\delta V - E}{R_a}. \tag{7.13}$$

Equation 7.13 suggests that the motoring operation takes place when $\delta > (E/V)$, and regenerative braking occurs when $\delta < (E/V)$. No-load operation is obtained when $\delta = (E/V)$.

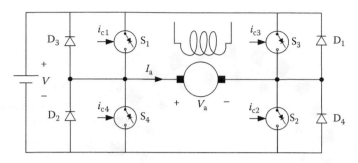

FIGURE 7.13
Class E four-quadrant chopper.

7.1.4.2 Four-Quadrant Operation

Four-quadrant operation can be obtained by combining two class C choppers (Figure 7.12a), as shown in Figure 7.13, which is referred to as a class E chopper. In this chopper, if S_2 is kept closed continuously, and S_1 and S_4 are controlled, a two-quadrant chopper is obtained, which provides positive terminal voltage (positive speed) and the armature current in either direction (positive or negative torque), giving a motor control in quadrants I and IV. Now if S_3 is kept closed continuously and S_1 and S_4 are controlled, one gets a two-quadrant chopper that can supply a variable negative terminal voltage (negative speed), and the armature current can be in either direction (positive or negative torque), giving a motor control in quadrants II and III.

This control method has the following features: the utilization factor of the switches is low due to the asymmetry in the circuit operation. Switches S_3 and S_2 should remain on for a long period. This may create commutation problems when the switches are using thyristors. The minimum output voltage depends directly on the minimum time for which the switch can be closed since there is always a restriction on the minimum time for which the switch can be closed, particularly in thyristor choppers.[7] The minimum available output voltage, and therefore the minimum available motor speed, is restricted.

To ensure that switches S_1 and S_4, or S_2 and S_3, are not on at the same time, some fixed-time interval must elapse between the turn-off for one switch and the turn-on of another switch. This restricts the maximum permissible frequency of operation. It also requires two switching operations during a cycle of the output voltage.

Dubey[6] provided other control methods to resolve the problems mentioned previously.

7.2 Induction Motor Drives

Commutatorless motor drives offer several advantages over conventional DC commutator motor drives for the electric propulsions of EVs and HEVs. At present, induction motor drives are the mature technology among commutatorless motor drives. Compared with DC motor drives, the AC induction motor drive has additional advantages such as its light-weight nature, small volume, low cost, and high efficiency. These advantages are particularly important for EV and HEV applications.

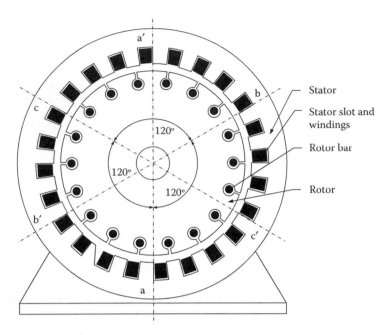

FIGURE 7.14
Cross section of induction motor.

There are two types of induction motors, wound-rotor and squirrel-cage motors. Because of the high cost, need for maintenance, and lack of sturdiness, wound-rotor induction motors are less attractive than their squirrel-cage counterparts, especially for electric propulsion in EVs and HEVs. Hence, squirrel-cage induction motors are loosely referred to as induction motors.

A cross section of a two-pole induction motor is shown in Figure 7.14. Slots in the inner periphery of the stator are inserted with three phase windings, a–a′, b–b′, and c–c′. The turns of each winding are distributed such that the current in the winding produces an approximate sinusoidally distributed flux density around the periphery of the air gap. The three windings are spatially arranged by 120°, as shown in Figure 7.14.

The most common types of induction motor rotors are the squirrel-cage motors in which aluminum bars are cast into slots in the outer periphery of the rotor. The aluminum bars are short-circuited together at both ends of the rotor by cast aluminum end rings, which also can be shaped as fans.

7.2.1 Basic Operation Principles of Induction Motors

Figure 7.15 shows, schematically, a cross section of the stator of a three-phase, two-pole induction motor. Each phase is fed with a sinusoidal AC current, which has a frequency of ω and a 120° phase difference between each other. Current i_{as}, i_{bs}, and i_{cs} in the three stator coils a–a′, b–b′, and c–c′ produce alternative magnetic motive forces (mmfs), F_{as}, F_{bs}, and F_{cs}, which are space vectors. The resultant stator mmf vector F_s^s constitutes a vector sum of the phase mmf vectors.

The mmfs produced by the phase currents can be written as:

$$F_{as} = F_{as} \sin \omega t, \tag{7.14}$$

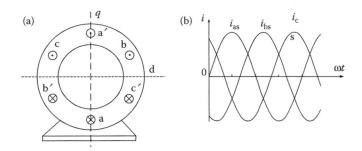

FIGURE 7.15
Induction motor stator and stator winding current: (a) spatially symmetric three-phase stator windings and (b) phase currents.

$$F_{bs} = F_{bs}\sin(\omega t - 120°), \tag{7.15}$$

$$F_{bs} = F_{cs}\sin(\omega t - 240°). \tag{7.16}$$

The resultant stator mmf vector, F_s^s, is expressed as

$$F_s^s = F_{as}^s e^{i0°} + F_{bs}^s e^{j120°} + F_{cs}^s e^{j240°}. \tag{7.17}$$

Assuming that the magnitudes of the three-phase mmfs are identical, equal to F_s, Equation 7.17 can be further expressed as

$$F_s^s = \frac{3}{2}F_s e^{(\omega t - 90°)}. \tag{7.18}$$

Equation 7.18 indicates that the resultant stator mmf vector is rotating with the frequency of the angle velocity of ω, and its magnitude is $3/2F_s$. Figure 7.16 graphically shows the stator mmf vectors at $\omega t = 0$ and $\omega t = 90°$; here, ωt is the angle in Equations 7.12 through 7.18, rather than the resultant stator mmf vector relative to the d-axis. If the ωt in Equations 7.14 through 7.16 is taken as the reference, the resultant stator mmf vector is a 90° delay to the phase a–a′ mmf.

The reaction between the rotating stator mmf and the rotor conductors induces a voltage in the rotor and electric current in the rotor. In turn, the rotating mmf produces a torque on the rotor, which is carrying the induced current. The induced current in the rotor is essential for producing torque, and in turn, the induced current depends on the relative movements between the stator mmf and the rotor. That is why there must exist a difference between the angular velocity of the rotating stator mmf and the angular velocity of the rotor.

The frequency ω, or angular velocity of the rotating stator mmf in the equation, depends only on the frequency of the alternative current of the stator; thus, it is referred to as electrical angular velocity. For a machine with two poles, the electrical angular velocity is identical to the mechanical angular velocity of the rotating stator mmf. However, for a machine with

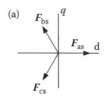

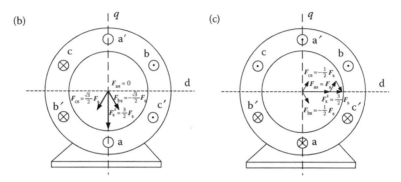

FIGURE 7.16
Stator mmf vectors: (a) positive direction of each phase mmf; (b) stator mmf vectors at $\omega t = 0$; (c) stator mmf vectors at $\omega t = 90°$.

more than two poles, the mechanical angular velocity differs from the electrical one, which can be expressed as

$$\omega_{ms} = \frac{2}{p}\omega = \frac{4\pi f}{p} \text{ rad/s,} \tag{7.19}$$

where f is the frequency of the alternative current or angular velocity of the rotating stator mmf in cycle/s. When the angular velocity of the rotor is equal to the mechanical angular velocity of the rotating stator mmf, there is no induced current in the rotor, and then no torque is produced. Thus, the mechanical angular velocity of the rotating stator mmf is also called synchronous speed.

If the rotor speed is ω_m in rad/s, then the relative speed between the stator rotating field and the rotor is given by

$$\omega_{sl} = \omega_{ms} - \omega_m = s\omega_{ms} \tag{7.20}$$

where ω_{sl} is called slip speed. The parameter s, known as slip, is given by

$$s = \frac{\omega_{ms} - \omega_m}{\omega_{ms}} = \frac{\omega_{sl}}{\omega_{ms}}. \tag{7.21}$$

Because of the relative speed between the stator field and the rotor, balanced three-phase voltages are induced in the rotor mentioned previously. The frequency of these voltages is proportional to the slip speed. Hence:

$$\omega_r = \frac{\omega_{sl}}{\omega_{ms}}\omega = s\omega, \tag{7.22}$$

where ω_r is the frequency of the rotor voltage induced.

For $\omega_m < \omega_{ms}$, the relative speed is positive; consequently, the rotor-induced voltages have the same phase sequence as the stator voltages. The three-phase current flowing through the rotor produces a magnetic field, which moves with respect to the rotor at the slip speed in the same direction as the rotor speed. Consequently, the rotor field moves in the space at the same speed as the stator, and a steady torque is produced. For $\omega_m = \omega_{ms}$, the relative speed between the rotor and the stator field becomes zero. Consequently, no voltages are induced, and no torque is produced by the motor. For $\omega_m > \omega_{ms}$, the relative speed between the stator field and the rotor speed reverses. Consequently, the rotor-induced voltages and currents also reverse and have a phase sequence opposite to that of the stator. Moreover, the developed torque has a negative sign, suggesting generator operation. (The generator is used to produce regenerative braking.)

7.2.2 Steady-State Performance

A per-phase equivalent circuit of an induction motor is shown in Figure 7.17a. The fields produced by the stator and the rotor are linked together by an ideal transformer. a_{T1} is the transformer factor, which is equal to n_s/n_r, where n_s and n_r are the number of turns of stator and rotor windings, respectively. For a squirrel-cage rotor, $n_r = 1$. The equivalent circuit can be simplified by referring the rotor quantities to the stator frequency and number of turns. The resultant equivalent circuit is shown in Figure 7.17b, where R'_r and X'_r are the rotor resistance and reactance referred to the stator, and is given by the following equation:

$$R'_r = a_{T1}^2 R_r \ \text{and} \ X'_r = a_{T1}^2 X_r. \tag{7.23}$$

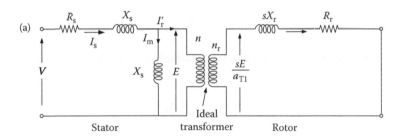

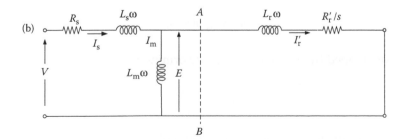

FIGURE 7.17
Per-phase equivalent circuit of an induction motor: (a) with a transformer between rotor and stator and (b) referring rotor quantities to stator.

The stator reactance, mutual reactance, and rotor reactance referred to the stator can be expressed by the stator frequency and their inductances, L_s, L_m, and L_r, as shown in Figure 7.17. The impedances of stator, field, and rotor can be expressed as

$$Z_s = R_s + jL_s\omega, \tag{7.24}$$

$$Z_m = jL_m\omega, \tag{7.25}$$

$$Z_r = \frac{R'_r}{s} + jL_r\omega. \tag{7.26}$$

The driving-point impedance of the circuit is

$$Z = Z_s + \frac{Z_m Z_r}{Z_m + Z_r}. \tag{7.27}$$

Hence, the current I_s and I'_r can be calculated as

$$I_s = \frac{V}{Z} \tag{7.28}$$

and

$$I'_r = \frac{Z_m}{Z_m + Z_r} I_s. \tag{7.29}$$

The total electrical power supplied to the motor for three phases is

$$P_{elec} = 3I'^2_r \frac{R'_r}{s}. \tag{7.30}$$

The mechanical power of the rotor can be obtained by subtracting the total power loss in the stator as

$$P_{mech} = P_{elec} - 3I'^2_r R'_r. \tag{7.31}$$

The angular velocity of the rotor, ω_m, is

$$\omega_m = \frac{2}{P}\omega(1-s). \tag{7.32}$$

The torque developed by the motor can be determined by

$$T = \frac{P_{mech}}{\omega_m}. \tag{7.33}$$

Figure 7.18 shows the torque–slip characteristics of an induction motor, which has fixed voltage and frequency. In the region of $0 < s < s_m$, where s_m is the rated slip of the motor, the torque increases approximately linearly with the increase of slip until reaching its maximum at $s = s_m$, then it decreases with the further increase of the slip. At $s = 1$, the rotor

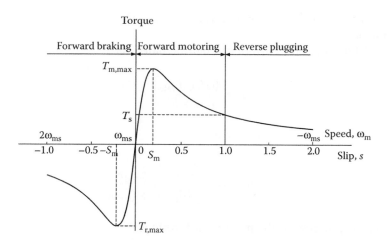

FIGURE 7.18
Torque–slip characteristics of an induction motor with fixed stator frequency and voltage.

speed is zero, and the corresponding torque is the starting torque, which is less than its torque at $s = s_m$. The region of $0 < s < 1$ is the forward motoring region. In the region of $s > 1$, the rotor torque is positive and decreases further with the increase of slip, and the rotor speed is negative, according to Equation 7.21. Thus, in this region, the operation of the motor is reverse braking. In the region of $s < 0$, that is, when the rotor speed is greater than the synchronous speed, the motor produces a negative torque.

The speed–torque characteristic of a fixed-voltage and fixed-frequency induction motor is not appropriate to vehicle traction applications. This is due to the low starting torque, limited speed range, and unstable operation in the range of $s > s_m$, in which any additional disturbing torque in the load leads the machine to stop as the torque decreases, with the speed decreasing characteristically. The high slip also results in high current, which may cause damage in the stator windings. The operation of the fixed voltage and frequency induction motor are usually operated in a narrow slip range of $0 < s < s_m$. Thus, for traction applications, an induction motor must be controlled to provide proper speed–torque characteristic, as mentioned in Chapters 2 and 4.

7.2.3 Constant Volt/Hertz Control

For traction application, the torque–speed characteristic of an induction motor can be varied by simultaneously controlling the voltage and frequency, which is known as the constant volt/hertz control. By emulating a DC motor at low speed, the flux may be kept constant. According to Figure 7.17b, the field current I_m should be kept constant and equal to its rated value. That is,

$$I_{mr} = \frac{E}{X_m} = \frac{E_{rated}}{\omega_r L_m},$$

(7.34)

where I_{mr} is the rated field current, and E_{rated} and ω_r are the rated mmf and frequency of the stator, respectively. To maintain a constant flux, the E/ω should be kept constant and equal to E_{rated}/ω_r. Ignoring the voltage drop in the stator impedance Z_s results in a constant V/ω

until the frequency and voltage reach their rated values. This approach is known as constant volt/hertz control.[6]

From Figure 7.17b, the rotor current can be calculated as

$$I'_r = \frac{(\omega/\omega_r)E_{rated}}{jL_r\omega + R'_r/s}.$$ (7.35)

The torque produced can be obtained as

$$T = \frac{3}{\omega}I'^2_r R'_r/s = \frac{3}{\omega}\left[\frac{(\omega/\omega)^2 E^2_{rated}R'_r/s}{(R'_r/s)^2 + (L_r\omega)^2}\right].$$ (7.36)

The slip s_m corresponding to the maximum torque is

$$s_m = \pm\frac{R'_r}{L_r\omega}.$$ (7.37)

And then the maximum torque is

$$T_{max} = \frac{3}{2}\frac{E^2_{rated}}{L_r\omega^2_r}.$$ (7.38)

Equation 7.38 indicates that with constant E/ω, the maximum torque is constant with varying frequency. Equation 7.37 indicates that $s_m\omega$ is constant, resulting in constant slip speed, ω_{sl}. In practice, due to the presence of stator impedance and the voltage drop, the voltage should be somewhat higher than that determined by constant E/ω, as shown in Figure 7.19.

When the motor speed is beyond its rated speed, the voltage reaches its rated value and cannot be increased with the frequency. In this case, the voltage is fixed to its rated value, and the frequency increases continuously with the motor speed. The motor goes into the field weakening operation. The slip s is fixed to its rated value corresponding to the rated frequency, and the slip speed ω_{sl} increases linearly with the motor speed. This control approach results in constant power operation, as shown in Figure 7.19.

In traction applications, speed control in a wide range is usually required, and the torque demand in the high-speed range is low. Control beyond constant power range is required. To prevent the torque from exceeding the breakdown torque, the machine is operated at a constant slip speed, and the machine current and power are allowed to decrease, as shown in Figure 7.19. Figure 7.20 shows a general block diagram where constant V/f control is implemented.

7.2.4 Power Electronic Control

As EV and HEV propulsion, an induction motor drive is usually fed with a DC source (e.g., battery, fuel cell), which has approximately constant terminal voltage. Thus, a variable-frequency and variable-voltage DC/AC inverter is needed to feed the induction motor. A general DC/AC inverter is constituted by power electronic switches and power diodes. The commonly used topology of a DC/AC inverter is shown in Figure 7.21a, which has three

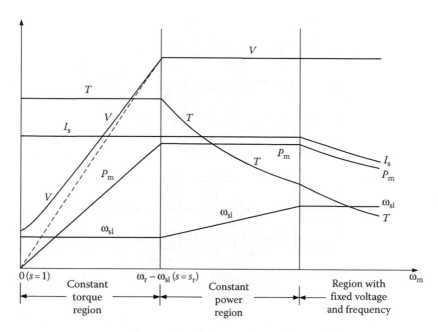

FIGURE 7.19
Operating variables varying with motor speed.

legs (S_1 and S_4, S_3 and S_6, and S_5 and S_2), feeding phases a, b, and c of the induction motor. When switches S_1, S_3, and S_5 are closed, S_4, S_6, and S_2 are opened, and phases a, b, and c are supplied with a positive voltage ($V_d/2$). Similarly, when S_1, S_3, and S_5 are opened and S_4, S_6, and S_2 are closed, phases a, b, and c are supplied with a negative voltage. All the diodes provide a path for the reverse current of each phase.

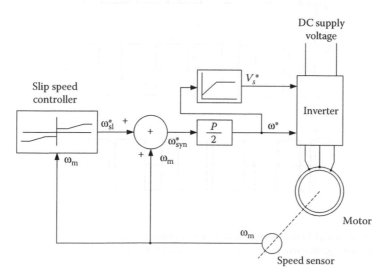

FIGURE 7.20
General configuration of constant V/f control.

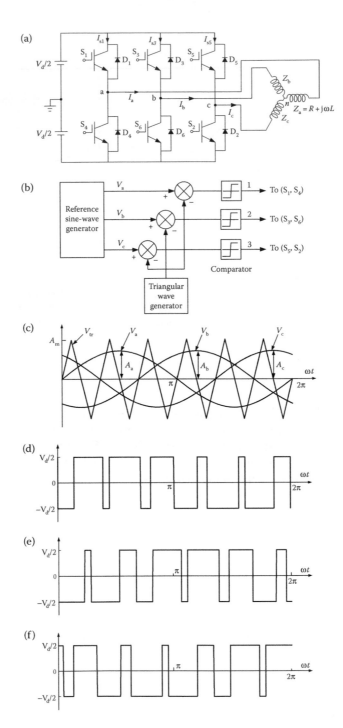

FIGURE 7.21
DC/AC inverter with sinusoidal pulse-width modulation: (a) inverter topology; (b) control signals; (c) three-phase reference voltage and triangular carrier waveforms; (d) voltage of phase a; (e) voltage of phase b; and (f) voltage of phase c.

For constant volt/hertz control of an induction motor, sinusoidal pulse width modulation (PWM) is used exclusively. Three-phase reference voltages V_a, V_b, and V_c of variable amplitudes A_a, A_b, and A_c are compared with a common isosceles triangular carrier wave V_{tr} of a fixed amplitude A_m, as shown in Figure 7.21c. The outputs of comparators 1, 2, and 3 form the control signals for the three legs of the inverter. When the sinusoidal reference voltages V_a, V_b, and V_c at a time t are greater than the triangular waved voltage, turn-on signals are sent to the switches S_1, S_3, and S_5 and turn-off signals to S_4, S_6, and S_6. Thus, the three phases of the induction motor have a positive voltage. On the other hand, when the reference sinusoidal voltage is smaller than the triangular wave voltage, turn-on signals are sent to switches S_1, S_3, and S_5 and turn-off signals to S_4, S_6, and S_2. The three phases of the induction motor then have a negative voltage. The voltages of the three phases are shown in Figure 7.21d–f.

The frequency of the fundamental component of the motor terminal voltage is the same as that of the reference sinusoidal voltage. Hence, the frequency of the motor voltage can be changed by changing the frequency of the reference voltage. The ratio of the amplitude of the reference wave to that of the triangular carrier wave, m, is called the modulation index; therefore

$$m = \frac{A}{A_m},\tag{7.39}$$

where A is the multitude of the reference sinusoidal voltage, V_a, V_b, or V_c, and A_m is the multitude of angular carrier voltage. The fundamental (rms) component in the phase waveform V_{ao}, V_{bo}, or V_{co} is given by

$$V_f = \frac{mV_d}{2\sqrt{2}}.\tag{7.40}$$

Thus, the fundamental voltage increases linearly with m until $m = 1$ (i.e., when the amplitude of the reference wave becomes equal to that of the carrier wave). For $m > 1$, the number of pulses in V_{ao}, V_{bo}, or V_{co} becomes less, and the modulation ceases to be sinusoidal.[6]

7.2.5 Field Orientation Control

The constant volt/hertz control of the induction motor is more suitably applied to motors that operate with relatively slow speed regulation. However, this approach shows poor response to frequent and fast speed variations and results in poor operation efficiency due to the poor power factor. In the last two decades, FOC or vector control technology has been successfully developed. This technology mostly overcomes the disadvantages of the constant volt/hertz control in AC motor drives.

7.2.5.1 Field Orientation Principles

The aim of FOC is to maintain the stator field perpendicular to the rotor field to always produce the maximum torque as in DC motors. However, for induction motors, phase voltages are the only accesses for control.

As mentioned in Section 7.2.1, when balanced three-phase sinusoidal currents flow through the three phases of the stator of an induction motor, the rotating field is developed;

current is induced in the rotor. In turn, the current induced in the rotor is also three-phase and produces a field, which rotates with the same angular velocity of the stator rotating field. The rotating fields of both stator and rotor can be described by two retorting vectors, referring to a common, stationary reference frame, d–q, as shown in Figure 7.16. The mmf of the stator field is expressed by Equation 7.17. For convenience, it is repeated as follows:

$$F_s^s = F_{as} e^{j0°} + F_{bs} e^{j120°} + F_{cs} e^{j240°}. \tag{7.41}$$

Similarly, the stator voltage, stator current, and stator flux can be expressed as vectors in the same way. That is,

$$v_s^s = v_{as}^s e^{j0°} + v_{bs}^s e^{j120°} + v_{cs}^s e^{j240°}, \tag{7.42}$$

$$i_s^s = i_{as}^s e^{j0°} + i_{bs}^s e^{j120°} + i_{cs}^s e^{j240°}, \tag{7.43}$$

$$\lambda_s^s = \lambda_{as}^s e^{j0°} + \lambda_{bs}^s e^{j120°} + \lambda_{cs}^s e^{j240°}. \tag{7.44}$$

The subscript s refers to the stator, and as, bs, and cs refer to phases a, b, and c of the stator. The superscript s refers to the variable that is referred to the stator fixed frame. Bold symbols stand for vector variables. The vectors of stator voltage, current, and flux can be also described by its components in *d* and *q* axes as follows:

$$\begin{bmatrix} v_{ds}^s \\ v_{qs}^s \end{bmatrix} = \begin{bmatrix} 1 & -\dfrac{1}{2} & -\dfrac{1}{2} \\ 0 & \dfrac{\sqrt{3}}{2} & -\dfrac{\sqrt{3}}{2} \end{bmatrix} \begin{bmatrix} v_{as} \\ v_{bs} \\ v_{cs} \end{bmatrix}, \tag{7.45}$$

$$\begin{bmatrix} i_{ds}^s \\ i_{qs}^s \end{bmatrix} = \begin{bmatrix} 1 & -\dfrac{1}{2} & -\dfrac{1}{2} \\ 0 & \dfrac{\sqrt{3}}{2} & -\dfrac{\sqrt{3}}{2} \end{bmatrix} \begin{bmatrix} i_{as} \\ i_{bs} \\ i_{cs} \end{bmatrix}, \tag{7.46}$$

$$\begin{bmatrix} \lambda_{ds}^s \\ \lambda_{qs}^s \end{bmatrix} = \begin{bmatrix} 1 & -\dfrac{1}{2} & -\dfrac{1}{2} \\ 0 & \dfrac{\sqrt{3}}{2} & -\dfrac{\sqrt{3}}{2} \end{bmatrix} \begin{bmatrix} \lambda_{as} \\ \lambda_{bs} \\ \lambda_{cs} \end{bmatrix}. \tag{7.47}$$

In a real induction motor, the rotor winding differs from the stator winding, that is, the effective number of turns per phase of the rotor winding, N_r, is not equal to that of the stator winding, N_s. Therefore, the turn ratio, $v = N_s/N_r$, must be taken into account. The vectors of rotor current, voltage, and magnetic flux can be described by i_r^r, v_r^r, and λ_r^r in the rotor frame. However, it is necessary to transform the vectors from the rotor frame to the

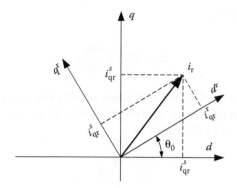

FIGURE 7.22
Transformation of rotor current vector from rotor frame to stator frame.

stator frame for easy analysis. The transformations of these vectors (Figure 7.22) are described by

$$i_r^s = \frac{e^{j\theta_0}}{v} i_{r'}^r \tag{7.48}$$

$$v_r^s = v e^{j\theta_0} v_{r'}^r \tag{7.49}$$

$$\lambda_r^s = v e^{j\theta_0} \lambda_r^r. \tag{7.50}$$

Using vector notation, either the stator or rotor windings can be represented by a simple resistive-plus-inductive circuit, using current, voltage, and magnetic flux space vectors, as illustrated in Figure 7.23.

Using the vector version of Kirchhoff's voltage law, the equation of the stator winding can be written

$$v_s^s = R_s i_s^s + \frac{d\lambda_s^s}{dt} \tag{7.51}$$

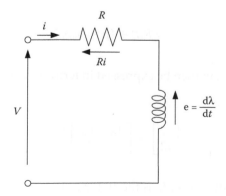

FIGURE 7.23
Resistive-plus-inductive equivalent circuit of either stator or rotor windings.

and that of the rotor winding

$$v_r^r = R_r^r i_r^r + \frac{d\lambda_r^r}{dt},$$ (7.52)

where R_s and R_r^r are the actual stator and rotor resistances per phase, respectively. As is known from the steady-state theory of induction machines, the relation between R_r^r and the rotor resistance referred to the stator is

$$R_r^r = \frac{1}{v^2} R_r.$$ (7.53)

Hence, from Equations 7.48 and 7.53, the first term of Equation 7.52 is

$$R_r^r i_r^r = \frac{e^{-j\theta_0}}{v} R_r i_r^s.$$ (7.54)

The second term, from Equation 7.50, gives

$$\frac{d\lambda_r^r}{dt} = \frac{e^{j\theta_0}}{v} \left(\frac{d\lambda_r^s}{dt} - j\omega_0 \right).$$ (7.55)

Finally, substituting Equations 7.54 and 7.55 into Equation 7.52 gives

$$v_r^s = R_r i_r^s + \frac{d\lambda_r^s}{dt} - j\omega_0 \lambda_r^s.$$ (7.56)

Introducing a differentiation operator $p \equiv d/dt$, the voltage equation of an induction motor can be written

$$v_s^s = R_s i_s^s + p\lambda_s^s,$$ (7.57)

$$v_r^s = R_r i_r^s + (p - j\omega_0)\lambda_r^s.$$ (7.58)

The flux vector λ_s^s and λ_r^s can then be expressed in terms of current vector i_s^s and i_r^s and the motor inductances as

$$\begin{bmatrix} \lambda_s^s \\ \lambda_r^s \end{bmatrix} = \begin{bmatrix} L_s & L_m \\ L_m & L_r \end{bmatrix} \begin{bmatrix} i_s^s \\ i_r^s \end{bmatrix},$$ (7.59)

where L_m is the mutual inductance, L_s is the stator inductance calculated as the sum of the stator leakage inductance L_{ls} and the mutual inductance L_m, and L_r is the rotor inductance calculated as the sum of the rotor leakage inductance L_{lr} and the mutual inductance L_m.

Finally, the voltage equation can be written in matrix format as

$$
\begin{bmatrix} v_{ds}^s \\ v_{qs}^s \\ v_{dr}^s \\ v_{qr}^s \end{bmatrix} = \begin{bmatrix} R_s & 0 & 0 & 0 \\ 0 & R_s & 0 & 0 \\ 0 & \omega_0 L_m & R_r & \omega_0 L_r \\ -\omega_0 L_m & 0 & -\omega_0 L_r & R_r \end{bmatrix} \begin{bmatrix} i_{ds}^s \\ i_{qs}^s \\ i_{dr}^s \\ i_{qr}^s \end{bmatrix} + \begin{bmatrix} L_s & 0 & L_m & 0 \\ 0 & L_s & 0 & L_m \\ L_m & 0 & L_r & 0 \\ 0 & L_m & 0 & L_r \end{bmatrix} \frac{d}{dt} \begin{bmatrix} i_{ds}^s \\ i_{qs}^s \\ i_{dr}^s \\ i_{dr}^s \end{bmatrix}. \quad (7.60)
$$

Because the rotor circuit of the induction motor is shorted, v_{dr}^s and v_{qr}^s are zero. At a given rotor speed, ω_0, the stator and rotor currents can be obtained by solving Equation 7.60. The torque developed by the motor can be expressed as

$$
T = \frac{P}{3} L_m \left(i_{qs}^s i_{dr}^s - i_{ds}^s i_{qr}^s \right) = \frac{P}{3} L_m \mathrm{Im}\left(i_s^s i_r^{s*} \right), \quad (7.61)
$$

where Im stands for the imaginary part of the production of vector i_s^s and conjugate vector of i_r^{s*}.

Transferring three-phase variables (voltage, current, and flux) into a stationary stator-based dq frame does not change the alternate characteristics of the variable with time. AC quantities are somewhat inconvenient for control purposes. For instance, control systems are usually represented by block diagrams in which the variables are time-varying DC signals. Therefore, another transformation is necessary, which allows the conversion of the AC dq components of the motor vectors into DC variables. To do this, a transformation is conducted from a stationary stator reference frame, dq, to the so-called excitation reference frame, DQ, which rotates at angular speed ω in the same direction as does mmf, F_s^s. As a result, in the steady state, coordinates of motor vectors in the new reference frame do not vary in time. This is illustrated in Figure 7.24, which shows the stator mmf vector in both reference frames.

The voltage vector of the stator in an excitation reference frame can be expressed as

$$
v_S^e = v_s^s e^{-j\omega t}. \quad (7.62)
$$

Considering $e^{-j\omega t} = \cos(\omega t) - \sin(\omega t)$, the components of stator voltage on DQ frame is

$$
\begin{bmatrix} v_{DS}^e \\ v_{QS}^e \end{bmatrix} = \begin{bmatrix} \cos(\omega t) & \sin(\omega t) \\ -\sin(\omega t) & \cos(\omega t) \end{bmatrix} \begin{bmatrix} v_{ds}^s \\ v_{qs}^s \end{bmatrix}. \quad (7.63)
$$

Thus, the motor equation in an excitation reference frame can be expressed as

$$
v_S^e = R_s i_s^e + (p + j\omega)\lambda_s^e, \quad (7.64)
$$

$$
v_R^e = R_r i_R^e + (p + j\omega - j\omega_0)\lambda_R^e = R_r i_R^e + (p + j\omega_r)\lambda_R^e \quad (7.65)
$$

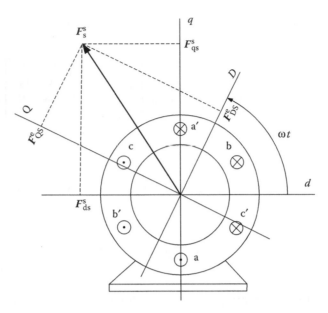

FIGURE 7.24
Stator mmf vector in stator and excitation reference frames.

where $\omega_r = \omega - \omega_0$ is the slip speed, and

$$\begin{bmatrix} \lambda_S^e \\ \lambda_R^e \end{bmatrix} = \begin{bmatrix} L_s & L_m \\ L_m & L_r \end{bmatrix} \begin{bmatrix} i_S^e \\ i_R^e \end{bmatrix}. \tag{7.66}$$

Again, the rotor voltage vector is normally assumed to be zero because of the shorted rotor winding.

The torque equation in the excitation reference frame is similar to that in the stator frame:

$$T = \frac{P}{3} L_m \left(i_{QS}^e i_{DR}^e - i_{DS}^e i_{QR}^e \right) = \frac{P}{3} L_m \mathrm{Im}\left(i_S^e i_e^{e*} \right). \tag{7.67}$$

In general, accurate control of instantaneous torque produced by a motor is required in high-performance drive systems, such as EV and HEV propulsions. The torque developed in the motor is a result of the interaction between current in the armature winding and the magnetic field produced in the stator field of the motor. The field should be maintained at a certain optimal level, sufficiently high to yield a high torque per unit of ampere, but not too high to result in excessive saturation of the magnetic circuit of the motor. With a fixed field, the torque is proportional to the armature current.

Independent control of the field and armature currents is desirable. In a similar manner to that of a DC motor, the armature winding in induction motors is also on the rotor, while the field is generated by currents in the stator winding. However, the rotor current is not directly derived from an external source but results from the EMF induced in the winding as a result of the relative motion of the rotor conductors with respect to the stator field. In the most commonly used squirrel-cage motors, only the stator current can be directly controlled, since the rotor winding is not accessible. Optimal torque production conditions are not inherent due to the absence of a fixed physical disposition between the stator and rotor fields, and the torque equation is nonlinear. FOC or vector control can realize the optimal control for transient operation of an induction drive. The FOC can decouple the field control from the torque control. A field-oriented induction motor emulates a separately excited DC motor in two aspects:

1. Both the magnetic field and the torque developed in the motor can be controlled independently.
2. Optimal conditions for torque production, resulting in the maximum torque per unit ampere, occur in the motor both in the steady state and in transient conditions of operation.

As mentioned in Section 7.1.1, the optimal torque production conditions are inherently satisfied in a DC motor (Figure 7.3). The armature current i_a, supplied through brushes, is always orthogonal to the flux vector (field flux), λ_f, produced in the stator and linking the rotor winding. In effect, the developed torque, T, is proportional both to the armature current and the field flux, that is,

$$T = K_T i_a \lambda_f, \tag{7.68}$$

where K_T is a constant depending on the physical parameters of the motor. Thus, the torque of separately excited DC motors can be controlled by independently controlling the armature current and flux, as mentioned in Section 7.1.2.

To emulate this independent armature and field control characteristic of a DC motor, the torque Equation 7.67 can be rearranged so that the torque is expressed in terms of the stator current and rotor flux. From Equation 7.66 the flowing equation can be obtained as

$$i_R^e = \frac{1}{L_r} \left(\lambda_R^e - L_m i_S^e \right). \tag{7.69}$$

Torque Equation 7.67 can be rewritten as

$$T = \frac{P}{3R_r} \frac{L_m}{\tau_r} \left(i_{QS}^e \lambda_{DR}^e - i_{DS}^e \lambda_{QR}^e \right), \tag{7.70}$$

where $\tau_r = L_r/R_r$ is the rotor time constant.

In Equation 7.70, if

$$\lambda_{QR}^e = 0, \tag{7.71}$$

then

$$T = \frac{P}{3R_r} \frac{L_m}{\tau_r} \lambda_{DR}^e i_{QS}^e. \tag{7.72}$$

Clearly, Equation 7.72 is analogous to Equation 7.68, describing a separately excited DC motor.

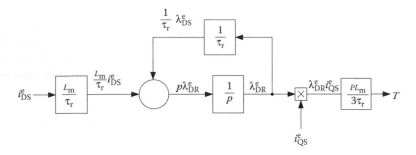

FIGURE 7.25
Block diagram of induction motor with $\lambda_{QR}^e = 0$.

Substituting $v_R^e = 0$ (shorted rotor winding) into Equation 7.65 yields

$$R_r i_R^e + (p + j\omega_r)\lambda_R^e = 0, \tag{7.73}$$

and substituting Equation 7.69 into Equation 7.73 yields

$$p\lambda_R^e = \frac{1}{\tau_r}\left[L_m i_S^e - (1 + j\omega_r\tau_r)\lambda_R^e\right]. \tag{7.74}$$

Thus,

$$p\lambda_{DR}^e = \frac{L_m}{\tau_r} i_{DS}^e - \frac{1}{\tau_r}\lambda_{DR}^e. \tag{7.75}$$

Equation 7.75 indicates that the flux λ_{DR}^e is produced by the current, i_{DS}^e. Thus, the torque produced can be represented by the block diagram, as shown in Figure 7.25.

Furthermore, Equation 7.75 can be expressed into a transfer function as

$$G(p) = \frac{\lambda_{DR}^e}{i_{DS}^e} = \frac{L_m}{\tau_r p + 1}. \tag{7.76}$$

Thus, the block diagram in Figure 7.25 can be further reduced, as shown in Figure 7.26.

If conditions in Equation 7.71 and $\lambda_{DR}^e = $ constant t are satisfied, that is, $\lambda_{QR}^e = 0$ and $p\lambda_{DR}^e = 0$, then Equation 7.64 yields $i_{DR}^e = 0$, that is, $i_R^e = ji_{QR}^e$. At the same time, $\lambda_R^e = \lambda_{DR}^e$. Consequently, vectors i_R^e and λ_R^e are orthogonal, which represents the optimal conditions for torque production, analogously to a DC motor. In an induction motor, the optimal torque-production conditions are always satisfied in the steady state. However, in transient operation, the motor needs delicate control to achieve this optimal torque production.

7.2.5.2 Control

As demonstrated in the previous section, the field orientation principle defines the conditions of optimal torque production. Orthogonality of the rotor current and stator flux vectors must be maintained at all times. This is inherently satisfied in the steady state when the rotor

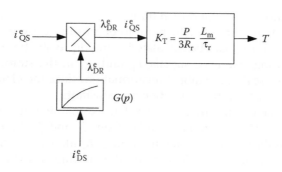

FIGURE 7.26
Block diagram of field-oriented induction motor.

settles down to such a speed that the developed torque matches the load torque. Under transient conditions, however, to meet the field orientation principle conditions, special techniques are required to provide an algorithmic equivalent of the actual physical disposition between the stator and rotor fields of the emulated DC motor.

A general block diagram of a vector control system for an induction motor drive is shown in Figure 7.27. A field orientation system produces reference signals, i_{as}^*, i_{bs}^*, and i_{cs}^*, of the stator currents, based on the input reference values, λ_r^* and T^*, of the rotor flux and motor torque, respectively, and the signals corresponding to selected variables of the motor. An inverter supplies the motor currents, i_{as}, i_{bs}, and i_{cs}, such that their waveforms follow the reference waveform, i_{as}^*, i_{bs}^*, and i_{cs}^*.

As shown in Figure 7.26, in a field-oriented induction motor, the i_{DS}^e and i_{QS}^e components of the stator current vector, i_S^e, in the excitation frame can be used for independent control of the motor field and torque, respectively. Hence, the field orientation system shown in Figure 7.27 first converts λ_r^* and T^* into the corresponding reference signals, i_{DS}^{e*} and i_{QS}^{e*},

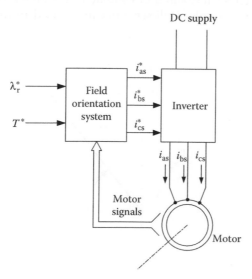

FIGURE 7.27
General block diagram of vector control system for induction motor.

of the vector of the stator current and then transfers these into the reference signals, i_{as}^*, i_{bs}^*, and i_{cs}^*, of the stator phase current, which are to be produced by the inverter. The stator phase currents, i_{as}^*, i_{bs}^*, and i_{cs}^*, can be calculated using dq to abc transformation (Equation 7.46) if the corresponding reference signals, i_{ds}^{s*} and i_{qs}^{s*}, in the stator reference frame are known. This is a simple scalar, or static, transformation, since the elements of the transformation matrix used to perform this operation are constant.

However, it can be seen from Equation 7.63 that a dynamic transformation, that is, one involving time, is required to determine i_{ds}^{s*} and i_{qs}^{s*} from i_{DS}^{e*} and i_{QS}^{e*}. Figure 7.24 does not indicate from which vector the excitation reference frame DQ aligns. Clearly, any one of the vectors can be used as a reference with which the excitation frame is to be aligned. Usually, it is the rotor flux vector, λ_r^s, along which the excitation frame is oriented. This method is usually referred to as the rotor flux orientation scheme,[8] as shown in Figure 7.28.

If the angular position of the rotor flux vector in the stator reference frame is denoted by θ_r, the DQ to dq transformation in the described scheme is expressed as

$$\begin{bmatrix} i_{ds}^{s*} \\ i_{qs}^{s*} \end{bmatrix} = \begin{bmatrix} \cos(\theta_r) & -\sin(\theta_r) \\ \sin(\theta_r) & \cos(\theta_r) \end{bmatrix} \begin{bmatrix} i_{DS}^{e*} \\ i_{QS}^{e*} \end{bmatrix}. \tag{7.77}$$

Note that this orientation of the orientation frame inherently satisfies the field orientation principle condition in Equation 7.71. The rotor flux is controlled by adjusting the i_{DS}^e component of the stator current vector—independently of the torque control, which is realized by means of the i_{QS}^e component. The only requirement for this scheme is an accurate identification of angle θ_r, that is, the position of λ_r^s. This can be done in either a direct or indirect way.

7.2.5.3 Direct Rotor Flux Orientation Scheme

In direct-field orientation systems, the magnitude and angular position (phase) of the reference flux vector, λ_r^e, are either measured or estimated from the stator voltage and current using flux observers. For example, Hall sensors can be used to measure magnetic fields.

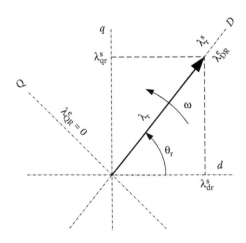

FIGURE 7.28
Orientation of excitation reference frame along rotor flux vector.

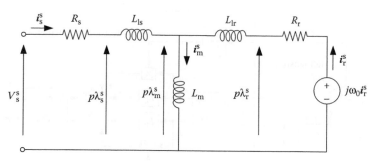

FIGURE 7.29
Dynamic T equivalent circuit of induction motor.

Placing the sensors in the air gap of the motor, on the d and q axes, makes it possible to determine the corresponding components of vector λ_m^s of the mutual flux (air gap flux). However, this air gap flux differs from the rotor flux, which is taken as the reference flux vector and needs derivation from the air gap flux λ_m^s. Referring to the dynamic T equivalent circuit shown in Figure 7.29, the flux appearing across the mutual inductance L_m is

$$\lambda_m^s = L_m i_m^s = L_m(i_s^s + i_r^s) \tag{7.78}$$

or

$$i_r^s = \frac{1}{L_m}\lambda_m^s - i_s^s. \tag{7.79}$$

Since λ_r^s differs from λ_m^s by only the leakage flux in the rotor, then

$$\lambda_r^s = \lambda_m^s + L_{1r}i_r^s = \lambda_m^s + L_{1r}\left(\frac{1}{L_m}\lambda_m^s - \lambda_s^s\right) = \frac{L_r}{L_m}\lambda_m^s - L_{1r}i_s^s. \tag{7.80}$$

A microprocessor-based rotor flux calculator is shown in Figure 7.30. It performs the algebraic operations as follows:

1. Signals i_{ds}^s and i_{qs}^s are calculated from the actual stator phase currents, i_{as}, i_{bs}, and i_{cs}, using the abc to dq transformation expressed in Equation 7.46.
2. Using Equation 7.80, signals λ_{dr}^s and λ_{qr}^s are calculated.
3. The magnitude λ_r and the phase θ_r of the rotor flux vector are determined using a rectangular to polar coordinate transformation.

It must be pointed out that the orthogonal spacing of the flux sensors in Figure 7.30 applies only to two-pole machines. In a P-pole machine, the sensors must be placed $180/P$ from each other.

Since $\lambda_{DR}^e = \lambda_r$ (Figure 7.28), the output variable, λ_r, of the rotor flux calculator can be used as a feedback signal in the field control loop. The same variables can also be used to calculate

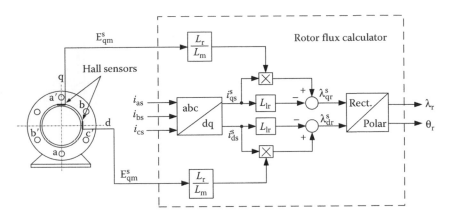

FIGURE 7.30
Determination of the magnitude and position of the rotor flux vector using a Hall sensor and a rotor flux calculator. (Adapted from D. C. Hanselman, *Brushless Permanent-Magnet Motor Design*, McGraw-Hill, New York, 1994.)

the developed torque, as shown in Figure 7.31. The torque calculator computes torque in the following steps:

1. The static abc to dq transformation is performed on the stator currents i_{as}, i_{bs}, and i_{cs} to obtain i_{ds}^s and i_{qs}^s.
2. Angle θ_r supplied by the rotor flux calculator is substituted into Equation 7.63 for ωt in order to transfer signals i_{ds}^s and i_{qs}^s into i_{DS}^e and i_{QS}^e components of the stator current vector in the excitation frame.
3. The magnitude, λ_r, of the rotor flux, also supplied by the rotor flux calculator and presumed equal to λ_{DS}^e, is multiplied by i_{QS}^e and by the torque constant K_T to calculate the developed torque.

Figure 7.31 shows the torque calculation process block diagram.

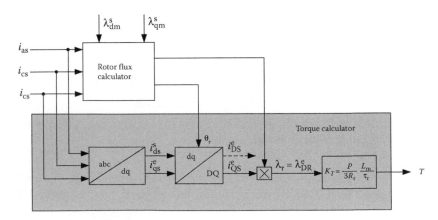

FIGURE 7.31
Torque calculator.

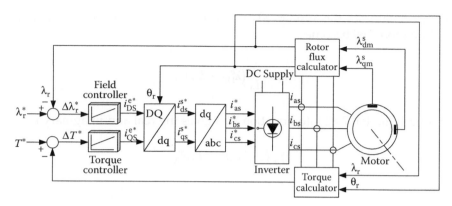

FIGURE 7.32
Vector control system for induction motor with direct rotor flux orientation.

Figure 7.32 illustrates an independent flux and torque control block diagram based on the vector control of an induction motor with direct rotor flux orientation. In the system, proportional-plus-integral (PI)-based field and torque controllers are used to generate the control signals i_{DS}^{e*} and i_{QS}^{e*} in the excitation frame by comparing the target rotor flux, λ_r^*, and target torque, T^*, with the actual rotor flux, λ_r, and torque, T. Then, i_{DS}^{e*} and i_{QS}^{e*} in the excitation frame are transformed into i_{ds}^{s*} and i_{qs}^{s*} in the stator reference frame using a rotor flux angle (Equation 7.63). Furthermore, i_{ds}^{s*} and i_{qs}^{s*} in the stator reference frame are transformed into phase current signals i_{as}^*, i_{bs}^*, and i_{cs}^* through static transformation (Equation 7.46). The phase current signals, as the reference signals, are used to control the power electronics of the inverter to generate a corresponding phase current i_{as}, i_{bs}, and i_{cs}.

In practice, the ratio of L_r to L_m, and the rotor leakage inductance, L_{ls}, which are used in the rotor flux calculator (Figure 7.30), are not significantly affected by changes in the operating conditions of the motor, such as the winding temperature or saturation of the magnetic circuit. Therefore, the field-orientation techniques described are considered to be the most robust and accurate. However, it requires the placement of vulnerable Hall sensors in the air gap of the motor, to the detriment of the cost and reliability of the drive system.

7.2.5.4 Indirect Rotor Flux Orientation Scheme

The presence of vulnerable Hall sensors in vector control with a direct rotor flux orientation would weaken the reliability and enhance the cost of the motor drive. The indirect approach is to obtain the rotor flux position by the calculation of the slip speed, ω_r, required for correct field orientation, and the imposition of this speed on the motor.

If the synchronous speed necessary to maintain the orthogonal orientation of vectors λ_R^e and i_R^e in the given operating conditions of the motor is denoted by ω^*, the θ_r angle can be expressed as

$$\theta_r = \int_0^t \omega^* \, dt = \int_0^t \omega_r^* \, dt + \int_0^t \omega_0 \, dt = \int_0^t \omega_r^* \, dt + \theta_0, \tag{7.81}$$

where ω^*, ω_r, and ω_0 are the synchronous speed, slip speed, and rotor speed, respectively, and θ_0 is the angular displacement of the rotor, which is easy to measure using a shaft position sensor.

The required value of the slip speed ω_r^* can be computed from Equation 7.69. Since $\lambda_R^e = \lambda_{DR}^e$, Equation 7.69 becomes

$$i_R^e = \frac{1}{L_r}\left(\lambda_{DR}^e - L_m i_S^e\right). \tag{7.82}$$

Substituting Equation 7.82 into Equation 7.73 gives the real and imaginary parts as

$$\lambda_{DR}^e\left(1 + \tau_r p\right) = L_m i_{DS}^e \tag{7.83}$$

and

$$\omega_r \tau_r \lambda_{DR}^e = L_m i_{QS}^e. \tag{7.84}$$

Replacing ω_r, λ_{DR}^e, and i_{QS}^e with ω_r^*, λ_r^*, and i_{QS}^{e*}, respectively, in Equation 7.84 yields

$$\omega_r^* = \frac{L_m}{\tau_r}\frac{i_{QS}^{e*}}{\lambda_r^*}. \tag{7.85}$$

Replacing λ_{Dr}^e and i_{DS}^e in Equation 7.83 with λ_r^* and i_{DR}^{e*} yields

$$i_{DS}^{e*} = \frac{1 + \tau_r p}{L_m}\lambda_R^*. \tag{7.86}$$

From the torque Equation 7.68, the signal i_{QS}^{e*} can be obtained as

$$i_{QS}^{e*} = \frac{T^*}{K_T \lambda_r^*}. \tag{7.87}$$

A vector control system for an induction motor based on an indirect rotor flux orientation scheme is shown in Figure 7.33. The rotor flux and developed torque are controlled in a feed-forward manner. Because of this, system performance strongly depends on an accurate knowledge of motor parameters, a requirement that is difficult to satisfy in practical

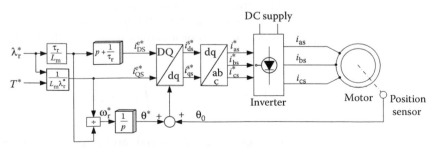

FIGURE 7.33
Vector control system for induction motor with indirect rotor flux orientation.

applications. On the other hand, a major advantage of such a system is that a standard motor can be used, whose rotor position is easily measured by an external sensor. Since the control scheme presented here constitutes an extension of the scalar torque control methods, the reference flux and torque values must satisfy the safe operation area condition described earlier.[9]

7.2.6 Voltage Source Inverter for FOC

The power electronic inverter for the FOC of induction motor drives has the same topology as shown in Figure 7.21a, which is again illustrated in Figure 7.34. The power switches in a given leg (a, b, or c) must never both be in the on state, since this would cause a short-circuit. On the other hand, if both switches on the same leg are in the off state, then the potential of the corresponding output terminal is unknown to the control system of the inverter. The circuit can be completed through either the upper or lower diode, and, consequently, the potential can be equal to that of either a positive bus (+) or a negative bus (−). Therefore, the inverter is controlled in such a way that, in a given leg, either the upper switch (SA, SB, or SC) is on, and the lower switch (SA′, SB′, or SC′) is off, or vice versa: the upper switch is off, and the lower switch is on.

Since only two combinations of states of switches in each leg are allowed, a switching variable can be assigned to each phase of the inverter. In effect, only eight logic states are permitted for the whole power circuit. The switching variables are defined as follows:

$$a = \begin{cases} 0 & \text{if SA is OFF and SA' is ON,} \\ 1 & \text{if SA is ON and SB' is OFF,} \end{cases} \tag{7.88}$$

$$b = \begin{cases} 0 & \text{if SB is OFF and SB' is ON,} \\ 1 & \text{if SB is ON and SB' is OFF,} \end{cases} \tag{7.89}$$

$$c = \begin{cases} 0 & \text{if SC is OFF and SC' is ON,} \\ 1 & \text{if SC is ON and SC' is OFF.} \end{cases} \tag{7.90}$$

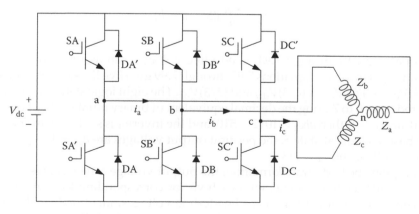

FIGURE 7.34
Circuit diagram of three-phase voltage source inverter.

The instantaneous values of the line-to-line output voltage of the inverter are given by

$$v_{ab} = V_{dc}(a - b),$$
(7.91)

$$v_{bc} = V_{dc}(b - c),$$
(7.92)

$$v_{ca} = V_{dc}(c - a),$$
(7.93)

where V_{dc} is the voltage of the DC supply of the inverter.

In a balanced three-phase system, the line-to-neutral voltage can be calculated from the line-to-line voltages as[9]

$$v_a = \frac{1}{3}(v_{ab} - v_{ca}),$$
(7.94)

$$v_b = \frac{1}{3}(v_{bc} - v_{ab}),$$
(7.95)

$$v_c = \frac{1}{3}(v_{ca} - v_{bc}).$$
(7.96)

Hence, after substituting Equations 7.88 through 7.90 into Equations 7.94 through 7.96, the line-to-neutral voltages are given by

$$v_a = \frac{V_{dc}}{3}(2a - b - c),$$
(7.97)

$$v_b = \frac{V_{dc}}{3}(2b - c - a),$$
(7.98)

$$v_c = \frac{V_{dc}}{3}(2c - a - b).$$
(7.99)

From Equations 7.91 through 7.93, line-to-line voltages can assume only three values: $-V_{dc}$, 0, and V_{dc}. However, Equations 7.97 through 7.99 give five line-to-neutral voltage values: $(-2/3)V_{dc}$, $(-1/3)V_{dc}$, 0, $(1/3)V_{dc}$, and $(2/3)V_{dc}$. The eight logic states of the inverter can be numbered from 0 to 7 using the decimal equivalent of binary number abc_2. For example, if $a = 1$, $b = 0$, and $c = 1$, then $abc_2 = 101_2 = 5_{10}$, and the inverter is said to be in state 5. Taking V_{dc} as the base voltage, at state 5, the per-unit output voltages are $v_{ab} = 1$, $v_{bc} = -1$, $v_{ca} = 0$, $v_a = 1/3$, $v_b = -2/3$, and $v_c = 1/3$.

Performing the abc to dq transformation, the output voltage can be represented as space vectors, the stator reference frame, with each vector corresponding to a given state of the inverter. The space vector diagrams of line-to-line voltages, identified by the superscript LTL, and line-to-neutral voltages, identified by the superscript LTN, of the voltage source inverter are shown in Figure 7.35. The vectors are presented in a per-unit format.

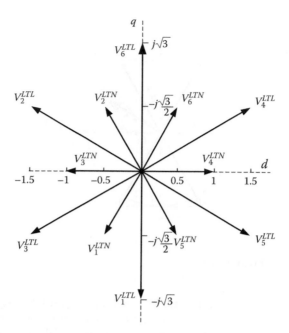

FIGURE 7.35
Space vectors of output voltage of voltage source inverter.

7.2.6.1 Voltage Control in Voltage Source Inverter

Many different PWM technologies have been developed and implemented in practical inverters. Currently, one of the most popular methods is based on the concept of space vectors of inverter voltages, as shown in Figure 7.35. This method is more suitable for application with the FOC of induction motor drives.

For a wye-connected induction motor, the load currents are generated by the line-to-neutral voltages of the inverter. Thus, the motor operation is controlled by the line-to-neutral voltage inverter voltages.

Space vectors of the line-to-neutral voltages are shown in Figure 7.36, together with an arbitrary vector v^*, to be generated by the inverter. In addition to showing six nonzero vectors (states 1 to 6), another two zero vectors, corresponding to states 0 and 7, are also shown. Clearly, only vectors v_0 to v_7, henceforth referred to as base vectors, can be produced at a given instant of time. Therefore, vector v^* represents an average rather than an instantaneous value, the average being taken over a period of switching or sampling interval, which, in practice, constitutes a small fraction of the cycle of the output frequency. The switching interval, at the center of which the reference vector is located, is shown in Figure 7.36 as the shaded segment.

The nonzero base vectors divided the cycle into six, 60°-wide sectors. The desired voltage vector v^*, located in a given sector, can be synthesized as a linear combination of the two adjacent base vectors, v_x and v_y, which frame the sector, and either one of the two zero vectors. That is,

$$v^* = d_x v_x + d_y v_y + d_z v_z, \tag{7.100}$$

where v_z is the zero vector, d_x, d_y, and d_z denote the duty ratios of the states of x, y, and z within the switching interval, respectively. For instance, the reference voltage vector v^*, in

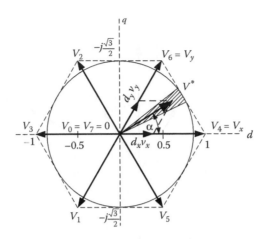

FIGURE 7.36
Illustration of the space vector PWM strategy.

Figure 7.36, is located within the first sector in which $v_x = v_4$ and $v_y = v_6$; hence, it can be produced by an appropriately timed sequence of states 4, 6, and 0 or 7 of the inverter.

The state duty ratio is defined as the ratio of the duration of the state to the duration of the switching interval. Therefore,

$$d_x + d_y + d_z = 1. \tag{7.101}$$

Under this condition, the locus of the maximum available vectors v^* constitutes the hexagonal envelope of the base vectors, as shown in Figure 7.36. To avoid low-order voltage harmonics, resulting from the noncircular shape of the envelope, the locus of the synthesized voltage vectors is, in practice, limited to the circle as shown in Figure 7.36. Consequently, the maximum available magnitude, V_{max}, of the resulting voltage is $(\sqrt{3}/2)V_{dc}$. With respect to vector v^*, in Figure 7.36, Equation 7.100 can be written

$$v^* = MV_{max}e^{j\alpha} = d_x v_4 + d_y v_6 + d_z v_z, \tag{7.102}$$

where M is the modulation index, adjustable within a 0 to 1 range, and α denotes the angular position of the vector v^* inside the sector, that is, the angular distance between vectors v^* and v_x. As seen in Figure 7.36, $v_x = v_4 = 1 + j0$ p.u., $v_y = v_6 = 1/2 + j\sqrt{3}/2$ p.u., and v_z (either v_0 or v_7) is zero, and $V_{max} = (\sqrt{3}/2)V_{dc}$. Equation 7.100 can be rewritten as

$$\frac{\sqrt{3}}{2}M\cos(\alpha) = d_x + \frac{1}{2}d_y, \tag{7.103}$$

and

$$\frac{\sqrt{3}}{2}M\sin(\alpha) = \frac{\sqrt{3}}{2}d_y. \tag{7.104}$$

Thus, d_x and d_y can be expressed as

$$d_x = M\sin(60° - \alpha), \tag{7.105}$$

$$d_y = M\sin(\alpha), \tag{7.106}$$

and

$$d_z = 1 - d_x - d_y. \tag{7.107}$$

The same equations can be applied to the other sectors.

The simple algebraic formulas (Equations 7.105 through 7.107) allow duty ratios of the consecutive logic states of an inverter to be computed in real time. Due to the freedom of choice of the zero vectors, various state sequences can be enforced in a given sector. A particularly efficient operation of the inverter is obtained when the state sequence in consecutive switching intervals is

$$|x - y - z|z - y - x|\ldots, \tag{7.108}$$

where $z = 0$ in the sectors $v_6 - v_2$, $v_3 - v_1$ and $v_5 - v_4$, and $z = 7$ in the remaining sectors. Figure 7.37 shows an example of switching signals and output voltages for a voltage source inverter in the previously described PWM mode with $M = 0.7$ and $20°$ width of the switching interval.[9]

7.2.6.2 *Current Control in Voltage Source Inverter*

Since the output currents of an inverter depend on load, feedforward current control is not feasible, and feedback from current sensors is required. There exist several different

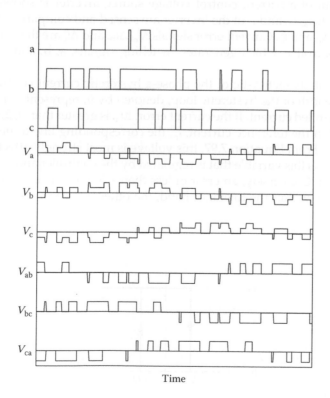

Time

FIGURE 7.37
Example switching signals and output voltage for voltage source inverter in PWM operation mode.

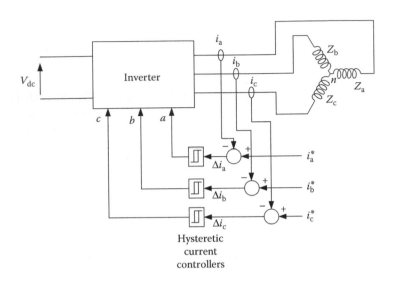

FIGURE 7.38
Block diagram of current-controlled voltage source inverter.

control technologies. The simplest one is the controller, based on the so-called hysteretic or bang-bang control.

A block diagram of a current control voltage source inverter is shown in Figure 7.38. The output currents i_a, i_b, and i_c of the inverter are sensed and compared with the reference current signals i_a^*, i_b^*, and i_c^*. Current error signals Δi_a, Δi_b, and Δi_c are then applied to the hysteresis current controller, which generates switching signals, a, b, and c, for the inverter switches.

The input–output characteristic of the phase-a hysteretic current controller is shown in Figure 7.39. The width of the hysteretic loop, denoted by h, represents the tolerance bandwidth for the controlled current. If the current error, Δi_a, is greater than $h/2$, that is, i_a is unacceptably lower than the reference current, i_a^*, the corresponding line-to-neutral voltage, v_a, must be increased. From Equation 7.97, this voltage is most strongly affected by the switch variable a; hence, it is this variable that is regulated by the controller and is set at 1 to a in the described situation. Conversely, an error of less than $-h/s$ results in $a = 0$; to decrease the current i_a that stays within the tolerance band, the other two controllers are operated in a similar manner.

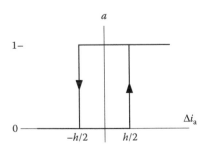

FIGURE 7.39
Input–output characteristics of hysteresis current controller.

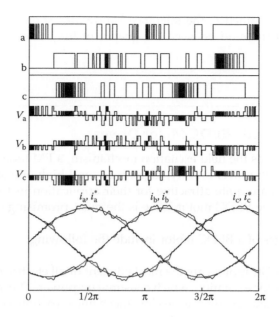

FIGURE 7.40
Current-controlled voltage source inverter (10% tolerance bandwidth).

The width, h, of the tolerance band affects the switching frequency of the inverter. The narrower the band, the more frequently the switching takes place, and the higher the quality of the current will be. This is illustrated in Figures 7.40 and 7.41, in which the switching variables, line-to-neutral voltages, and currents for an inverter supplying a resistive-plus-inductive load are at values of $h = 10$ and 5% of the amplitude of the reference

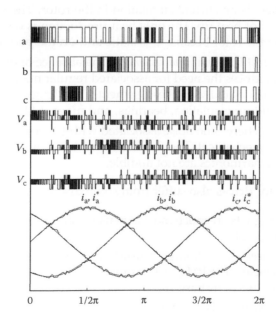

FIGURE 7.41
Current-controlled voltage source inverter (5% tolerance bandwidth).

current, respectively. In practice, the tolerance bandwidth should be set to a value that represents an optimal trade-off between the quality of the currents and the efficiency of the inverter.

7.3 Permanent Magnetic BLDC Motor Drives

Using high-energy PMs as the field excitation mechanism, a PM motor drive can be potentially designed with high power density, high speed, and high operation efficiency. These prominent advantages are quite attractive for their application in EVs and HEVs. Of the family of PM motors, the BLDC motor drive is the most promising candidate for EV and HEV applications.[4]

The major advantages of a BLDC motor include the following:

- *High efficiency*: BLDC motors are the most efficient of all electric motors. This is due to the use of PMs for excitation, which consume no power. The absence of mechanical commutators and brushes means low mechanical friction losses and, therefore, higher efficiency.

- *Compactness*: The recent introduction of high-energy-density magnets (rare-earth magnets) has made it possible to achieve very high-flux densities in BLDC motors. This accordingly makes it possible to achieve high torques, which in turn allows one to make the motor small and light.

- *Ease of control*: A BLDC motor can be controlled as easily as a DC motor because the control variables are easily accessible and constant throughout the operation of the motor.

- *Ease of cooling*: There is no current circulation in the rotor. Therefore, the rotor of a BLDC motor does not heat up. The only heat production is on the stator, which is easier to cool than the rotor because it is static and on the periphery of the motor.

- *Low maintenance, great longevity, and reliability*: The absence of brushes and mechanical commutators reduces the need for associated regular maintenance and reduces the risk of failure associated with these elements. The longevity is therefore only a function of the winding insulation, bearings, and magnet life length.

- *Low noise emissions*: There is no noise associated with the commutation because it is electronic and not mechanical. The driving converter switching frequency is high enough so that the harmonics are not audible.

However, BLDC motor drives also suffer from some disadvantages as follows:

- *Cost*: Rare-earth magnets are much more expensive than other magnets and result in an increased motor cost.

- *Limited constant power range*: A large constant power range is crucial to achieving high vehicle efficiencies. The PM BLDC motor is incapable of achieving a maximum speed greater than twice the base speed.

- *Safety*: Large rare-earth PMs are dangerous during the construction of the motor because flying metallic objects are attracted to them. There is also a danger in the case of vehicle wreck if the wheel spins freely: the motor is still excited by its magnets, and high voltage is present at the motor terminals, which could endanger passengers or rescuers.

- *Magnet demagnetization*: Magnets can be demagnetized by large opposing magneto-motive forces and high temperatures. The critical demagnetization force is different for each magnet material. Great care must be taken in cooling the motor, especially if it is compact.

- *High-speed capability*: Surface-mounted PM motors cannot reach high speeds because of the limited mechanical strength of the assembly between the rotor yoke and the PMs.

- *Inverter failures in BLDC motor drives*: Because of the PMs on the rotor, BLDC motors present major risks in the case of short-circuit failures of the inverter. Indeed, the rotating rotor is always energized and constantly induces an EMF in the short-circuited windings. A very large current circulates in those windings, and an correspondingly large torque tends to block the rotor. The dangers of blocking one or several wheels of a vehicle are nonnegligible. If the rear wheels are blocked while the front wheels are spinning, the vehicle will spin uncontrollably. If the front wheels are blocked, the driver will have no directional control over the vehicle. If only one wheel is blocked, it will induce a yaw torque that will tend to spin the vehicle, which will be difficult to control. In addition to the dangers to the vehicle, it should be noted that the large current resulting from an inverter short-circuit poses a risk of demagnetizing and destroying the PMs.

Open-circuit faults in BLDC motor drives are no direct threat to vehicle stability. The impossibility of controlling a motor due to an open circuit may, however, pose problems in terms of controlling the vehicle. Because the magnets are always energized and cannot be controlled, it is difficult to control a BLDC motor to minimize the fault. This is a particularly important issue when the BLDC motor is operated in its constant power region. Indeed, in this region, a flux is generated by the stator to oppose the magnet flux and allow the motor to rotate at higher speeds. If the stator flux disappears, the magnet flux induces a large EMF in the windings, which can be harmful to the electronics or passengers.

7.3.1 Basic Principles of BLDC Motor Drives

A BLDC motor drive consists mainly of the BLDC machine, the digital signal processor (DSP)-based controller, and the power-electronics-based power converter, as shown in Figure 7.42. Position sensors H_1, H_2, and H_3 sense the position of the machine rotor. The rotor position information is fed to the DSP-based controller, which in turn supplies gating signals to the power converter by turning on and off the proper stator pole windings of the machine. In this way, the torque and speed of the machines are controlled.

7.3.2 BLDC Machine Construction and Classification

BLDC machines can be categorized by the position of the rotor PM, geometrically, according to the way in which the magnets are mounted on the rotor. The magnets can either be surface mounted or interior mounted.

Figure 7.43a shows a surface-mounted PM rotor. Each PM is mounted on the surface of the rotor. It is easy to build, and specially skewed poles are easily magnetized on this

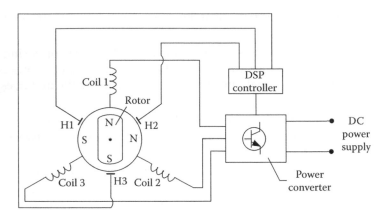

FIGURE 7.42
BLDC motor.

surface-mounted type to minimize cogging torque, but there is a possibility that it will fly apart during high-speed operation.

Figure 7.43b shows an interior-mounted PM rotor. Each PM is mounted inside the rotor. It is not as common as the surface-mounted type, but it is a good candidate for high-speed operation. Note that there is inductance variation for this type of rotor because the PM part is equivalent to air in the magnetic circuit calculation.

In the case of the stator windings, there are two major classes of BLDC motor drives, both of which can be characterized by the shapes of their respective back EMF waveforms, trapezoidal and sinusoidal.

The trapezoidal-shaped back EMF BLDC motor is designed to develop trapezoidal back EMF waveforms. It has the following ideal characteristics:

- Rectangular distribution of magnet flux in the air gap
- Rectangular current waveform
- Concentrated stator windings

Excitation waveforms take the form of quasisquare current waveforms with two 60° electrical intervals of zero current excitation per cycle. The nature of the excitation waveforms

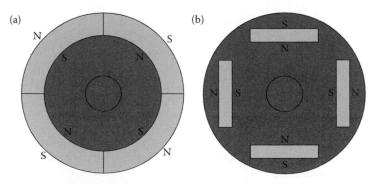

FIGURE 7.43
Cross-sectional view of PM rotor: (a) surface-mounted PM rotor and (b) interior-mounted PM rotor.

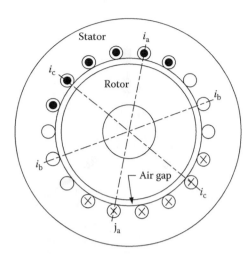

FIGURE 7.44
Winding configuration of trapezoidal-shaped back EMF BLDC.

for trapezoidal back EMF permits some important system simplifications compared to sinusoidal back EMF machines. The resolution requirements for the rotor position sensor are much lower, since only six commutation instants are necessary per electrical cycle. Figure 7.44 shows the winding configuration of the trapezoidal-shaped back EMF BLDC machine.

Figure 7.45a shows an equivalent circuit, and Figure 7.45b shows a trapezoidal back EMF, current profiles, and Hall sensor signals of the three-phase BLDC motor drive. The voltages seen in this figure, e_a, e_b, and e_c, are the line-to-neutral back EMF voltages, the result of the PM flux crossing the air gap in a radial direction, cutting the coils of the stator at a rate proportional to the rotor speed. The coils of the stator are positioned in the standard three-phase full-pitch, concentrated arrangement, and thus the phase trapezoidal back EMF waveforms are displaced by 120° electrical degrees. The current pulse generation is a "120°-on and 60°-off" type, meaning each phase current is flowing for two-thirds of an electrical 360° period, 120° positively and 120° negatively. To drive the motor with maximum and constant torque per ampere, it is desirable for the line current pulses to be synchronized with the line-neutral back EMF voltages of the particular phase.

A sinusoidal-shaped back EMF BLDC motor is designed to develop sinusoidal back EMF waveforms. It has the following ideal characteristics:

1. Sinusoidal distribution of magnet flux in the air gap,
2. Sinusoidal current waveforms,
3. Sinusoidal distribution of stator conductors.

The most fundamental aspect of the sinusoidal-shaped back EMF motor is that the back EMF generated in each phase winding by the rotation of the magnet should be a sinusoidal wave function of the rotor angle. The drive operation of the sinusoidal-shaped back EMF BLDC machine is similar to the AC synchronous motor. It has a rotating stator MMF wave like a synchronous motor and, therefore, can be analyzed with a phasor diagram. Figure 7.46 shows the winding configuration of the sinusoidal-shaped back EMF BLDC machine.

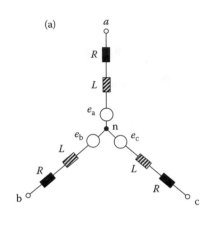

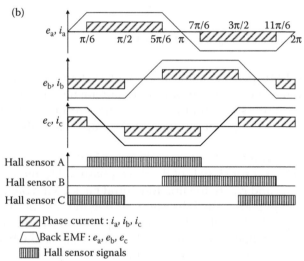

FIGURE 7.45
(a) Three-phase equivalent circuit and (b) back EMFs, currents, and Hall sensor signals of a BLDC motor.

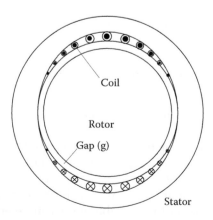

FIGURE 7.46
Winding configuration of sinusoidal-shaped back EMF BLDC.

7.3.3 Properties of PM Materials

There are three classes of PMs currently used for electric motors:

1. Alnicos (Al, Ni, Co, Fe).
2. Ceramics (ferrites), for example, barium ferrite (BaO × 6Fe$_2$O$_3$) and strontium ferrite (SrO × 6Fe$_2$O$_3$).
3. Rare-earth materials, that is, samarium–cobalt (SmCo) and neodymium–iron–boron (NdFeB).

Demagnetization curves of the preceding PM materials are shown in Figure 7.47.[10]

7.3.3.1 Alnico

The main advantages of Alnico are its high magnetic remanent flux density and low temperature coefficients. The temperature coefficient of its remanent magnetic flux density B_r, or remanence, is 0.02%/°C, and the maximum service temperature is 520°C. These advantages allow quite a high air gap flux density and high operating temperature. Unfortunately, coercive force is very low, and the demagnetization curve is extremely nonlinear. Therefore, it is very easy not only to magnetize but also to demagnetize Alnico. Alnico magnets have been used in motors having ratings in a range of a few watts to 150 kW. Alnicos dominated the PM industry from the mid-1940s to about 1970, when ferrites became the most widely used materials.[10]

7.3.3.2 Ferrites

Barium and strontium ferrites were invented in the 1950s. A ferrite has a higher coercive force than Alnico, but at the same time, it has a lower remanent magnetic flux density.

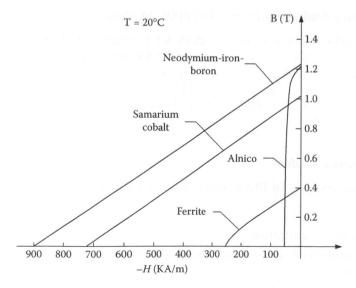

FIGURE 7.47
Demagnetization curves for different PM materials.

Temperature coefficients are relatively high, that is, the coefficient of B_r is 0.20%/°C and the coefficient of coercive field strength, H_c, or coercivity is 0.27%/°C. The maximum service temperature is 400°C. The main advantages of ferrites are their low cost and very high electric resistance, which means no eddy-current losses in the PM volume.

7.3.3.3 Rare-Earth PMs

During the last three decades, greater progress regarding available energy density $(BH)_{max}$ has been achieved with the development of rare-earth PMs. The first generation of the rare earth PMs based on the composition of samarium–cobalt ($SmCo_5$) was invented in the 1960s and has been commercially produced since the early 1970s. Today, it is a well-established hard magnetic material. $SmCo_5$ has the advantages of high remanent flux density, high coercive force, high-energy product, linear demagnetization curve, and low temperature coefficient. The temperature coefficient of B_r is 0.03%–0.045%/°C, and the temperature coefficient of H_c is 0.14%–0.40%/°C. The maximum service temperature is 250–300°C. It is well suited to build motors with low volume and, consequently, high specific power and low moment of inertia. The cost is the only drawback. Both Sm and Co are relatively expensive due to their supply restrictions.

With the discovery in recent years of a second generation of rare-earth magnets based on inexpensive neodymium (Nd) and iron, remarkable progress with regard to lowering raw material costs has been achieved. NdFeB magnets, which are now produced in increasing quantities, have better magnetic properties than those of SmCo, but only at room temperature. The demagnetization curves, especially the coercive force, are strongly temperature dependent. The temperature coefficient of B_r is 0.095%–0.15%/°C, and the temperature coefficient of H_c is 0.40%–0.7%/°C. The maximum service temperature is 150°C, and the Curie temperature is 310°C.

The latest grades of NdFeB have better thermal stability, enabling an increase in working temperature by 50°C, and offer greatly improved resistance to corrosion.[10]

7.3.4 Performance Analysis and Control of BLDC Machines

Speed–torque performance is most important for traction and other applications. As any other electric machine, the torque is produced by the interaction of the magnetic field and current. The magnetic field is produced in BLDC by the PM, and the current depends on the source voltage, control, and the back EMF, which is determined by the magnetic field and speed of the machine. To obtain the desired torque and speed at a given load, the current needs to be controlled.

7.3.4.1 Performance Analysis

The performance analysis of BLDC machines is based on the following assumption for simplification:

1. The motor is not saturated.
2. Stator resistances of all the windings are equal, and self- and mutual inductances are constant.
3. Power semiconductor devices in the inverter are ideal.
4. Iron losses are negligible.

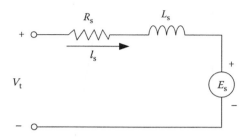

FIGURE 7.48
Simplified equivalent circuit of BLDC motor.

A simplified equivalent circuit of one phase is shown in Figure 7.48, where V_t is the voltage of the power supply, R_s is the resistance of the winding, L_s is the leakage inductance ($L_s = L - M$, where L is the self-inductance of the winding and M is the mutual inductance), and E_s is the back EMF induced in the winding by the rotating rotor.

Based on the equivalent circuit of Figure 7.48, the performance of the BLDC motor can be described by

$$V_t = R_s I_s + L_s \frac{dI_s}{dt} + E_s, \tag{7.109}$$

$$E_s = k_E \omega_r, \tag{7.110}$$

$$T_e = k_T I_s, \tag{7.111}$$

$$T_e = T_L + J \frac{d\omega_r}{dt} + B\omega_r, \tag{7.112}$$

where k_E is the back EMF constant, which is associated with the PMs and rotor structure, ω_r is the angular velocity of the rotor, k_T is the torque constant, T_L is the load torque, and B is the viscous resistance coefficient. For steady-state operation, Equations 7.109 through 7.111 can be simply reduced to

$$T_e = \frac{(V_t - k_E \omega_r) k_T}{R_s}. \tag{7.113}$$

The speed–torque performance with constant voltage supply is shown in Figure 7.49. It can be seen from Equation 7.113 and Figure 7.49 that at low speed, especially while starting, very high torque is produced, which results in very high current due to the low back EMF. This very high current would damage stator windings.

With a variable voltage supply, the winding current can be restricted to its maximum by actively controlling the voltage; thus, a maximum constant torque can be produced, as shown in Figure 7.50.

For dynamic or transient operation, the performance of the BLDC machine is described by Equations 7.109 through 7.112. However, a Laplace transform is helpful in simplifying the

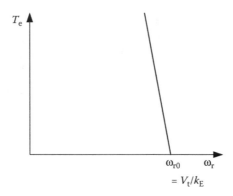

FIGURE 7.49
Speed–torque curve in steady state with constant voltage.

analysis. Equations 7.109 through 7.112 can be expressed by their Laplace forms as follows:

$$V_t(s) = E_s(s) + (R_s + sL_s)I_s(s), \qquad (7.114)$$

$$E_s(s) = k_E\omega_r(s), \qquad (7.115)$$

$$T_e(s) = k_T I_s(s), \qquad (7.116)$$

$$T_e(s) = T_L(s) + (B + sJ)\omega_r(s). \qquad (7.117)$$

Thus, the transfer function of the BLDC motor drive system is

$$\omega_r(s) = \frac{k_T}{(R_s + sL_s)(sJ + B) + k_T k_E} V_t(s) - \frac{R_s + sL_s}{(R_s + sL_s)(sJ + B) + k_T k_E} T_L(s). \qquad (7.118)$$

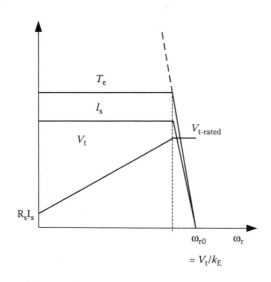

FIGURE 7.50
Speed–torque curve in steady state with variable voltage supply.

L_s and J in Equation 7.118 represent the electrical and mechanical delay in transient operation. L_s determines how quickly the armature current builds up in response to a step change in the terminal voltage, where the rotor speed is assumed to be constant. J determines how quickly the speed builds up in response to a step change in the terminal voltage.

7.3.4.2 Control of BLDC Motor Drives

In vehicle traction applications, the torque produced is required to follow the torque desired by the driver and commanded through the accelerator and brake pedals. Thus, torque control is the basic requirement.

Figure 7.51 shows a block diagram of a torque control scheme for a BLDC motor drive. The desired current I^* is derived from the commanded torque T^* through a torque controller. The current controller and the commutation sequencer receive the desired current I^*, position information from the position sensors, and perhaps the current feedback through current transducers and then produce gating signals. These gating signals are sent to the three-phase inverter (power converter) to produce the desired phase current to the BLDC machine.

In traction applications, speed control may be required, cruising control operation, for example (Figure 7.52). Many high-performance applications include current feedback for torque control. At a minimum, a DC bus current feedback is required to protect the drive

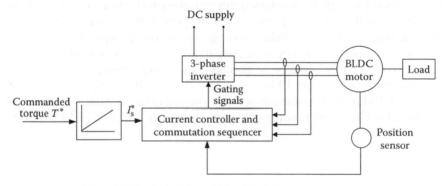

FIGURE 7.51
Block diagram of torque control of BLDC motor.

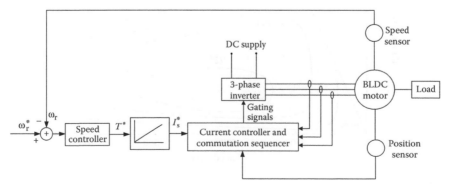

FIGURE 7.52
Block diagram of speed control of BLDC motor.

and the machine from over-currents. The controller blocks, "speed controller" may be any type of classical controller such as a proportional-integral (PI) controller or a more advanced controller such as an artificial intelligence control. The current controller and commutation sequencer provide the properly sequenced gating signals to the three-phase inverter while comparing sensed currents to a reference to maintain a constant peak current control by hysteresis (current chopping) or with a voltage source (PWM)-type current control. Using position information, the commutation sequencer causes the inverter to electronically commutate, acting as the mechanical commutator of a conventional DC machine. The commutation angle associated with a brushless motor is normally set so that the motor commutates around the peak of the torque angle curve. Considering a three-phase motor, connected in Delta or wye, commutation occurs at electrical angles, which are plus or minus 30° (electrical) from the peaks of the torque–angle curves. When the motor position moves beyond the peaks by an amount equal to 30° (electrical), then the commutation sensors cause the stator phase excitation to switch to move the motor suddenly to −30° relative to the peak of the next torque–angle curve.[11]

7.3.5 Extend Speed Technology

As discussed previously, PM BLDC motors inherently have a short constant power range due to their rather limited field-weakening capability. This is a result of the presence of the PM field, which can only be weakened through the production of a stator field component, which opposes the rotor magnetic field. The speed ratio, x, is usually less than 2.[12]

Recently, the use of additional field windings to extend the speed range of PM BLDC motors has been adopted.[1] The key is to control the field current in such a way that the air gap field provided by PMs can be weakened during high-speed, constant-power operation. Due to the presence of both PMs and field windings, these motors are known as PM hybrid motors. A PM hybrid motor can achieve a speed ratio of around 4. The optimal efficiency profiles of a PM hybrid motor drive are shown in Figure 7.53.[1] However, the PM

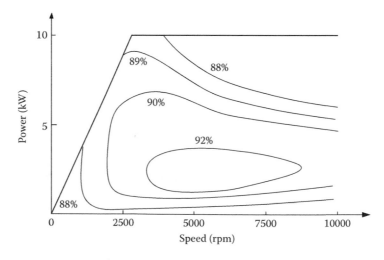

FIGURE 7.53
Optimal efficiency profiles of a PM hybrid motor drive. (Adapted from C. C. Chan and K. T. Chau, *Modern Electric Vehicle Technology*, Oxford University Press, Oxford, 2001.)

hybrid motors have the drawback of a relatively complex structure. The speed ratio is still not enough to meet the vehicle performance requirement, especially for an off-road vehicle. Thus, a multigear transmission is required.

7.3.6 Sensorless Techniques

As mentioned previously, the operation of BLDC motor drives relies mostly on position sensors for obtaining the rotor position information so as to properly perform the turn-on or turn-off of each phase properly.[8] The position sensor is usually either a three-element Hall-effect sensor or an optical encoder. These position sensors are expensive, fragile elements. Thus, its presence not only increases the cost of the motor drive but also seriously lowers its reliability and limits its application in some environments, such as in the military. Position sensorless technology can effectively continue the operation of a system in case the position sensors lose their function. This is crucial in some applications, such as military vehicles.

Several sensorless technologies have been developed. Most them are based on the voltage, current, and back EMF detection. These techniques can be primarily grouped into four categories:

1. Those using measured currents, voltages, fundamental machine equations, and algebraic manipulations.
2. Those using observers.
3. Those using back EMF methods.
4. Those with novel techniques not falling into the previous three categories.

7.3.6.1 Methods Using Measurables and Math

The method consists of two subtypes: (1) those that calculate the flux linkages using measured voltages and currents and (2) those that utilize a model's prediction of a measurable voltage or current, compare the model's value with the actual measured voltage or current, and calculate the change in position, which is proportional to the difference between the measured and the actual voltage or current.

The first subtype is seen in some studies.[13–20] The fundamental idea is to calculate the flux linkage from the measured voltage and current:

$$\psi = \int_0^t (V - Ri)\mathrm{d}\tau. \tag{7.119}$$

With the knowledge of the initial position, machine parameters, and the flux linkages' relationship with the rotor position, the rotor position can be estimated. By determining the rate of change of the flux linkage from the integration results, the speed can also be determined. An advantage of the flux-calculating method is that line–line voltages may be used in the calculations, and thus, no motor neutral is required.[8] This is beneficial, as the most common BLDC configuration is Y-connected with no neutral.

The second subtype is seen in some other studies.[21–24] This method consists of first developing an accurate d–q model of the machine. Utilizing the measured currents and a d–q transformation, the output voltages of the model are compared to the measured and transformed voltages. The difference is proportional to the difference in angular reference between the model's coordinate system and the actual coordinate system, which is the rotor position with reference to the actual coordinate system's reference. Conversely, measured voltages have been used to find current differences. In either case, the difference between the measured (and transformed) and the calculated is used as the multiplier in an updated equation for the rotor position.

7.3.6.2 Methods Using Observers

These methods determine the rotor position and/or speed using observers. The first of these considered are those that use the well-known Kalman filter as a position estimator.[25–30] One of the first of these to appear in the literature was by M. Schroedl in 1988. In his many publications, Schroedl utilized various methods of measuring system voltages and currents, which could produce rough estimates of the angular rotor position. The Kalman filtering added the additional refinements to the first estimates of position and speed. Other observer-based systems include those utilizing nonlinear,[31–33] full-order,[13,34,35] and sliding-mode observers.[15,22,36]

7.3.6.3 Methods Using Back EMF Sensing

Using back EMF sensing is the main approach in sensorless control technology of a BLDC motor drive. This approach consists of several methods, such as (1) the terminal voltage sensing method, (2) the third-harmonic back EMF sensing method, (3) freewheeling diode conduction, and (4) back EMF integration.

Terminal voltage sensing: In the normal operation of a BLDC motor, the flat part of a phase back EMF is aligned with the phase current. The switching instants of the converter can be obtained by knowing the zero crossing of the back EMF and a speed-dependent period of time delay.[37]

Since a back EMF is zero at rest and proportional to speed, it is not possible to use the terminal voltage sensing method to obtain a switching pattern at low speeds. As the speed increases, the average terminal voltage increases, and the frequency of excitation increases. The capacitive reactance in the filters varies with the frequency of excitation, introducing a speed-dependent delay in switching instants. This speed-dependent reactance disturbs the current alignment with the back EMF and field orientation, which causes problems at higher speeds. In this method, a reduced speed operating range is normally used, typically around 1000–6000 rpm. This method is a good method for the steady state; however, phase differences in the circuits used due to speed variations do not allow optimal torque per ampere over a wide speed range.

Third-harmonic back EMF sensing: Rather than using the fundamentals of the phase back EMF waveform as in the previous technique, the third harmonic of the back EMF can be used in the determination of the switching instants in the wye-connected 120° current conduction operating mode of a BLDC motor.[38] This method is not as sensitive to phase delay as the zero-voltage crossing method since the frequency to be filtered is three times as high. The reactance of the filter capacitor in this case dominates the phase angle output of the filter more so than at the lower frequency. This method provides a wider speed range than the zero-crossing method, does not introduce as much phase delay as the zero-crossing method, and requires less filtering.

Freewheeling diode conduction: This method uses indirect sensing of the zero crossing of the phase back EMF to obtain the switching instants of a BLDC motor.[39] In a 120°

conducting Y-connected BLDC motor, one of the phases is always open-circuited. For a short period after opening the phase, the phase current remains flowing via a free-wheeling diode due to the inductance of the windings. This open-phase current becomes zero in the middle of the commutation interval, which corresponds to the point where the back EMF of the open phase crosses zero. The largest drawback of this method is the requirement of six additional isolated power supplies for the comparator circuitry for each free wheeling diode.

Back EMF *integration*: In this method, position information is extracted by integrating the back EMF of the unexcited phase.[40-43] The integration is based on the absolute value of the open phase's back EMF. Integration of the voltage divider scaled-down back EMF starts when the open phase's back EMF crosses zero. A threshold is set to stop the integration that corresponds to a commutation instant. As the back EMF is assumed to vary linearly from positive to negative (trapezoidal back EMF assumed), and this linear slope is assumed to be speed insensitive, the threshold voltage is kept constant throughout the speed range. If desired, current advance can be implemented by the change of the threshold. Once the integrated value reaches the threshold voltage, a reset signal is imported to zero the integrator output.

This approach is less sensitive to switching noise and automatically adjusts to speed changes, but the low-speed operation is poor. With this type of sensorless operation scheme, up to 3600 rpm has been reported.[43]

7.3.6.4 Unique Sensorless Techniques

The following sensorless methods are completely original and unique. These range from artificial intelligence methods to variations in the machine structure. The first of the novel methods to be considered are those utilizing artificial intelligence, that is, artificial neural networks (ANNs) and fuzzy logic. Peters and Harth[43] utilized a neural network using a back propagation training algorithm (BPN) to act as a nonlinear function implementation between measured phase voltages and currents, which were inputs, and rotor position, which was the output. Using the equations in this method, the flux linkage can be calculated using the measured voltages, currents, and system parameters.

Utilizing fuzzy logic, Hamdi and Ghribi[44] proposed two fuzzy logic subsystems in an application. Using the conventional equations of phase voltages and currents, the rotor position can be calculated.[8] With the knowledge of the relationships between these measurables and the rotor position, a fuzzy Mamdani -type system was developed to produce rotor position estimates. It was noted that this could have been accomplished just as easily with lookup tables; however, for the desired resolution, the size of the lookup tables becomes unmanageably large. The second fuzzy system used took as input the estimated rotor position and produced reference current values for two different drive strategies: unity power factor and maximum torque per ampere.

In Hesmondhalgh et al.,[45] an additional stator lamination with equally spaced slots around the periphery is added to the end of the machine. Each of the slots contains a small sensing coil. The local magnetic circuit variations for each of the sensing coils are affected by the PM rotor's position. A 20-kHz signal is injected through the coils. The signal distortions are analyzed at the terminals of the sensing coils, the second harmonic yielding position information. An artificial saliency was created in Hesmondhalgh et al.[45] by attaching small pieces of aluminum to the surface of the PMs. The flow of eddy currents in the aluminum acts to increase the reluctance of the various windings' magnetic circuits, causing changes in the winding's inductances with rotor position.

7.4 SRM Drives

The SRM drive is considered an attractive candidate for variable-speed motor drives due to its low cost, rugged structure, reliable converter topology, high efficiency over a wide speed range, and simplicity in control.[46,47] These drives are suitable for EVs, HEV traction applications, aircraft starter/generator systems, mining drives, washing machines, door actuators, and so on.[48–51]

The SRM has a simple, rugged, and low-cost structure. It has no PM or winding on the rotor. This structure not only reduces the cost of the SRM but also offers high-speed operation capability for this motor. Unlike the induction and PM machines, the SRM is capable of high-speed operation without the concern of mechanical failures that result from high-level centrifugal force. In addition, the inverter of the SRM drive has a reliable topology. The stator windings are connected in series with the upper and lower switches of the inverter. This topology can prevent the shoot-through fault that exists in the induction and permanent motor drive inverter. Moreover, high efficiency over a wide speed range and control simplicity are known merits of the SRM drive.[46,47]

A conventional SRM drive system consists of the SRM, power inverter, sensors such as voltage, current, and position sensors, and control circuitry such as the DSP controller and its peripherals, as shown in Figure 7.54. Through proper control, high performance can be achieved in the SRM drive system.[46,47] The SRM drive inverter is connected to a DC power supply, which can be derived from the utility lines through a front-end diode rectifier or from batteries. The phase windings of the SRM are connected to the power inverter, as shown in Figure 7.55. The control circuit provides a gating signal to the switches of the inverter according to particular control strategies and the signals from various sensors.

7.4.1 Basic Magnetic Structure

The SRM has salient poles on both the stator and the rotor. It has concentrated windings on the stator and no winding or PM on the rotor. There are several configurations for SRM depending on the number and the size of the rotor and stator poles. The configurations of the 8/6 and 6/4 SRMs, which are more common, are shown in Figure 7.56.

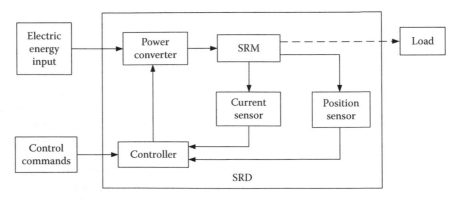

FIGURE 7.54
SRM drive system.

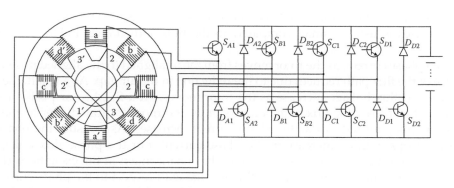

FIGURE 7.55
SRM and its power supply.

Due to its double saliency structure, the reluctance of the flux path for a phase winding varies with the rotor position. Also, since the SRM is commonly designed for high-degree saturation at high phase current, the reluctance of the flux path also varies as the phase current. As a result, the stator flux linkage, phase bulk inductance, and phase incremental inductance all vary with the rotor position and phase current.

The phase voltage equation of the SRM (Figure 7.55) is given by

$$V_j = Ri_j + \frac{d}{dt}\sum_{k=1}^{m} \lambda_{jk},$$ (7.120)

where m is the total number of phases, V_j is the applied voltage in phase j, ij is the current in phase j, R is the winding resistance per phase, λ_{jk} is the flux linkage of phase j due to the current of phase k, and t is the time. The phase flux linkage λ_{jk} is given by

$$\lambda_{jk} = L_{jk}(i_{k,\theta}, \theta)i_k,$$ (7.121)

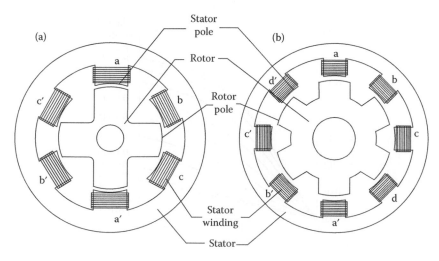

FIGURE 7.56
Cross section of common SRM configurations: (a) 6/4 SRM and (b) 8/6 SRM.

where L_{jk} is the mutual inductance between phase k and phase j. The mutual inductance between phases is usually small compared to the bulk inductance and is neglected in equations.

At a fixed phase current, as the rotor moves from the unaligned to the aligned position, the reluctance of the flux path reduces due to the reduction in the air gap. As a result, the phase inductance and flux linkage increase as the rotor moves. At a fixed rotor position, as the phase current increases, the flux path becomes more and more saturated. Hence, the reluctance of the flux path reduces as the phase current increases. As a result, the phase bulk inductance drops with an increase in the phase current, but the phase flux linkage still increases as the phase current increases due to the enhancement in the excitation. The variations of the phase bulk inductance and flux linkage with respect to the phase current and rotor position for an 8/6 SRM are shown in Figures 7.57 and 7.58, respectively. In those figures, $\theta = -30°$ and $\theta = 0°$ represent the unaligned and aligned rotor positions of the referenced SRM, respectively.

Substituting Equation 7.121 into Equation 7.120, one can have:

$$
\begin{aligned}
V_j &= Ri_j + \frac{\mathrm{d}}{\mathrm{d}t} \sum_{k=1}^{m} \lambda_{jk} = Ri_j + \sum_{k=1}^{m} \left\{ \frac{\partial \lambda_{jk}}{\partial i_k} \frac{\mathrm{d}i_k}{\mathrm{d}t} + \frac{\partial \lambda_{jk}}{\partial \theta} \frac{\mathrm{d}\theta}{\mathrm{d}t} \right\} \\
&= Ri_j + \sum_{k=1}^{m} \left\{ \frac{\partial(L_{jk}i_k)}{\partial i_k} \frac{\mathrm{d}i_k}{\mathrm{d}t} + \frac{\partial(L_{jk}i_k)}{\partial \theta} \omega \right\} \\
&= Ri_j + \sum_{k=1}^{m} \left\{ \left(L_{jk} + i_k \frac{\partial L_{jk}}{\partial i_k} \right) \frac{\mathrm{d}i_k}{\mathrm{d}t} + i_k \frac{\partial L_{jk}}{\partial \theta} \omega \right\}.
\end{aligned}
\tag{7.122}
$$

When only one phase is energized in an operation, Equation 7.122 can be written as

$$
V_j = Ri_j + \left(L_{jj} + i_j \frac{\partial L_{jj}}{\partial i_j} \right) \frac{\mathrm{d}i_j}{\mathrm{d}t} + i_j \frac{\partial L_{jj}}{\partial \theta} \omega.
\tag{7.123}
$$

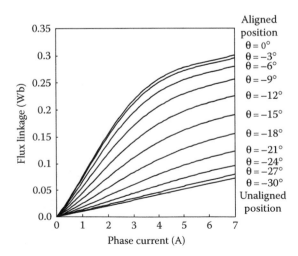

FIGURE 7.57
Variation of phase flux linkage with rotor position and phase current.

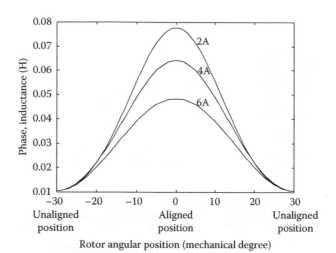

FIGURE 7.58
Variation of phase bulk inductance with rotor position and phase current.

The third term on the right-hand side of Equation 7.123 is the back EMF. The phase incremental inductance is defined as the derivative of the phase flux linkage against the phase current as

$$\ell_{jj} = \frac{\partial \lambda_{jj}}{i_j} = L_{jj} + i_j \frac{\partial L_{jj}}{\partial i_j}, \tag{7.124}$$

where $\ell_{jj}(i, \theta)$ and $L_{jj}(i, \theta)$ are the phase incremental inductance and bulk inductance, respectively. Figure 7.57 shows a typical example of flux linkage varying with rotor position, θ, and phase current, i, of an SRM. Figure 7.58 shows typical variation of phase bulk inductance with rotor position and phase current.

When the magnetic flux is not saturated, the flux linkage varies linearly with the phase current. The incremental inductance can be viewed as equal to the phase bulk inductance. However, if the machine is saturated at a certain phase current and rotor position, the phase incremental inductance no longer equals the phase bulk inductance. The variation of the phase incremental inductance with respect to the phase current and the rotor position can be derived from the variation of the phase linkage with respect to the phase current and the rotor position. The variation of the phase incremental inductance with respect to the phase current and the rotor position for an 8/6 SRM is shown in Figure 7.59.

7.4.2 Torque Production

Torque in SRM is produced by the tendency of the rotor to get into alignment with the excited stator poles. The analytical expression of the torque can be derived using the derivative of the coenergy against the rotor position at a given current.

For a phase coil with current i linking a flux λ, the stored field energy W_f and the coenergy W'_f are indicated as shaded regions in Figure 7.60. Co-energy can be found from the definite integral

$$W'_f = \int_0^i \lambda \, di. \tag{7.125}$$

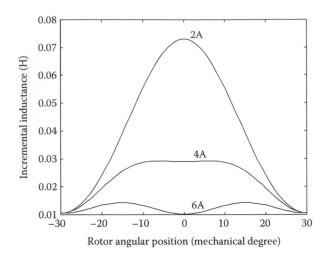

FIGURE 7.59
Variation of phase incremental inductance with rotor position and phase current for a typical 8/6 SRM.

The torque produced by one phase coil at any rotor position is given by

$$T = \left[\frac{\partial W_f'}{\partial \theta}\right]_{i=\text{constant}}. \tag{7.126}$$

When flux is linear with current, for example, an unsaturated field, the magnetization curve in Figure 7.60 will be a straight line, and the coenergy will be equal to the stored

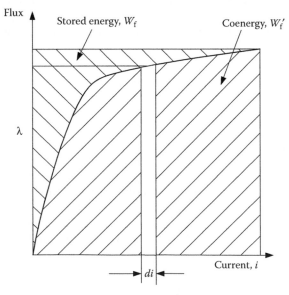

FIGURE 7.60
Stored field energy and coenergy.

field energy. The instantaneous torque can be given as

$$T = \frac{1}{2}i^2\frac{dL(\theta)}{d\theta},$$

(7.127)

where L is the unsaturated phase bulk inductance.

In a saturated phase, the torque cannot be calculated by a simple algebraic equation; instead, an integral equation such as

$$T = \int_0^i \frac{\partial L(\theta, i)}{\partial \theta} i \, di$$

(7.128)

is used.

From Equations 7.127 and 7.128 it can be seen that in order to produce positive torque (motoring torque) in SRM, the phase has to be excited when the phase bulk inductance increases as the rotor rotates. It can also be observed from Equation 7.127 and 7.128 that the phase current can be unidirectional for motoring torque production. Hence, low-cost and reliable inverter topology introduced in a later section can be employed for the SRM drive. Figure 7.61 shows the ideal phase inductance, current, and torque of the SRM. Positive (motoring) torque is produced if the phase is excited when the phase inductance increases as the rotor rotates. Negative torque is generated if the phase is excited when the phase inductance decreases as the rotor moves.[52,53] This implies that the position information is necessary for control of an SRM drive.

The output torque of an SRM is the summation of the torque in all phases:

$$T_m = \sum_{i=1}^{N} T(i, \theta),$$

(7.129)

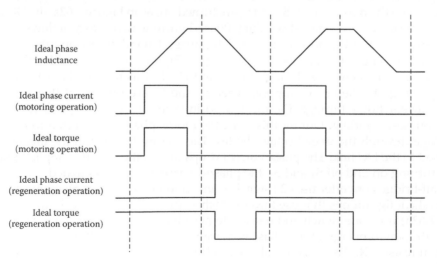

FIGURE 7.61
Idealized inductance, current, and torque profiles of SRM.

where T_m and N are the output torque and phase number of the motor, respectively. The relation between the motor torque and the mechanical load is usually given by

$$T_m - T_1 = J\frac{d\omega}{dt} + B\omega, \tag{7.130}$$

where J, B, and T_1 are the moment of inertia, viscous friction, and load torque, respectively. The relation between position and speed is given by

$$\omega = \frac{d\theta}{dt}. \tag{7.131}$$

7.4.3 SRM Drive Converter

It can be seen from Figure 7.61 that the torque developed by the motor can be controlled by varying the amplitude and the timing of the current pulses in synchrony with the rotor position. To control the amplitude and pulse width of the phase current, a certain type of inverter should be used.

The input to the SRM drive is DC voltage, which is usually derived from the utility through a front-end diode rectifier or from batteries. Unlike other AC machines, the currents in SRMs can be unidirectional. Hence, conventional bridge inverters used in AC motor drives are not used in SRM drives. Several configurations have been proposed for an SRM inverter in the literature,[54,55] some of the most commonly used ones are shown in Figure 7.62.

The most commonly used inverter uses two switches and two freewheeling diodes per phase and is called a classic converter. The configuration of the classic converter is shown in Figure 7.62a. The main advantage of the classic converter is the flexibility in control. All phases can be controlled independently, which is essential for very high-speed operation, where there will be considerable overlap between the adjacent phase currents.[56]

The operation of the classic converter is shown in Figure 7.63 by taking phase-1 as an example. When the two switches S_1 and S_2 are turned on, as in Figure 7.62a, the DC bus voltage, V_{dc}, is applied to the phase-1 winding. Phase-1 current increases as it flows through the path consisting of the V_{dc} positive terminal, S_1, phase-1 winding, S_2, and the V_{dc} negative terminal. By turning off S_1 and holding on S_2 (i.e., Figure 7.63b), when the phase is energized, the current freewheels through the S_2 and D_1. In this mode, phase-1 is not getting or giving energy to the power supply. When S_1 and S_2 are turned off (Figure 7.63c), the phase-1 current flows through D_2, the V_{dc} positive terminal, the V_{dc} negative terminal, D_1, and phase-1 winding. During this time, the motor phase is subjected to negative DC bus voltage through the freewheeling diodes. The energy trapped in the magnetic circuit is returned to the DC link. The phase current drops due to the negative applied phase voltage. By turning on and off S_1 and S_2, the phase-1 current can be regulated.

The half-bridge converter uses $2n$ switches and $2n$ diodes for an n-phase machine. There are several configurations that use fewer switches; for example, the R-dump-type inverter (Figure 7.62b) uses one switch and one diode per phase. This drive is not efficient; during turn-off, the stored energy of the phase is charging capacitor C to the bus voltage and dissipating in resistor R. Also, a zero voltage mode does not exist in this configuration.

An alternative configuration is an $(n+1)$ switch inverter. In this inverter, all phases share a switch and diode so that an overlapping operation between phases is not possible, which is

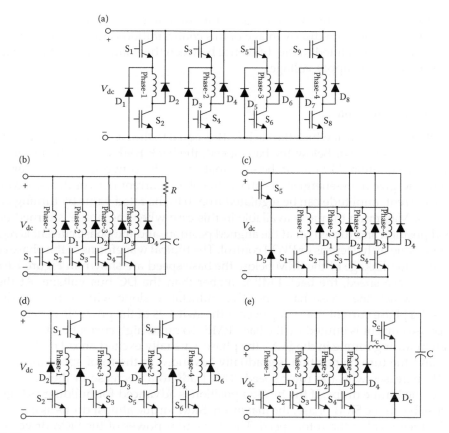

FIGURE 7.62
Different inverter topologies for SRM drives: (a) classic half-bridge converter; (b) R-dump; (c) $n+1$ switches (Miller converter); (d) $1.5n$ switch converter; and (e) C-dump.

inevitable in high-speed operations of this motor. This problem has been solved by sharing switches of each couple nonadjacent phases, as shown in Figure 7.62d. This configuration is limited to an even number of phases of SRM drives.

One of the popular inverter configurations is a C-dump (Figure 7.62e), which has the advantage of fewer switches and allows independent phase current control. In this configuration, during the turn-off time, the stored magnetic energy charges capacitor C, and if the

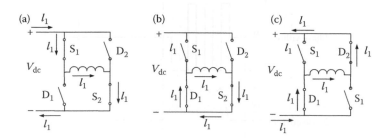

FIGURE 7.63
Modes of operation for classic converter: (a) turn-on mode, (b) zero voltage mode, and (c) turn-off mode.

voltage of the capacitor reaches a certain value, for example V_c, it is transferred to the supply through switch S_c. The main disadvantage of this configuration is that the negative voltage across the phase coil is limited to the difference between the voltage across the capacitor V_c and the system power supply voltage.

7.4.4 Modes of Operation

For SRM, there is a speed at which the back EMF is equal to the DC bus voltage. This speed is defined as the base speed. Below the base speed, the back EMF is lower than the DC bus voltage. From Equation 7.125 it can be seen that when the converter switches are turned on or off to energize or de-energize the phase, the phase current rises or drops accordingly. The phase current amplitude can be regulated from 0 to the rated value by turning on or off the switches. Maximum torque is available in this case when the phase is turned on at an unaligned position and turned off at the aligned position, and the phase current is regulated at the rated value by hysteresis or PWM control. The typical waveforms of the phase current, voltage, and flux linkage of the SRM below the base speed are shown in Figure 7.64.

Above the base speed, the back EMF is higher than the DC bus voltage. At the rotor position—at which the phase has a positive inductance slope with respect to the rotor position—the phase current may drop even if the switches of the power inverter are turned on. The phase current is limited by the back EMF. To build high current and therefore produce high motoring torque in the SRM, the phase is usually excited ahead of the unaligned position, and the turn-on position is gradually advanced as the rotor speed increases. The back EMF increases with the rotor speed. This leads to a decrease in the phase current, and hence, the torque drops. If the turn-on position is advanced for building as high a current as possible in the SRM phase, the maximum SRM torque almost drops as a linear function of the reciprocal of the rotor speed. The maximum power of the SRM drive is almost constant. The typical waveforms at high-speed operation are shown in Figure 7.65.

The advancing of the phase turn-on position is limited to the position at which the phase inductance has a negative slope with respect to the rotor position. If the speed of the rotor further increases, no phase advancing is available for building a higher current in the phase,

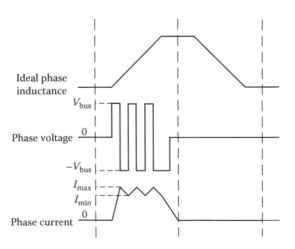

FIGURE 7.64
Low-speed (below the base speed) operation of SRM.

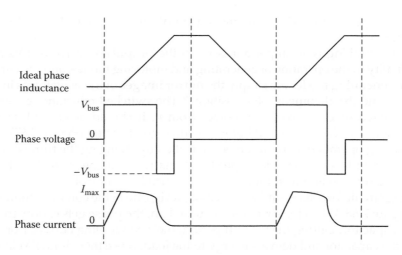

FIGURE 7.65
High-speed (above the base speed) operation of SRM.

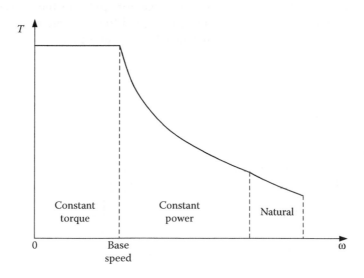

FIGURE 7.66
Torque–speed characteristic of SRM.

and the torque of the SRM drops significantly.[11] The mode is referred to as the natural mode operation. The torque–speed characteristic of the SRM is shown in Figure 7.66.

7.4.5 Generating Mode of Operation (Regenerative Braking)

Torque in SRMs is created based on the principle of reaching the minimum reluctance for the excited phase. Therefore, if the rotor pole is approaching the excited phase, which means the bulk inductance is increasing, the torque produced is in the direction of the rotor, and it is in motoring mode, but if the rotor pole is leaving the stator phase—which means a negative slope of the bulk inductance—the stator tries to keep it in alignment; the torque produced is then in the opposite direction of the movement of the rotor, and the SRM works in the generating mode.

Regenerative braking is an important issue in the propulsion drive of EVs and HEVs. There is a duality in the operation of generating and motoring modes, and the current waveforms in the generating modes are simply the mirror images of the waveforms in the motoring region around the alignment rotor position.[57] The switched reluctance generator (SRG) is a singly excited machine, so to get power from it, it should be excited near the rotor aligned position and then turned off before the unaligned region (Figure 7.67).

As in motoring operation, current can be controlled by changing the turn-on and turn-off angles and current level while at low speed. Alternatively, at speeds higher than the base speed, only the turn-on and turn-off angles can be used for control.

The driving circuit for SRG is similar to that of SRMs; one of the common configurations is shown in Figure 7.68. When the switches are turned on, the phase gets energy from the supply and the capacitor. During the turn-off period, the freewheeling current from the motor charges up the capacitor and delivers energy to the load. Since there is no PM in this motor, during the start-up and initial condition, it needs an external source such as a battery to deliver energy to the phase; after taking transient time, the capacitor is then charged up to the output voltage. Depending on the output voltage during phase-on time, both the capacitor and the external source, or just the capacitor, provide the current to the load and the phase coil. The external source can be designed to be charged, or it can be disconnected from the system after the system reaches its operating point.

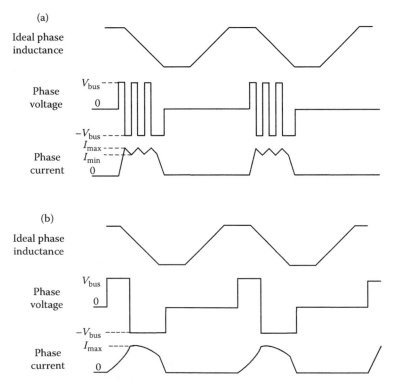

FIGURE 7.67
Low- and high-speed operation in generating mode: (a) low-speed operation and (b) high-speed operation.

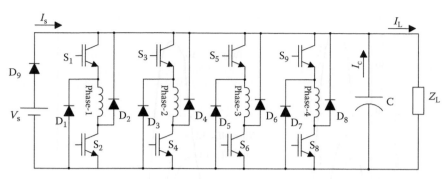

FIGURE 7.68
Driving circuit example for SRG.

In the generating region, the back EMF is negative, so it helps the phase to be charged very fast; then during turn-off, the back EMF opposes the negative supply voltage, and it decreases slowly:

$$V_C - e = L\frac{di}{dt} + Ri, \quad e > 0 \text{ (during phase-on period).} \tag{7.132}$$

$$-V_C - e = L\frac{di}{dt} + Ri, \quad e > 0 \text{ (during phase-off period).} \tag{7.133}$$

In Equations 7.132 and 7.133, V_c is the bus voltage of the inverter or, equivalently, the voltage of the bus capacitor, and e is the back EMF voltage.

In certain conditions such as high speed and high loads, the back EMF voltage is greater than the bus supply voltage, so the current increases even after turning off the phase. In addition to uncontrollable torque, this necessitates an oversized converter, thereby adding to the cost and overall size of the system. Due to variations in the speed of the prime mover, the power electronic converter should be designed for the worst possible case. This will magnify the additional cost and size issues. By properly selecting the turn-off angle, this maximum generating current can be coaxed into the safe region.[58] Figure 7.69 shows the effect of turn-off angle in the maximum generating current.

7.4.6 Sensorless Control

Excitation of the SRM phases needs to be properly synchronized with the rotor position for effective control of speed, torque, and torque pulsation. A shaft position sensor is usually used to provide the rotor position. However, these discrete position sensors not only add complexity and cost to the system, but they also tend to reduce the reliability of the drive system and restrict their application in some specific environments, such as military applications. Position sensorless technology can effectively continue the operation of the system, in case the position sensors lose their function. This is crucial in some applications, such as military vehicles.

Several sensorless control methods have been reported in the literature over the past two decades.[15–28] Most of these techniques are based on the fact that the magnetic status of the SRM is a function of the angular rotor position. As the rotor moves from the unaligned position toward the aligned position, the phase inductance increases from the minimum value to

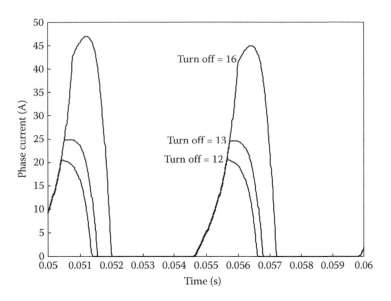

FIGURE 7.69
Effect of turn-off angle at maximum current level in generating mode in an 8/6 switched reluctance machine.

the maximum value. It is obvious that if the phase bulk inductance can be measured and the functional relation between the phase bulk inductance and the rotor position is known, the rotor position can be estimated according to the measured phase bulk inductance.[59]

Some sensorless techniques do not use the magnetic characteristic and voltage equation of the SRM directly to sense the rotor position. Instead, these sensorless control methods are based on the observer theory or synchronous operation method similar to that applied to conventional AC synchronous machines.

Generally, the existing sensorless control methods can be classified as follows:

1. Phase flux linkage-based method.
2. Phase inductance-based method.
3. Modulated signal injected methods.
4. Mutually induced voltage-based method.
5. Observer-based methods.

7.4.6.1 Phase Flux Linkage-Based Method[60]

This method uses the phase voltage and current data of the active phases to estimate the rotor position. The basic principle of this method is to use the functional relation between the phase flux linkage, the phase current, and the rotor position for rotor position detection. From Figure 7.57 it can be observed that if the flux linkage and the phase current are known, the rotor position can be estimated accordingly, as shown in Figure 7.70.

The problem with this sensorless control method is the inaccurate estimation of the phase flux linkage at low speed. At high speed (above the base speed), the phase voltage keeps its positive polarity until the phase is turned off. The V term dominates in $V–Ri$, and integration of $V–Ri$ in a relatively short period does not lead to a huge error in flux estimation. However, at low speed (below the base speed), the phase voltage changes its polarity from one

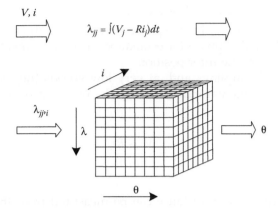

FIGURE 7.70
Flux linkage-based rotor position estimation method.

hysteresis cycle to the next hysteresis cycle. When $V–Ri$ is integrated in a relatively long period, the phase voltage term cancels itself due to the excursions, while the Ri term keeps its polarity during the integration period—and becomes significant after a long time of integration. The error in R or i may lead to a huge error in the flux estimation in this case. Therefore, this sensorless control method is only suitable for high-speed operation of SRM.

7.4.6.2 Phase Inductance-Based Method

Similar to the phase flux linkage, the phase bulk and incremental inductances are both functions of the phase current and the rotor position. Hence, they can also be used for rotor position estimation.

7.4.6.2.1 Sensorless Control Based on Phase Bulk Inductance

Using the phase flux linkage obtained as shown in Figure 7.70,[61] the phase inductance can be obtained as

$$L_{jj} = \frac{\lambda_{jj}}{i_j}.$$ (7.134)

The estimated phase bulk inductance and the measured phase current can be input to a prestored lookup table storing the functional relation between the phase bulk inductance, the phase current, and the rotor position to find the corresponding rotor position. Instead of using a lookup table, one can also use an analytical model to represent the functional relation between phase bulk inductance, phase current, and rotor position.[61]

Like the flux linkage-based method, since integration of $V–Ri$ is used for phase inductance estimation, this method is only suitable for high-speed operation. Some sensorless control methods that can work both at standstill and low speed, such as the open-loop method, must be used to start the SRM and bring the rotor speed to a certain level. After the rotor speed has reached a threshold, the phase flux linkage and/or the inductance are calculated using the integration method, and the rotor position is estimated according to the calculated phase flux linkage and inductance.

7.4.6.2.2 Sensorless Control Based on Phase Incremental Inductance

The position estimation method using the phase incremental inductance utilizes the current and voltage data of the active phase for estimation of the incremental inductance of this phase and, consequently, of the rotor position.

Neglecting the mutual couplings and, at very low speeds [neglectable motional-EMF term $i_j(\partial L_{jj}/\partial\theta)\omega$], the incremental inductance can be obtained from Equations 7.123 and 7.124 as

$$\ell_{jj} = \frac{V_j - Ri_j}{\mathrm{d}i_j/\mathrm{d}t}. \tag{7.135}$$

Thus, the phase incremental inductance can be measured from the phase voltage and current. If the relation between the phase incremental inductance and the rotor position is known, the rotor position can be estimated according to the estimated phase incremental inductance.

At low phase current—and therefore in an unsaturated phase—the phase incremental inductance can be viewed as equal to the phase bulk inductance, and it monotonically increases as the rotor moves from the unaligned position to the aligned position. The incremental inductance has a one-to-one relation with the rotor position in this case. However, at high phase current, and therefore in a saturated phase, when the rotor moves from the unaligned position to the aligned position, the phase incremental inductance may be the same value at two or more rotor positions.[62]

Even though the phase incremental inductance does not have a one-to-one relation with the rotor position at high phase current, it can still be used for rotor position estimation. Some SRMs are designed with a high degree of saturation such that the phase incremental inductance at the aligned position has its minimum value at high currents. In this case, the phase incremental inductance at the aligned position is unique; hence, it can be used to detect the aligned rotor position. This rotor position estimation technique gives one rotor position at one electrical cycle.

This method does not require any extra sensing circuitry. However, it is applicable only for very low speeds, less than 10% of the base speed, because the back EMF term is neglected for calculation of the phase incremental inductance.

7.4.6.3 Modulated Signal Injection Methods

These methods are to apply a voltage to the idle phase winding and measure the resultant phase current to detect the phase inductance. This derived phase inductance provides the rotor position information. Both an extra-low-amplitude voltage source and the power converter can be used to apply a voltage to the phase winding. When an extra voltage source is used, a sinusoidal voltage is usually used for sensing the phase inductance. The phase angle and the amplitude of the resultant phase current contain the phase inductance; hence, the rotor position information can be obtained. This is the idea behind the amplitude modulation (AM) and phase modulation (PM) methods. When the power converter is used for sensing purposes, a short period voltage pulse is usually applied to the idle phase, and a triangular current is induced in the corresponding phase. The changing rate of the phase current contains the phase inductance and, hence, the rotor position information. This is the basic idea of the diagnostic pulse-based method.

7.4.6.3.1 Frequency Modulation Method

This method is used to first generate a train of square wave voltage whose frequency is in reverse proportion to the instantaneous inductance of the idle phase.[49,63] The circuitry used for generating a square wave voltage train whose frequency is in reverse proportion to the inductance is referred to as an *L–F* converter.

To detect the frequency of this square wave voltage train, and hence the phase inductance, the timer of a microcontroller can be used to count the frequency of the square wave voltage train. Another approach is to use a frequency-to-voltage converter (*F–V* converter) to obtain a voltage proportional to the frequency of the square wave voltage train and sample this voltage using an A/D converter. To connect the phase winding, which is in the power circuit, to the sensing circuitry, which is in the control circuit, two photovoltaic BOSFET switches are used for each phase.

Since the signal used for position estimation in this method is an inductance-encoded frequency signal, this method is referred to as the frequency modulation (FM) method. This method is easy to implement and is robust. However, at high-speed operation, a phase energizing current exists—even when the phase inductance is decreasing as the rotor moves. This restricts the signal injection to the SRM phases. Another problem with this method is that it requires additional circuitry for implementation. The cost associated with this additional circuitry may be a concern in some applications. Furthermore, it is very sensitive to mutual coupling, since the current in the active phase induces voltage in the unenergized phases, which strongly distorts the probing pulses.

7.4.6.3.2 AM and PM Methods

The PM and AM techniques are based on the phase and amplitude variations, respectively, of the phase current due to the time-varying inductance when a sinusoidal voltage is applied to the phase winding in series with a resistance R.[64] The current flowing through the circuit in response to the applied voltage is a function of the circuit impedance. Since the coil inductance varies periodically, the phase angle between the current and the applied voltage also varies in a periodic manner. With a large inductance, the lagging angle of the current wave behind the voltage wave is large, and the peak current is small. The PM encoder technique measures the instantaneous phase angle on a continuous basis, while the AM encoder technique measures the amplitude. These instantaneous measurements contain the phase inductance information that can be obtained after passing the signals through a demodulator. The demodulator generates a signal that represents the phase inductance as a function of the rotor position. Using an inverse function or a conversion table, the rotor position can be estimated.

Since the PM and AM methods need the injection of a low-amplitude signal to one of the idle phases, photovoltaic BOSFET switches are needed to connect the phase winding to the sensing circuitry.

Like the FM method, signal injection to one of the idle phases is restricted at high-speed operation where the torque producing current occupies most of the electrical cycle and makes signal injection impossible. Another disadvantage of these methods is that they require additional hardware for indirect position sensing. As stated previously, it is very sensitive to mutual coupling.

7.4.6.3.3 Diagnostic Pulse-Based Method

Instead of using an additional voltage source to inject the sensing signal to the idle phase, the power converter of the SRM drive can be used to provide a short period voltage pulse to the idle phase, and low-amplitude current is produced.[65] Therefore, the back EMF, saturation effect, and voltage drop on the winding resistance can all be neglected. From Equations 7.123 and 7.124, the changing rate of the phase current is given as

$$\frac{\mathrm{d}i_j}{\mathrm{d}t} = \frac{V}{L_{jj}}. \tag{7.136}$$

Equation 7.136 indicates that the phase current changing rate contains the phase inductance and, hence, the rotor position information.

Similar to the case of switches being turned on, when the switches connected to the phase are turned off, the phase current freewheels through the diodes. The phase voltage equals the negative DC bus voltage; the change rate of the current has the same expression as Equation 7.136 but with a negative sign.

Either the current growing rate or the dropping rate can be used for sensing the phase inductance. When the current changing rate is found to exceed a threshold that is dictated by the phase inductance at the commutation position, the phase can be commutated. This method does not require additional hardware for indirect rotor position sensing. However, at high-speed operation, the phase excitation current occupies the majority of one electrical cycle and restricts the injection of the testing signal, and, like the FM, AM, and PM methods, it is very sensitive to mutual coupling.

7.4.6.4 Mutually Induced Voltage-Based Method

The idea of this method is based on measuring the mutually induced voltage in an idle phase, which is either adjacent or opposite to the energized phase of an SRM.[66] The mutual voltage in the "off" phase, induced due to the current in the active phase, varies significantly with respect to the rotor position. This mutually induced voltage variation can be sensed by a simple electronic circuit. If the functional relation between the mutually induced voltage in the inactive phase due to the current in the active phase and the rotor position is known, the rotor position information can be extracted from the mutually measured induced voltage in the inactive phase. This method is only suitable for low-speed operation. Furthermore, it is very sensitive to noise since the ratio between induced voltage and system noise is small.

7.4.6.5 Observer-Based Methods

In this method, state-space equations are used to describe the dynamic behavior of the SRM drive.[67] An observer is then developed based on these nonlinear state-space differential equations for estimation of the rotor position. The input and output of this observer are the phase voltage and phase current, respectively. The state variables of this observer are the stator flux linkage, rotor position angle, and rotor speed. The phase current, flux linkage, rotor position, and rotor speed can be estimated using this observer. The phase current estimated by this observer is compared to the actual phase current of the SRM, and the resultant current errors are used to adjust the parameters of the observer. When the current estimated by the observer matches the actual current, the observer is considered a correct representation of the dynamic behavior of the actual SRM drive, and the rotor position estimated by the observer is used to represent the actual rotor position.

The main disadvantages of these methods are real-time implementation of complex algorithms, which require a high-speed DSP and a significant amount of stored data. This increases the cost and the speed limitations by the DSP. However, high resolution in detecting rotor position and applicability to the whole speed range are some merits of these methods.

7.4.7 Self-Tuning Techniques of SRM Drives

As discussed in previous sections, the SRM drive has a simple and rugged construction—favorable characteristics for traction application—but its control is very complicated due to the nonlinearity of its magnetic circuit and the fact that control depends heavily on the mechanical and electrical parameters, such as air gap, resistance, and so on.[53] In mass production and real-world operation, it is important for these parameters to have exact values and remain unchanged. For example, the air gap would be changed due to mechanical vibration wearing, and the resistance in windings and inductance would vary with temperature. These parameter variations would cause significant degradation of the drive performance if the control system cannot "know" these variations and implement corresponding corrections in the control process. Self-tuning techniques are referred to as methods of updating the control strategy in a control system.

The major purpose of self-tuning control for the SRM drive is to update the control variables in the presence of motor parameters' variations to optimize the torque per ampere.[68] There are two approaches to this problem: the arithmetic mean method and the neural network-based method.

7.4.7.1 Self-Tuning with Arithmetic Method

To optimize the SRM drive performance, it is necessary to maximize torque per ampere through real-time optimization. The SRM drive control variables are the phase current, turn-on angle, and turn-off angle. In a low-speed region, hysteresis-type current control is used to keep the commanded current constant. The chopping current band must be optimally chosen, as there is a trade-off between the width of the band and the chopping frequency. Assuming the selected band is optimal, maximum torque per ampere can be obtained by aptly tuning the turn-on angle (θ_{on}) and the turn-off angle (θ_{off}) of the phase current excitation. Computer simulations, based on a simple mathematical model, have been performed to prove the existence of a unique (θ_{on}, θ_{off}) optimal pair, which gives the maximum torque per ampere for a given current and speed.[31] It has been shown that the optimal values of θ_{on} and θ_{off} are bounded within the following limits:

$$\theta_{on}^{min} < \theta_{on} < 0°, \tag{7.137}$$

$$\theta_{off}^{max} < \theta_{off} < 180°, \tag{7.138}$$

where θ_{on}^{min} is the turn-on angle such that the current reaches the desired value at $0°$, and θ_{off}^{max} is the turn-off angle such that the current reduces to zero at $180°$.

The intuitive selections for control angles are such as to turn on each phase exactly at its unaligned position and turn off the phase just before its aligned position. The optimal θ_{on} is not very susceptible to the change in inductance due to the parameter variations because of the large air gap at the unaligned position.[46] Hence, optimal θ_{on} calculated off-line based on the linear model is sufficient to give the optimal torque per ampere. Therefore, the optimization problem reduces to calculation of θ_{off} on-line that gives maximum torque per ampere.

7.4.7.1.1 Optimization with Balanced Inductance Profiles

To minimize the phase current for a given torque and speed, a heuristic search algorithm for finding the optimum turn-off angle[69] can be used in which both the reference current and the turn-off angle are varied, while the PI controller maintains the speed.

The optimization algorithm is explained as follows. Initially, the default turn-off angle is used to reach the commanded speed. Then, the turn-off angle is reduced in steps. With the turn-off angle variation, the torque either decreases or increases, and the PI controller adjusts the phase reference current accordingly to a new value so that the speed remains at its set value. If the current reduces with a change in the turn-off angle, then the direction of search is correct and is continued until the current starts increasing with a further change in the turn-off angle. The step size for the turn-off angle can be a function of the operating point itself.

Once the optimization is completed for a given operating point, the optimum values of control variables—that is, the reference current and the control angles—can be stored in lookup tables so that the controller can directly pick up these values if the same operating point is to be reached in the future. This saves some time and effort.

7.4.7.1.2 Optimization in the Presence of Parameter Variations

Initially, when optimization is performed, the reference currents for all the phases are kept the same, assuming that the phase inductances are balanced. If there are parameter variations, then different phases will have different optimal reference currents and turn-off angles. To address this problem, once the general optimization is complete, the control variables for the individual phases are tuned separately, that is, the reference current and turn-off angle for only one of the phases are varied at a time, while these parameters for the other phases are kept fixed. Finally, when all the phases are tuned, the optimum reference current and turn-off angle for different phases are different if there is any parameter variation. The main advantage of using this method is that the optimization algorithm does not require any information about the degree of imbalance present in the inductance profile, which may change considerably over a period of time.

Figure 7.71 shows the current waveform of the SRM phase with default values of turn-on and turn-off angles, and Figure 7.72 shows the current waveform after applying the self-tuning algorithm, which is running almost at the same operating point. By comparing the two figures, there is a considerable reduction in both the amplitude and the width of the phase current and, hence, its rms value. The operating speeds are also the same before and after optimization for both cases.

7.4.7.2 Self-Tuning Using an ANN

ANNs with highly nonlinear and adaptive structures have been used in many applications. ANNs have an inherent interpolation property, so they are an ideal candidate for storing turn-on and turn-off angles instead of storing them in look-up tables. Figure 7.73 shows a three-layer feedforward neural network with two inputs—current and speed—and one output—the optimum turn-off angle.[70]

The proposed self-tuning control technique incorporates a heuristic search method along with an adaptive-type ANN-based method. The weights of the ANN are initially set to default values. The control technique incorporates a periodic heuristic search of optimal θ_{off} to verify the accuracy of the θ_{off} obtained from the ANN. If there is a variation in the inductance profile due to parameter drift, the optimal θ_{off} obtained from the ANN is no longer valid. This prompts the controller to activate the heuristic search by modifying the θ_{off} in

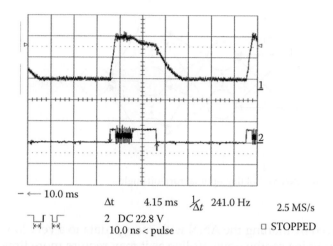

FIGURE 7.71
Phase current and gating pulse without optimization; terminal voltage: 50 V, load: 120 W, reference current: 5.5 A, speed: 1200 rpm, conduction angle: 30°.

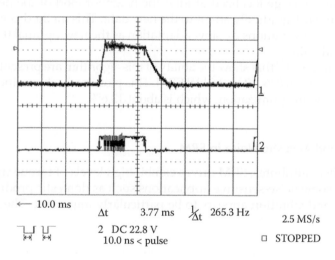

FIGURE 7.72
Phase current and gating pulse with optimization; terminal voltage: 50 V, load: 120 W, reference current: 4.7 A, speed: 1200 rpm, conduction angle: 27.25°.

small steps until the current reaches its minimum value. This new optimal θ_{off} at that particular operating point is now used to adapt the weights of the ANN. Hence, this novel ANN-based control technique coupled with the heuristic search learns and adapts to any parameter drift to give the optimal θ_{off}.

ANNs have been successfully used for many applications in control systems, but the ANN learning algorithm shows great performance when used off-line. This means that they must be fully trained before being applied. Neural networks with incremental learning capability and stable adaptation of network parameters are essential for on-line adaptive control. The adaptive learning assumes that the ANN to start with is well trained in such a way that it can perform input/output mapping for the initial training set with a high degree of accuracy.

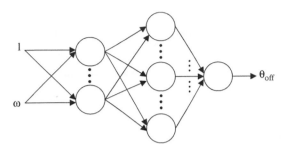

FIGURE 7.73
A three-layer feedforward ANN for holding optimal turn-off angles.

This can be achieved by training the ANN with enough data to a very low error rate. In this application, this training can be done off-line as it may require more time.

Now when new training data are obtained, the already trained ANN is used to generate additional examples. These additional examples with the newly obtained training data are then used to retrain the current ANN. This ensures that the original ANN mapping is retained with only a change localized around the neighborhood of the new training data. This makes the network gradually adapt to the new data. This method ensures the stability of the network weight variations by slow adaptation as the new optimal θ_{off} is in the neighborhood of the old value.

Some simulation results[70] that show the ability of this algorithm are presented in Figure 7.74. These results belong to an 8/6, 12 V, 0.6 kW SRM. This plot clearly shows the improvement in torque per ampere with optimization, which is about 13.6%.

7.4.8 Vibration and Acoustic Noise in SRM

Despite the excellent attributes, SRM drives exhibit high levels of torque ripple and audible noise. Indeed, in some noise-sensitive applications such as domestic products, the problem of acoustic noise and vibration appears to be particularly important. The acoustic noise in

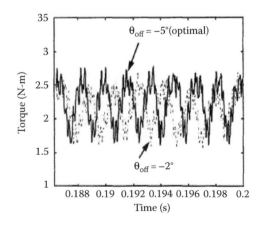

FIGURE 7.74
Developed torque before and after optimization.

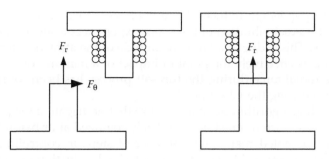

FIGURE 7.75
Static profile of radial component of force.

the SRM is mainly due to the varying magnetic forces between the stator and the rotor poles, as shown in Figures 7.75 and 7.76.[71,72] The tangential and radial components of the electromagnetic force density in the air gap are given by

$$F_\theta = v_0 \int B_\theta B_r \, d\theta, \tag{7.139}$$

$$F_r = v_0 \int (B_r^2 - B_\theta^2) \, d\theta, \tag{7.140}$$

where v_0, B_θ, B_r, and θ stand for reluctivity of the air, tangential, and radial components of the flux density, and rotor position, respectively.

The varying magnetic forces, especially the radial force, cause the deformation of the stator and, therefore, radial vibrations of the stator and acoustic noise.

The results of a structural study of SRMs show that back iron is the most significant parameter in the dynamic behavior of the stator deformations.[60,73] Increasing the back-iron length results in larger natural frequencies and smaller deformation, which consequently reduce the chance of a mechanical resonance even at high speeds.

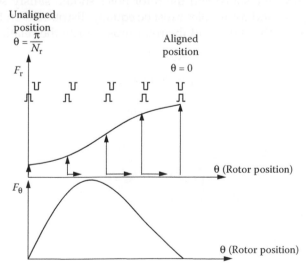

FIGURE 7.76
Distribution of radial (F_r) and tangential (F_θ) forces.

Increasing the air gap length reduces the radial forces.[74] However, it vitiates the performance of the SRM. Radial vibration of the stator experiences a severe acceleration during the turn-off process. This is due to a large magnitude of the attraction forces and their fast rate of change. This is effectively the point of impact of a hammer on the stator structure. Smoothing of the radial force during the turn-off process has been found to be the most direct method for reducing the vibration.[75]

The current profiling algorithm must make sure that no negative torque is generated. In other words, the phase current must be completely removed at or before the aligned position. Also, it must be noted that a high number of steps in controlling the tail current increases the switching losses. Moreover, unconstrained reduction of vibration using this method vitiates the performance of the machine. Therefore, study of other objectives such as efficiency and torque ripple under the proposed control method is an essential step.[71]

Fahimi and Ehsani[76] showed that, for practical implementation, two levels of current provide a smooth variation in radial force. Therefore, the turn-off instant, the position at which the second current limit is assigned and the position at which final hard chopping of the phase current occurs, is considered a controlled variable. These parameters are computed at various operating points using the analytical model of the SRM drive.[77]

7.4.9 SRM Design

SRM has a simple construction. However, this does not mean its design is simple. Due to the double-salient structure, continuously varying inductance and high saturation of pole tips, and the fringing effect of poles and slots, the design of SRMs suffers great difficulty in using the magnetic circuit approach. In most cases, electromagnetic finite-element analysis is used to determine the motor parameters and performances. Typical electromagnetic field distributions of an 8/6 SRM are shown in Figure 7.77. Nevertheless, there are some basic criteria to initialize the design process of SRM for EVs and HEVs.[78,79]

7.4.9.1 Number of Stator and Rotor Poles

For continuous rotation, the stator and the rotor poles should satisfy some special conditions, that is, stator poles and rotor poles must be equally distributed on the circumferences, and the pole numbers of the stator and the rotor must satisfy the relationship as follows:

$$N_s = 2mq, \tag{7.141}$$

$$N_r = 2(mq \mp 1), \tag{7.142}$$

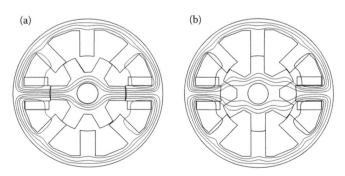

(a)   (b)

FIGURE 7.77
Typical electromagnetic field distribution of an 8/6 SRM: (a) aligned position and (b) unaligned position.

TABLE 7.1

Common Combination of q, N_s, and N_r

				N_r		
		$m = 1$			$m = 2$	
q	N_s	"−"	"+"	N_s	"−"	"+"
3	6	4	8	12	10	14
4	8	6	10	16	14	18
5	10	8	12	20	18	22

where N_s and N_r are the pole numbers of the stator and rotor, respectively; q is the phase number of the machine; and m is equal to 1 or 2. To reduce the switch frequency and the minimum inductance, the rotor pole number is less than the stator pole number, that is, a minus sign is used in Equation 7.142. The most common combination of q, m, N_s, and N_r is listed in Table 7.1.

Four-phase 8/6 and three-phase 6/4 configurations are the most commonly used SRM structures. A three-phase 6/4 configuration has the advantage of having more room for phase advancing in high-speed applications. In addition, compared to an 8/6 configuration, this structure minimizes the effects of mutual coupling between adjacent phases. However, it results in more torque pulsation due to its torque–angle characteristics that contain large dead zones. Furthermore, starting torque can be a problem associated with this configuration. On the other hand, an 8/6 structure can be used to reduce the torque ripple and improve the starting torque.[11] However, by selecting an 8/6 machine, the cost of the silicon increases. By increasing the number of poles per phase (12/8 and 16/12 configurations), one can minimize the demerits of a 6/4 machine while maintaining the same cost of silicon. In this design study an 8/6 configuration has been selected.

7.4.9.2 Stator Outer Diameter

The stator outer diameter is mainly designed based on the available space given in desired specifications. In fact, the main compromise must be made between the length of the machine and its outer diameter. A pancake (the length of the machine is less than the stator outer radius) type of design is subject to three-dimensional effects of coil endings,[80] whereas a very long structure faces cooling and rotor bending issues, which are of particular importance for large machines.

7.4.9.3 Rotor Outer Diameter

The relationship between the torque developed by the SRM and machine parameters can be represented by the following equation:

$$T \propto D_r^2 l (N_i)^2, \tag{7.143}$$

where D_r, N_i, and l are the outer diameter of the rotor, equivalent ampere-turn of one phase, and length of the machine. Once the outer diameter of the SRM is fixed, any increase in rotor outer diameter results in a reduction of Ni, thereby reducing the torque developed by SRM. Because of this and because SRM is highly saturated, rotor boring should be equal to or slightly larger than the stator outer radius. It must be noted that the rotor geometry enhances the moment of inertia and vibrational modes of the machine.

7.4.9.4 Air Gap

Air gaps have an important impact on the generated torque and dynamic behavior of SRMs. In fact, by reducing the air gap, inductance at the aligned position increases, resulting in higher torque density. On the other hand, a very small air gap causes severe saturation in stator and rotor pole areas.[74] In addition, the mechanical manufacture of a very small air gap might not be feasible. The following empirical formula can be used as a reference for selecting air gaps in large machines[81]:

$$\delta(\text{mm}) = 1 + \frac{D_s}{1000},\tag{7.144}$$

where δ and D_s are the air gap and the stator outer boring in (mm) and (m), respectively. By investigating the value of the flux density (B) in the stator and rotor poles, the level of saturation and, consequently, air gap can be finalized.

7.4.9.5 Stator Arc

Since the developed torque depends on the area available for the coils, it is important to design the stator arc in such a way that maximum space for inserting the coils is provided. A very narrow stator arc results in the tangential vibration of the stator pole. Moreover, it reduces the effective region in the torque–angle characteristics, which increases the torque ripple and reduces the average torque. An optimal value for the stator arc can be chosen using the following inequality[77]:

$$0.3\frac{\pi D_R}{N_S} \leq \lambda_S \leq 0.35\frac{\pi D_R}{N_S},\tag{7.145}$$

where D_R, N_S, and λ_S are, respectively, the rotor diameter, number of stator poles, and stator arc.

7.4.9.6 Stator Back Iron

For designing the back iron, the following constraints must be considered:

1. Radial vibration of the stator body must be minimized.
2. There should be enough space for cooling of the stator.
3. The back iron should be capable of carrying half of the flux existing in the stator poles without getting saturated.
4. The area available for inserting the coils should not be reduced.

7.4.9.7 Performance Prediction

Clearly, most performance requirements are related to the dynamic performance of the drive and, hence, call for an overall modeling of the drive system, including control and power electronics considerations. However, to predict the dynamic performance of the drive, the static characteristics of the machine (phase inductance and torque–angle profiles) should be available.

The improved magnetic equivalent circuit (IMEC)[82] approach is a shortcut method that gives an approximation of the steady-state parameters of the SRM. Indeed, by replacing all magnetomotive sources (ampere-turn) with voltage sources and various parts of the

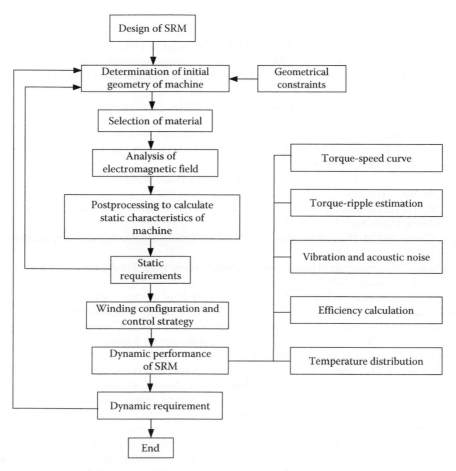

FIGURE 7.78
Basic design strategy.

magnetic structure with their equivalent reluctances, one can perform a magnetic analysis. Furthermore, by dividing the stator and rotor poles into several smaller portions, the accuracy of this method can be arbitrarily improved. It must be noted that finite-element analysis of SRMs is a time-consuming procedure. Therefore, the IMEC method is more appropriate for developing first design examples.

Figure 7.78 depicts a general design strategy for SRM drives.

Bibliography

1. C. C. Chan and K. T. Chau, *Modern Electric Vehicle Technology*, Oxford University Press, Oxford, 2001.
2. D. W. Novotny and T. A. Lipo, *Vector Control and Dynamics of AC Drives*, Oxford Science Publications, Oxford, 1996. ISBN: 0-19-856439-2.
3. D. C. Hanselman, *Brushless Permanent-Magnet Motor Design*, McGraw-Hill, New York, 1994.

4. F. Huang and D. Tien, A neural network approach to position sensorless control of brushless DC motors, In *Proceedings of the IEEE 22nd International Conference on Industrial Electronics, Control, and Instrumentation*, Vol. 2, pp. 1167–1170, August 1996.

5. S. Vukosavic, L. Peric, E. Levi, and V. Vuckovic, Sensorless operation of the SR motor with constant dwell, In *Proceedings of the 1990 IEEE Power Electronics Specialists Conference*, pp. 451–454, 1990.

6. G. K. Dubey, *Power Semiconductor Controlled Drives*, Prentice-Hall, Englewood Cliffs, NJ, 1989.

7. S. R. MacMinn and J. W. Sember, Control of a switched-reluctance aircraft starter-generator over a very wide speed range, In *Proceedings of the Intersociety Energy Conversion Engineering Conference*, pp. 631–638, 1989.

8. M. Ehsani, *Method and Apparatus for Sensing the Rotor Position of a Switched Reluctance Motor*, U.S. Patent No. 5,410,235, April 1995.

9. A. M. Trzynadlowski, *The Field Orientation Principle in Control of Induction Motor*, Kluwer Academic Publishers, Dordrecht, 1994.

10. J. F. Gieras and M. Wing, *Permanent Magnet Motor Technology*, Design and Applications, Marcel Dekker, New York, 1997.

11. I. Husain, Minimization of torque ripple in SRM drives, *IEEE Transactions on Industrial Electronics* 49(1), February 2002: 28–39.

12. K. M. Rahman and M. Ehsani, Performance analysis of electric motor drives for electric and hybrid electric vehicle application, *IEEE Power Electronics in Transportation*, 1996: 49–56.

13. T. Senjyu and K. Uezato, Adjustable speed control of brushless DC motors without position and speed sensors, In *Proceedings of the IEEE/IAS Conference on Industrial Automation and Control: Emerging Technologies*, pp. 160–164, 1995.

14. A. Consoli, S. Musumeci, A. Raciti, and A. Testa, Sensorless vector and speed control of brushless motor drives, *IEEE Transactions on Industrial Electronics*, 41, February 1994: 91–96.

15. P. Acarnley, Sensorless position detection in permanent magnet drives, In *IEE Colloquium on Permanent Magnet Machines and Drives*, pp. 10/1–10/4, 1993.

16. T. Liu and C. Cheng, Adaptive control for a sensorless permanent-magnet synchronous motor drive, *IEEE Transactions on Aerospace and Electronic Systems*, 30, July 1994: 900–909.

17. R. Wu and G. R. Slemon, A permanent magnet motor drive without a shaft sensor, *IEEE Transactions on Industry Applications*, 27, September/October 1991: 1005–1011.

18. T. Liu and C. Cheng, Controller design for a sensorless permanent magnet synchronous drive system, *IEE Proceedings—B*, 140, November 1993: 369–378.

19. N. Ertugrul, P. P. Acarnley, and C. D. French, Real-time estimation of rotor position in PM motors during transient operation, In *IEE Fifth European Conference on Power Electronics and Applications*, pp. 311–316, 1993.

20. N. Ertugrul and P. Acarnley, A new algorithm for sensorless operation of permanent magnet motors, *IEEE Transactions on Industry Applications*, 30, January/February 1994: 126–133.

21. T. Takeshita and N. Matsui, Sensorless brushless DC motor drive with EMF constant identifier, In *IEEE International Conference on Industrial Electronics, Control, and Instrumentation*, Vol. 1, pp. 14–19, 1994.

22. N. Matsui and M. Shigyo, Brushless DC motor control without position and speed sensors, *IEEE Transactions on Industry Applications*, 28, January/February 1992: 120–127.

23. N. Matsui, Sensorless operation of brushless DC motor drives, In *Proceedings of the IEEE International Conference on Industrial Electronics, Control, and Instrumentation*, Vol. 2, pp. 739–744, November 1993.

24. N. Matsui, Sensorless PM brushless DC motor drives, *IEEE Transactions on Industrial Electronics*, 43, April 1996: 300–308.

25. M. Schrodl, Digital implementation of a sensorless control algorithm for permanent magnet synchronous motors, In *Proceedings of the International Conference "SM 100"*, ETH Zurich, Switzerland, pp. 430–435, 1991.

26. M. Schrodl, Operation of the permanent magnet synchronous machine without a mechanical sensor, In *IEE Proceedings on the International Conference on Power Electronics and Variable Speed Drives*, pp. 51–56, July 1990.

27. M. Schrodl, Sensorless control of permanent magnet synchronous motors, *Electric Machines and Power Systems*, 22, 1994: 173–185.

28. B. J. Brunsbach, G. Henneberger, and T. Klepsch, Position controlled permanent magnet excited synchronous motor without mechanical sensors, In *IEE Conference on Power Electronics and Applications*, Vol. 6, pp. 38–43, 1993.

29. R. Dhaouadi, N. Mohan, and L. Norum, Design and implementation of an extended Kalman filter for the state estimation of a permanent magnet synchronous motor, *IEEE Transactions on Power Electronics*, 6, July 1991: 491–497.

30. A. Bado, S. Bolognani, and M. Zigliotto, Effective estimation of speed and rotor position of a PM synchronous motor drive by a Kalman filtering technique, In *Proceedings of the 23rd IEEE Power Electronics Specialist Conference*, Vol. 2, pp. 951–957, 1992.

31. K. R. Shouse and D. G. Taylor, Sensorless velocity control of permanent-magnet synchronous motors, In *Proceedings of the 33rd Conference on Decision and Control*, pp. 1844–1849, December 1994.

32. J. Hu, D. M. Dawson, and K. Anderson, Position control of a brushless DC motor without velocity measurements, *IEE Proceedings on Electronic Power Applications*, 142, March 1995: 113–119.

33. J. Solsona, M. I. Valla, and C. Muravchik, A nonlinear reduced order observer for permanent magnet synchronous motors, *IEEE Transactions on Industrial Electronics*, 43, August 1996: 38–43.

34. R. B. Sepe and J. H. Lang, Real-time observer-based (adaptive) control of a permanent-magnet synchronous motor without mechanical sensors, *IEEE Transactions on Industry Applications*, 28, November/December 1992: 1345–1352.

35. L. Sicot, S. Siala, K. Debusschere, and C. Bergmann, Brushless DC motor control without mechanical sensors, In *Proceedings of the IEEE Power Electronics Specialist Conference*, pp. 375–381, 1996.

36. T. Senjyu, M. Tomita, S. Doki, and S. Okuma, Sensorless vector control of brushless DC motors using disturbance observer, In *Proceedings of the 26th IEEE Power Electronics Specialists Conference*, Vol. 2, pp. 772–777, 1995.

37. K. Iizuka, H. Uzuhashi, and M. Kano, Microcomputer control for sensorless brushless motor, *IEEE Transactions on Industry Applications* IA-27, May–June 1985: 595–601.

38. J. Moreira, Indirect sensing for rotor flux position of permanent magnet AC motors operating in a wide speed range, *IEEE Transactions on Industry Applications Society*, 32, November/December 1996: 401–407.

39. S. Ogasawara and H. Akagi, An approach to position sensorless drive for brushless DC motors, *IEEE Transactions on Industry Applications* 27, September/October 1991: 928–933.

40. T. M. Jahns, R. C. Becerra, and M. Ehsani, Integrated current regulation for a brushless ECM drive, *IEEE Transactions on Power Electronics* 6, January 1991: 118–126.

41. R. C. Becerra, T. M. Jahns, and M. Ehsani, Four-quadrant sensorless brushless ECM drive, In *Proceedings of the IEEE Applied Power Electronics Conference and Exposition*, pp. 202–209, March 1991.

42. D. Regnier, C. Oudet, and D. Prudham, Starting brushless DC motors utilizing velocity sensors, In *Proceedings of the 14th Annual Symposium on Incremental Motion Control Systems and Devices*, Champaign, IL, Incremental Motion Control Systems Society, pp. 99–107, June 1985.

43. D. Peters and J. Harth, I.C.s provide control for sensorless DC motors, *EDN* April 1993: 85–94.

44. M. Hamdi and M. Ghribi, A sensorless control scheme based on fuzzy logic for AC servo drives using a permanent-magnet synchronous motor, In *IEEE Canadian Conference on Electrical and Computing Engineering*, pp. 306–309, 1995.

45. D. E. Hesmondhalgh, D. Tipping, and M. Amrani, Performance and design of an electromagnetic sensor for brushless DC motors, *IEE Proceedings*, Vol. 137, pp. 174–183, May 1990.

46. T. J. E. Miller, *Switched Reluctance Motors and Their Control*, Oxford Science Publications, London, 1993.
47. P. J. Lawrenson, J. M. Stephenson, P. T. Blenkinsop, J. Corda, and N. N. Fulton, Variable-speed switched reluctance motors, *Proceedings of IEE*, 127(Part B, 4), July 1980: 253–265.
48. E. Richter, J. P. Lyons, C. A. Ferreira, A. V. Radun, and E. Ruckstadter, Initial testing of a 250-kW starter/generator for aircraft applications, In *Proceedings of the SAE Aerospace Atlantic Conference Expo.*, Dayton, OH, April 18–22, 1994.
49. M. Ehsani, *Phase and Amplitude Modulation Techniques for Rotor Position Sensing in Switched Reluctance Motors*, U.S. Patent No. 5,291,115, March 1994.
50. D. A. Torrey, Variable-reluctance generators in wind-energy systems, In *Proceedings of the IEEE PESC '93*, pp. 561–567, 1993.
51. J. M. Kokernak, D. A. Torrey, and M. Kaplan, A switched reluctance starter/alternator for hybrid electric vehicles, In *Proceedings of the PCIM '99*, pp. 74–80, 1999.
52. J. T. Bass, M. Ehsani, and T. J. E. Miller, Simplified electronics for torque control of sensorless switched reluctance motor, *IEEE Transactions on Industrial Electronics*, 34(2), 1987.
53. M. Ehsani, *Self-Tuning Control of Switched Reluctance Motor Drives System*, U.S. Patent Pending, File Number 017575.0293.
54. M. Ehsani, *Switched Reluctance Motor Drive System*, U.S. Patent Pending, Serial Number 60/061,087, Filing Date: January 1997.
55. R. Krishnan, *Switched Reluctance Motors Drives: Modeling, Simulation Analysis, Design and Applications*, CRC Press, Boca Raton, FL, 2001.
56. N. Mohan, T. M. Undeland, and W. P. Robbins, *Power Electronics—Converters, Applications, and Design*, John Wiley & Sons, New York, 1995, ISBN: 0-471-58408-8.
57. A. Radun, Generating with the switched-reluctance motor, In *Proceedings of the IEEE APEC '94*, pp. 41–47, 1994.
58. B. Fahimi, A switched reluctance machine based starter/generator for more electric cars, In *Proceedings of the IEEE Electric Machines and Drives Conference*, pp. 73–78, 2000.
59. H. Gao, F. R. Salmasi, and M. Ehsani, Sensorless control of SRM at standstill, In *Proceedings of the 2000 IEEE Applied Power Electronics Conference*, Vol. 2, pp. 850–856, 2000.
60. J. P. Lyons, S. R. MacMinn, and M. A. Preston, Discrete position estimator for a switched reluctance machine using a flux-current map comparator, U.S. Patent 5140243, 1991.
61. G. Suresh, B. Fahimi, K. M. Rahman, and M. Ehsani, Inductance based position encoding for sensorless SRM drives, In *Proceedings of the 1999 IEEE Power Electronics Specialists Conference*, Vol. 2, pp. 832–837, 1999.
62. H. Gao, *Sensorless Control of the Switched Reluctance Motor at Standstill and Near-Zero Speed*, PhD dissertation, Texas A&M University, December 2001.
63. M. Ehsani, *Position Sensor Elimination Technique for the Switched Reluctance Motor Drive*, U.S. Patent 5072166, 1990.
64. M. Ehsani, I. Husain, S. Mahajan, and K. R. Ramani, New modulation encoding techniques for indirect rotor position sensing in switched reluctance motors, *IEEE Transactions on Industry Applications*, 30(1), 1994: 85–91.
65. G. R. Dunlop and J. D. Marvelly, Evaluation of a self-commuted switched reluctance motor, In *Proceedings of the 1987 Electric Energy Conference*, pp. 317–320, 1987.
66. M. Ehsani and I. Husain, Rotor position sensing in switched reluctance motor drives by measuring mutually induced voltages, In *Proceedings of the 1992 IEEE Industry Application Society Annual Meeting*, Vol. 1, pp. 422–429, 1992.
67. A. Lumsdaine and J. H. Lang, State observer for variable reluctance motors, *IEEE Transactions on Industrial Electronics*, 37(2), 1990: 133–142.
68. K. Russa, I. Husain, and M. E. Elbuluk, A self-tuning controller for switched reluctance motors, *IEEE Transactions on Power Electronics*, 15(3), May 2000: 545–552.
69. P. Tandon, A. Rajarathnam, and M. Ehsani, Self-tuning of switched reluctance motor drives with shaft position sensor, *IEEE Transactions on Industry Applications*, 33(4), July/August 1997: 1002–1010.

70. A. Rajarathnam, B. Fahimi, and M. Ehsani, Neural network based self-tuning control of a switched reluctance motor drive to maximize torque per ampere, In *Proceedings of the IEEE Industry Applications Society Annual Meeting*, Vol. 1, pp. 548–555, 1997.
71. B. Fahimi, *Control of Vibration in Switched Reluctance Motor Drive*, PhD dissertation, Texas A&M University, May 1999.
72. D. E. Cameron, J. H. Lang, and S. D. Umans, The origin and reduction of acoustic noise and doubly salient variable reluctance motor, *IEEE Transactions on Industry Applications*, IA-28(6), November/December 1992: 1250–1255.
73. H. Gao, B. Fahimi, F. R. Salmasi, and M. Ehsani, Sensorless control of the switched reluctance motor drive based on the stiff system control concept and signature detection, In *Proceedings of the 2001 IEEE Industry Applications Society Annual Meeting*, pp. 490–495, 2001.
74. B. Fahimi and M. Ehsani, Spatial distribution of acoustic noise caused by radial vibration in switched reluctance motors: Application to design and control, In *Proceedings of the 2000 IEEE Industry Application Society Annual Meeting*, Rome, Italy, October 2000.
75. B. Fahimi, G. Suresh, K. M. Rahman, and M. Ehsani, Mitigation of acoustic noise and vibration in switched reluctance motor drive using neural network based current profiling, In *Proceedings of the 1998 IEEE Industry Application Society Annual Meeting*, Vol. 1, pp. 715–722, 1998.
76. B. Fahimi and M. Ehsani, *Method and Apparatus for Reducing Noise and Vibration in Switched Reluctance Motor Drives*, U.S. Patent pending.
77. G. S. Buja and M. I. Valla, Control characteristics of the SRM drives—part I: Operation in the linear region, *IEEE Transactions on Industrial Electronics*, 38(5), October 1991: 313–321.
78. B. Fahimi, G. Suresh, and M. Ehsani, Design considerations of switched reluctance motors: Vibration and control issues, In *Proceedings of the 1999 IEEE Industry Application Society Annual Meeting*, Phoenix, AZ, October 1999.
79. J. Faiz and J. W. Finch, Aspects of design optimization for switched reluctance motors, *IEEE Transactions on Energy Conversion*, 8(4), December 1993: 704–712.
80. A. M. Michaelides and C. Pollock, Effect of end core flux on the performance of the switched reluctance motor, *IEE on Electronic Power Applications* 141(6), November 1994: 308–316.
81. G. Henneberger, *Elektrische Maschinen I, II, III, RWTH Aachen*, Manuscripts at Institut fuer Elektrische Maschinen, 1989.
82. B. Fahimi, G. Henneberger, and M. Moallem, Prediction of transient behavior of SRM drive using improved equivalent magnetic circuit method, In *PCIM Conference Records*, pp. 285–291, 1995.
83. Y. Yang et al., Design and comparison of interior permanent magnet motor topologies for traction applications, *IEEE Transactions on Transportation Electrification* 3(1), 2017: 86–97.
84. G. Pellegrino et al., Performance comparison between surface-mounted and interior PM motor drives for electric vehicle application, *IEEE Transactions on Industrial Electronics* 59(2), 2012: 803–811.
85. Z. Yang et al., Comparative study of interior permanent magnet, induction, and switched reluctance motor drives for EV and HEV applications, *IEEE Transactions on Transportation Electrification* 1(3), 2015: 245–254.
86. B. Bilgin, A. Emadi, and M. Krishnamurthy, Comprehensive evaluation of the dynamic performance of a 6/10 SRM for traction application in PHEVs, *IEEE Transactions on Industrial Electronics* 60(7), 2013: 2564–2575.
87. H. Nasiri, A. Radan, A. Ghayebloo, and K. Ahi, Dynamic modeling and simulation of transmotor based series-parallel HEV applied to Toyota Prius 2004, In *2011 10th International Conference on Environment and Electrical Engineering (EEEIC), IEEE*, pp. 1–4, May 2011.
88. A.Y. Yeksan, N. Ershad, and M. Ehsani, Dual-shaft electrical machine for vehicle applications, *IEEE Transactions on Energy Conversion* 2017 (under review).
89. A. Ghayebloo and A. Radan, Superiority of dual-mechanical-port-machine-based structure for series–parallel hybrid electric vehicle applications, *IEEE Transactions on Vehicular Technology* 65(2), 2016: 589–602.

90. L. Xu, A new breed of electric machines-basic analysis and applications of dual mechanical port electric machines, In *Electrical Machines and Systems, 2005. ICEMS 2005. Proceedings of the Eighth International Conference on*, Vol. 1, pp. 24–31, 2005. IEEE.

91. L. Xu, Y. Zhang, and X. Wen, Multioperational modes and control strategies of dual-mechanical-port machine for hybrid electrical vehicles, *IEEE Transactions on Industry applications* 45(2), 2009: 747–755.

92. K. Ji, S. Huang, J. Zhu, Y. Gao, and C. Zeng, A novel brushless dual-mechanical-port electrical machine for hybrid electric vehicle application, In *2012 15th International Conference on Electrical Machines and Systems (ICEMS)*, pp. 1–6, 2012. IEEE.

93. S. Cui, S. Han, X. Zhang, and Y. Cheng, Design optimization for unified field permanent magnet dual mechanical ports machine, In *Vehicle Power and Propulsion Conference (VPPC), 2014 IEEE*, pp. 1–6, 2014.

94. F. Zhao, Z. Xingming, X. Wen, G. Qiujian, Z. Li, and Z. Guangzhen, A control strategy of unified field permanent magnet dual mechanical port machine, In *2013 International Conference on Electrical Machines and Systems (ICEMS)*, pp. 685–690, 2013. IEEE.

95. G. Prajapati et al., Development of a P3 5-speed hybrid AMT, SAE Technical Paper 2017-26-0090, 2017, doi: 10.4271/2017-26-0090.

96. Y. Suzuki et al., Development of new plug-in hybrid transaxle for compact class vehicles, SAE Technical Paper 2017-01-1151, 2017, doi: 10.4271/2017-01-1151.

97. Y. Yang, X. Hu, H. Pei, and Z. Peng, Comparison of power-split and parallel hybrid powertrain architectures with a single electric machine: Dynamic programming approach, *Applied Energy* 168, 2016: 683–690. http://dx.doi.org/10.1016/j.apenergy.2016.02.023.

8

Design Principle of Series (Electrical Coupling) Hybrid Electric Drivetrain

The concept of a series hybrid electric drivetrain was developed from the electric vehicle (EV) drivetrain.[1] As mentioned in Chapter 4, EVs, compared with conventional gasoline- or diesel-fueled vehicles, have the advantages of zero mobile pollutant emissions, multi-energy sources, and high efficiency. However, EVs using present technologies have some disadvantages: a limited drive range due to the shortage of energy storage in the on-board batteries, limited payload and volume capacity due to heavy and bulky batteries, and long battery charging time. The initial objective of developing a series HEV was aimed at extending the drive range by adding an engine/alternator system to charge the batteries on board.

A typical series hybrid electric drivetrain configuration is shown in Figure 8.1. The vehicle is propelled by a traction motor. The traction motor is powered by a battery pack and/or an engine/generator unit. The powers of both power sources are merged together in a power electronics-based and controllable electrical coupling device. Many operation modes are available to choose from, according to the power demands of the driver and the operational status of the drivetrain system.

Vehicle performance (in terms of acceleration, gradeability, and maximum speed) is completely determined by the size and the characteristics of the traction motor drive. Motor power capability and transmission design are the same as in the EV design discussed in Chapter 4. However, the drivetrain control is essentially different from the pure electric drivetrain due to the involvement of the additional engine/generator unit. This chapter will focus on the design principles of the engine/alternator system, the drivetrain control, and the energy and power capacity of the battery pack. In this chapter, the term "peak power source" will replace "battery pack" because in HEVs the major function of the batteries is to supply peaking power, and they can be replaced with other kinds of sources such as ultracapacitors, flywheels, or combinations.

8.1 Operation Patterns

In series hybrid electric drivetrains, the engine/generator system is mechanically decoupled from the drive wheels, as shown in Figure 8.1. The speed and torque of the engine are independent of vehicle speed and traction torque demand, and they can be controlled at any operating point on its speed–torque plane.[2,3] Generally, the engine should be controlled in such a way that it always operates in its optimal operation region, where fuel consumption and emissions of the engine are minimized (Figure 8.2). Due to the mechanical decoupling of the engine from the drive wheels, this optimal engine operation is realizable. However, it heavily depends on the operating modes and the control strategy of the drivetrain.

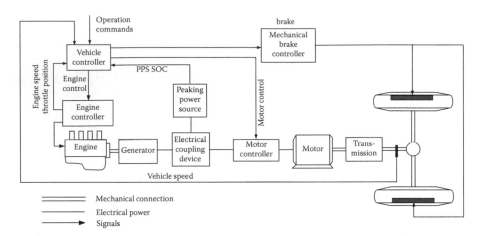

FIGURE 8.1
Configuration of a typical series hybrid electric drivetrain.

The drivetrain has several operating modes that can be used selectively according to the driving conditions and wishes of the driver. These operating modes are as follows:

1. *Hybrid traction mode*: When a large amount of power is demanded, that is, the driver steps hard on the gas pedal, both the engine/generator and peaking power source (PPS) supply their power to the electric motor drive. In this case, the engine should be controlled to operate in its optimal region for efficiency and emission

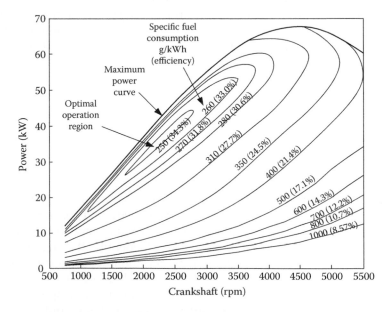

FIGURE 8.2
Example of engine characteristics and optimal operating region.

reasons as shown in Figure 8.2. The PPS supplies additional power to meet the traction power demand. This operation mode can be expressed as

$$P_{\text{demand}} = P_{\text{e/g}} + P_{\text{pps}}, \tag{8.1}$$

where P_{demand} is the power demanded by the driver, $P_{\text{e/g}}$ is the engine/generator power, and P_{pps} is the PPS power.

2. *Peak power source-alone traction mode*: In this operating mode, the PPS alone supplies its power to meet the power demand:

$$P_{\text{demand}} = P_{\text{pps}}. \tag{8.2}$$

3. *Engine/generator-alone traction mode*: In this operating mode, the engine/generator alone supplies its power to meet the power demand:

$$P_{\text{demand}} = P_{\text{e/g}}. \tag{8.3}$$

4. *PPS charge from engine/generator*: When the energy in the PPS decreases to some bottom threshold, the PPS must be charged. This can be done by regenerative braking or by the engine/generator. Usually, engine/generator charging is needed, since regenerative braking charging is insufficient. In this case, the engine/generator power is divided into two parts: one to propel the vehicle and the other to charge the PPS:

$$P_{\text{demand}} = P_{\text{e/g}} + P_{\text{pps}}. \tag{8.4}$$

It should be noted that the operation mode is only effective when the power of the engine/generator is greater than the load power demand. It should be noted that the PPS power is given a negative sign when it is being charged.

5. *Regenerative braking mode*: When a vehicle brakes, the traction motor can be used as a generator, converting part of the kinetic energy of the vehicle mass into electric energy to charge the PPS.

As shown in Figure 8.1, the vehicle controller governs the operation of each component according to the traction power (torque) command from the driver, the feedback from each of the components, and also the drivetrain and the preset control strategy. The control objectives are to (1) meet the power demand of the driver, (2) operate each component with optimal efficiency, (3) recapture braking energy as much as possible, and (4) maintain the state of charge (SOC) of the PPS in a preset window.

8.2 Control Strategies

A control strategy is a control rule that is preset in the vehicle controller and governs the operation of each component. The vehicle controller receives operation commands from the driver and feedback from the drivetrain and all the components, and it then makes decisions to use the proper operational modes. Obviously, the performance of the drivetrain relies mainly on control quality, in which control strategy plays a crucial role.

In practice, there are several control strategies that can be employed in a drivetrain for vehicles with different mission requirements. In this chapter, two typical control strategies are introduced: (1) maximum SOC of PPS (Max. SOC-of-PPS) and (2) engine turn-on and turn-off (engine on/off) or thermostat control strategies.[4]

8.2.1 Max. SOC-of-PPS Control Strategy

The object of this control strategy is to meet the power demand of the driver and, at the same time, maintain the SOC of the PPS at a high level. The engine/generator is the primary power source, and the PPS is the secondary source. This control strategy is the proper design for vehicles in which performance (e.g., speed, acceleration, gradeability) is the first concern, such as vehicles with frequent stop–go driving patterns and military vehicles in which carrying out their mission is the most important objective. A high SOC level in the PPS will guarantee a high performance of vehicles at any time.

The Max. SOC-of-PPS control strategy is depicted in Figure 8.3, in which points A, B, C, and D represent the power demands that the driver commands in either traction mode or braking mode. Point A represents the demanded traction power that is greater than the power that the engine/generator can produce. In this case, the PPS must produce its power to make up for the power shortage of the engine/generator. Point B represents the commanded power that is less than the power that the engine/generator produces when operating in its optimal operation region (Figure 8.2). In this case, two operating modes may be used, depending on the SOC level of the PPS. If the SOC of the PPS is below its top line, such as less than 70%, the engine/generator is operated with a full load. (The operating point of the engine/generator with a full load depends on the engine/generator design. For details, see the next section.) Part of its power goes

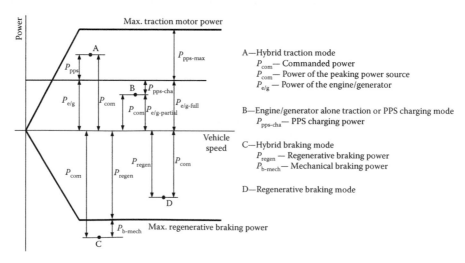

FIGURE 8.3
Illustration of the maximum PPS SOC control strategy.

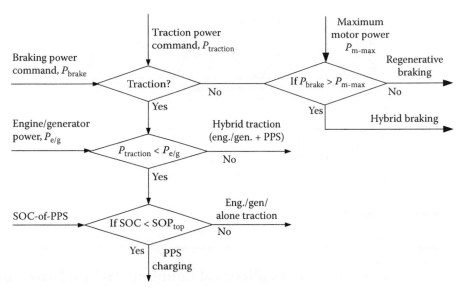

FIGURE 8.4
Control flowchart of Max. SOC-of-PPS control strategy.

to the traction motor to propel the vehicle, and the other part goes to the PPS to increase the energy level. On the other hand, if the SOC of the PPS reaches its top line, the engine/generator traction mode alone is supplied, that is, the engine/generator is controlled to produce power equal to the demanded power, and the PPS is set at idle. Point C represents the commanded braking power that is greater than the braking power the motor can produce (maximum regenerative braking power). In this case, a hybrid braking mode is used, in which the electric motor produces its maximum braking power, and the mechanical braking system produces the remaining braking power. Point D represents the commanded braking power that is less than the maximum braking power that the motor can produce. In this case, only regenerative braking is used. The control flowchart of the Max. SOC-of-PPS is illustrated in Figure 8.4.

8.2.2 Engine On–Off or Thermostat Control Strategy

The Max. SOC-of-PPS control strategy emphasizes maintaining the SOC of the PPS at a high level. However, in some driving conditions, such as driving for a long time (with a low load) on a highway at constant speed, the PPS can be easily charged to its full level, and the engine/generator is forced to operate at a power output smaller than its optimum. Hence, the efficiency of the drivetrain is reduced. In this case, the engine on–off or thermostat control strategy would be appropriate. This control strategy is illustrated in Figure 8.5. The operation of the engine/generator is completely controlled by the SOC of the PPS. When the SOC of the PPS reaches its preset top line, the engine/generator is turned off, and the vehicle is propelled only by the PPS. On the other hand, when the SOC of the PPS reaches its bottom line, the engine/generator is turned on. The PPS obtains its charging from the engine/generator. In this way, the engine can be always operated within its optimal deficiency region.

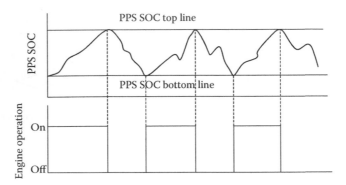

FIGURE 8.5
Illustration of thermostat control.

8.3 Design Principles of a Series (Electrical Coupling) Hybrid Drivetrain

Successful design of the drivetrain system means ensuring the vehicle is capable of achieving the desired performance, such as acceleration, gradeability, high speed, and high operating efficiency. The traction motor drive, engine/generator unit, PPS, and electrical coupling device are the major design components of concern. Their design should primarily be considered at the system level with ensuring that all the components work harmoniously.

8.3.1 Electrical Coupling Device

As mentioned previously, the electrical coupling device is the sole linkage point for combining the three sources of power together: engine/generator, PPS, and traction motor. Its major function is to regulate the power (electric current) flow between these power sources and sinks. The power (current) regulation is carried out based on the proper control of the terminal voltages. The simplest structure is to connect the three terminals together directly, as shown in Figure 8.6.

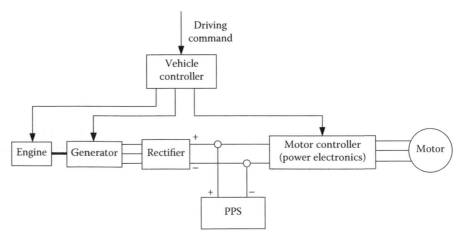

FIGURE 8.6
Directly connected power source and sink.

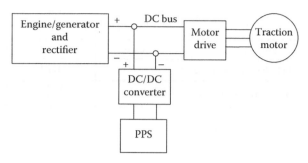

FIGURE 8.7
Configuration with DC/DC converter on PPS side.

This configuration is the simplest and has the lowest cost. Its major feature is that the bus voltage is equal to the rectified voltage of the generator and that of the PPS. The bus voltage is determined by the minimum of the two voltages mentioned previously. The power flow is solely controlled by the voltage of the generator. To deliver its power to the traction motor and/or the PPS, the open circuit voltage (zero current) of the generator, rectified, must be higher than the PPS voltage. This can be done by controlling the engine throttle and/or the magnetic field of the generator. When the engine/generator is controlled to generate the rectified terminal voltage equal to the open circuit voltage of the PPS, the PPS does not deliver power, and the engine/generator alone powers the electric motor. When the rectified voltage of the engine/generator is lower than the PPS voltage, the PPS alone powers the electric motor. In regenerative braking, the generated bus voltage by the traction motor must be higher than the PPS voltage. However, the voltage generated by the traction motor is usually proportional to the rotational speed of the motor. Therefore, the regenerative braking capability at a low speed will be rather limited for this design. It is also obvious that this simple design requires the engine/generator and the PPS to have the same rated voltage. This constraint may result in a heavy PPS due to the high voltage.

Adding a DC/DC converter, and thus releasing the voltage constraints, may significantly improve the performance of the drivetrain.[5,6] Two alternative configurations are shown in Figures 8.7 and 8.8. In the configuration of Figure 8.7, the DC/DC converter is placed between the PPS and the DC bus, and the engine–generator–rectifier is connected directly to the DC bus. In this configuration, the PPS voltage can be different from the DC bus voltage, and the rectified voltage of the engine/generator is always equal to the DC bus voltage. In the configuration of Figure 8.8, the DC/DC converter is placed between the engine–

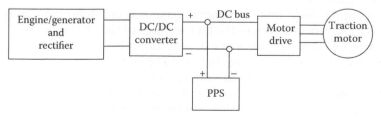

FIGURE 8.8
Configuration with DC/DC converter on engine/generator side.

Energy flow	PPS discharging	PPS charging
PPS traction	Boost	—
PPS charging from e/g	—	Buck
Regenerative braking	—	Buck or buck/boost

FIGURE 8.9
Basic functions of DC/DC converter.

generator–rectifier, and the DC bus and the PPS are directly connected to the DC bus. Contrary to the configuration of Figure 8.7, the DC/DC converter conditions the rectified voltage of the engine/generator, and the voltage of the PPS is always equal to the DC bus voltage.

Of these two configurations, the one in Figure 8.7 seems to be more appropriate. Its advantages over the other one are mainly the following: (1) changes in the voltage of PPS do not affect the DC bus voltage, (2) the energy in the PPS can be fully used, (3) the voltage of the DC bus can be maintained by controlling the engine throttle and/or the magnetic field of the generator, (4) a low PPS voltage can be used, which may lead to a small and light PPS pack and less cost, and (5) the charging current of PPS can be regulated during regenerative braking and charging from the engine/generator.

It is obvious that the DC/DC converter in this configuration must be bidirectional. In the case of the rated voltage of the PPS being lower than the DC bus voltage, the DC/DC converter must boost the PPS voltage to the level of the DC bus to deliver its power to the DC bus and buck the DC bus voltage to the level of the PPS charging voltage to charge the PPS. In regenerative braking, if the voltage generated by the traction motor at a given low speed is still higher than the voltage of the PPS, the buck DC/DC converter in the PPS charging direction is still usable. However, if the voltage generated by the traction motor at the given low speed is lower than the terminal voltage of the PPS, the DC/DC converter may need to boost the DC bus voltage to charge the battery. In this case, a buck/boost (step down/step up) converter is needed. The basic functions of the DC/DC required converter are summarized in Figure 8.9.

Figure 8.10 shows a bidirectional DC/DC converter connected between the low voltage of the PPS and the high voltage of the DC bus, boosting for PPS discharging (traction) and bucking for PPS charging from the engine/generator or from regenerative braking.[6] In the PPS discharging (traction) mode, switch S_1 is turned off, and switch S_2 is turned on and off periodically. In the on period of S_2, the inductor L_d is charged with energy from the PPS, and the load is powered by capacitor C, as shown in Figure 8.11a. In the off period of S_2, both the PPS and the inductor supply energy to the load and the charging of capacitor C, as shown in Figure 8.11b.

In the PPS charging mode from the engine/generator or traction motor in regenerative braking, the DC/DC converter bucks the high voltage of the DC bus to the low voltage of the PPS. Switch S_1 and diode D_2 serve as a unidirectional buck converter. The current flows during the on and off periods of S_1 are shown in Figure 8.12.

As mentioned previously, when the voltage generated by the traction motor at low speed, in regenerative braking, is lower than the voltage of the PPS, a bidirectional boost/buck DC/DC converter is required. Such a bidirectional buck/boost DC/DC converter is shown in Figure 8.13. Its basic operations in PPS discharging and charging modes are as follows.

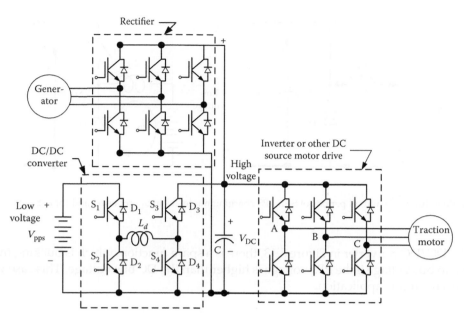

FIGURE 8.10
Bidirectional DC/DC with low-voltage PPS and high-voltage DC bus.

In the PPS discharging mode, that is, boosting the PPS voltage to the DC bus level, switch S_1 is always on, S_2 and S_3 are always off, and S_4 is turned on and off periodically in the same manner as S_2 in Figure 8.11. In the PPS charging mode with the DC bus voltage higher than the PPS voltage in regenerative braking or engine/generator charging mode, that is, bucking the DC bus voltage to PPS level, switches S_1, S_2, and S_4 are turned off, and S_3 is turned on and off periodically in the same manner as shown in Figure 8.12. In the PPS charging mode with the DC bus voltage lower than the PPS voltage (regenerative braking at low speed), that is, boosting the DC bus voltage to the PPS level, switches S_1 and S_4 are kept off, S_3 on, and S_2 is turned on and off periodically. In the on period of S_2, the inductor Ld is charged by the DC bus through S_3 and S_2. In the off period of S_2, both the DC bus and the inductor charge the PPS through S_3 and D_1.

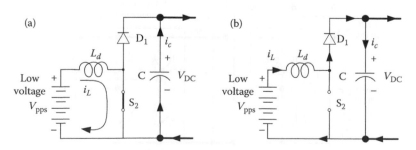

FIGURE 8.11
Current flow during on and off periods of S_2 in PPS discharging mode: (a) in S_2 on period and (b) in S_2 off period.

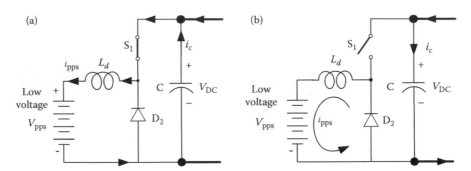

FIGURE 8.12
Current flow during on and off periods of S_1 in PPS charging mode: (a) in S_1 on period and (b) in S_2 off period.

In the DC/DC converter in Figure 8.13, there is a spare function, which is bucking the PPS voltage to bus voltage if the PPS voltage is higher than the DC bus voltage. This case would never occur in this application.

8.3.2 Power Rating Design of Traction Motor

Similar to the pure EV discussed in Chapter 4, the power rating of the electric motor drive in series HEVs is completely determined by the vehicle acceleration performance requirement, motor characteristics, and transmission characteristics (Chapter 4). At the beginning of the design, the power rating of the motor drive can be estimated, according to the acceleration performance (time used to accelerate the vehicle from zero speed to a given speed),

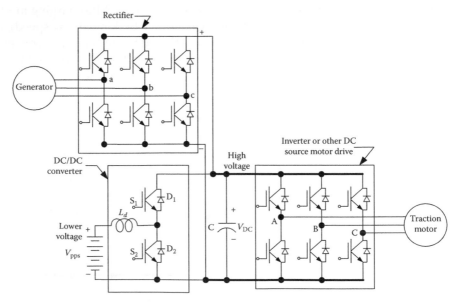

FIGURE 8.13
A boost/buck DC/DC converter.

using the following equation:

$$P_t = \frac{\delta M}{2t_a}\left(V_f^2 + V_b^2\right) + \frac{2}{3}Mgf_rV_f + \frac{1}{5}\rho_aC_DA_fV_f^3, \tag{8.5}$$

where M is the total vehicle mass in kilograms, t_a is the expected acceleration time in s, V_b is the vehicle speed in m/s, corresponding to the motor-based speed (Figure 8.14), V_f is the final speed of the vehicle during acceleration in m/s, g is the gravity acceleration in 9.80 m/s², f_r is the tire rolling resistance coefficient, ρ_a is the air density in 1.202 kg/m³, A_f is the front area of the vehicle in m², and C_D is the aerodynamic drag coefficient.

The first term in Equation 8.5 represents the power used to accelerate the vehicle mass, and the second and third terms represent the average power for overcoming the tire rolling resistance and aerodynamic drag.

Figure 8.14 shows the tractive effort and the tractive power versus vehicle speed with a two-gear transmission. During acceleration, starting from a low gear, the tractive effort follows the sequence a–b–d–e–f. At point f, the electric motor reaches its maximum speed, and the transmission must be shifted to high gear for further acceleration. In this case, the base speed of the vehicle in Equation 8.5 is $V_{b1.}$ However, when a single-gear transmission is used, that is, only a high gear is available, the tractive effort follows the sequence c–d–e–f–g, and $V_b = V_{b2}$.

It is obvious that for a given final speed during acceleration, such as 100 km/h at point e, the vehicle with a two-gear transmission will have a short acceleration time because the tractive effort at low speed in low gear, represented by a–b–d, is larger than that in a higher gear, represented by a–d.

Figure 8.15 shows an example of the power rating of the motor versus speed ratio, which is defined as the ratio of maximum speed to base speed, as shown in Figure 8.14.

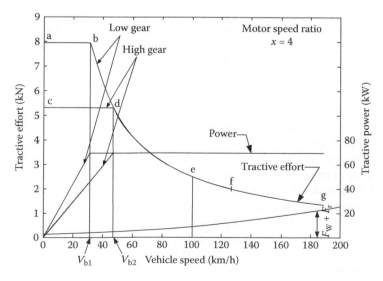

FIGURE 8.14
Speed–torque (power) characteristics of an electric motor.

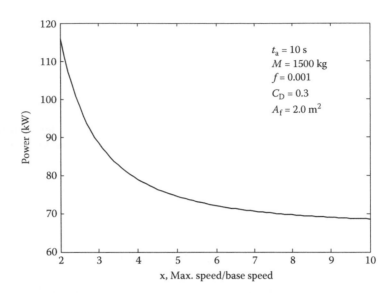

FIGURE 8.15
Power rating of traction power versus speed ratio of a drivetrain.

It should be noted that the rated motor power determined by Equation 8.5 is only an estimate for meeting the acceleration performance specifications. In some special applications, such as in off-road military vehicles, cross-country operation may be the primary concern. In this case, the traction motor must be powerful enough to overcome the required maximum grade of an off-road trail. The traction power in hill climbing can be expressed as

$$P_{grade} = \left(Mgf_r \cos\alpha + \frac{1}{2}\rho_a C_D A_f V^2 + Mg\sin\alpha \right)V(W), \tag{8.6}$$

where α is the ground slope angle, and V is the vehicle speed in m/s, specified by the gradeability requirement. When the off-road vehicle is climbing its required maximum slope, 60% or 31°, for example, at a speed of 10 km/h in real operation, the ground is usually unpaved, and the rolling resistance is much larger than those of paved roads because of the road surface deformations. Therefore, in the calculation of motor power required for gradeability, additional resistance power should be added to reflect this situation.

Based on the specified gradeability requirement of 60% or 31° at 10 km/h, the tractive efforts versus vehicle speeds of a 10-ton military vehicle, with different extended speed ratios and motor power ratings, can be calculated using Equation 8.6, as shown in Figure 8.16. Larger extended-speed ratios can effectively reduce the power rating requirement of the traction motor to meet the gradeability requirement. However, the speed on the maximum slope will be smaller. The large extended-speed ratio can be implemented either by the motor itself or by a multigear transmission.

To ensure that the vehicle meets an acceleration requirement, for example, 8 s from 0 to 48 km/h, the motor power rating requirement with different extended-speed ratios on paved roads is also calculated using Equation 8.5. Figure 8.17 shows the calculation results. It is obvious that the motor power rating is determined by gradeability performance. This means that the power rating designed to meet the gradeability will naturally meet the acceleration

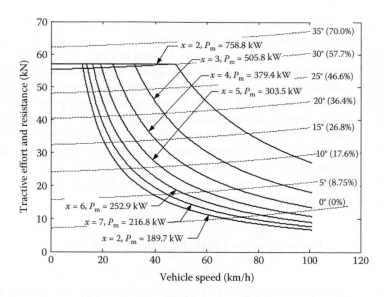

FIGURE 8.16
Tractive effort versus vehicle speed with different speed ratios and motor power.

requirement. In engineering design, trade-offs may need to be made between motor power rating and system complexity to design an appropriate motor extended-speed ratio.

8.3.3 Power Rating Design of Engine/Generator

As discussed in Chapter 6, the engine/generator in a series hybrid drivetrain is used to supply steady-state power to prevent the PPS from being discharged completely. In the design

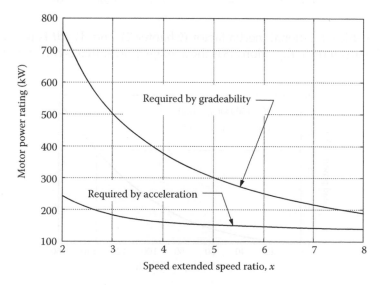

FIGURE 8.17
Motor power required by acceleration and gradeability along with extended-speed ratios.

of the engine/generator, two driving conditions should be considered: driving for a long time at constant speed, such as highway driving and off-road driving on a soft road, and driving with a frequent stop–go pattern, such as driving in cities. With the first driving pattern, the vehicle should not rely on the PPS to support the operation at high speeds, for example, 130 km/h for on-road vehicles and 60 km/h for cross-country driving in off-road vehicles. The engine/generator should be able to produce sufficient power to support vehicle speed. For a frequent stop–go driving pattern, the engine/generator should produce sufficient power to maintain the energy store of the PPS at a certain level, so that enough power can be drawn to support vehicle acceleration and hill climbing. As mentioned previously, the energy consumption of the PPS is closely related to the control strategy.

At a constant speed and on a flat road, the power output from the power source (engine/generator and/or the PPS) can be expressed as

$$P_{e/g} = \frac{V}{1000\eta_t\eta_m}\left(Mgf_r + \frac{1}{2}\rho_a C_D A_f V^2\right)(\text{kW}),\tag{8.7}$$

where η_t and η_m are the efficiencies of transmission and traction motor, respectively. Figure 8.18 shows an example of the load power (not including η_t and η_m, curve versus vehicle speed) for a 1500-kg passenger car. It indicates that the power demand at constant speed is much less than that for acceleration (Figure 8.15). In this example, about 35 kW (including losses in the transmission and traction motor, $\eta_t = 0.9$, $\eta_m = 0.8$, for example) is needed at a constant speed of 130 km/h.

When the vehicle is driving in a stop-and-go pattern in urban areas, the power that the engine/generator produces should be equal to or slightly greater than the average load power to maintain balanced PPS energy storage. The average load power can be expressed as

$$P_{ave} = \frac{1}{T}\int_0^T \left(Mgf_r + \frac{1}{2}\rho_a C_D A_f V^2\right)V dt + \frac{1}{T}\int_0^T \delta M\frac{dV}{dt} dt,\tag{8.8}$$

where δ is the vehicle rotational inertia factor (Chapter 2), and dV/dt is the acceleration of the vehicle. The first term in Equation 8.8 is the average power that is consumed to overcome

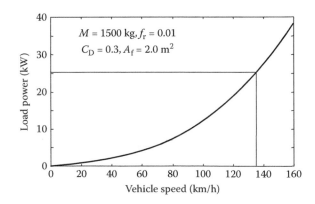

FIGURE 8.18
Load power of 1500-kg passenger car at constant speed.

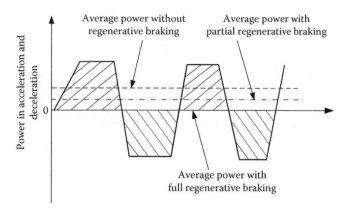

FIGURE 8.19
Average power consumed in acceleration and deceleration with full, partial, and zero regenerative braking.

the tire rolling resistance and aerodynamic drag. The second term is the average power consumed in acceleration and deceleration. When the vehicle can recover all the kinetic energy of the vehicle, the average power consumed in acceleration and deceleration is zero. Otherwise, it will be greater than zero, as shown in Figure 8.19.

In the design of an engine/generator system, the power capability should be greater than, or at least not less than, the power that is needed to support a vehicle driving at a constant speed (highway driving) and at average power when driving in urban areas. In actual design, some typical urban drive cycles must be used to predict the average power of the vehicle, as shown in Figure 8.20.

In the engine/generator size design, the operating point at which the engine/generator produces the aforementioned power should be determined. In fact, there are two possible designs. One approach is to design the engine operating point at its most efficient point, as shown by point a in Figure 8.21. At this operating point, the engine produces the needed power, as discussed previously. This design leads to a somewhat larger engine since its maximum power will not be used most of the time. This design has the advantage of making more power available for special situations. For instance, when the PPS is completely discharged or has failed, the engine/generator can be operated at a higher power (point b) to ensure that the vehicle performance has not suffered too much. The larger engine power can also be used to quickly charge the PPS. Shifting of the operating point from a to b, as shown in Figure 8.21, causes the bus voltage to increase due to the increase in speed. With a properly designed traction motor control, a higher voltage does not affect the operation of the traction motor. On the contrary, the higher DC bus voltage enables the motor to produce more power.

Another design approach is to design the operating point at point b, that is, close to the engine's maximum power, to meet the acceleration and gradeability requirements, as discussed previously. This design leads to a smaller engine. However, its operating efficiency is somewhat lower than the former design with no excess power to support the vehicle.

8.3.4 Design of PPS

The PPS must be capable of delivering sufficient power to the traction motor at any time. At the same time, the PPS must store sufficient energy to avoid failure of power delivery due to too-deep discharge.

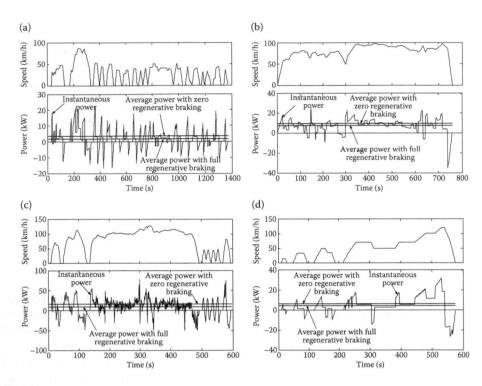

FIGURE 8.20
Instantaneous power and average power with full and zero regenerative braking in typical drive cycles. (a) FTP75 urban, (b) FTP75 highway, (c) US06, and (d) ECE-15.

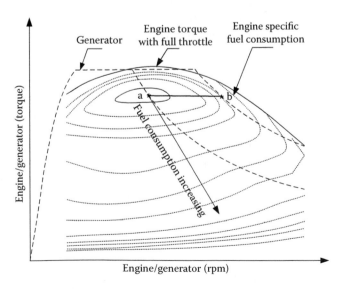

FIGURE 8.21
Operating point of engine/generator.

8.3.4.1 Power Capacity of PPS

To fully utilize the traction motor power capacity, the total power of the engine/generator and PPS should be greater than, or at least equal to, the rated maximum power of the electric motor. Thus, the power capacity of the PPS can be expressed as

$$P_{pps} \geq \frac{P_{m,max}}{\eta_m} - P_{e/g}, \tag{8.9}$$

where $P_{m,max}$ is the maximum rated power of the motor, η_m is the efficiency of the motor, and $P_{e/g}$ is the power of the engine/generator system at its designed operating point.

8.3.4.2 Energy Capacity of PPS

In some driving conditions, a frequent accelerating/decelerating driving pattern would result in a low SOC in the PPS, thus losing its delivery power. To properly determine the energy capacity of the PPS, the energy variations in some typical drive cycles must be known. The energy variation in the PPS can be expressed as

$$\Delta E = \int_0^T P_{pps} \, dt, \tag{8.10}$$

where P_{pps} is the power of the PPS. A positive P_{pps} represents charging power, and a negative P_{pps} represents discharging power. It is obvious that the energy variation in the PPS is closely associated with the control strategy. Figure 8.22 shows an example in which the energy changes in the PPS vary with driving time in the FTP75 urban drive cycle with the maximum SOC control strategy. Figure 8.22 also shows the maximum amount of energy changes, ΔE_{max}, in the whole drive cycle, if the SOC of the PPS is allowed in an operating range between SOC_{top} and SOC_{bott}. The whole energy capacity of the PPS can be determined by Equation 8.11. The operating range of the SOC of the PPS depends on the operating characteristics of the PPS. For example, for efficiency reasons, chemical batteries would have an optimal operating range in the middle (0.4–0.7), and for limited voltage variation reasons, ultracapacitors would only have a very limited energy change range (0.8–1.0):

$$E_{cap} = \frac{\Delta E_{max}}{SOC_{top} - SOC_{bott}}. \tag{8.11}$$

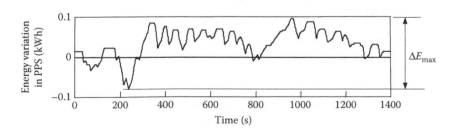

FIGURE 8.22
Energy variations in FTP75 urban drive cycle with Max. SOC control strategy.

8.4 Design Example

Design specification

  Parameters

   Vehicle total mass: 1500 kg
   Rolling resistance coefficient: 0.01
   Aerodynamic drag coefficient: 0.3
   Front area: 2.0 m^2
   Transmission efficiency (single gear): 0.9

  Performance specification

   Acceleration time (from 0 to 100 km/h): 10 ± 1 s
   Maximum gradeability: >30% at low speed and >5% at 100 km/h
   Maximum speed: 160 km/h

8.4.1 Design of Traction Motor Size

Using Equation 8.5 and assuming the motor drive has a speed ratio of $x = 4$, the motor drive power rating can be obtained as 82.5 kW for the specified acceleration time of 10 s from 0 to 100 km/h. Figure 8.23 shows the speed–torque and speed–power profiles of the traction motor.

8.4.2 Design of Gear Ratio

The gear ratio is designed so that the vehicle reaches its maximum speed at the motor's maximum speed:

$$i_g = \frac{\pi n_{m,max} r}{30 V_{max}},$$

(8.12)

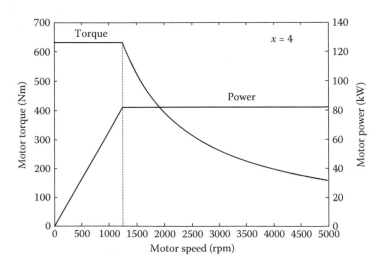

FIGURE 8.23
Characteristics of traction motor versus motor rpm.

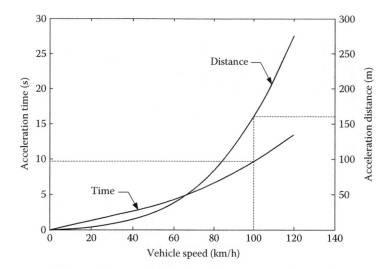

FIGURE 8.24
Acceleration time and distance versus vehicle speed.

where $n_{m,max}$ is the maximum motor rpm, V_{max} is the maximum speed of the vehicle in m/s, and r is the radius of the tire. Suppose $n_{m,max} = 5000$ rpm, $V_{max} = 44.4$ m/s (160 km/h or 100 mph), and $r = 0.2794$ m (11 in.); $i_g = 3.29$ is obtained.

8.4.3 Verification of Acceleration Performance

Based on the torque–speed profile of the traction motor, gear ratio, and vehicle parameters, and using the calculation method discussed in Chapters 2 and 4, vehicle acceleration performance (acceleration time and distance versus vehicle speed) can be obtained, as shown in Figure 8.24. If the acceleration time obtained does not meet the design specification, the motor power rating should be redesigned.

8.4.4 Verification of Gradeability

Using the motor torque–speed profile, gear ratio, and vehicle parameters and the equations described in Chapters 2 and 4, the tractive effort and resistance versus vehicle speed can be calculated, as shown in Figure 8.25a. Further, the gradeability of the vehicle can be calculated, as shown in Figure 8.25b. Figure 8.25 indicates that the gradeability calculated is much greater than that specified in the design specification. This result implies that for a passenger car the power needed for acceleration performance is usually larger than that needed for gradeability; the former determines the power rating of the traction motor.

8.4.5 Design of Engine/Generator Size

The power rating of the engine/generator is designed to be capable of supporting a vehicle at a regular highway speed (130 km/h or 81 mph) on a flat road. Figure 8.26 shows that the engine power needed at 130 km/h or 81 mph is 32.5 kW, in which energy losses in transmission (90% of efficiency), motor drive (85% of efficiency), and generator (90% of efficiency) are

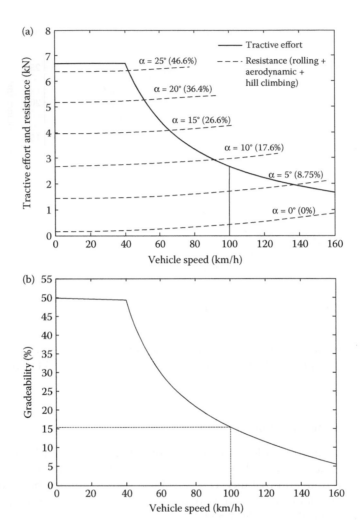

FIGURE 8.25
Tractive effort, resistance, and gradeability of vehicle versus speed: (a) tractive effort and resistance and (b) gradeability.

involved. Figure 8.26 also indicates that 32.5 kW of engine power can be capable of supporting a vehicle driving at 78 km/h (49 mph) on a 5% grade road.

Another consideration in the design of the power rating of the engine/generator is the average power when driving with some typical stop-and-go driving patterns, as illustrated in Figure 8.20. The typical data in these drive cycles are listed in Table 8.1.

Compared with the power needed in Figure 8.25, the average power in these driving cycles is smaller. Hence, 32.5 kW of engine power can meet the power requirements in these drive cycles. Figure 8.27 shows the engine characteristics. The engine would need to supply additional power to support the continuous nontraction loads, such as lights, entertainment, ventilation, air conditioning, power steering, brake boosting, and so on. In summary, the engine needs to produce about 35 kW of power to support the vehicle at 130 km/h on a flat road, without the need for power assistance from the PPS. This power can sufficiently meet the average power requirements for the stop-and-go driving pattern in urban areas.

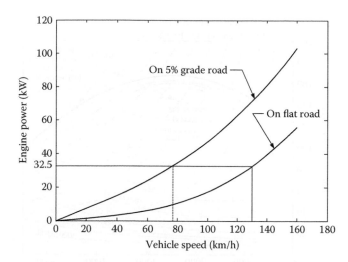

FIGURE 8.26
Engine power versus vehicle constant speed on flat road and 5% grade road.

Figure 8.27a shows the engine fuel consumption map and the minimum fuel consumption operating point (point a) at which 35 kW of power is produced. The maximum engine power is about 63 kW at point b. Another design of the engine power is shown in Figure 8.27b, in which the engine operating point is designed close to its maximum power to produce a power demand of 35 kW. The engine size in this design is smaller than the former design, but the fuel consumption is higher than the former design at 35 kW power level. As discussed previously, this power is for 130 km/h of constant speed on a flat road. At lower speeds or driving in urban areas in which the average load power is much less, the latter design may not show higher fuel consumption than the former.

8.4.6 Design of Power Capacity of PPS

The sum of the output power of the engine/generator and PPS should be greater than, or at least equal to, the input power of the traction motor:

$$P_{\text{pps}} = \frac{P_{\text{motor}}}{\eta_{\text{motor}}} - P_{\text{e/g}} = \frac{82.5}{0.85} - 32.5 = 64.5\,\text{kW}, \tag{8.13}$$

where 32.5 kW is the power of the engine/generator for traction.

TABLE 8.1

Typical Data of Different Drive Cycles

	Max. Speed (km/h)	Average Speed (km/h)	Average Power with Full Regen. Braking (kW)	Average Power without Regen. Braking (kW)
FTP75 urban	86.4	27.9	3.76	4.97
FTP75 highway	97.7	79.6	12.6	14.1
US06	128	77.4	18.3	23.0
ECE-1	120	49.8	7.89	9.32

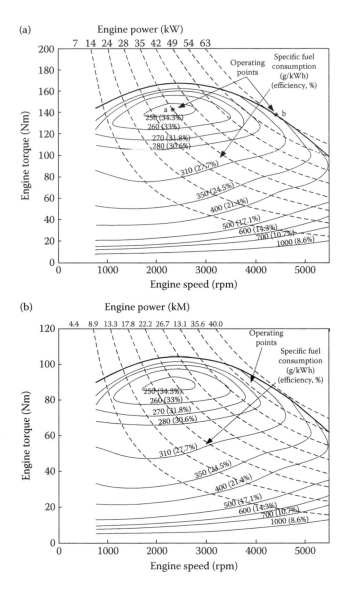

FIGURE 8.27
Engine characteristics and operating points: (a) operating with best efficiency and (b) operating at close to maximum power.

8.4.7 Design of Energy Capacity of PPS

The energy capacity of the PPS depends heavily on the drive cycle and the overall control strategy. In this design, because the power capacity of the engine/generator is much greater than the average load power (Figure 8.20), the engine on–off (thermostat) control strategy is appropriate.

Figure 8.28 shows the simulation results of the previously mentioned vehicle with an engine on–off control strategy in the FTP75 urban drive cycle. In the simulation, regenerative braking is involved (Chapter 14, Regenerative Braking). In the control, the allowed maximum energy variation in the PPS is 0.5 kWh. Suppose that the PPS can operate in an SOC

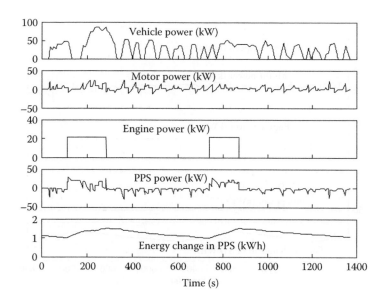

FIGURE 8.28
Simulation results in FTP75 urban drive cycle.

range of 0.2. Using batteries as the PPS, operating in a range of 0.4–0.6 of SOC will have optimal efficiency. Using ultracapacitors, a 0.2 variation in the SOC will limit the terminal voltage to 10%. The total storage energy in the PPS can be calculated as

$$E_{pps} = \frac{\Delta E_{max}}{\Delta SOC} = \frac{0.5}{0.2} = 2.5 \, kWh. \tag{8.14}$$

The weight and volume of the PPS are determined by its power capability or energy capability, depending on the power and energy density ratings of the PPS. For batteries, power density is usually the determining factor, whereas for ultracapacitors, energy density is usually the determining factor. A hybrid PPS, both battery and ultracapacitors, would be much smaller and lighter than using any one of the above. For details, refer to Chapter 13, Energy Storage.

8.4.8 Fuel Consumption

The fuel consumption in various drive cycles can be calculated by simulation. In the simulation in this example, the engine shown in Figure 8.27b is used. When the engine is on, its power output is around 20 kW, corresponding to its best fuel efficiency operating point. In the FTP75 urban drive cycle (Figure 8.28), the vehicle has a fuel economy of 5.57 L/100 km or 42.4 miles per gallon (mpg), and in the FTP highway drive cycle (Figure 8.29), 5.43 L/100 km or 43.5 mpg. It is clear that a hybrid vehicle with a performance similar to that of a conventional vehicle is much more efficient, especially in a frequent stop-and-go environment. The main reasons are the high operating efficiency of the engine and the significant amount of braking energy recovered by regenerative braking. Regenerative braking techniques are described in Chapter 14.

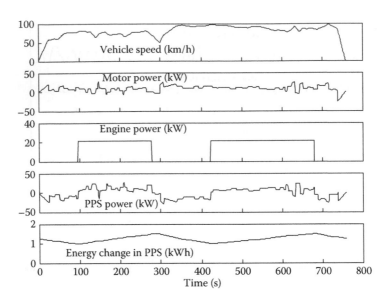

FIGURE 8.29
Simulation results in FTP75 highway drive cycle.

Bibliography

1. C. C. Chan and K. T. Chau, *Modern Electric Vehicle Technology*, Oxford University Press, New York, 2001.
2. C. G. Hochgraf, M. J. Ryan, and H. L. Wiegman, Engine control strategy for a series hybrid electric vehicle incorporating load-leveling and computer controlled energy management, *Society of Automotive Engineers (SAE) Journal*, Paper No. 960230, Warrendale, PA, 2002.
3. M. Ender and P. Dietrich, Duty cycle operation as a possibility to enhance the fuel economy of an SI engine at part load, *Society of Automotive Engineers (SAE) Journal*, Paper No. 960227, Warrendale, PA, 2002.
4. M. Ehsani, Y. Gao, and K. Butler, Application of electric peaking hybrid (ELPH) propulsion system to a full-size passenger car with simulation design verification, *IEEE Transactions on Vehicular Technology*, 48(6), November 1999.
5. C. C. Chan, The state of the art of electric and hybrid, and fuel cell vehicles, *Proceedings of the IEEE, Special issue on Electric, Hybrid and Fuel Cell Vehicles*, 95(4), April 2007.
6. J.-S. Lai and D. J. Nelson, Energy management power converters in hybrid electric and fuel cell vehicles, *Proceedings of the IEEE, Special issue on Electric, Hybrid and Fuel Cell Vehicles*, 95(4), April 2007.
7. B. Sarlioglu et al. Driving toward accessibility: A review of technological improvements for electric machines, power electronics, and batteries for electric and hybrid vehicles. *IEEE Industry Applications Magazine*, 23(1), 2017: 14–25.
8. O. Veneri (ed.) *Technologies and Applications for Smart Charging of Electric and Plug-in Hybrid Vehicles*, Springer, 2017.
9. K. T. Chau, C. C. Chan, and C. Liu, Overview of permanent-magnet brushless drives for electric and hybrid electric vehicles. *IEEE Transactions on Industrial Electronics*, 55(6), June 2008, 2246.
10. W. J. Bradley, M. K. Ebrahimi, and M. Ehsani. A general approach for current-based condition monitoring of induction motors. *Journal of Dynamic Systems, Measurement, and Control*, 136(4), 2014: 041024.

11. N. Denis, R. D. Maxime, and D. Alain. Fuzzy-based blended control for the energy management of a parallel plug-in hybrid electric vehicle. *IET Intelligent Transport Systems*, 9(1), 2014: 30–37.

12. S. J. Kim, K. Kyung-Soo, and K. Dongsuk. Feasibility assessment and design optimization of a clutchless multimode parallel hybrid electric powertrain. *IEEE/ASME Transactions on Mechatronics*, 21(2), 2016: 774–786.

13. M. Pourabdollah et al. Optimal sizing of a parallel PHEV powertrain. *IEEE Transactions on Vehicular Technology*, 62(6), 2013: 2469–2480.

14. M. Cacciato et al. Energy management in parallel hybrid electric vehicles exploiting an integrated multi-drives topology. Electrical and Electronic Technologies for Automotive, 2017 International Conference of IEEE, 2017.

15. H. I. Dokuyucu and M. Cakmakci. Concurrent design of energy management and vehicle traction supervisory control algorithms for parallel hybrid electric vehicles. *IEEE Transactions on Vehicular Technology*, 65(2), 2016: 555–565.

16. M. Saikyo et al. Optimization of energy management system for parallel hybrid electric vehicles using torque control algorithm. Society of Instrument and Control Engineers of Japan (SICE), 2015 54th Annual Conference of the IEEE, 2015.

9

Parallel (Mechanically Coupled) Hybrid Electric Drivetrain Design

Unlike the series hybrid drivetrain, the parallel or mechanically coupled hybrid drivetrain has features that allow both the engine and the traction motor to apply their mechanical power in parallel directly to the drive wheels. As mentioned in Chapter 6, mechanical coupling has two forms: torque and speed couplings. When using conventional IC engines as the primary power source, torque coupling is more appropriate since the IC engine is essentially a torque source.

The major advantages of a torque-coupling parallel configuration over a series configuration are (1) the nonnecessity of a generator, (2) a smaller traction motor, and (3) only part of the engine power going through multipower conversion. Hence, the overall efficiency can be higher than in the series hybrid.[1] However, control of the parallel hybrid drivetrain may be more complex than that of the series hybrid drivetrain because of the simultaneous mechanical coupling between the engine and the drive wheels.

There are many possibile configurations in a parallel hybrid drivetrain, as mentioned in Chapter 6. The design methodology for one configuration may not be applicable to others. Each particular configuration may be only applicable to the specified operation environment and mission requirement. This chapter will focus on the design methodology of parallel drivetrains with torque coupling, which operates with the electrically peaking principle. That is, the engine supplies its power to meet the base load (operating at a given constant speed on flat and mild-grade roads, or the average of the load of a stop-and-go driving pattern), and the electric motor supplies the power to meet the peak load requirement. Other options, such as a mild hybrid drivetrain, are discussed in Chapter 12.

9.1 Drivetrain Configuration and Design Objectives

The drivetrain structure of a parallel (torque coupling) hybrid vehicle is shown in Figure 9.1. The control system of the drivetrain consists of a vehicle controller, an engine controller to control the engine power, an electric motor controller, and, perhaps, a mechanical brake controller and a clutch controller. The vehicle controller is the highest-level controller. It receives the operation command from the driver through the accelerator and brake pedals, and other operating variables of the vehicle and its components, which includes vehicle speed, engine speed and throttle position, SOC of the PPS, and so on. By processing all signals received, based on the embedded drivetrain control algorithm, the vehicle controller generates control commands and sends the commands to the corresponding component controllers. The component controllers control the corresponding components to carry out the commands coming from the vehicle controller. Since the torque coupler is uncontrollable, the power flow in the drivetrain can only be regulated by controlling the power sources, that is, the engine, traction motor, clutch, and mechanical brake.

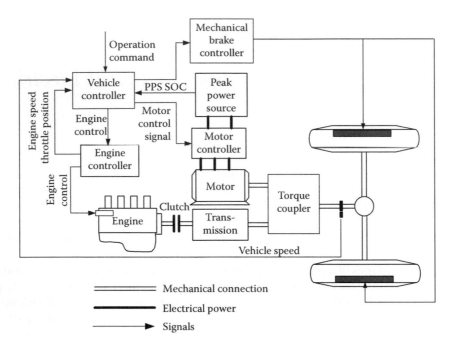

FIGURE 9.1
Configuration of parallel torque-coupling hybrid drivetrain.

In the drivetrain design, the important factors are the power of the engine; motor, PPS, and its energy capacity; transmission; and, more importantly, the control strategy of the drive-train. The design objectives are as follows: (1) satisfying the performance requirements (gradeability, acceleration, and maximum cruising speed), (2) achieving high overall efficiency whenever possible, (3) maintaining the SOC-of-PPS at reasonable levels while driving on highways and in urban areas without the need of charging the PPS from outside the vehicle, and (4) recovering as much brake energy as possible.

9.2 Control Strategies

The available operation modes in a parallel torque-coupling hybrid drivetrain, as mentioned in Chapter 6, mainly include (1) engine-alone traction, (2) electric-alone traction, (3) hybrid traction (engine plus motor), (4) regenerative braking, and (5) PPS charging from the engine. During operation, proper operation modes should be used to meet the traction torque requirement, achieve high overall efficiency, maintain a reasonable level of SOC of the PPS, and recover as much braking energy as possible.[2–6]

The overall control scheme is schematically shown in Figure 9.2. It consists of the vehicle controller, engine controller, electric motor controller, and mechanical brake controller. The vehicle controller is at the highest level. It collects data from the driver and all the components, such as the desired torque from the driver, vehicle speed, SOC of PPS, engine speed and throttle position, electric motor speed, and so on. Based on these data, the component characteristics, and the preset control strategy, the vehicle controller sends its control signals

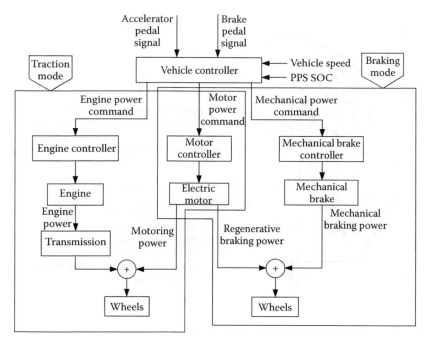

FIGURE 9.2
Overall control scheme of parallel torque-coupling hybrid drivetrain.

to each component controller. Each component controller then regulates the operation of the corresponding component to meet the requirement of the drivetrain.

The vehicle controller plays a central role in the operation of the drivetrain. The vehicle controller should fulfill various operation modes, according to the data collected from components and the driver's command, and should send the correct control command to each component controller. Hence, the control strategy in the vehicle controller is the key to the success of the drivetrain operation.

9.2.1 Max. SOC-of-PPS Control Strategy

When a vehicle is operating in a stop-and-go driving pattern, the PPS must deliver its power to the drivetrain frequently. Consequently, the PPS tends to be discharged quickly. In this case, maintaining a high SOC in the PPS is necessary to ensure that the PPS can deliver sufficient power to the drivetrain to support the vehicle's frequent acceleration. The basic rules in this control strategy are using the engine as the primary power source as much as possible and charging the PPS whenever the engine has excess power over that required for propulsion, without the PPS SOC exceeding its full charge limit.[3]

The maximum control strategy can be explained by Figure 9.3. In this figure, the maximum power curves for hybrid traction (engine plus electric motor), engine-alone traction, electric motor-alone traction, and regenerative braking are plotted against vehicle speed. Power demands in different conditions are also plotted, represented by points A, B, C, and D.

The operation modes of the drivetrain are explained as follows.

Motor-alone propelling mode: When the vehicle speed is less than a preset value V_{eb}, which is the bottom line of the vehicle speed below which the engine cannot operate stably, or

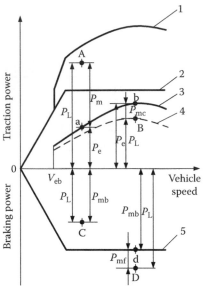

1: Maximum power in hybrid mode
2: Maximum power with electric alone traction
3: Engine power on its optimum operating line
4: Engine power with partial load
5: Maximum generative power of electric motor

P_L : Load power, traction or braking
P_e : Engine power
P_m : Motor traction power
P_{mb}: Motor braking power
P_{mc}: PPS charging power
P_{mf}: Mechanical braking power
V_{eb} : Vehicle speed corresponding to engine minimum rpm

FIGURE 9.3
Demonstration of various operating modes based on power demand.

operates with high fuel consumption or high emissions, the electric motor alone propels the vehicle. Meanwhile the engine is shut down or idles with the clutch open. The engine power, electric traction power, and PPS discharge power can be written as

$$P_e = 0, \tag{9.1}$$

$$P_m = P_L, \tag{9.2}$$

$$P_{\text{pps-d}} = \frac{P_m}{\eta_m}, \tag{9.3}$$

where P_e is the engine power output, P_L is the propelling power demanded by the driver from the accelerator pedal, P_m is the power output of the electric motor, $P_{\text{pps-d}}$ is the PPS discharge power, and η_m is the motor efficiency.

Hybrid propelling mode: When the propelling power, P_L, demanded by the driver, represented by point A in Figure 9.3, is greater than the power that the engine can produce, both the engine and the electric motor must deliver their power to the drive wheels at the same time. In this case, engine operation is set at its optimum operation line (point a) by controlling the engine throttle to produce power P_e. The remaining power demand is supplied by the electric motor. The motor power output and PPS discharge power are

$$P_m = P_L - P_e, \tag{9.4}$$

$$P_{\text{pps-d}} = \frac{P_m}{\eta_m}. \tag{9.5}$$

PPS charge mode: When the demanded propelling power, P_L, represented by point B in Figure 9.3, is less than the power that the engine can produce while operating on its optimum operation line, and the SOC of the PPS is below its top line, the engine is operated on its optimum operating line (point b), producing its power P_e. In this case, the electric motor is controlled by its controller to function as a generator, powered by the remaining power of the engine. The input power to the electric motor and PPS charge power are

$$P_m = (P_e - P_L)\eta_{t,e,m}, \tag{9.6}$$

$$P_{pps-c} = P_m\eta_m, \tag{9.7}$$

where $\eta_{t,e,m}$ is the transmission efficiency from the engine to the electric motor.

Engine-alone propelling mode: When the demanded propelling power, represented by point B in Figure 9.3, is less than the power that the engine can produce while operating on its optimum operation line, and the SOC of the PPS has reached its top line, the engine-alone propelling mode is used. In this case, the electric system is shut down, and the engine is operated to supply its power that meets the load power demanded. The power output curve of the engine with a partial load is represented by the dashed line in Figure 9.3. The engine power, electric power, and battery power can be expressed by

$$P_e = P_L, \tag{9.8}$$

$$P_m = 0, \tag{9.9}$$

$$P_{pps} = 0. \tag{9.10}$$

Regenerative-alone brake mode: When a vehicle experiences braking, and the demanded braking power is less than the maximum regenerative braking power that the electric system can supply (point C in Figure 9.3), the electric motor is controlled to function as a generator to produce its braking power, which equals the demanded braking power. In this case, the engine is shut down or set idling. The electric power output from the motor and PPS charging power is

$$P_{mb} = P_L\eta_m, \tag{9.11}$$

$$P_{pps-c} = P_{mb}. \tag{9.12}$$

Hybrid braking mode: When the demanded braking power is greater than the maximum regenerative braking power that the electric system can supply (point D in Figure 9.3), the mechanical brake must be applied. In this case, the electric motor should be controlled to produce its maximum regenerative braking power, and the mechanical brake system handles the remaining portion. The motor output power, PPS charging power, and mechanical braking power are

$$P_{mb} = P_{mb,max}\eta_m, \tag{9.13}$$

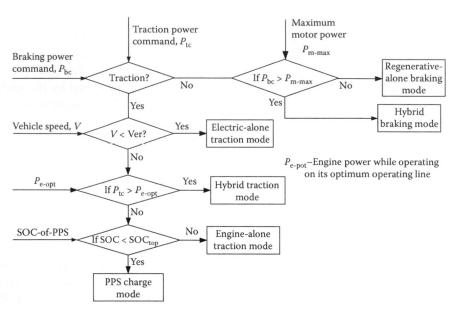

FIGURE 9.4
Flowchart of Max. SOC-of-PPS control strategy.

$$P_{pps-c} = P_{mb}, \tag{9.14}$$

$$P_{mf} = P_L - P_{mb}. \tag{9.15}$$

It should be noted that for good braking performance, the braking forces on the front and rear wheels should be proportional to their normal load on the wheels. Thus, braking power control will not be exactly as mentioned previously (for more details, see Chapter 14, "Regenerative Braking"). The control flowchart of the Max. SOC-of-PPS is illustrated in Figure 9.4.

The major objective of this control strategy is to use the engine as the vehicle's primary mover as much as possible, with no pure electric traction, when the vehicle speed is higher than a preset value. This control strategy minimizes the part of the engine's energy that cycles through the electric motor and the PPS. This may reduce the engine's energy transmission losses. However, when the PPS is fully charged and the vehicle load is small, the engine is throttled down to meet the small load power. In this case, the engine suffers low operating efficiency.

9.2.2 Engine On–Off (Thermostat) Control Strategy

When a vehicle operates in a state in which the load power is less than the power that the engine produces at optimal operating efficiency and the PPS is fully charged, the Max. SOC-of-PPS control strategy forces the engine to operate away from its optimal operating region. Consequently, the overall efficiency of the vehicle suffers. In this situation, the engine on–off (thermostat) control strategy may be used. In the engine on–off control strategy, the operation of the engine is controlled by the SOC of the PPS, as shown in Figure 9.5.

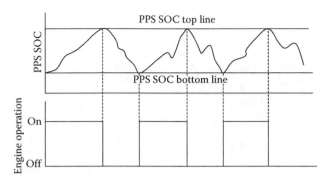

FIGURE 9.5
Illustration of engine on–off control strategy.

During the engine's on period, the control is Max. SOC-of-PPS strategy, in which the engine is always operated on its optimal curve. When the SOC of the PPS reaches its top line, the engine is turned off, and the vehicle is propelled only by the electric motor. When the SOC of the PPS reaches its bottom line, the engine is turned on, and the control again goes into Max. SOC-of-PPS.

This control strategy uses the electric motor as the primary power source. The engine either operates in its optimal region or stops. Thus, the average operating efficiency of the engine is optimized. However, contrary to the Max. SOC-of-PPS control strategy, the engine energy that goes through the electric motor and the PPS is also maximized, which may cause more energy losses in energy conversion.

It should be noted that with this control strategy, the electric motor must have sufficient power to meet the vehicle's peak power during the engine's off periods.

9.2.3 Constrained Engine On–Off Control Strategy

The constrained engine on–off control strategy is the trade-off between the Max. SOC-of-PPS and engine on–off control strategies. The principle of this control strategy is to add the engine on-and-off operation in the Max. SOC of-PPS control strategy. The control will be exactly the Max. SOC-of-PPS, when the vehicle speed is less than V_{eb}, and the demanded traction powers are at points A, C, and D, as shown in Figure 9.3. However, when the demanded traction power is at point B in Figure 9.3, that is, less than the engine's power at optimal efficiency, the engine can be operated with optimal throttle, partial throttle, or turned off, depending on the SOC of the PPS. This control strategy is explained using the diagram in Figure 9.6.

The engine torque or power is divided into three special regions—large torque area, TL, medium torque area, TM, and small torque area, TS—as shown in Figure 9.6a. These three torque areas are separated by the torque curves T_{e-l}, T_{e-m}, and T_{e-s}. These three curves may be generated by three special throttle openings. In Figure 9.6a, the isofuel consumption curves are also plotted. Similarly, the SOC of the PPS is also divided into three regions—high, medium, and low—as shown in Figure 9.6b. The engine control is based on the real-time demanded traction torque T_L and the SOC of the PPS. The suggested control strategy is illustrated in Figure 9.7.

When the commanded traction torque is in the TL area, as shown by point A in Figure 9.6a, and if the SOC of the PPS is in the medium or high region, the engine is

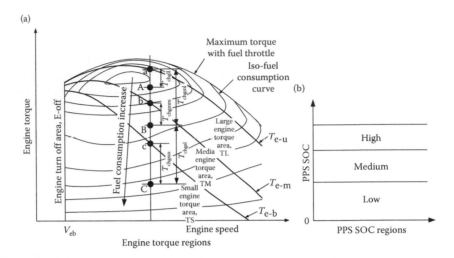

FIGURE 9.6
Illustration of the constrained engine on and off control strategy: (a) engine operation regions and (b) PPS SOC regions.

controlled to produce its torque equal to the demanded traction torque, and no additional torque is produced to charge the PPS. However, if the SOC of the PPS is in the low region, the engine needs to produce more torque to charge the PPS. In this case, the engine is controlled to operate on its optimal efficiency line, represented by point a. The charging torque is T_{chgll}, as shown in Figure 9.6a.

When the demanded traction torque is in the TM area, as shown by point B in Figure 9.6a, and if the SOC of the PPS is in the high region, the engine is controlled to produce its

Commanded torque, T_L	PPS SOC		
	Low	Medium	High
In small area (point C)	$T_e = T_b$ $T_{chgsl} = T_b - T_c$	$T_e = T_c$ $T_{chgsm} = T_c - T_c$	$T_e = 0$ $T_{chgsh} = 0$
In medium area (point B)	$T_e = T_a$ $T_{chgml} = T_a - T_B$	$T_e = T_b$ $T_{chgmm} = T_b - T_B$	$T_e = T_B$ $T_{chgmh} = 0$
In large area (point A)	$T_e = T_a$ $T_{chgll} = T_a - T_A$	$T_e = T_A$ $T_{chglh} = 0$	$T_e = T_A$ $T_{chglh} = 0$

T_A, T_B, T_C — Demanded traction torques in large, medium, and low torque areas, corresponding to points A, B, and C in Figure 9.6a
T_a, T_b, T_c — Torques that engine is controlled to produce, corresponding to points a, b, and c
T_{chgxx} — PPS charging torque, footnote xx = lh— large torque, low SOC, and so on
T_e — Engine torque

FIGURE 9.7
Engine torque control strategy with different demanded traction torque and PPS SOC.

torque equal to the demanded traction torque with no additional engine torque to charge the PPS. Otherwise, if the SOC of the PPS is in the medium region, the engine torque is controlled on the top boundary line of this area, as shown by point b in Figure 9.6a. The PPS charging torque is T_{chgmm}. However, if the SOC of the PPS is in the low region, and to quickly bring the SOC of the PPS to the medium level, the engine is controlled to operate at the optimal efficiency line, as shown by point a in Figure 9.6a. The PPS charging torque is T_{chgml}.

When the demanded traction torque is in the TS area as shown by point C in Figure 9.6a, and if the SOC of the PPS is in the high region, the engine is shut down, and the electric motor alone propels the vehicle. If the SOC of the PPS is in the medium region, the engine is controlled to operate at the upper boundary line of this area, as shown by point c in Figure 9.6a. The PPS charging torque is T_{chgsm}. However, if the SOC of the PPS is in the low region, the engine is controlled to operate at point a, as shown in Figure 9.6a, to quickly bring the SOC of the PPS to the medium region. The PPS charging torque is T_{chgsl}.

Figure 9.7 summarizes engine control in all the commanded engine traction torque areas and SOC regions of the PPS.

9.2.4 Fuzzy Logic Control Technique

The aforementioned engine and electric motor control strategy can be further developed by using fuzzy logic control methods, based on the demanded traction torque and SOC of the PPS. In fuzzy logic language, input variables of the demanded traction torque and SOC of the PPS are described by linguistic values as high (H), medium (M), and low (L). The output variables of the demanded engine torque are described as high (H), medium (M), low (L), and a crisp value zero (Z). Similarly, the electric motor torques are described as negative high (NH), negative medium (NM), negative low (NL), zero (Z, a crisp value), positive low (PL), positive medium (PM), and positive high (PH). Positive torque is for traction, and negative is for generating. The control rules are very similar to those described in Figure 9.7. The block diagram of fuzzy logic control is shown in Figure 9.8.

In Figure 9.8, only the engine torque is determined by the fuzzy logic control rules, based on the SOC of the PPS and the demanded traction torque. The motor torque is obtained from the demanded traction torque, and the engine torque is obtained from fuzzy logic, that is,

$$T_m = T_{ct} - T_e.$$

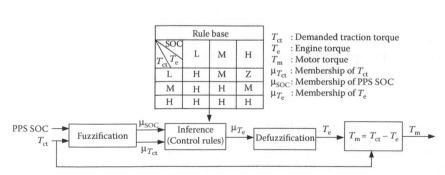

FIGURE 9.8
Block diagram of fuzzy logic control.

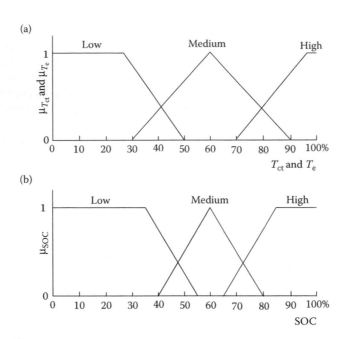

FIGURE 9.9
Membership functions of (a) T_{ct}, T_e and (b) SOC.

Since this control strategy is for the operation mode in which the demanded traction torque, T_{ct}, is smaller than the maximum torque that the engine can produce with its optimal efficiency curve, as shown by point B in Figure 9.3, the demanded traction torque, T_{ct}, and the engine torque, T_e, have the same boundary, that is, from zero to maximum. Thus, T_{ct} and T_e have the same membership function. A possible function is shown in Figure 9.9a, and a possible membership function of the SOC-of-PPS is shown in Figure 9.9b. The standard procedure to solve fuzzy logic control problems is not discussed here. Readers should consult the associated references.[7–11]

It must be noted that the threshold or fuzzy logic control strategies discussed previously are based on experience and knowledge about the drivetrain operation. Real-time operation is the most reliable way of obtaining this experience and knowledge. However, it requires a lot of time and involves high cost. Another effective way is to use simulation technologies to tune the control parameters iteratively until the optimal parameters are achieved for the specified operation environment. This work may be time consuming. It is easier to use off-line optimization techniques than iteratively running rules/fuzzy logic-based simulation to obtain the optimal control parameters. Dynamic programming is one of these techniques.[12,13]

9.2.5 Dynamic Programming Technique

The basic idea of the control algorithm using the dynamic programming technique considers the dynamic nature of the HEV system when performing the optimization. Furthermore, the optimization is with respect to the time horizon, rather than for an instant in time,[12–14] that is, for the whole drive cycle.

Contrary to the rule-based fuzzy logic control algorithm, the dynamic optimization approach usually relies on a drivetrain model to compute the best control strategy. The model can be either analytical or numerical. For a given drive cycle, the optimal operating strategy to achieve the best fuel economy can be obtained by solving a dynamic optimization problem. The problem formulation is described in what follows.[12,13]

In the discrete-time format, a model of the HEV can be expressed as

$$x(k+1) = f(x(k), u(k)), \tag{9.16}$$

where $u(k)$ is the vector of control variables such as engine throttle opening, desired motor torque, gear shift command of the transmission, and so on, and $x(k)$ is the vector of state variables of the system, which is the response to control variables $u(k)$. The goals of the optimization are to find the optimal control input $u(k)$ to minimize the total fuel consumption, or the combination of total consumption and total emission over a given driving cycle.[14] The total fuel consumption or the combined fuel consumption and emission is defined as a cost function to be minimized. In the following expression, only the total fuel consumption is included in the cost function:

$$J = \text{Fuel} = \sum_{k=0}^{N-1} L(x(k), u(k)), \tag{9.17}$$

where N is the time length of the drive cycle, and L is the instantaneous fuel consumption rate, which is a function of the system state x and input u.

In the minimizing procedure of Equation 9.17, some constraints must be imposed to ensure that all the operating parameters are within their valid ranges. These constraints include the following:

$$\omega_{e\text{-min}} \leq \omega_e \leq \omega_{e\text{-max}}, \tag{9.18}$$

$$0 \leq T_e \leq T_{e\text{-max}}, \tag{9.19}$$

$$0 \leq \omega_m \leq \omega_{m\text{-max}}, \tag{9.20}$$

$$T_{m\text{-min}} \leq T_m \leq T_{m\text{-max}}, \tag{9.21}$$

$$\text{SOC}_{\text{min}} \leq \text{SOC} \leq \text{SOC}_{\text{max}}, \tag{9.22}$$

where ω_e is the engine angular velocity; $\omega_{e\text{-min}}$ and $\omega_{e\text{-max}}$ are the specified minimum and maximum engine angular velocities, respectively; T_e is the engine torque, which must be greater than or equal to zero and smaller than or equal to its maximum torque at the corresponding angular velocity; ω_m is the motor angular velocity, which is defined in a range of zero to its maximum; T_m is the motor torque; $T_{m\text{-min}}$ is the minimum motor torque, which may be the maximum generating torque (negative); $T_{m\text{-max}}$ is the maximum motor torque in traction; and SOC is the SOC of the PPS, which is constrained to the range of its bottom SOC_{min} and top SOC_{max} levels. To sustain the charge of the PPS (PPS SOC at the end of the drive cycle is no lower than that at the beginning of the drive cycle), a final state constraint for the SOC of the PPS should be imposed. Thus, a soft terminal constraint on the PPS SOC (quadratic penalty function) is added to the cost function as follows:[12,13]

$$J = \text{Fuel} = \sum_{k=0}^{N-1} L(x(k), u(k)) + G(x(N)), \qquad (9.23)$$

where $G(x(N)) = \alpha(\text{SOC}(N) - \text{SOC}_f)^2$ represents the penalty associated with the error in the SOC at the end of the drive cycle; SOC_f is the desired SOC at the end of the drive cycle, which may be set equal to the SOC at the beginning of the drive cycle; and α is a weight factor.

Dynamic programming is well known for requiring computations that grow exponentially with the number of states.[12,15] Therefore, a simplified vehicle and component model is necessary. The engine, electric motor, PPS, transmission, and so on may need to be reduced to static models with lookup tables for I/O mapping and efficiencies.

Standard procedures to solve the preceding optimization problem based on Bellman's principle of optimality are given by Lin et al.[12] and Betsekas.[15] The dynamic programming algorithm is presented as follows:

Step $N - 1$:

$$J^*_{N-1}(x(N-1)) = \min_{u(N-1)} [L(x(N-1), u(N-1)) + G(x(N))]. \qquad (9.24)$$

Step k, for $0 \leq k < N - 1$:

$$J^*_k(x(k)) = \min_{u(k)} [L(x(k), u(k)) + J^*_{k+1}(x(k+1))]. \qquad (9.25)$$

The recursive equation is solved backward from step $N - 1$ to zero. Each minimization in a given drive cycle is performed subject to the constraints imposed by Equations 9.18 through 9.22.

The standard method to solve a dynamic programming problem numerically is to use quantization and interpolation.[12,13,16] The states and control values are first quantized into finite grids. At each step of the dynamic programming algorithm, the function $J_k(x(k))$ is evaluated only at the grid points. If the next states, $x(k+1)$, do not fall exactly on a quantized value, function interpolation is used to determine the value of $J^*_{k+1}(x(k+1))$ in Equation 9.25 as well as $G(x(N))$ in Equation 9.24.

The dynamic programming procedure produces an optimal, time-varying, state-feedback control policy that is stored in a table for each of the quantized states and time stages, that is, $u^*(x(k), k)$. This function is then used as a state-feedback controller in the simulations. It should be noted that dynamic programming creates a family of optimal paths for all possible initial conditions. In this application, once the initial SOC is given, the optimal policy finds an optimal way of bringing the final SOC back to the terminal value (SOC_f) while achieving the minimal fuel consumption.

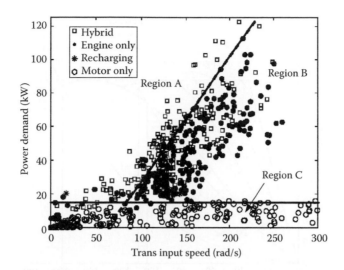

FIGURE 9.10
Operating points of dynamic programming optimization over UDDSHDC cycle. (C.-C. Lin et al., *IEEE Transactions on Control System Technology*, 11(6), November 2003.)

Although the dynamic programming approach provides an optimal solution for minimizing fuel consumption, the resulting control policy is not implementable in real driving conditions because the optimal policy requires a knowledge of the future speed and load profiles of the vehicle. Nevertheless, analytic optimal policies determined through dynamic programming can provide insight into how fuel economy improvement is achieved. Figure 9.10 shows an example of the operating points of different operating modes that can be used to refine the rules/fuzzy logic-based control strategy, as discussed previously.

9.3 Parametric Design of a Drivetrain

Parameters of a parallel (torque coupling) hybrid drivetrain such as engine power, electric motor power, gear ratios of the transmission, and power and energy capacity of the PPS are key parameters and exert considerable influence on vehicle performance and operational efficiency. However, as initial steps in the drivetrain design, these parameters should be estimated based on vehicle performance requirements. Such parameters should also be refined by more accurate simulations.

In the following sections, the parameters of a passenger car are used in calculations. These parameters are vehicle mass, $M = 1500$ kg; rolling resistance coefficient, $f_r = 0.01$; air density, $\rho_a = 1.205$ kg/m^3; front area, $A_f = 2.0$ m^2; aerodynamic drag coefficient, $CD = 0.3$; radius of drive wheels, $r = 0.2794$ m; transmission efficiency from engine to drive wheels, $\eta_{t,e} = 0.9$; and transmission efficiency from motor to drive wheels, $\eta_{t,m} = 0.95$.

9.3.1 Engine Power Design

The engine should be able to supply sufficient power to support vehicle operation at normal constant speeds on both a flat and a mild-grade road without the help of the PPS. The engine

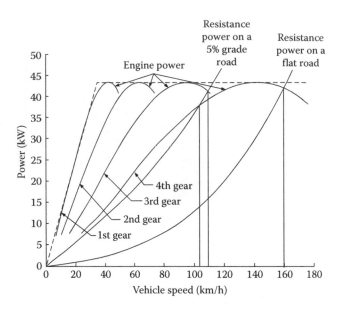

FIGURE 9.11
Engine power required at constant speed on a flat road and a 5% grade road.

should also be able to produce an average power that is larger than the average load power when the vehicle operates in a stop-and-go operating pattern.

As a requirement of normal highway driving at a constant speed on a flat or a mild-grade road, the power needed is expressed as

$$P_e = \frac{V}{1000\eta_{t,e}} \left(Mgf_r + \frac{1}{2}\rho_a C_D A_f V^2 + Mgi \right) \text{ (kW)}. \tag{9.26}$$

Figure 9.11 shows the load powers of a 1500-kg example passenger car, along with vehicle speed, on a flat road and a road with a 5% grade. It is seen that on a flat road, a speed of 160 km/h (100 mph) needs a power of 43 kW. For a comprehensive analysis, the power curves of a 43-kW engine with a multigear transmission are also plotted in Figure 9.11. From Figure 9.11 it can also be seen that on a 5% grade road, the vehicle can reach maximum speeds of about 103 and 110 km/h in fourth gear and third gear, respectively.

Figure 9.12 is the same diagram as in Figure 9.11, with the addition of the engine fuel consumption map at each gear. This figure can be used to analyze the influence of transmission gears on vehicle performance, such as vehicle acceleration, gradeability, and fuel consumption (for more details, refer to the section on transmission design).

The previously discussed engine power should be evaluated so that it meets the average power requirement while driving in a stop-and-go pattern. In a drive cycle, the average load power of a vehicle can be calculated by

$$P_{ave} = \frac{1}{T} \int_0^T \left(Mgf_r V + \frac{1}{2}\rho_a C_D A_f V^3 + \delta M V \frac{dV}{dt} \right) dt. \tag{9.27}$$

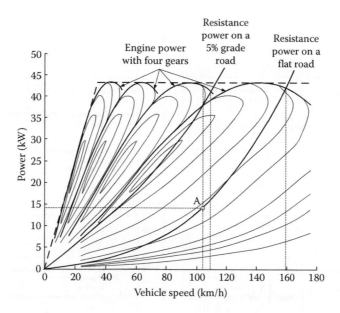

FIGURE 9.12
Engine power required at constant speed on flat road and 5% grade road with engine fuel consumption map at each gear.

The average power varies with the degree of regenerative braking. The two extreme cases are the full and zero regenerative braking cases. Full regenerative braking recovers all the energy consumed in braking, and the average power is calculated by Equation 9.27, where negative dV/dt (deceleration) tends to reduce the average power, P_{ave}. However, when the vehicle has no regenerative braking, the average power is larger than that with full regenerative braking, which can be calculated from Equation 9.27 in such a way that when the instantaneous power is less than zero, it is given a zero.

Figure 9.13 shows vehicle speed, instantaneous load power, and average powers with full regenerative braking and zero regenerative braking in some typical drive cycles for a 1500-kg passenger car.

In the engine power design, the average power that the engine produces must be greater than the average load power, as shown in Figure 9.13. In a parallel drivetrain, the engine is mechanically coupled to the drive wheels. Hence, engine rotating speed varies with vehicle speed. On the other hand, engine power with full throttle varies with engine rotating speed. In other words, engine power at full throttle is associated with vehicle speed. Thus, the determination of engine power to meet the average power in a drive cycle is not as straightforward as in a series hybrid, in which the engine operating point can be fixed. The average power that the engine produces at full throttle can be calculated by

$$P_{max\text{-}ave} = \frac{1}{T} \int_{0}^{T} P_e(V) \, dt, \qquad (9.28)$$

where T is the total time in the drive cycles, and $P_e(V)$ is the engine power at full throttle, which is a function of vehicle speed when the gear ratio of the transmission is given, as shown in Figures 9.11 and 9.12.

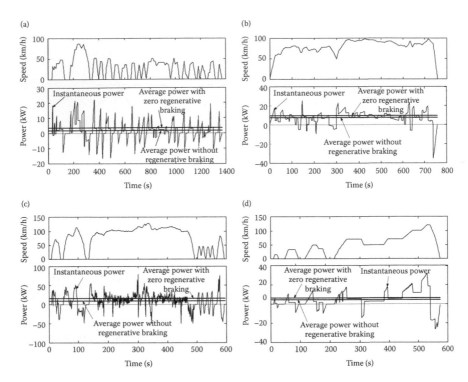

FIGURE 9.13
Instantaneous power and average power with full and zero regenerative braking in typical drive cycles: (a) FTP75 urban, (b) FTP75 highway, (c) US06, and (d) ECE-15.

The possible operating points of the engine at full throttle and the maximum possible average powers in some typical drive cycles are shown in Figure 9.14, in which the maximum engine power is 43 kW and the transmission is single gear (fourth gear only in Figures 9.11 and 9.12). If a multigear transmission is used, the maximum average power of the engine is greater than that used in a single-gear transmission (Figures 9.11 and 9.12). Comparing the maximum possible average powers to the load average powers in the typical drive cycles as shown in Figure 9.13, it is concluded that the engine power design is sufficient to support a vehicle operating in these typical drive cycles.

9.3.2 Transmission Design

Multigear transmissions can effectively increase the tractive effort of drive wheels from the engine torque, especially at a low to medium speed range (Figures 9.11 and 9.12). The direct benefit of increased tractive effort is a reduction of acceleration time and enhanced gradeability at a given motor power rating. In other words, the motor power rating can be reduced to meet a given acceleration performance and gradeability. Another benefit is the large remaining engine torque for charging the PPS, in addition to propelling the vehicle. Thus, the SOC of the PPS can be brought back to a high level quickly. However, a multigear transmission adds complexity to the drivetrain, especially for a control system in which a shifting control module must be added.

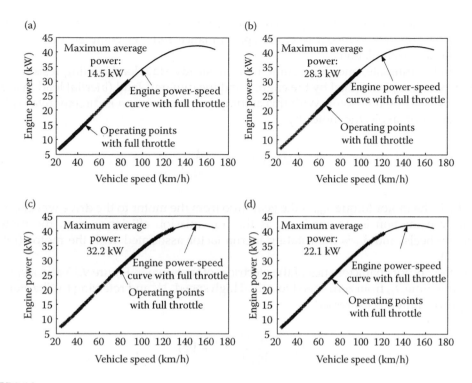

FIGURE 9.14
Maximum possible operating points of engine and maximum average power in typical drive cycles: (a) FTP75 urban, (b) FTP75 highway, (c) US06, and (d) ECE-15.

In real engineering, careful analysis is needed to make decisions about using a single-gear or multigear transmission. In the example mentioned previously, the maximum average engine power in the typical drive cycle with a single-gear transmission, as shown in Figure 9.14, is significantly greater than the average load power even without regenerative braking, as shown in Figure 9.13. Thus, the engine is considered powerful enough with a single-gear transmission if it meets the sustainability of the SOC-of-PPS. Of course, if a small engine must be used, a multigear transmission is necessary.

The operating efficiency of the engine is not expected to be significantly improved from using a multigear transmission. As shown in Figure 9.12, at most engine speeds, the engine has higher operating efficiency in the highest gear (fourth gear in Figure 9.12) than in lower gears. However, in the highest gear, the low speed range, in which the engine cannot stably operate and pure electric traction is used, is higher than that in other gears.

9.3.3 Electric Motor Drive Power Design

In HEVs, the major function of the electric motor is to supply peak power to the drivetrain. In motor power capacity design, acceleration performance and peak load power in typical drive cycles are the major concerns.[17]

It is difficult to directly design the motor power from the acceleration performance specified. This is because we have two power sources and their maximum power relationship with vehicle speed. An effective approach is to initially make an estimate of the motor power capacity based on the specified acceleration performance and then complete the final design through iterative simulations.

As an initial estimate, one can assume that the steady-state load (rolling resistance and aerodynamic drag) is handled by the engine, and the dynamic load (inertial load in acceleration) is handled by the motor. With this assumption, acceleration is directly related to the torque output of an electric motor by

$$\frac{T_m i_{tm} \eta_{tm}}{r} = \delta_m M \frac{dV}{dt}, \tag{9.29}$$

where T_m is the motor torque, i_{tm} is the gear ratio from the motor to the drive wheels, where a single transmission is assumed; η_{tm} is the transmission efficiency from the motor to the drive wheels; and δ_m is the rotating inertia factor associated with the motor (refer to Chapter 2).

Using the output characteristics of the electric motor shown in Figure 9.15 and a specified acceleration time, t_a, from zero speed to a final high speed, V_f, and referring to Chapter 4, the motor power rating is expressed as

$$P_m = \frac{\delta_m M}{2\eta_{tm} t_a} (V_f^2 + V_b^2). \tag{9.30}$$

For a 1500-kg passenger car with a maximum speed of 160 km/h, a base speed of 50 km/h, a final acceleration speed of 100 km/h, acceleration time $t_a = 10$ s, and $\delta_m = 1.04$, the power rating of the electric motor is about 74 kW.

It should be noted that the motor power obtained above is somewhat overestimated. The engine has some remaining power to help the motor accelerate the vehicle, as shown in Figures 9.11 and 9.12. This is also seen in Figure 9.16, in which vehicle speed, engine power at full throttle, and resistance power (rolling resistance, aerodynamic drag, and power losses

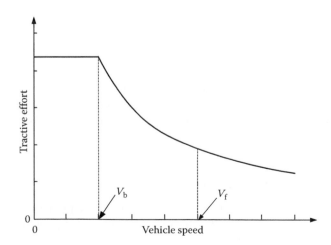

FIGURE 9.15
Tractive effort versus vehicle speed of electric-motor-driven vehicle.

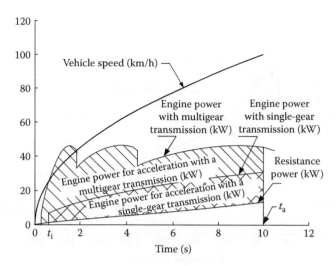

FIGURE 9.16
Vehicle speed, engine power, and resistance power versus acceleration time.

in transmission) with a multigear and single-gear transmission are plotted along acceleration time. The average remaining power of the engine, used to accelerate the vehicle, can be expressed as

$$P_{e,a} = \frac{1}{t_a - t_i} \int_{t_i}^{t_a} (P_e - P_r)\,dt, \tag{9.31}$$

where P_e and P_r are engine power and resistance power, respectively. It should be noted that the engine power transmitted to the drive wheels is associated with the gear number and gear ratios of the transmission. It is obvious from Figures 9.11 and 9.12 that a multigear transmission effectively increases the remaining power at the drive wheels, thereby reducing the motor power required for acceleration.

Figure 9.16 shows the engine's remaining power over that required for overcoming the rolling resistance and aerodynamic drag with a multigear and single-gear transmission during acceleration. This figure indicates that around 17- and 22-kW engines with a single-gear and multigear transmission, respectively, are available for assisting the motor in acceleration. Finally, the motor power rating is $74 - 17 = 57$ kW for a single-gear transmission and $74 - 22 = 52$ kW for a multigear transmission.

When the power ratings of the engine and electric motor as well as the transmission are initially designed, a more accurate calculation needs to be performed to evaluate vehicle performance, mainly maximum speed, gradeability, and acceleration. Maximum speed and gradeability can be obtained from the diagram of tractive effort and resistance versus vehicle speed. This diagram can be created using the methods discussed in Chapter 2.

The diagram (Figure 9.17) shows the design results of an example passenger car with a single-gear transmission. It indicates that the maximum gradeability of the vehicle is about 42% or 22.8° at a vehicle speed of about 40 km/h (point A). At a vehicle speed of 100 km/h, the gradeability of the vehicle with full hybrid, motor-alone, and engine-alone traction is 18.14% or 10.28° (point B), 10.36% or 5.91° (point C), and 4.6% or 2.65° (point D),

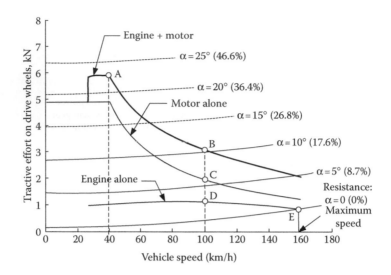

FIGURE 9.17
Tractive effort and resistance on a slope versus vehicle speed.

respectively. The maximum speed of the vehicle is around 160 km/h with engine-alone traction, which is dictated by the engine power point (E). However, if the engine and the motor top speed can extend beyond this vehicle speed, the vehicle maximum speed in hybrid mode and motor-alone mode can be extended further.

Figure 9.18 shows the acceleration performance for the example passenger car with a single-gear transmission. It indicates that 10.7 s are used, and 167 m are covered for accelerating the vehicle from zero speed to 100 km/h.

The calculation results for vehicle performance shown in Figures 9.17 and 9.18 indicate that the design of the engine and motor power capacities are appropriate.

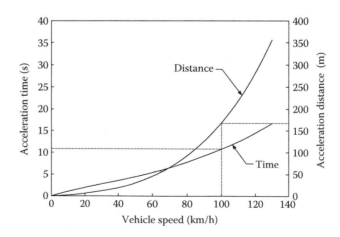

FIGURE 9.18
Acceleration time and distance versus vehicle speed.

9.3.4 PPS Design

The PPS design mainly includes the design for power capacity and energy capacity. The power capacity design is somewhat straightforward. The terminal power of the PPS must be greater than the input electric power of the electric motor, that is,

$$P_s \geq \frac{P_m}{\eta_m},$$ (9.32)

where P_m and η_m are the motor power rating and efficiency, respectively.

The energy capacity design of the PPS is closely associated with the electrical energy consumption in various driving patterns—mainly the full load acceleration and typical urban drive cycles.

During the acceleration period, the energies drawn from the PPS, and the engine can be calculated along with the calculation of acceleration time and distance by

$$E_{pps} = \int_0^{t_a} \frac{P_m}{\eta_m}\,dt$$ (9.33)

and

$$E_{engine} = \int_0^{t_a} P_e\,dt,$$ (9.34)

where E_{pps} and E_{engine} are the energies drawn from the PPS and the engine, respectively, and P_m and P_e are the powers drawn from the motor and the engine, respectively. Figure 9.19 shows the energies drawn from the PPS and the engine in a period of full acceleration along

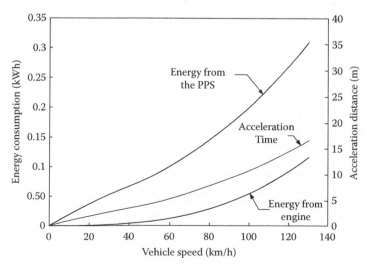

FIGURE 9.19
Energies drawn from PPS and engine during full acceleration period.

with the vehicle speed for the example passenger car. At an end speed of 120 km/h, about 0.3 kWh energy is drawn from the PPS.

The energy capacity of the PPS must also meet the requirement while driving in a stop-and-go pattern in typical drive cycles. In other words, the energy in the PPS will not be fully discharged. The energy changes in the PPS can be obtained by

$$E_{var} = \int_0^t (P_{pps-ch} - P_{pps-disch})\, dt, \tag{9.35}$$

where P_{pps-ch} and $P_{pps-disch}$ are the instantaneous charging and discharging power of the PPS. With a given control strategy, the charging and discharging power of the PPS can be obtained by a drivetrain simulation in a typical drive cycle (Section 9.4).

Figure 9.20 shows the simulation results of the example passenger car in an FTP75 urban drive cycle with the maximum SOC control strategy. The maximum energy change in the PPS is about 0.11 kWh, which is less than that under full load acceleration (0.3 kWh). Thus, the energy consumption in full load acceleration determines the energy capacity of the energy storage. This conclusion is only valid for the Max. SOC control strategy and FTP75 urban driving cycle. When other control strategies and drive cycles are used, the conclusion may be different.

In fact, not all the energy stored in the PPS can be fully used to generate sufficient power for the drivetrain. If batteries are used as the PPS, a low SOC severely limits their power output and at the same time leads to low efficiency due to an increase in internal resistance. If ultracapacitors are used as the PPS, a low SOC results in low terminal voltage that affects the

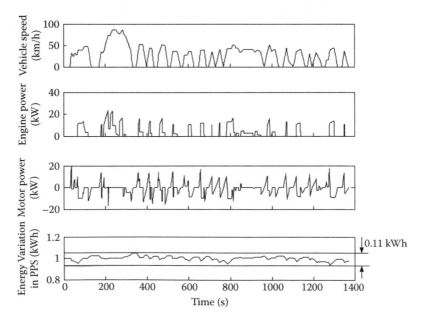

FIGURE 9.20
Vehicle speed, engine power, motor power, and energy variation in PPS storage in FTP75 urban drive cycle with Max. SOC control strategy.

performance of the traction motor. Similarly, when a flywheel is used, a low SOC means the low flywheel speed at which the terminal voltage of the electric machine, functioning as the energy exchange port, is low. Thus, only part of the energy stored in the PPS can be available for use, which can be represented by the SOC or state-of-energy. The energy capacity of the PPS can be obtained from

$$E_{c\text{-pps}} = \frac{E_{\text{dis-max}}}{SOC_t - SOC_b},$$
(9.36)

where $E_{\text{dis-max}}$ is the allowed maximum energy discharging from the PPS, and SOC_t and SOC_b are the top and bottom lines of the SOC of the PPS. In the example, $E_{\text{dis-max}} = 0.3$ kWh and we assume that 30% of the total energy of the PPS can be used; then the minimum energy capacity of the PPS is 1 kWh.

9.4 Simulations

When all the major components have been designed, the drivetrain should be simulated using a simulation program. The simulation in typical drive cycles can produce a great deal of useful information about the drivetrain, such as engine power, motor power, energy changes in the PPS, engine operating points, motor operating points, fuel consumption, and so on.

Figure 9.20 shows the vehicle speed, engine power, motor power, and energy changes in the PPS along with the driving time for the example passenger car in the FTP75 urban drive cycle. Figures 9.21 and 9.22 show the engine and motor operating points, respectively. The simulation results of the fuel economy of the example passenger car are 4.66 L per 100 km or

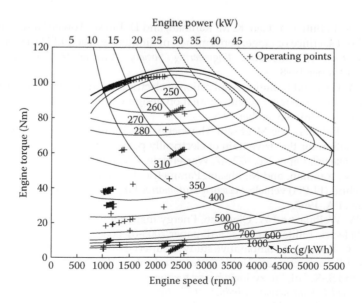

FIGURE 9.21
Engine operating points overlap with its fuel consumption map in an FTP75 urban drive cycle with maximum SOC control strategy.

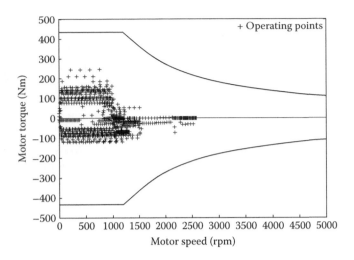

FIGURE 9.22
Motor operating points in an FTP75 urban drive cycle with maximum SOC control strategy.

50.7 miles per gallon (mpg) when the engine is turned off during the period of standstill and braking and 5.32 L per 100 km or 44.4 mpg when the engine is sitting idle during the period of standstill and braking.

Bibliography

1. M. Ehsani, K. L. Butler, Y. Gao, K. M. Rahman, and D. Burke, Toward a sustainable transportation without sacrifice of range, performance, or air quality: The ELPH car concept, In *International Federation of Automotive Engineering Society Automotive Congress*, Paris, France, September/October 1998.
2. M. Ehsani, Y. Gao, and K. Butler, Application of electric peaking hybrid (ELPH) propulsion system to a full-size passenger car with simulation design verification, *IEEE Transactions on Vehicular Technology*, 48(6), November 1999.
3. Y. Gao, K. M. Rahman, and M. Ehsani, The energy flow management and battery energy capacity determination for the drive train of electrically peaking hybrid, *Society of Automotive Engineers (SAE) Journal*, Paper No. 972647, Warrendale, PA, 1997.
4. Y. Gao, K. M. Rahman, and M. Ehsani, Parametric design of the drive train of an electrically peaking hybrid (ELPH) vehicle, *Society of Automotive Engineers (SAE) Journal*, Paper No. 970294, Warrendale, PA, 1997.
5. C. Liang, W. Weihua, and W. Qingnian, Energy management strategy and parametric design for hybrid electric military vehicle, SAE paper 2003-01-0086.
6. P. Pisu and G. Rizzoni, A comparative study of supervisory control strategies for hybrid electric vehicles, *IEEE Transaction on Control Systems Technology*, 15(3), May 2007.
7. H.-D. Lee and S.-K. Sul, Fuzzy-logic-based torque control strategy for parallel-type hybrid electric vehicle, *IEEE Transaction on Industrial Electronics*, 45(4), August 1998.
8. G. Shi, Y. Jing, A. Xu, and J. Ma, Study and simulation of based-fuzzy-logic parallel hybrid electric vehicles control strategy, In *Proceedings of the Sixth International on Intelligent Systems Design and Application (ISDA'06)*, 2006.

9. R. Langari and J.-S. Won, Intelligent energy management agent for a parallel hybrid vehicle—part I: System architecture and design of the driving situation identification process, *IEEE Transactions on Vehicular Technology*, 54(3), May 2005.

10. R. Langari and J.-S. Won, Intelligent energy management agent for a parallel hybrid vehicle—part II: Torque distribution, charge sustenance strategies, and performance results, *IEEE Transactions on Vehicular Technology*, 54(3), May 2005.

11. T. Heske and J. N. Heske, *Fuzzy Logic for Real World Design*, Annabooks, ISBN: 0-929392-24-8, 1996.

12. C.-C. Lin, H. Peng, J. W. Grizzle, and J.-M. Kang, Power management strategy for a parallel hybrid electric truck, *IEEE Transactions on Control System Technology*, 11(6), November 2003.

13. C.-C. Lin, J.-M. Kang, J. W. Grizzle, and H. Peng, Energy management strategy for a parallel hybrid electric truck, In *Proceedings of the American Control Conference*, Arlington, VA, June 25–27, 2001.

14. C.-C. Lin, H. Peng, S. Jeon, and J. M. Lee, Control of a hybrid electric truck based on driving pattern recognition, In *Proceedings of the 2002 Advanced Vehicle Control Conference*, Hiroshima, Japan, September 2002.

15. D. P. Betsekas, *Dynamic Programming and Optimal Control*, Athena Scientific, 1995.

16. C.-C. Lin, Z. Filipi, L. Louca, H. Peng, D. Assanis, and J. Stein, Modeling and control of a medium-duty hybrid electric truck, *International Journal of Heavy Vehicle Systems*, 11(3/4), 2004.

17. Y. Gao, H. Moghbelli, and M. Ehsani, Investigation of proper motor drive characteristics for military vehicle propulsion, *Society of Automotive Engineers (SAE) Journal*, Paper No. 2003-01-2296, Warrendale, PA, 2003.

18. H-T. Ngo, K-B. Sheu, Y-C. Chen, Y-C. Hsueh, H-S. Yan, Design and analysis of a novel series-parallel hybrid transmission, In *The Proceedings of the JSME International Conference on Motion and Power Transmissions*, http://doi.org/10.1299/jsmeimpt.2017.10-04.

19. X. Hu, N. Murgovski , L. Johannesson, B. Egardt. Energy efficiency analysis of a series plug-in hybrid electric bus with different energy management strategies and battery sizes. *Appl Energy*, 111, 2013, 1001–1009. http://dx.doi.org/10.1016/j.apenergy.2013.06.056.

20. V. Sezer, M. Gokasan, and S. Bogosyan. A novel ECMS and combined cost map approach for high-efficiency series hybrid electric vehicles. *IEEE Transactions on Vehicular Technology*, 60.8, 2011: 3557–3570.

21. R. M. Patil, Z. Filipi, and H. K. Fathy. Comparison of supervisory control strategies for series plug-in hybrid electric vehicle powertrains through dynamic programming. *IEEE Transactions on Control Systems Technology*, 22.2, 2014: 502–509.

22. W. Shabbir, and S. A. Evangelou. Exclusive operation strategy for the supervisory control of series hybrid electric vehicles. *IEEE Transactions on Control Systems Technology*, 24.6, 2016: 2190–2198.

23. S. Di Cairano, et al. Power smoothing energy management and its application to a series hybrid powertrain. *IEEE Transactions on Control Systems Technology*, 21.6, 2013: 2091–2103.

24. H. Borhan, et al. MPC-based energy management of a power-split hybrid electric vehicle. *IEEE Transactions on Control Systems Technology*, 20.3, 2012: 593–603.

25. M. Roche, W. Shabbir, and S. A. Evangelou. Voltage control for enhanced power electronic efficiency in series hybrid electric vehicles. *IEEE Transactions on Vehicular Technology*, 66.5, 2017: 3645–3658.

10

Design and Control Methodology of Series–Parallel (Torque and Speed Coupling) Hybrid Drivetrain

As discussed in Chapter 6, the series–parallel, or more accurately the torque/speed-coupling hybrid drivetrain, has some advantages over the series (electrical coupling) and parallel (single torque or speed coupling) drivetrains. The torque and speed couplings in this drivetrain free the engine from the drive wheels in the torque and speed constraints. Consequently, the instantaneous engine torque and speed can be independent of the load torque and speed of the vehicle. Therefore, the engine can be operated in its high-efficiency region in a similar way as that of the series (electrical coupling) drivetrain. On the other hand, part of the engine power is directly delivered to the drive wheels without experiencing multiform conversion.[1-3] This feature is more similar to the parallel (torque or speed coupling) drivetrain.

As discussed in Chapter 6, the series–parallel hybrid drive train can be composed of speed-coupling devices such as planetary gears and transmotors, as shown in Figures 6.22 through 6.24. All these configurations have similar features, designs, and control principles. This chapter focuses on the design and control principles of the configuration, which uses the planetary gear unit as its speed-coupling device, as shown in Figure 6.22. For a more detailed example of a commercial series–parallel vehicle, see the appendix in this book.

10.1 Drivetrain Configuration

10.1.1 Speed-Coupling Analysis

A series–parallel hybrid drivetrain can be formed using both torque and speed coupling. The well-known torque-coupling devices are mostly a gear set, sprocket-chain set, or pulley-belt set.[4,5] However, speed-coupling devices are less familiar to the reader and more complex. The operating characteristics of planetary gear functioning as a speed-coupling device are discussed in detail as follows.

A mechanical planetary unit has the structure shown in Figure 10.1. It consists of a sun gear, labeled s in Figure 10.1, a ring gear, labeled r, several planetary gears, labeled p (usually three or four for force balance), and a yoke, labeled y, which is hinged to the centers of the planetary gears. As discussed in Chapter 6, the speeds in rpm of the sun gear, n_s, ring gear, n_r, and yoke, n_y, have the relationship

$$n_y = \frac{1}{1 + i_g} n_s + \frac{i_g}{1 + i_g} n_r, \tag{10.1}$$

where i_g is the gear ratio defined as R_r/R_s as shown in Figure 10.1. The speeds n_s, n_r, and n_y are defined as positive in the direction shown in Figure 10.1. Defining $k_{ys} = (1 + i_g)$ and

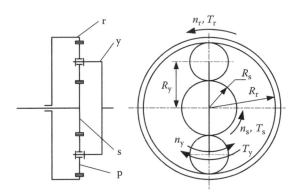

FIGURE 10.1
Planetary gear unit used as a speed coupling.

$k_{yr} = (1 + i_g)/i_g$, Equation 10.1 can be further expressed as

$$n_y = \frac{1}{k_{ys}} n_s + \frac{1}{k_{yr}} n_r. \tag{10.2}$$

Neglecting the energy losses in steady-state operation, the torques acting on the sun gear, ring gear, and yoke have the relationship

$$T_y = -k_{ys} T_s = -k_{yr} T_r. \tag{10.3}$$

Equation 10.3 indicates that the torques acting on the sun gear, T_s, and ring gear, T_r, always have the same sign; in other words, they always have to be in the same direction. However, the torque acting on the yoke, T_y, is always in the opposite direction of T_s and T_r. Equation 10.3 also indicates that with $i_g > 1$, which is the general case since $R_r > R_s$, T_s is the smallest, T_y is the largest, and T_r is in between. This means that the torque acting on the yoke is balanced by torques acting on the sun gear and ring gear.

When one element among the sun gear, ring gear, and yoke is locked to the vehicle frame, that is, one degree of freedom of the unit is constrained, the unit becomes a single-gear transmission (one input and one output); the speed and torque relationship, with one element fixed, is shown in Figure 10.2.

Element fixed	Speed	Torque
Sun gear	$n_y = \dfrac{1}{k_{yr}} n_r$	$T_y = -k_{yr} T_r$
Ring gear	$n_y = \dfrac{1}{k_{ys}} n_s$	$T_y = -k_{ys} T_s$
Yoke	$n_s = -\dfrac{k_{ys}}{k_{yr}} n_r$	$T_s = \dfrac{k_{yr}}{k_{ys}} T_r$

FIGURE 10.2
Speed and torque relationships while one element is fixed.

In composing a hybrid drivetrain with the planetary gear unit as a speed coupling, there are many options, as shown in Figure 10.3. To reduce the torque capacity requirement, thereby reducing the motor/generator physical size and weight, connecting the motor/generator to the sun gear of the planetary gear unit may be the appropriate choice. The engine may be connected to either the yoke or to the ring gear, as shown in Figure 10.3a and b. In the former design (Figure 10.3a), the torques of the engine and motor/generator have the relationship

$$T_e = -k_{ys}T_{m/g},$$

(10.4)

where T_e and $T_{m/g}$ are the torques acting on the yoke and sun gear, produced by the engine and the motor/generator. T_e and $T_{m/g}$ have opposite directions. The engine operates in the first quadrant, and the motor operates in the third and fourth quadrants, as shown in Figure 10.4. Since the motor/generator must produce its torque to balance the engine maximum torque at any speed, the motor/generator must have a constant maximum torque in its whole speed range, as shown in Figure 10.4.

In the latter design (Figure 10.3b), the engine torque and motor/generator torque have the relationship

$$T_e = \frac{k_{yr}}{k_{ys}}T_{m/g}.$$

(10.5)

The engine and motor/generator operating areas are shown in Figure 10.5.

Comparing Figures 10.4 and 10.5, with the same maximum engine torque, the former design (Figures 10.3a and 10.4) results in a smaller motor/generator (smaller motor/generator torque). However, the torque from the ring gear, delivered to the drive wheel through the gearbox, is smaller than the engine torque (similar to the overdrive gear in the conventional transmission). Nevertheless, the gear ratio of the gearbox can be designed to meet the tractive torque requirement. Further discussion in the following section will be based on this design (Figures 10.3a and 10.4).

10.1.2 Drivetrain Configuration

Figure 10.6 shows the detailed configuration of a series–parallel (torque/speed coupling) drivetrain.[6] The planetary gear unit constitutes the speed coupling that connects an engine and a motor/generator together. The engine and the motor/generator are connected to the yoke and the sun gear, respectively. The ring gear of the planetary gear is connected to the drive wheels through gears of Z_1, Z_2, Z_4, Z_5, and a differential. A traction motor is connected to the drive wheels through gears of Z_3, Z_2, Z_4, Z_5, and the differential, which couples the output torques of the ring gear and the traction motor together. In this configuration, one clutch and two locks are used. The clutch serves for connecting or disconnecting the engine to or from the yoke of the planetary gear unit. Lock 1 is used to lock or release the sun gear and the shaft of the generator/motor to or from the stationary frame of the vehicle. Lock 2 is used to lock or release the yoke to or from the stationary frame of the vehicle. By controlling the clutch, locks, engine, motor/generator, and the traction motor, many operation modes are available to be used as follows.

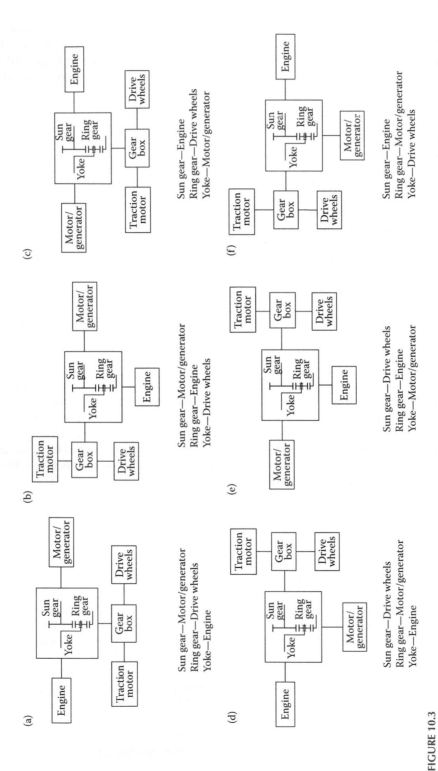

FIGURE 10.3

Possible configurations of torque- and speed-coupling hybrid drivetrain with connections of (a) motor/generator to sun gear, drive wheel to ring gear, and engine to yoke; (b) motor/generator to sun gear, engine to ring gear, and drive wheel to yoke; (c) engine to sun gear, drive wheel to ring gear, and motor/generator to yoke; (d) drive wheel to sun gear, motor/generator to ring gear, and engine to yoke; (e) drive wheel to sun gear, engine to ring gear, and motor/generator to yoke; and (f) engine to sun gear, motor/generator to ring gear, and drive wheel to yoke.

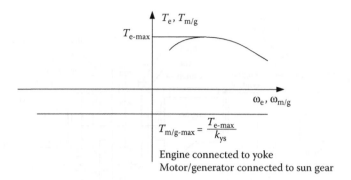

FIGURE 10.4
Operating areas of engine and motor/generator.

1. *Speed-coupling mode*: In this mode, the traction motor is de-energized. There are three submodes:

   1.1. *Engine-alone traction*: The clutch is engaged to connect the engine to the yoke, lock 1 locks the sun gear to the vehicle stationary frame, and the motor/generator is de-energized. Lock 2 releases the yoke from the vehicle stationary frame. The energy flow route is shown in Figure 10.7.

   In this case, the engine alone delivers its torque to the drive wheels. The speed relationship between the engine and the drive wheels is

   $$n_{dw} = \frac{k_{yr} n_e}{i_{rw}}, \tag{10.6}$$

   where n_{dw} and n_e are the speeds of the drive wheel and the engine, and i_{rw} is the gear ratio from the ring gear to the drivetrain wheels, which is expressed as

   $$i_{rw} = \frac{Z_5 Z_2}{Z_1 Z_4}, \tag{10.7}$$

   where Z_1, Z_2, Z_4, and Z_5 are the tooth numbers of the gears Z_1, Z_2, Z_4, and Z_5.

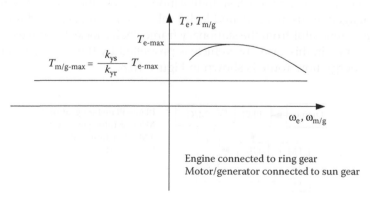

FIGURE 10.5
Operating areas of engine and motor/generator.

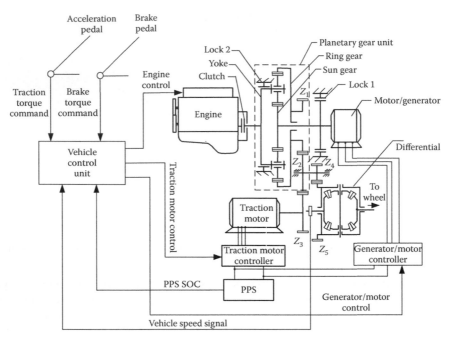

FIGURE 10.6
Drivetrain configuration.

The torque relationship between the drive wheels and the engine is

$$T_{dw} = \frac{i_{rw}\eta_{yr}\eta_{rw}T_e}{k_{yr}},$$ (10.8)

where T_{dw} is the torque developed on the drive wheels by the engine torque T_e, η_{yr} is the efficiency from the yoke to the ring gear, and η_{rw} is the efficiency from the ring gear to the drive wheels.

1.2. *Motor/generator-alone traction*: In this mode, the engine is shut down; the clutch is engaged or disengaged, and lock 1 releases the sun gear and the shaft of the motor/generator from the stationary frame; lock 2 locks the yoke to the stationary frame. In this case, the vehicle is propelled by the motor/generator alone. The energy flow route is shown in Figure 10.8.

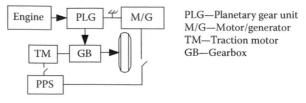

PLG—Planetary gear unit
M/G—Motor/generator
TM—Traction motor
GB—Gearbox

FIGURE 10.7
The traction energy flow route is from the engine alone.

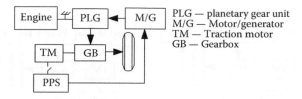

FIGURE 10.8
Energy flow route in mode of motor/generator traction.

The speed and torque relationship between the generator/motor and the drive wheels is

$$n_{dw} = -\frac{k_{yr}}{k_{ys}i_{rw}} n_{m/g} \tag{10.9}$$

and

$$T_{dw} = \frac{k_{ys}i_{rw}\eta_{sr}\eta_{rw}}{k_{yr}} T_{m/g}, \tag{10.10}$$

where T_{dw} is the tractive torque on the drive wheels developed by the motor/generator torque $T_{m/g}$, and η_{sr} is the efficiency from the sun gear to the ring gear.

It should be noted that the motor/generator must be operated in the third quadrant, that is, negative angular velocity and negative torque as defined in Figure 10.1.

1.3. *Engine and motor/generator with speed-coupling traction*: In this mode, the clutch is engaged. Locks 1 and 2 are released from the stationary frame. From Equation 10.2, the angular velocities of the drive wheels, engine, and motor/generator have the relationship

$$n_{dw} = \frac{k_{yr}}{i_{rw}}\left(n_e - \frac{1}{k_{ys}}n_{m/g}\right), \tag{10.11}$$

and the torques have the relationship

$$T_{dw} = \frac{i_{rw}\eta_{yr}\eta_{rw}}{k_{yr}} T_e = \frac{k_{ys}\eta_{sr}^{b}\eta_{rw}}{k_{yr}} T_{m/g}, \tag{10.12}$$

where b is an index when the power flows from the motor/generator to the sun gear, that is, $n_{m/g} < 0$, $b = 1$, otherwise $b = -1$. Equation 10.11 implies that, at a given vehicle speed, the engine speed can be adjusted by the motor/generator speed. Equation 10.12 indicates that the engine torque, motor/generator torque, and load torque on the drive wheels always have a fixed relationship. This implies that a change in one torque will cause a change in the other two

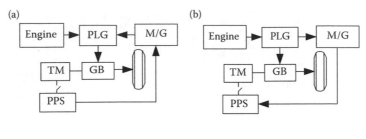

PLG—Planetary gear unit
M/G—Motor/generator
TM—Traction motor
GB—Gearbox

FIGURE 10.9
Energy flow route in speed-coupling mode: (a) M/G motoring and (b) M/G generating.

torques, causing the operating points of the engine and motor/generator to change. For a detailed discussion, refer to the next section for drivetrain control.

The energy flow routes are shown in Figure 10.9.

2. *Torque-coupling mode*: When the traction motor is energized, its torque can be added to the torque output of the ring gear to constitute the torque-coupling mode. Corresponding to the three modes (1.1), (1.2), and (1.3), when the traction motor is controlled to operate in motoring and generating modes, six basic operation modes are constituted.

   2.1. *Engine alone in mode (1.1) plus traction motor motoring*: This mode is the same as the general parallel hybrid traction mode. The energy flow route is shown in Figure 10.10.

   2.2. *Engine alone in mode (1.1) plus traction motor generating*: This mode is the same as the PPS charging from the engine mode in the general hybrid drivetrain. The energy flow route is shown in Figure 10.11.

   2.3. *Motor/generator-alone mode (1.2) plus traction motor motoring*: This mode is similar to mode (2.1), but the engine is replaced by the motor/generator. The energy flow route is shown in Figure 10.12.

   2.4. *Motor/generator alone in mode (1.2) plus traction motor generating*: This mode is similar to mode (2.2), but the engine is replaced by the motor/generator. This mode may never be used because part of the motor/generator energy

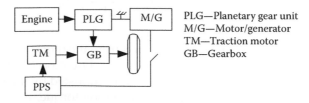

PLG—Planetary gear unit
M/G—Motor/generator
TM—Traction motor
GB—Gearbox

FIGURE 10.10
Energy flow route in parallel traction mode.

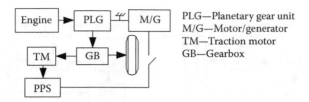

FIGURE 10.11
Energy flow route in parallel PPS charging.

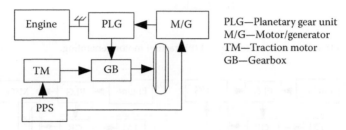

FIGURE 10.12
Energy flow routes in the two-mode traction in parallel.

circles from the PPS and finally to the PPS through the motor/generator and traction motor, as shown in Figure 10.13.

2.5. *Speed-coupling traction in mode (1.3) plus traction motor motoring*: This mode uses the full functions of speed and torque coupling. There are two operating states of the motor/generator: motoring and generating, as shown in Figure 10.14. The operating states of the motor/generator in motoring (Figure 10.14a) may be used at high vehicle speeds. In this case, the engine speed may be limited to somewhat lower than its medium speed to avoid too high an engine speed where its operating efficiency may be low. The motor generator contributes its speed to the drivetrain for supporting the high vehicle speed, as shown in Figure 10.14a. Similarly, the operating states in Figure 10.14b may be used in the case of low vehicle speed. In this case, the engine can be operated at speeds somewhat lower than its medium-speed to avoid too low a speed operation, where its operating efficiency may be low. The motor/generator absorbs part of the engine speed.

2.6. *Speed-coupling traction in mode (1.3) plus traction motor generating*: Similar to mode (2.5), the engine and the motor/generator operate in speed-coupling mode, but the traction motor operates in generating mode, as shown in Figure 10.15.

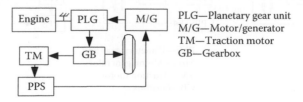

FIGURE 10.13
Energy flow route in mode of motor/generator traction and PPS charging.

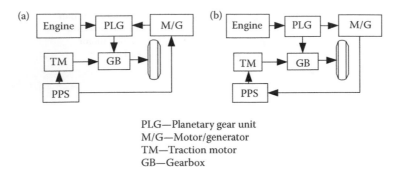

FIGURE 10.14
Energy flow route: (a) traction motor motoring and (b) traction motor generating.

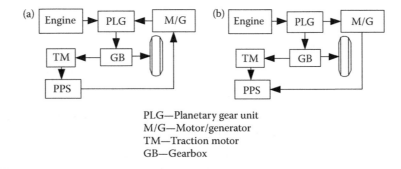

FIGURE 10.15
Energy flow route: (a) M/G motoring and (b) M/G generating.

3. *Regenerative braking*: When a vehicle is braking, the traction motor, motor/generator, or both can produce braking torque and recapture part of the braking energy to charge the PPS. In this case, the engine is shut down with the clutch opened. The possible energy flow is shown in Figure 10.16.

As discussed previously, several operating modes are available for use. In a control scheme design, perhaps not all the operating modes are really used, depending on the drive-train design, driving conditions, operating characteristics of the major components, and so on.

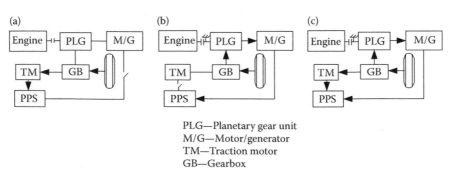

FIGURE 10.16
Energy flow in regenerative braking: (a) traction motor alone, (b) M/G alone, and (c) both traction motor and M/G.

10.2 Drivetrain Control Methodology

10.2.1 Control System

The control system of the drivetrain is shown in Figure 10.6. The vehicle controller unit (VCU) receives the traction or braking torque commands from the driver through the accelerator or brake pedals and other necessary operating information, such as the SOC of the PPS and vehicle speed. Based on the real-time information received and the control logic preset in the VCU, the VCU generates control signals to control the engine, motor/generator, and traction motor, as well as clutch and locks, through the engine throttle actuator, motor/generator controller, traction motor controller, clutch, and lock actuators.

10.2.2 Engine Speed Control Approach

Equation 10.11 indicates that the engine speed, n_e, can be adjusted by controlling the motor/generator speed, $n_{m/g}$, at a given wheel speed, n_{dw}. However, this control activity must be carried out by controlling the engine throttle and the motor/generator torque, as shown in Figure 10.17. The control procedure is as follows.

Suppose the engine is operating at point a at a speed of $n_{e,a}$, producing torque $T_{e,1}$, with a throttle angle of 60° as shown in Figure 10.17; the motor/generator has to produce its torque $T_{m/g,1} = T_{e,1}/k_{ys}$ (where the losses are ignored) to balance the engine torque. With a fixed motor/generator torque, and thus a fixed engine torque, increasing the engine throttle opening causes the engine speed to increase, to point b at $\theta = 70°$, for example. Similarly, reducing the engine throttle opening causes the engine speed to decrease, to point c at $\theta = 50°$, for example. The engine speed can also be changed by changing the

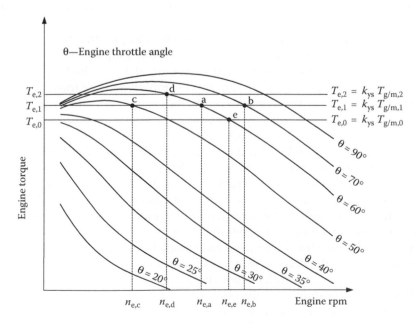

FIGURE 10.17
Engine speed controlled by engine throttle and motor/generator torque.

motor/generator torque, as shown in Figure 10.17. With a fixed engine throttle, reducing the motor/generator torque (and thus the engine torque) causes the engine speed to increase from point a to point e, or increasing the motor/generator torque causes the engine speed to decrease from point a to point d. Thus, the engine speed can be potentially controlled within its optimal speed range by instantaneously controlling the engine throttle and/or the motor/generator torque.

10.2.3 Traction Torque Control Approach

The traction torque on the drive wheels is the sum of the torques transmitted from the ring gear of the planetary gear units and the traction motor. The traction torque on the drive wheels can be expressed as

$$T_{tdw} = i_{rw}\eta_{rw}T_{ring} + i_{mw}\eta_{mw}T_{tm}, \qquad (10.13)$$

where T_{tdw} is the total tractive effort on the drive wheels; T_{ring} is the torque output from the ring gear of the planetary gear unit, which is generated by the engine and the motor/generator; i_{rw} and η_{rw} are the gear ratio and transmission efficiency from the ring gear to the drive wheels, respectively; T_{tm} is the traction motor torque; and η_{mw} and i_{mw} are the transmission efficiency and gear ratio from the traction motor to the drive wheels, where $i_{mw} = (Z_2Z_5)/(Z_3Z_4)$ (Z_2, Z_3, Z_4, and Z_5 are the tooth numbers of the corresponding gears, as shown in Figure 10.6).

The total traction torque request on the drive wheels, which is demanded by the driver through the accelerator pedal, can be met by the torque outputs from the ring gear, T_{ring}, and the traction motor, T_{tm}. As mentioned previously, T_{ring} can be obtained by controlling the engine throttle and the motor/generator torque to operate the engine with high efficiency. The contributions of T_{ring} and T_{tm} to the total depend on the control strategy of the drivetrain, which is discussed in the next section.

Figure 10.18 illustrates the simulation results of an example drivetrain with full engine throttle opening and full traction motor load (maximum torque versus motor speed). In the simulation, the engine rpm is controlled such that, at low vehicle speeds, the engine operates with a constant speed (1200 rpm in this example), and the motor/generator operates at positive speeds. At medium vehicle speeds, the motor/generator is locked to the vehicle frame, and the engine speed linearly increases with vehicle speed (pure parallel or pure torque-coupling operation). At high vehicle speeds, the engine again operates at a constant speed (3500 rpm in this example), and the motor/generator operates with negative speed (rotating in the opposite direction of the engine). With the previous engine speed control, the engine operating speeds are constrained in the medium range in which the engine efficiency may be higher. It is noted that the motor/generator is de-energized in the medium vehicle speed range for using the high engine torque and closing the energy flow through the motor/generator, which may cause more energy loss.

Similar to the series (electrical coupling) and parallel (mechanical coupling) drivetrains, the maximum torque on the drive wheels, corresponding to full accelerator pedal position, full engine throttle opening, and full traction motor load, dictates vehicle performance, such as acceleration and gradeability. On the other hand, with a partially depressed accelerator pedal (partial load request), the engine, traction motor, or both must reduce their torques to meet the traction torque demand. Thus, a control strategy is needed to properly allocate the total load power to the power sources.

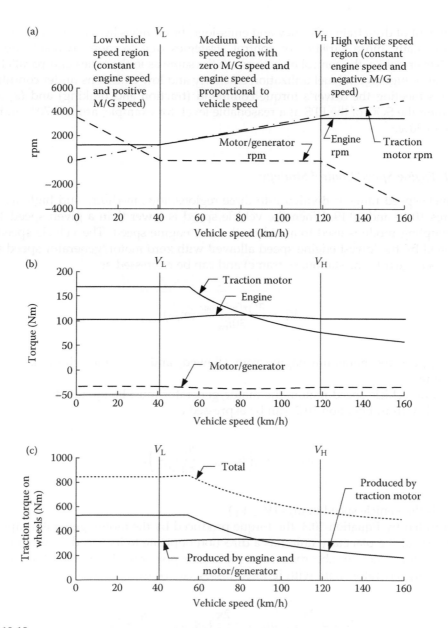

FIGURE 10.18
Torques and speeds of engine, traction motor, motor/generator, and drive wheels with full engine throttle opening and full loading of traction motor along vehicle speed: (a) speeds, (b) torques of power plants, and (c) torques on drive wheels.

10.2.4 Drivetrain Control Strategies

The distinct property of the foregoing hybrid drivetrain is that the engine speed and torque can be decoupled completely or partially from the drive wheels through speed coupling and torque coupling. It also has much more flexibility than series or parallel hybrid drivetrains in the choice of the active operation mode. Thus, this drivetrain has more potential for the improvement of drivetrain efficiency and emissions, but it relies heavily on system control. There are many more varieties of control strategies due to more available operation modes. Nevertheless, the control objectives are the same as in the series and parallel drive-trains, that is, high overall fuel utilization efficiency and low emissions under conditions of (1) always meeting the driver's torque command (traction and braking) and (2) always maintaining the SOC of the PPS at a reasonable level, for example, around 70% and never lower than 30%.

10.2.4.1 Engine Speed Control Strategy

The vehicle speed range is divided into three regions, low, medium, and high, as shown in Figures 10.18 and 10.19. When the vehicle speed is lower than a given speed V_L, the speed-coupling mode is used to avoid too low an engine speed. The vehicle speed V_L is determined by the lowest engine speed allowed with zero motor/generator speed (lock 1 locks the sun gear to the stationary frame) and can be expressed as

$$V_L = \frac{\pi k_{yr} r_w n_{e\text{-min}}}{30 i_{rw}} \text{ (m/s)},\tag{10.14}$$

where $n_{e\text{-min}}$ is the minimum engine rpm allowed, and r_w is the wheel radius of the vehicle in m.

In this low vehicle speed region, the motor/generator must be operated with a positive speed, which, from Equation 10.2, can be expressed as

$$n_{m/g} = k_{ys}\left(n_{e\text{-min}} - \frac{30 i_{rw} V}{\pi k_{yr} r_w}\right),\tag{10.15}$$

where V is the vehicle speed in m/s ($V \le V_L$).

As indicated by Equation 10.3, the torque produced by the motor/generator, applied to the sun gear of the planetary unit, has a direction opposite to its speed. Therefore, in this case, the motor/generator absorbs part of the engine power to charge the PPS. The power on the motor/generator shaft, with ignored losses, can be expressed as

$$P_{m/g} = \frac{2\pi}{60} T_{m/g} n_{m/g} = \frac{2\pi}{60} T_e n_{e\text{-min}} - \frac{i_{rw}}{k_{yr} r_w} T_e V.\tag{10.16}$$

The first term on the right-hand side is the power that the engine produces, and the second term is the power that is delivered to the drive wheels.

When the vehicle speed is higher than V_L, but lower than a given speed V_H, the motor/generator is de-energized, and the sun gear (the shaft of the motor/generator) is locked to the stationary frame of the vehicle. The drivetrain operates in the torque-coupling mode. The engine speed is proportional to the vehicle speed. Speed V_H is determined by the maximum engine speed allowed $n_{e\text{-max}}$, beyond which the engine operating efficiency may be reduced. When the vehicle speed is higher than V_H, the engine speed is kept constant at $n_{e\text{-max}}$, and the motor/generator starts working again at a negative speed to compensate

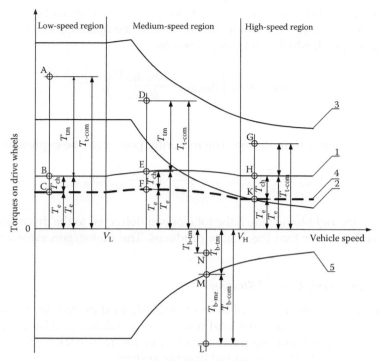

1—Traction torque developed by maximum
   torque with optimal throttle
2—Traction torque developed by maximum
   traction motor torque
3—Maximum traction torque developed by both
   engine and traction motor
4—Traction torque developed by engine at
   partial throttle
5—Maximum braking torque developed by the
   traction motor

$T_{t\text{-com}}$—Traction torque demanded by driver

$T_{t\text{-e}}$—Traction torque developed by engine torque

$T_{t\text{-tm}}$—Traction torque developed by traction motor

T_{ch}—Equivalent traction torque for PPS charging

$T_{b\text{-com}}$—Demanded braking torque

$T_{b\text{-tm}}$—Braking torque produced by traction motor

$T_{b\text{-me}}$—Braking torque produced by mechanical brake

FIGURE 10.19
Schematic illustration of Max. SOC control strategy.

for the engine speed. V_H can be expressed as

$$V_H = \frac{\pi k_{yr} r_w n_{e\text{-max}}}{30 i_{rw}} \, (\text{m/s}),\qquad(10.17)$$

where $n_{e\text{-max}}$ is the maximum engine rpm allowed.

In this medium-speed region, all the engine power is delivered to the drive wheels.

When the vehicle speed is higher than V_H, to keep the engine speed below the maximum engine speed allowed $n_{e\text{-max}}$, the motor/generator must be operated in the direction opposite to the engine speed, which can be expressed as

$$n_{m/g} = k_{ys}\left(n_{e\text{-max}} - \frac{30k_{ys}i_{rw}V}{\pi k_{yr}r_w}\right),\tag{10.18}$$

where $V \geq V_H$.

The motor/generator is in motoring. The motoring power can be expressed as

$$P_{m/g} = \frac{2\pi}{60}T_{m/g}n_{m/g} = \frac{i_{rw}}{k_{yr}r_w}T_eV - \frac{2\pi}{60}T_e n_{e\text{-max}}.\tag{10.19}$$

The first term on the right-hand side is the total power delivered to the drive wheels, and the second term is the power that the engine produces. The motor/generator accepts power from the PPS.

10.2.4.2 Traction Torque Control Strategy

Similar to the torque (power) control in the parallel hybrid drivetrain, Figure 10.19 conceptually shows the allocation of the total traction torque demanded by the driver for the engine (motor/generator) and the traction motor, or the total braking torque demanded for the traction motor and the mechanical braking system.

10.2.4.2.1 In Low Vehicle Speed Region

As mentioned previously, when the vehicle speed is lower than V_L, the engine is operated at a specified speed, $n_{e\text{-min}}$. The engine torque labeled 1 in Figure 10.19 is produced with an engine throttle at which the engine has maximum fuel utilization efficiency at this speed. This engine throttle is possibly near its full opening point.

Point A represents a traction torque demanded by the driver, which is larger than the torque that the engine can produce with the optimal engine throttle, as shown in Figure 10.16. In this case, the engine alone cannot handle this demanded torque and needs the help of the traction motor. In this case, the engine should be controlled at its optimal throttle, as shown by point B in Figure 10.19. However, the torque that the traction motor produces depends on the energy level of the PPS. When the SOC of the PPS is lower than a specified value SOC_L (30% for example), the PPS should not be further discharged. In this case, the maximum power of the traction motor is the power generated by the motor/generator described by Equation 10.16. Neglecting the losses, the traction motor torque can be expressed as

$$T_{mt} = \frac{60}{2\pi}\frac{P_{m/g}}{n_{tm}} = \left(\frac{n_{e\text{-min}}}{n_{tm}} - \frac{i_{rw}}{k_{yr}i_{mw}}\right)T_e = \left(\frac{2\pi r_w}{60i_{mw}}\frac{n_{e\text{-min}}}{V} - \frac{i_{rw}}{i_{mw}k_{yr}}\right)T_e,\tag{10.20}$$

where i_{mw} is the gear ratio from the traction motor to the drive wheels described by $i_{mw} = (Z_2Z_5)/(Z_3Z_4)$, as shown in Figure 10.6.

In this case, the planetary gear unit, the motor/generator, and the traction motor together function as an electric vehicle traction (EVT) because no energy goes into or comes out of the PPS.

When the SOC is higher than the bottom line, SOC_L, that is, the PPS has sufficient energy to support the traction motor, the traction motor should be controlled to produce its torque,

T_{tm}, to meet the demanded traction torque, as shown in Figure 10.19. In this case, the PPS supplies its power to the traction motor.

When the demanded torque, $T_{\text{t-com}}$, is smaller than the engine torque produced at optimal throttle, as shown by point B in Figure 10.19, there are several options in choosing the engine and traction motor operations: (1) With the SOC of the PPS lower than SOCL, the engine may be operated at a speed of $n_{\text{e-min}}$ and optimal throttle (point B in Figure 10.19). The PPS is charged by the motor/generator by a power of $P_{\text{m/g}}$ (Equation 10.16) and the traction motor torque, T_{ch}, as shown in Figure 10.19. (2) With the SOC of the PPS in between the SOC_L and SOC_H ($\text{SOC}_\text{L} < \text{SOC} < \text{SOC}_\text{H}$), the engine and the motor/generator may be controlled so that the engine operates at a speed of $n_{\text{e-min}}$ and produces the torque that meets the demanded traction torque. The traction motor is idling (de-energized). The PPS is charged only by the motor/generator. (3) With the SOC of the PPS higher than SOC_H, the engine is shut down, and the traction motor alone produces its torque to meet the traction torque demand.

10.2.4.2.2 In Medium Vehicle Speed Region

When the vehicle speed is in a range higher than V_L but lower than V_H as shown in Figures 10.18 and 10.19, only the torque-coupling (traditionally parallel) mode is employed, that is, lock 1 locks the sun gear of the planetary gear unit (the shaft of the motor/generator) to the stationary vehicle frame. In this mode, engine speed is proportional to vehicle speed. The engine and traction motor control strategy, based on the demanded traction torque and SOC of the PPS, is exactly the same as that discussed in Chapter 9.

10.2.4.2.3 In High Vehicle Speed Region

When the vehicle speed is higher than V_H, as shown in Figures 10.18 and 10.19, the engine speed is controlled at its top $n_{\text{e-max}}$. In this case, the motor/generator works in the motoring mode, taking energy from the PPS and delivering it to the drivetrain. The motoring power is described by Equation 10.19. The torques of the engine and the traction motor are controlled based on the demanded traction torque and energy level of the PPS.

When the demanded traction torque, $T_{\text{t-com}}$ (point G in Figure 10.19), is larger than the torque that the engine can produce with its optimal throttle at the speed of $n_{\text{e-max}}$, and the SOC of the PPS is lower than SOC_L, that is, the PPS can no longer be discharged to support the motoring operation of the motor/generator and the traction motor, the engine is forced to operate with a speed higher than the specified speed, $n_{\text{e-max}}$, to develop greater power. In this case, there are two options: one is to use the engine-alone mode with only torque coupling, which is the same as in the medium vehicle speed range; the other is to control the engine to operate at a speed somewhat higher than the speed that corresponds to the vehicle speed in the torque-coupling mode. That is,

$$n_e > \frac{30 i_{\text{rw}} V}{\pi k_{\text{yr}} r_{\text{w}}}. \qquad (10.21)$$

The term on the right-hand side is the engine speed that corresponds to the vehicle speed V in the torque-coupling mode. In this way, the motor/generator can be operated in its generating mode. The generating power from the motor/generator can feed the traction motor to generate additional traction torque. This operating mode is the EVT mode, as discussed previously.

If the SOC of the PPS is at its medium and high levels, that is, SOC > SOCL, the engine is controlled at its specified speed, $n_{\text{e-max}}$, at the optimal engine throttle (point H in Figure 10.19). The traction motor produces its torque, together with the engine torque, to meet the demanded traction torque.

In case the demanded traction torque is smaller than the engine torque at the optimal throttle, as shown by point K in Figure 10.19, and the SOC of the PPS is below SOC_L, the engine is operated at point K, and the traction motor works in its generating mode to charge the PPS. If the SOC is in the medium region ($SOC_L < SOC < SOC_H$), the traction motor may be de-energized, and the engine alone propels the vehicle (point K). If the SOC of the PPS is at a high level ($SOC > SOC_H$), the engine may be shut down, and the traction motor alone propels the vehicle.

10.2.4.3 Regenerative Braking Control

Similar to the parallel drivetrain control, when the demanded braking torque is larger than the maximum torque that the motor can produce in generating mode, both regenerative braking by the traction motor and mechanical braking are applied. Otherwise, only regenerative braking is applied. For a detailed discussion, refer to Chapter 14.

It should be noted that the control strategies discussed previously are only for guidance in a real control strategy design. More careful and insightful studies are necessary based on the special design constraints, design objectives, component characteristics, operation environments, and so on. More complicated and subtle approaches may be employed, such as fuzzy logic, dynamic programming, and so on. Further, computer simulations are very useful in designing a good control strategy.

10.3 Drivetrain Parameter Design

The design principles of drivetrain parameters, such as engine power, motor power, and the power and energy capacity of the PPS, are very similar to those in series and parallel drivetrains discussed in Chapters 8 and 9. Therefore, they are not discussed further in this chapter. However, the torque and power capacity design of the motor/generator may need further discussion.

From Equations 10.3 and 10.4 and Figure 10.5, it can be seen that the motor/regenerator torque is required to balance the engine torque at nearly full throttle, in the speed regions of lower than the minimum speed, $n_{\text{e-min}}$, and higher than the maximum speed, $n_{\text{e-max}}$. Thus, the torque capacity of the motor/generator is determined by the maximum engine torque in the low-speed and high-speed regions. However, for purposes of safety, the torque capacity of the motor/generator would be designed to be able to balance the maximum torque of the engine in its entire speed range. Figure 10.5 also indicates that this maximum motor torque should be available in its entire speed range, rather than a special point. Thus, the ideal torque–speed characteristic is a constant torque in its entire speed range, which can be expressed as

$$T_{\text{m/g-max}} = \frac{T_{\text{e-max}}}{k_{\text{ys}}}. \tag{10.22}$$

Here, $T_{\text{e-max}}$ is the maximum engine torque with fully open throttle.

From Equation 10.16, it is obvious that the generating power of the motor/generator is maximized at zero vehicle speed, that is, all the power produced by the engine goes to the motor/generator. Thus, the maximum generating power of the motor/generator can be determined by

$$P_{m/g\text{-max}} = \frac{2\pi}{60} T_{e\text{-max}} n_{e\text{-min}}. \tag{10.23}$$

Similarly, the maximum motoring power of the motor/generator occurs at the maximum vehicle speed, V_{max}, as indicated by Equation 10.19, which can be expressed as

$$P_{m/g\text{-max}} = \frac{i_{rw}}{k_{yr}r_w} T_{e\text{-max}} V_{max} - \frac{2\pi}{60} T_{e\text{-max}} n_{e\text{-max}}. \tag{10.24}$$

10.4 Simulation of an Example Vehicle

Based on the design and control principles discussed in previous sections, a 1500-kg passenger car was simulated in FTP75 urban and highway drive cycles. The parameters of the vehicle simulated are listed in Table 10.1.

Figure 10.20 shows the simulation results of vehicle speed, engine power, motor/ generator power, traction motor power, and SOC of the PPS in an FTP urban drive cycle. The motor/generator always works in the generating mode (negative power) because of the low vehicle speeds. Through the regenerative braking and charging from the engine by the motor/generator, the PPS SOC can be easily maintained at a high level, which ensures that the PPS is always able to supply sufficient power to the drivetrain for acceleration.

Figure 10.21 shows the engine operating points on the engine fuel consumption map. This figure indicates that the engine, most of the time, operates in its high-efficiency area. Engine-alone traction with light load and high SOC of the PPS causes some engine operating points away from its high-efficiency area. The fuel consumption of the vehicle in an FTP urban drive cycle obtained from the simulation is 5.88 L/100 km or 40.2 mpg.

Figures 10.22 and 10.23 show the simulation results while driving in an FTP75 highway cycle. The generator/motor power is zero, except in a short time period of cycle start. This means that the drivetrain, most of the time, worked with a pure torque coupling (the

TABLE 10.1

Vehicle Parameters

Vehicle mass	1500 kg
Engine power	28 kW
Traction motor power	40 kW
Generator motor power	15 kW
Tire rolling resistance coefficient	0.01
Aerodynamic drag coefficient	0.3
Front area	2.2 m^2

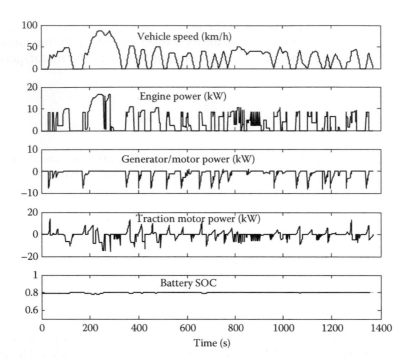

FIGURE 10.20
Vehicle speed, engine power, generator/motor power, traction motor power, and battery SOC in an FTP75 urban drive cycle.

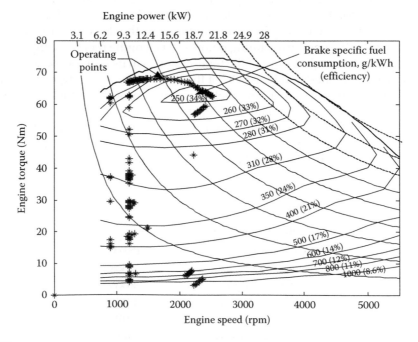

FIGURE 10.21
Engine operating points on its fuel consumption map in an FTP75 urban drive cycle.

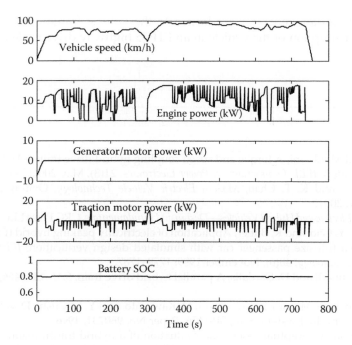

FIGURE 10.22
Vehicle speed, engine power generator/motor power, traction motor power, and battery SOC in an FTP75 highway drive cycle.

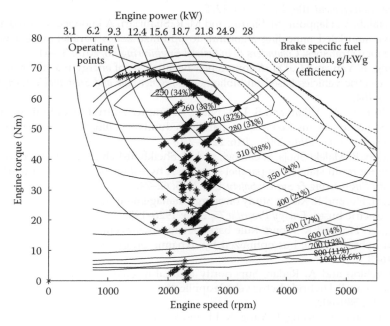

FIGURE 10.23
Engine operating points on its fuel consumption map in FTP75 highway drive cycle.

sun gear and its motor/generator are locked to the vehicle frame). Simulation indicates that the fuel consumption of the vehicle in an FTP75 highway drive cycle is 4.96 L/100 km or 47.7 mpg.

Bibliography

1. Y. Gao and M. Ehsani, A torque and speed coupling hybrid drive train—Architecture, control, and simulation. *IEEE Transactions on Power Electronics*, 21(3), May 2006: 741–748.
2. C. C. Chan and K. T. Chau, *Modern Electric Vehicle Technology*, Oxford University Press, New York, 2001.
3. I. Husani, *Electric and Hybrid Vehicles—Design and Fundamentals*, CRC Press LLC, New York, 2003.
4. M. Ehsani, Y. Gao, and K. Butler, Application of electrically peaking hybrid (ELPH) propulsion system to a full-size passenger car with simulated design verification, *IEEE Transaction On Vehicular Technology*, 48(6), November 1999: 1779–1787.
5. M. W. Nedunadi and D. Dardalis, A parallel hybrid drive train, *SAE*, SP-1466, Paper No. 1999-01-2928.
6. K. Yamaguchi, S. Moroto, K. Kobayashi, M. Kawamto, and Y. Miyaishi, Development of a new hybrid system-duel system, *SAE*, SP-1156, Paper No. 960231, 1996.
7. Y. Chen et al., Conceptual design and evaluation of a hybrid transmission with power-split, series, and two parallel configurations, *SAE International Journal of Alternative Powertrains* 6(1), 2017: 122–135.
8. S. Lin, S. Chang, and B. Li, Gearshift control system development for direct-drive automated manual transmission based on a novel electromagnetic actuator, *Mechatronics*, 24(8), 2014: 1214–1222.
9. E. Wang, D. Guo, and F. Yang, System design and energetic characterization of a four-wheel-driven series–parallel hybrid electric powertrain for heavy-duty applications, *Energy Conversion and Management* 106, 2015: 1264–1275.
10. A. Zaretalab, V. Hajipour, M. Sharifi, and M. R. Shahriari, A knowledge-based archive multi-objective simulated annealing algorithm to optimize series–parallel system with choice of redundancy strategies, *Computers & Industrial Engineering* 80, 2015: 33–44.
11. P. Zhang, F. Yan, and C. Du, A comprehensive analysis of energy management strategies for hybrid electric vehicles based on bibliometrics, *Renewable and Sustainable Energy Reviews* 48, 2015: 88–104.
12. S. M. Mousavi, N. Alikar, and S. T. A. Niaki, An improved fruit fly optimization algorithm to solve the homogeneous fuzzy series–parallel redundancy allocation problem under discount strategies, *Soft Computing* 20(6), 2016: 2281–2307.
13. A. Azadeh, B. Maleki Shoja, S. Ghanei, and M. Sheikhalishahi, A multi-objective optimization problem for multi-state series-parallel systems: A two-stage flow-shop manufacturing system, *Reliability Engineering & System Safety* 136, 2015: 62–74.
14. J. Peng, H. He, and R. Xiong, Study on energy management strategies for series-parallel plug-in hybrid electric buses, *Energy Procedia* 75, 2015: 1926–1931.
15. J. Peng, H. He, and R. Xiong, Rule based energy management strategy for a series–parallel plug-in hybrid electric bus optimized by dynamic programming, *Applied Energy* 185, 2017: 1633–1643.
16. A. Ghayebloo and A. Radan, Superiority of dual-mechanical-port-machine-based structure for series–parallel hybrid electric vehicle applications, *IEEE Transactions on Vehicular Technology* 65(2), 2016: 589–602.
17. X. Sun, C. Shao, G. Wang, L. Yang, X. Li, and Y. Yue, Research on electrical brake of a series-parallel hybrid electric vehicle, *2016 World Congress on Sustainable Technologies (WCST)* 2016: 70–75.

18. K. Huang, C. Xiang, and R. Langari, Model reference adaptive control of a series–parallel hybrid electric vehicle during mode shift, *Proceedings of the Institution of Mechanical Engineers, Part I: Journal of Systems and Control Engineering* 231(7), 2017: 541–553.
19. H. Wang, Y. Huang, A. Khajepour, and Q. Song, Model predictive control-based energy management strategy for a series hybrid electric tracked vehicle, *Applied Energy* 182, 2016: 105–114.
20. H. Zhang, Y. Zhang, and C. Yin, Hardware-in-the-loop simulation of robust mode transition control for a series–parallel hybrid electric vehicle, *IEEE Transactions on Vehicular Technology* 65(3), 2016: 1059–1069.

11

Design and Control Principles of Plug-In Hybrid Electric Vehicles

As discussed in previous chapters, in the PPS charge sustained hybrid drivetrain, the net energy consumption in the PPS in a complete drive cycle is zero, that is, the energy level in the PPS at the beginning of the drive cycle is equal to the energy level at the end of the drive cycle. All the propulsion energy comes from the primary energy source: gasoline or diesel for internal combustion (IC) engines, hydrogen or hydrogen-based fuel for fuel cells. During operation, the energy in the PPS fluctuates in a narrow window. The PPS size is determined by power rather than energy capacity. The energy-to-power ratio is in the range of 0.05–0.1 kWh/kW. With a given power capacity, the energy storage in the PPS is considered to be sufficient if it can sustain 0.05–0.1 h with the given power. Thus, the PPS is more an energy buffer than form of energy storage. This is also the origin of the name PPS (peak power source). At present and in the immediate future, ultracapacitors and high-power batteries or their combination are the most promising candidates as the PPS of the PPS charge sustained hybrid electric vehicles (HEVs). For details, refer to Chapter 13, "Peaking Power Sources and Energy Storages."

With the development and maturing of advanced battery technologies, the energy storage capacity of batteries has significantly improved. Obviously, using high-energy batteries only as a PPS is a waste.

The plug-in hybrid electric drivetrain is designed to fully or partially use the energy of the energy storage to displace part of the primary energy source, such as gasoline, diesel, hydrogen, and so on.

All configurations discussed in Chapter 6 can be employed in plug-in hybrid electric drivetrains. Most of the differences from the PPS sustained hybrid drivetrain are in the drivetrain control strategy, energy storage design, and, perhaps, slightly different electric motor power design. This chapter concentrates on these three topics.

11.1 Statistics of Daily Driving Distance

Charging the energy in the energy storage device from the utility grid to displace part of the petroleum fuel is the major feature of plug-in hybrid electric vehicles (PHEVs). The amount of petroleum fuel displaced by the utility electricity depends mainly on the amount of electrical energy per recharge, that is, the energy capacity of the energy storage; total driving distance between recharges, that is, usual daily driving distance; and electrical power usage profiles, that is, the drive cycle features and control strategies. To achieve optimal design, especially for the energy storage system, understanding the daily driving distance in a typical environment is very helpful.

Figure 11.1 is a histogram showing the daily driving distance distribution and the cumulative frequency derived from the 1995 National Personal Transportation Survey data.[1,2]

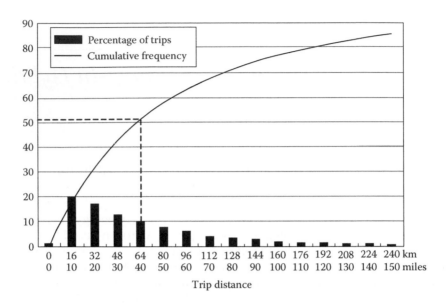

FIGURE 11.1
Daily driving distance distribution and cumulative factor.

The cumulative frequency or utility factor in reference[1] represents the percentages of the total driving time (days) during which the daily driving distances are less than or equal to said distance on the horizontal axis. Figure 11.1 reveals the fact that about half of the daily driving distance is less than 64 km (40 miles). If a vehicle is designed to have 64 km (40 miles) of pure EV range, that vehicle will have half of its total driving distance from the pure EV mode. Even if the daily traveling distance is beyond this 60 km (40 miles) pure EV range, a large amount of the petroleum fuel can be displaced by electricity because the pure EV mode takes a large portion of the daily travel. Research also shows that even if the pure EV range is less than 64 km (40 miles), such as 32 km (20 miles), still a large amount of petroleum can be displaced in normal daily driving.[1]

11.2 Energy Management Strategy

First, some definitions about PHEV are introduced:

- *Charge-depleting (CD) mode*: An operating mode in which the SOC of the energy storage may fluctuate, but on average, it decreases while driving.
- *Charge-sustaining (CS) mode*: An operating mode in which the SOC of the energy storage may fluctuate but on average is maintained at a certain level while driving.
- *All electric range (AER)*: After a full recharge, the total miles (kilometers) driven electrically (engine off) before the engine turns on for the first time.
- *Electric vehicle miles (EVM) or kilometers (EVKM)*: After a full recharge, the cumulative miles or kilometers driven electrically (engine off) before the vehicle reaches CS mode.

- *Charge-depleting range (CDR)*: After a full recharge, the total miles or kilometers driven before the vehicle reaches CS mode. It should be noted that EVM or EVKM indicates pure electric driving. However, CDR may include engine propulsion, but the on-average SOC of the energy storage decreases till the sustaining level.

- *PHExx*: A PHEV with useable energy storage equivalent to *xx* miles of driving energy on a reference driving cycle, where *xx* stands for the mileage number. For example, $PHEV_20$ can displace petroleum energy equivalent to 20 miles of driving on the reference drive cycle with off-board electricity. A similar definition can be made in kilometers. It should be noted that $PHEV_20$, for example, does not imply that the vehicle will achieve 20 miles of AER, EVM, or CDR on the reference cycle, or any other cycle. Operating characteristics also depend on the power ratings of components, the power train control strategy, and the nature of the driving cycle.[1]

11.2.1 AER-Focused Control Strategy

The idea of this control strategy is to use the energy of the energy storage intensively in the AER.[3,4] One possibility is to allow the driver to manually select between a CS mode and a full EV operating mode. This design could be useful for vehicles that may be used in the region where combustion engine use is restricted. This design provides flexibility for the driver to determine the times that the pure EV mode is used. For example, in a trip that includes places where pure EV operation is required, the driver can select the pure EV operating mode just prior to entering this area to have sufficient range. In other places, the vehicle may be operated in pure EV mode or CS mode, depending on the energy status of the energy storage and the power demand. In normal conditions, where the trip does not include mandatory pure EV operation, the driver could select the pure EV mode at the start of the trip to fully use the energy of the energy storage to displace the petroleum fuel until the energy of the energy storage reaches its specified level, at which time the CS mode starts automatically.

This energy management approach clearly divides the whole trip into pure EV and CS modes. Thus, the design and control techniques developed for EV and HEV in previous chapters can be used. When a series hybrid configuration is used, the power rating designs of the motor, engine, and energy storage are almost the same as in the CS hybrid. The motor power guarantees the acceleration and gradeability performance, the engine/generator power supports the vehicle driving at a constant speed on flat or mild grades, and the energy storage power is larger (or at least not smaller) than the motor power minus the engine/generator power. However, the energy storage must be designed so that its useable energy can meet the requirements of the pure EV range. When a parallel or series-parallel configuration is used, the motor power should be designed to meet the peaking power requirements of the reference driving cycles. Otherwise, the vehicle will not be able to follow the speed profile of the drive cycle and will be somewhat sluggish compared to the driver's expectation.

The traction power computations in typical drive cycles were discussed in detail in previous chapters. However, for the reader's convenience, they are repeated below.

The traction power on the drive wheels includes the rolling resistance, aerodynamic drag, inertial force of acceleration, and grade resistance, which can be expressed as

$$P_t = \frac{V}{1000}\left(Mgf_r + \frac{1}{2}\rho_a C_D A_f V^2 + M\delta\frac{dV}{dt} + Mgi\right)(kW),\qquad(11.1)$$

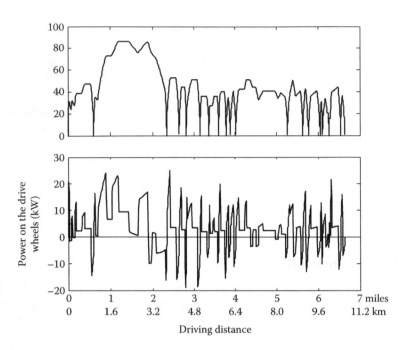

FIGURE 11.2
Vehicle speed and traction power in an FTP75 urban drive cycle.

where M is the vehicle mass in kg, V is the vehicle speed in m/s, g is the gravity acceleration, 9.81 m/s^2, ρ_a is the air mass density, 1.205 kg/m^3, C_D is the aerodynamic drag coefficient of the vehicle, A_f is the front area of the vehicle in m^2, δ is the rotational inertia
factor, dV/dt is the acceleration in m/s^2, and i is the road grade. In standard drive cycles,
flat roads are used.

Figure 11.2 is a diagram showing the vehicle speed and the traction power, on the drive
wheels, versus the traveling distance in the FTP75 urban drive cycle. The vehicle parameters
used in this computation are listed in Table 11.1. Figure 11.2 indicates that the peaking traction power on the drive wheels is about 25 kW. However, there are power losses in the path
from the energy storage to the drive wheels. To meet the power requirements, the motor
output power should be designed to account for the power losses from the motor shaft to
the drive wheels. Suppose that the efficiency from the motor shaft to the drive wheels is
90%; then the motor shaft power rating is about 28 kW. It should be noted that this required

TABLE 11.1

Vehicle Parameters Used in Power Computation

Vehicle mass (kg)	1700
Rolling resistance coefficient	0.01
Aerodynamic drag coefficient	0.3
Front area (m^2)	2.2
Rotational inertia factor	1.05

TABLE 11.2

Powers of Motor and Energy Storage in Typical Drive Cycles

Cycle Power (kW) Item	FTP75 Urban	FTP75 Highway	LA92	US06
Motor at vehicle speed	28 at 50 km/h (31 mph)	32 at 72 km/h (57 mph)	55 at 57 km/h (36 mph)	98 at 117 km/h (73 mph)
Energy storage	35.7	39	68.5	121

motor power is also related to the vehicle speed at which this peak power occurs. For example, the peak power in Figure 11.2 occurs at a vehicle speed of 50 km/h (31.25 mph). In the motor power design, we must be sure that the motor can produce this peak power at this vehicle speed. Similarly, the peaking power of the energy storage should include the losses in the electric motor, the power electronics, and the transmission. Suppose that the efficiencies of the motor and the power electronics are 0.85 and 0.95, respectively; then the power capacity of the energy storage is about 34.7 kW in this example. Table 11.2 lists the motor power and the energy storage power in FTP75 urban, FTP75 highway, LA92, and US06 drive cycles.

Integrating Equation 11.1 over the driving time in a driving cycle can give the energy consumption by the drive wheels, as shown in Figure 11.3. Here, no regenerative braking is included. When including energy losses in the power electronics, the motor, and the transmission, the useable energy in the energy storage for 32 km (20 miles) and 64 km (40 miles) of pure EV driving in typical drive cycles is listed in Table 11.3.

In vehicle design, an appropriate reference drive cycle should be selected. An aggressive drive cycle, such as US06, needs a large motor drive and energy storage but also gives good vehicle acceleration and gradeability performance. In contrast, a mild driving cycle,

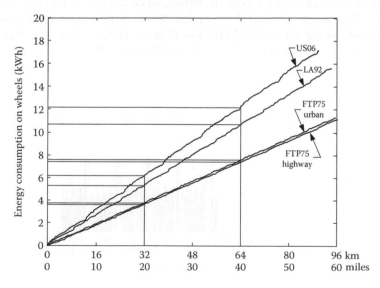

FIGURE 11.3

Energy consumption by drive wheels versus driving distance in typical drive cycles.

TABLE 11.3

Energy Consumption in Typical Drive Cycles

Cycle Energy (kWh) Distance	FTP75 Urban	FTP75 Highway	LA92	US06
32 km (20 miles)	5.2	5.14	7.29	8.4
64 km (40 miles)	10.4	10.28	14.58	16.8

such as FTP75, leads to a small motor drive and energy storage but also sluggish vehicle performance.

The following figures show simulation results of the drivetrain in the reference driving cycle, FTP75 urban. The vehicle parameters listed in Table 11.1 were used. The total energy in the energy storage, fully charged, is 10 kWh. The simulation ran nine sequential cycles, and the pure EV mode was started at the beginning of the simulation until the SOC reached about 30%, beyond which the CS mode was started. The control strategy in the CS mode employed the constrained engine on–off control strategy, discussed in Section 9.2.3. In the simulation, 400 W of constant auxiliary power was added at the terminal of the energy storage.

Figures 11.4 and 11.5 show the engine power and the motor power. Figure 11.6 shows the SOC of the energy storage and the remaining energy in the energy storage versus the traveling distance. The pure EV mode range is about 32 km (20 miles). Figure 11.7 shows the engine operating points overlapping its brake-specific fuel consumption map.

Figures 11.8 and 11.9 show the fuel and electric energy consumption scenarios in metric and English units, respectively. When the traveling distance is less than four drive cycles (42.5 km or 26.6 miles), the vehicle can completely displace the petroleum fuel with electricity in pure EV mode. The total electric energy consumed is about 7.1 and 15.5 kWh per 100 km, or 4.05 miles/kWh (Figure 11.9). With increasing total traveling distance, the percentage of the fuel displacement decreases, since the CS modes take larger percentages of the trip. For nine sequential drive cycles (96 km or 60 miles), the fuel and electrical energy consumptions are about 3.2 L/100 km (Figure 11.8) or 74 mpg (Figure 11.9) and 7.42 kWh/100 km (Figure 11.8) or 8.43 mile/kWh (Figure 11.9).

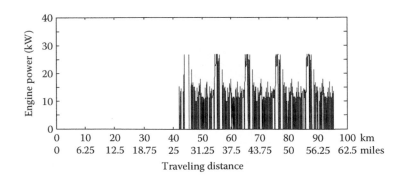

FIGURE 11.4

Engine power versus traveling distance in FTP75 urban drive cycle in AER mode.

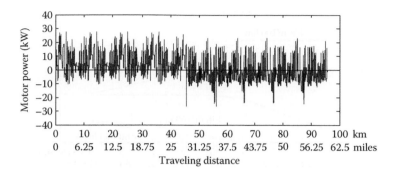

FIGURE 11.5
Motor power versus traveling distance in FTP75 urban driving cycle in AER mode.

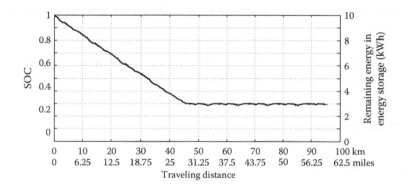

FIGURE 11.6
SOC and the remaining energy in the energy storage versus traveling distance in FTP75 urban driving cycle with AER mode.

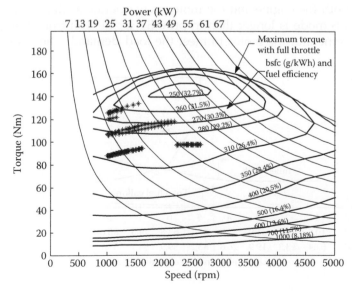

FIGURE 11.7
Engine operating points overlapping its fuel consumption map in FTP75 urban drive cycle in AER mode.

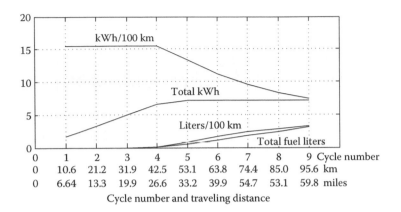

FIGURE 11.8
Fuel and electric energy consumption versus number of FTP75 urban drive cycle and traveling distance in AER mode in metric units.

A simulation of the same design in the LA92 drive cycle was also performed. The results are illustrated in the following figures (Figures 11.10 through 11.15). Comparing the two drive cycles, the LA92 drive cycle has higher vehicle speed and larger acceleration rate. The pure EV range is shorter, and the fuel and electric energy consumptions are higher than in FTP75 urban drive cycles.

11.2.2 Blended Control Strategy

Unlike the AER-focused control strategy in which a pure EV range is designed, the blended control strategy uses both the engine and the motor for traction, in CD mode, until the SOC of the energy storage reaches the specified low threshold, beyond which the vehicle operates in CS mode.

In CD mode, both the engine and the motor may operate at the same time. The range before entering CS mode is longer than in pure EV mode. Control strategies are needed to control the engine and the motor to meet the load demand. There are many possible

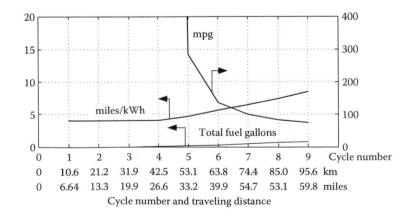

FIGURE 11.9
Fuel and electric energy consumption versus number of FTP75 urban drive cycle and traveling distance in AER mode in English units.

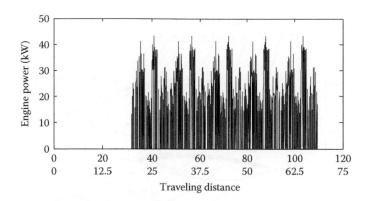

FIGURE 11.10
Engine power versus traveling distance in LA92 drive cycle in AER mode.

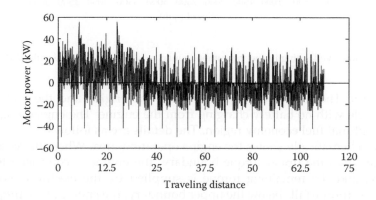

FIGURE 11.11
Motor power versus traveling distance in LA92 drive cycle in AER mode.

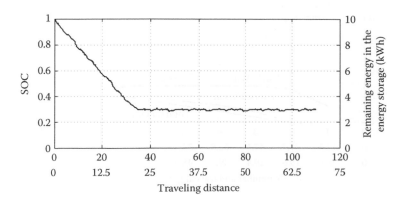

FIGURE 11.12
SOC and remaining energy in energy storage versus traveling distance in LA92 drive cycle in AER mode.

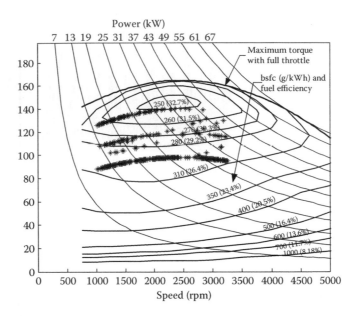

FIGURE 11.13
Engine operating points overlapping its fuel consumption map in LA92 driving cycle with AER mode.

control strategies. The following is one in which the engine and the motor alternately propel the vehicle with no battery charging from the engine. The engine is constrained to operate in its optimal fuel economy region. The details are as follows.

Figure 11.16 schematically shows the engine operating area. When the requested engine torque is larger than the upper torque boundary, the engine is controlled to operate on this boundary, and the remaining torque is supplied by the electric motor. When the requested engine torque falls below the upper boundary, the engine alone propels the vehicle. When the requested engine torque is below the lower torque boundary, the engine is shut down, and the electric motor alone propels the vehicle. In this way, the engine

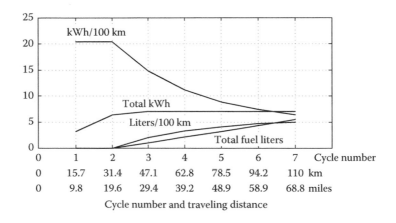

FIGURE 11.14
Fuel and electric energy consumption versus number of LA92 drive cycle and traveling distance in AER mode in metric units.

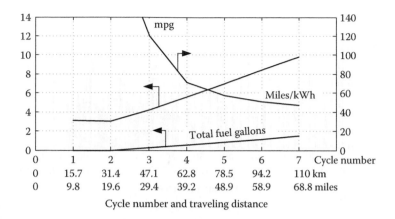

FIGURE 11.15
Fuel and electric energy consumption versus number of LA92 drive cycle and traveling distance in AER mode in English units.

operation is constrained within its optimal region. Due to the absence of battery charging from the engine, the battery energy level continuously falls to its specified lower level. Then the drivetrain goes into CS mode.

The example vehicle mentioned was simulated with the control strategy discussed previously in nine sequences of the FTP75 urban drive cycle. The results are shown in Figures 11.17 through 11.22.

Similarly, simulation results in seven sequences of the LA92 drive cycle are shown in Figures 11.23 through 11.28.

It should be noted that the pure EV range is mostly determined by the capacity of the energy storage and its SOC level, at which the CS mode started. The range in CD mode is

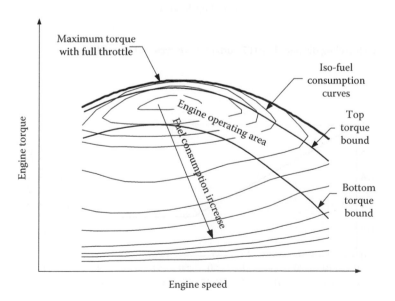

FIGURE 11.16
Operation area of engine in CD mode.

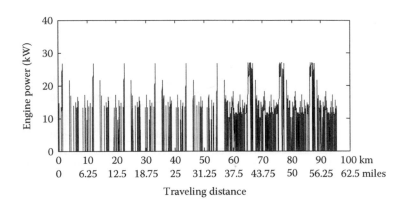

FIGURE 11.17
Engine power versus traveling distance in FTP75 urban drive cycle in CD mode.

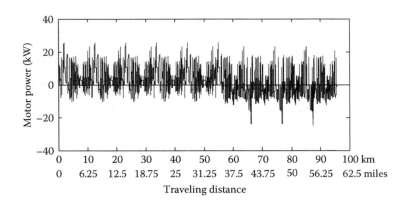

FIGURE 11.18
Motor power versus traveling distance in FTP75 urban drive cycle in CD mode.

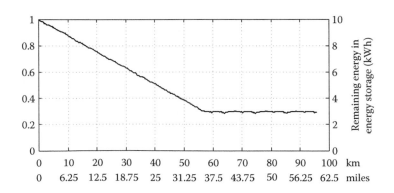

FIGURE 11.19
SOC and remaining energy in energy storage versus traveling distance in FTP75 urban drive cycle in CD mode.

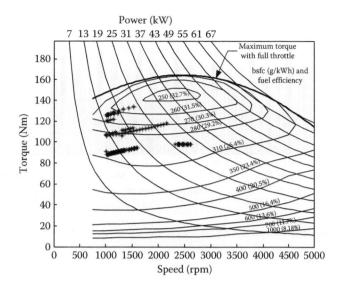

FIGURE 11.20
Engine operating points superimposed on its fuel consumption map in FTP75 urban drive cycle in CD mode.

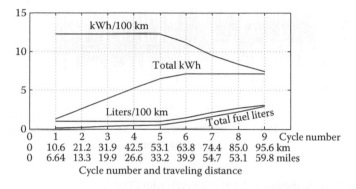

FIGURE 11.21
Fuel and electric energy consumption versus number of FTP75 urban drive cycle and traveling distance in CD mode in metric units.

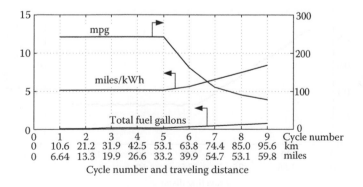

FIGURE 11.22
Fuel and electric energy consumption versus number of FTP75 urban drive cycle and traveling distance in CD mode in English units.

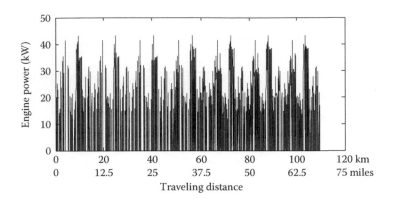

FIGURE 11.23
Engine power versus traveling distance in LA92 drive cycle in CD mode.

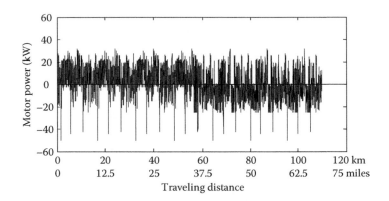

FIGURE 11.24
Motor power versus traveling distance in LA92 drive cycle with CD mode.

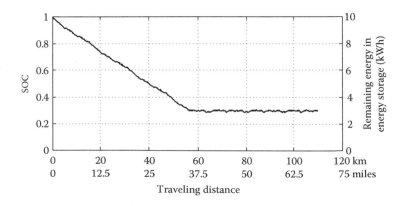

FIGURE 11.25
SOC and remaining energy in energy storage versus traveling distance in LA92 drive cycle in CD mode.

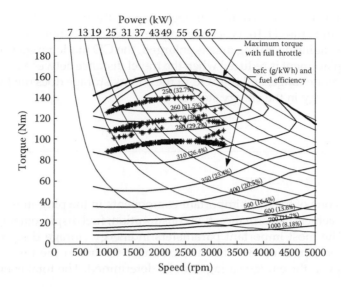

FIGURE 11.26
Engine operating points superimposed on its fuel consumption map in LA92 drive cycle with CD mode.

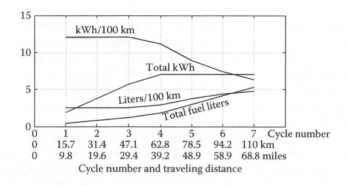

FIGURE 11.27
Fuel and electric energy consumption versus number of LA92 drive cycle and traveling distance in CD mode in metric units.

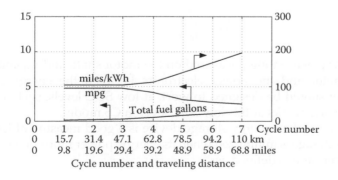

FIGURE 11.28
Fuel and electric energy consumption versus number of LA92 drive cycle and traveling distance in CD mode in English units.

also related to the drivetrain control strategy, especially the specified engine operating region, as shown in Figure 11.16. When the engine lower torque boundary is moved downward, that is, the engine operating area is enlarged, the CD mode range is increased. However, the fuel displacement is reduced when the travel distance between chargings is shorter than the full CD mode range, and the SOC of the energy storage does not hit its lower limit, leaving useable energy in the energy storage.

11.3 Energy Storage Design

Energy storage is one of the most important components in the plug-in hybrid vehicle. It is closely related to vehicle performance, fuel consumption, fuel displacement, initial cost, and operation cost. The most important parameters in energy storage design are the storage energy and power capacities. By simulation, similar to that in the preceding discussion, the useable energy in the energy storage can be determined. The total energy capacity can be obtained by

$$E_c = \frac{E_{\text{usable}}}{\text{SOC}_{\text{top}} - \text{SOC}_{\text{bottom}}}, \tag{11.2}$$

where E_{usable} is the useable energy in the energy storage, consumed in pure EV or CD mode; SOC_{top} is the top SOC with fully charged energy storage (which is usually equal to 1); and $\text{SOC}_{\text{bottom}}$ is the SOC of the energy storage at which the operation mode is switched from pure EV or CD mode to CS mode. In the previous example, the useable energy is about 7 kWh (Figures 11.6, 11.8, 11.12, 11.14, 11.19, 11.21, 11.25, and 11.27), and the SOC operating window is 0.7 (from 1 to 0.3). The total energy capacity of the energy storage is about 10 kWh.

It should be noted that the depth of discharge (DOD) of batteries is closely related to battery life. Figure 11.29 illustrates the battery cycle life with the DOD.[5,6] If one deep discharge per day is assumed, a total of 4000+ deep charges would be required for a 10- to 15-year lifetime. With the characteristics shown in Figure 11.29, a 70% DOD for NiMH and a 50% DOD for Li-ion batteries may be the proper designs.

The power requirement of the energy storage is completely determined by the electric motor power rating, which can be expressed as

$$P_{\text{es}} \geq \frac{P_{\text{m}}}{\eta_{\text{m}} \eta_{\text{pe}}}, \tag{11.3}$$

where P_{m} is the motor power rating, measured on the motor shaft, and η_{m} and η_{pe} are the efficiencies of the motor and the power electronics between the energy storage and the motor. This power should be designed to work at low SOC levels, such as 30%, since the energy storage always works at this low SOC level in CS mode.

The energy/power ratio of an energy storage device is a good measure of its suitability. The size of the energy storage is minimized when its energy/power ratio equals the required one. The energy/power ratio is defined as

$$R_{\text{e/p}} = \frac{\text{total energy}}{\text{power at operating SOC}}. \tag{11.4}$$

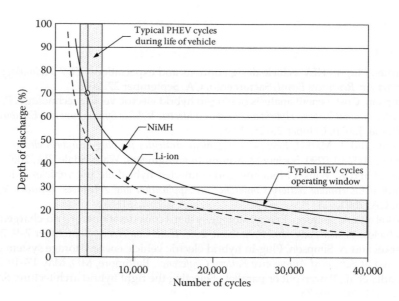

FIGURE 11.29
Cycle life characteristics of VARTA energy storage technologies.

In the example vehicle simulated previously, the total energy required is about 10 kWh, and the power required is about 60 kW, which is defined at 30% of the battery SOC, yielding an $R_{e/p}$ of 0.167 h at 30% battery SOC. Figure 11.30 shows a typical energy/power ratio versus the specific power of energy storage technologies.[5] If 0.2 h of the energy/power ratio is used in the design, Cobasys' NiMH battery yields a total weight of 129 kg (60/0.465), which carries 12 kWh of total energy (0.2 × 60). However, when SAFT's Li-ion battery is used, the total weight is about 56 kg (60/1.08), carrying the same amount of energy of 12 kWh.

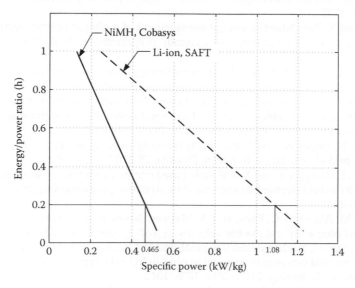

FIGURE 11.30
Typical energy/power ratios versus specific power.

Bibliography

1. T. Markel, Plug-in HEV vehicle design options and expectations, *ZEV Technology Symposium, California Air Resources Board*, Sacramento, CA, September 27, 2006.
2. A. Simpson, Cost–benefit analysis of plug-in hybrid electric vehicle technology, *Presented at the 22nd International Battery, Hybrid and Fuel Cell Electric Vehicle Symposium and Exhibition (EVS-22)*, Yokohama, Japan, October 23–28, 2006.
3. J. Gonder and T. Markel, *Energy management strategies for plug-in hybrid electric vehicles*, SAE Paper no. 2007-01-0290, Society of Automotive Engineers, Warrendale, PA, 2007.
4. T. Markel and K. Wipke, Modeling grid-connected hybrid electric vehicles using ADVISOR, *The 16th IEEE Annual Battery Conference on Application and Advances*, January 9, 2001, Long Beach, CA, 2001.
5. T. Markel and A. Simpson, Energy storage systems considerations for grid-charged hybrid electric vehicles, *Vehicle Power and Propulsion, 2005 IEEE Conference*, September 7–9, 2005.
6. T. Markel and A. Simpson, Plug-in hybrid electric vehicle energy storage system design, to be presented at *Advanced Automotive Battery Conference*, Baltimore, MD, May 17–19, 2006.
7. J. Kapadia et al., Powersplit or parallel - Selecting the right hybrid architecture, *SAE Int. J. Alt. Power.*, 6(1), 2017: 68–76.
8. S. Zhang, R. Xiong, and F. Sun, Model predictive control for power management in a plug-in hybrid electric vehicle with a hybrid energy storage system. *Appl Energy*, 185, 2015: 1654–1662. http://dx.doi.org/10.1016/j.apenergy.2015.12.035.
9. X. Wang, H. Hongwen, S. Fengchun, and Z. Jieli, Application study on the dynamic programming algorithm for energy management of plug-in hybrid electric vehicles. *In Energies*, 8(4), 2015: 3225.
10. M. F. M. Sabri, K. A. Danapalasingam, and M. F. Rahmat, A review on hybrid electric vehicles architecture and energy management strategies. *In Renewable and Sustainable Energy Reviews*, 53, 2016: 1433–1442.
11. C. Zheng, X. Bing, Y. Chenwen, and C. M. Chunting, A novel energy management method for series plug-in hybrid electric vehicles. *In Applied Energy*, 145, 2015: 172–179.
12. M. Montazeri-Gh and M.-K. Mehdi, An optimal energy management development for various configuration of plug-in and hybrid electric vehicle. *In Journal of Central South University*, 22(5), 2015: 1737–1747.
13. Z. Shuo and X. Rui, Adaptive energy management of a plug-in hybrid electric vehicle based on driving pattern recognition and dynamic programming. *In Applied Energy*, 155, 2015: 68–78.
14. Z. Chen, X. Rui, W. Kunyu, and J. Bin, Optimal energy management strategy of a plug-in hybrid electric vehicle based on a particle swarm optimization algorithm. *Energies*, 8(5), 2015: 3661.
15. C.-A. Andrea, O. Simona, and R. Giorgio, A control-oriented lithium-ion battery pack model for plug-in hybrid electric vehicle cycle-life studies and system design with consideration of health management. *In Journal of Power Sources*, 279, 2015: 791–808.
16. P. Jiankun, H. Hongwen, and X. Rui, Study on energy management strategies for series-parallel plug-in hybrid electric buses. *In Energy Procedia*, 75, 2015: 1926–1931.
17. P. Jiankun, H. Hongwen, and X. Rui, Rule based energy management strategy for a series–parallel plug-in hybrid electric bus optimized by dynamic programming. *In Applied Energy*, 185, 2017: 1633–1643.
18. J. Carroll, M. Alzorgan, C. Page, and A. Mayyas, Active battery thermal management within electric and plug-in hybrid electric vehicles. *In SAE Technical Paper*, 2016.
19. X. Hu, S. J. Moura, N. Murgovski, B. Egardt, and D. Cao. Integrated optimization of battery sizing, charging, and power management in plug-in hybrid electric vehicles. *IEEE Transactions on Control Systems Technology*, 24(3), 2016: 1036–1043.
20. K. Dominik, S. Vadim, and J. Jeong, Fuel saving potential of optimal route-based control for plug-in hybrid electric vehicle. *In IFAC-PapersOnLine*, 49(11), 2016: 128–133.

12

Mild Hybrid Electric Drivetrain Design

Full hybrid electric vehicles (HEVs) with parallel, series, or series–parallel configurations can significantly reduce fuel consumption by operating the engine optimally and using effective regenerative braking.[1-3] However, a high electric power demand requires a bulky and heavy energy storage pack. This causes difficulties for packing the drivetrain under the hood and reduces the loading capacity, and it also increases the energy losses in the rolling tires. Full hybrid drivetrains have structures totally different from conventional drivetrains. To transition totally from conventional drivetrains to full hybrids, a huge investment of time and money is needed. A compromise is to develop an intermediate product that is easier to convert from the current products and yet is more efficient than those products. One solution is to put a small electric motor behind the engine to constitute the so-called mild or soft hybrid electric drivetrain. This small electric motor operates as an engine starter as well as an electrical generator. It can also add additional power to the drivetrain when high power is demanded and can convert part of the braking energy into electric energy. This small motor can potentially replace the clutch or the torque converter, which is inefficient when operating with a high slip ratio.

A mild hybrid electric drivetrain does not require high-power energy storage due to the small power rating of the electric motor. A 42-V electrical system may be able to meet requirements. Other subsystems of the conventional vehicle, such as engine, transmission (gearbox), and brake, do not require many changes.

This chapter introduces two typical configurations of mild hybrid drivetrains. Their control and parametric designs are explained along with a design example.

12.1 Energy Consumed in Braking and Transmission

As indicated in Chapter 14, a significant amount of energy is consumed in braking, especially when driving in urban areas. Chapter 14 also indicates that the braking power in normal driving is not large (Figure 14.6).[1,4] Thus, a small motor is able to recover most of the braking energy.

Another source of energy loss in conventional vehicles is the transmission. Conventional vehicles are usually equipped with automatic transmissions, especially in North America. In an automatic transmission, the dynamic hydraulic torque converter is the basic element and has low efficiency when operating at a low speed ratio (high-speed slip), as shown in Figure 12.1.

When the vehicle is operating in a stop-and-go driving pattern in urban areas, the frequent accelerating of the vehicle leads to a low speed ratio in the torque converter, resulting in low operation efficiency. Figure 12.2 shows the operating efficiency of a typical automatic transmission in an FTP75 urban drive cycle. In this drive cycle, the average efficiency is about 0.5.[5,6]

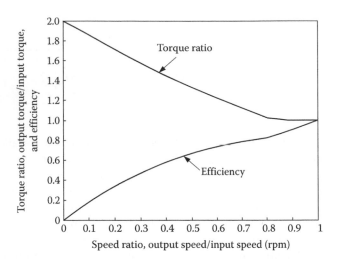

FIGURE 12.1
Characteristics of a typical dynamic hydraulic torque converter.

In addition, when driving in urban areas, the engine idling time during standstill and braking is significant. In the FTP75 drive cycle, the percentage of engine idling time reaches 44%, and in New York City, it reaches about 57%. When the engine is idling, not only does the engine itself consume energy, but also the energy is needed to drive the transmission. For instance, about 1.7 kW of engine power is needed to drive an automatic transmission when a vehicle is at a standstill.

Using a small electric motor to replace the torque converter and then constitute a mild hybrid electric drivetrain is an effective approach to saving the energy losses in an automatic transmission and in braking and engine idling operations.

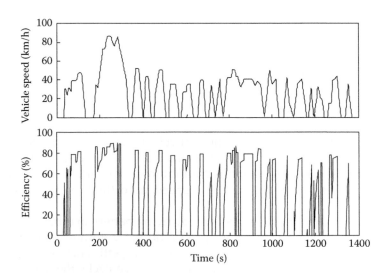

FIGURE 12.2
Vehicle speed and operating efficiency of an automatic transmission while driving in an FTP75 urban drive cycle.

12.2 Parallel Mild Hybrid Electric Drivetrain

12.2.1 Configuration

A parallel connected mild hybrid electric drivetrain is shown in Figure 12.3. A small electric motor, which can function as engine starter, generator, and traction motor, is placed between the engine and the automatically shifted multigear transmission (gearbox). The clutch is used to disconnect the gearbox from the engine when required, such as during gear shifting and low vehicle speed. The power rating of an electric motor may be in a range of about 10% of the engine power rating. An electric motor can be smoothly controlled to operate at any speed and torque; thus, isolation between electric motor and transmission is not necessary. The operation of the drivetrain and each individual component is controlled by the drivetrain controller and the component controllers.

12.2.2 Operating Modes and Control Strategy

The drivetrain has several operating modes, depending on the operation of the engine and the electric motor.

Engine-alone traction mode—In this mode, the electric motor is de-energized, and the vehicle is propelled by the engine alone. This mode may be used when the SOC

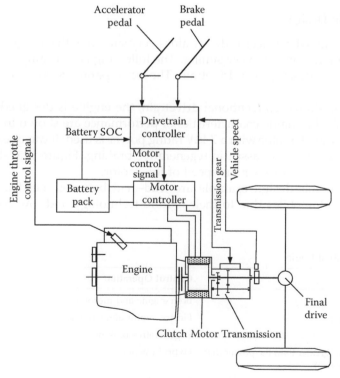

FIGURE 12.3
Configuration of parallel connected mild hybrid electric drivetrain.

of the batteries is in the high region and the engine alone can handle the power demand.

Motor-alone traction mode—In this mode, the engine is shut down, and the clutch is disengaged (open). The vehicle is propelled by the electric motor-alone. This operating mode may be used at low vehicle speed, less than 10 km/h, for example.

Battery charge mode—In this mode, the electric motor operates as a generator and is driven by the engine to charge the batteries.

Regenerative braking mode—In this mode, the engine is shut down, and the clutch is disengaged. The electric motor is operated to produce a braking torque to the drivetrain. Part of the kinetic energy of the vehicle mass is converted into electric energy and stored in the batteries.

Hybrid traction mode—In this mode, both the engine and the electric motor deliver traction power to the drivetrain.

Which of the operating modes is used in real operation depends on the power demand, which is commanded by the driver through the accelerator or the brake pedal, the SOC of the batteries, and vehicle speed.

The control strategy is the preset control logic in the drivetrain controller. The drivetrain controller receives the real-time signals from the driver and each individual component (refer to Figure 12.3) and then commands the operation of each component, according to the preset control logic. A proposed control logic is illustrated in Table 12.1 and Figure 12.4.[5]

12.2.3 Drivetrain Design

The design of a mild hybrid electric drivetrain is very similar to the design of a conventional drivetrain because the two are very similar. The following is an example of the systematic design of a 1500-kg passenger car drivetrain. The major parameters of the vehicle are listed in Table 12.2.

Referring to the similar conventional drivetrain, the engine is designed to have a peak power of 108 kW. The engine characteristics of performance are shown in Figure 12.5.

In this design, a small motor with a 7-kW rating power is used; it can operate as an engine starter and an alternator and assist in regenerative braking. Figure 12.6 shows the torque and the power characteristics versus speed of this motor.

The batteries in this design example are lead–acid batteries. Lead–acid batteries are widely used in automobiles due to their mature technology and low cost. They have a

TABLE 12.1

Illustration of Control Logic

Driving Condition	Control Operation
Standstill	Both engine and motor are shut down
Low-speed (<10 km/h)	Electric motor-alone traction
Braking	Regenerative braking
High power demand (greater than the power that the engine can produce)	Hybrid traction
Medium and low power demand	Battery charge mode or engine-alone traction mode, depending on battery SOC (Figure 12.4)

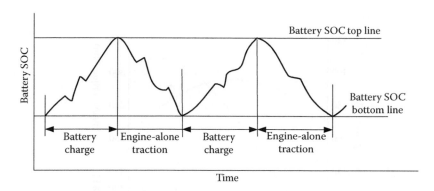

FIGURE 12.4
Battery charge and engine-alone traction, depending on battery SOC.

relatively high power density compared to other kinds of common batteries.[7] Thus, they are considered to be the proper selection for mild HEVs, in which power density is more important than energy density.

A cell of a lead–acid battery has the characteristics shown in Figure 12.7. The terminal voltage varies with discharging current and time, which in turn represent the SOC of the battery. These characteristics may be simply modeled as shown in Figure 12.8.

In the discharging process, the battery terminal voltage can be expressed as

$$V_t = V_o(SOC) - [R_i(SOC) + R_c]I, \tag{12.1}$$

where $V_o(SOC)$ and $R_i(SOC)$ are the open circuit voltage and internal resistance of the battery, respectively, which are functions of the battery SOC, and R_c is the conductor resistance. The discharging power at the terminals can be expressed as

$$P_t = I V_o(SOC) - [R_i(SOC) + R_c]I^2. \tag{12.2}$$

TABLE 12.2

Major Parameters of Mild Hybrid Electric Drivetrain

Vehicle Mass	**1500 kg**
Rolling resistance coefficient	0.01
Aerodynamic drag coefficient	0.28
Front area	$2.25 \, \text{m}^2$
Four-gear transmission	
Gear ratio:	
First gear	2.25
Second gear	1.40
Third gear	1.00
Fourth gear	0.82
Final gear ratio	3.50

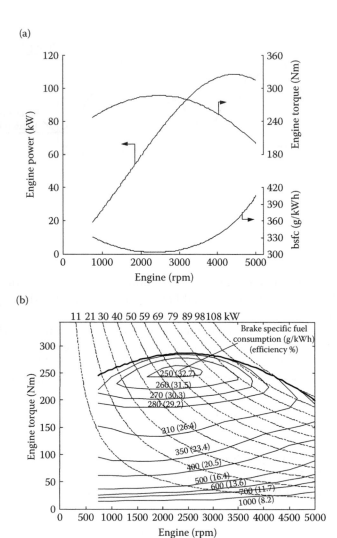

FIGURE 12.5
Performance of engine: (a) performance with full throttle and (b) fuel consumption map.

The maximum power that the load can attain at the terminals can be expressed as

$$P_{t\,max} = \frac{V_o^2(SOC)}{4[R_i(SOC)+R_c]}.$$

(12.3)

This maximum power is attained when the discharging current is

$$I = \frac{V_o}{2[R_i(SOC)+R_c]}.$$

(12.4)

Figure 12.9a shows the terminal voltages and currents of 36- and 12-V batteries with a current capacity of 100 Ah versus load power (discharge power). This indicates that for the 36-V

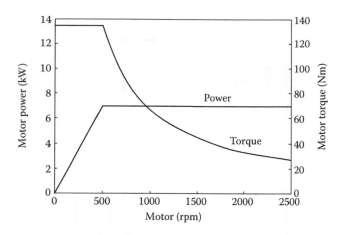

FIGURE 12.6
Power and torque of electric motor versus motor speed.

battery, the maximum power that the battery can supply is about 8.5 kW, but for the 12-V battery, it is less than 3 kW. Figure 12.9b shows that the 36-V battery has a discharge efficiency of over 70% at a power of less than 7 kW. For the 12-V battery, it is less than 2.5 kW. Thus, for the mild hybrid electric drivetrain proposed in this chapter, a 42-V electric system (36-V battery) can support the operation of the electric motor (rated power of 7 kW).

12.2.4 Performance

Because there are few differences from a conventional drivetrain (e.g., engine, transmission), a mild hybrid electric drivetrain is expected to have similar acceleration and gradeability performance. Figure 12.10 shows the performance of an example 1500-kg mild hybrid passenger car.

Figure 12.11 shows the simulation results of a 1500-kg hybrid passenger car in an FTP75 urban cycle. Figure 12.11b indicates that a mild hybrid electric drivetrain with a small motor

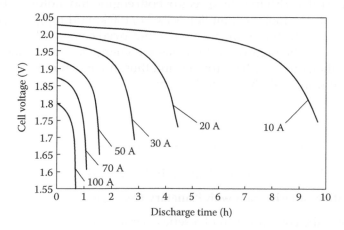

FIGURE 12.7
Discharge characteristics of lead–acid battery.

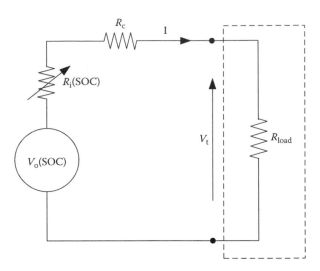

FIGURE 12.8
Battery model.

cannot significantly improve engine operating efficiency because most of the time the engine still operates in a low-load region. However, because of the elimination of engine idling and of the inefficient torque converter and utilization of regenerative braking, fuel economy in urban driving is significantly improved. The simulation shows that for the 1500-kg passenger car mentioned previously, fuel consumption is 7.01 L/100 km (33.2 mpg). The simulated fuel consumption for a similar conventional vehicle is 10.7 L/100 km (22 mpg), whereas the Toyota Camry (1445 kg curb weight, four-cylinder, 2.4 L, 157 hp, or 117 kW maximum engine power, automatic transmission) has a fuel economy of about 10.3 L/100 km (23 mpg).[8] With mild hybrid technology, fuel consumption can be reduced by more than 30%. Figure 12.11c shows the motor efficiency map and the operating points. It indicates that the electric motor operates as a generator more than a traction motor to support the electric load of auxiliaries and maintain the battery SOC balance.

Figure 12.12 shows the simulation results of the same vehicle on an FTP75 highway drive cycle. Compared to urban driving, the speeds of both engine and motor are higher due to the higher vehicle speed. The fuel consumption is 7.63 L/100 km (31 mpg) (Toyota Camry: 7.38 L/100 km or 32 mpg).[8] The fuel economy has not improved compared to conventional vehicles. The reason is that the highway vehicle has less energy loss in engine idling, braking, and transmission than during urban driving, and thus, not much room exists for fuel economy improvement using mild hybrid technology.

12.3 Series–Parallel Mild Hybrid Electric Drivetrain

12.3.1 Configuration of Drivetrain with Planetary Gear Unit

Figure 12.13 shows the configuration of a series–parallel (speed coupling and torque coupling) mild hybrid electric drivetrain that uses a planetary gear unit to connect the engine, motor, and transmission (gearbox) together. This configuration is very similar to that shown

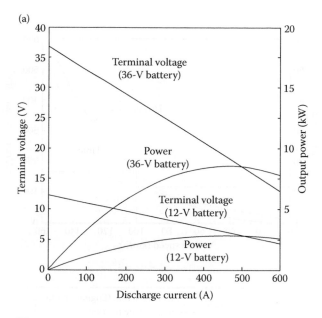

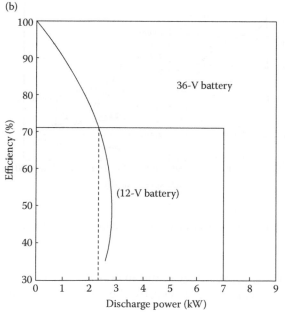

FIGURE 12.9
Battery performance with 36- and 12 V-rated voltages: (a) battery power and terminal voltage versus discharge current and (b) battery discharge efficiency.

in Figure 10.6 in Chapter 10, but it removes the traction motor and replaces it with a multi-gear transmission. The engine is connected to the yoke of the planetary gear unit through clutch 1, which is used to couple or decouple the engine from the yoke. Lock 2 is used to lock the yoke of the planetary gear unit to the vehicle frame. The electric motor is connected to the sun gear. Clutch 2 is used to couple or decouple the sun gear (electric motor) to or from

(a)

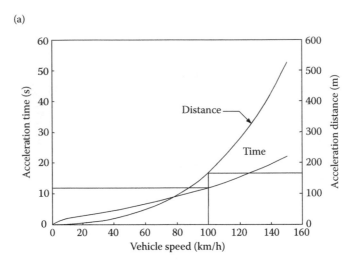

(b)

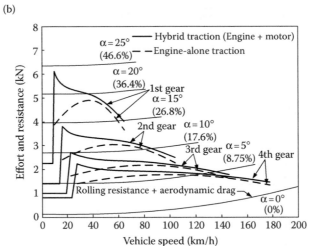

FIGURE 12.10
Performance of the hybrid electric drivetrain: (a) acceleration and (b) tractive effort versus vehicle speed.

the yoke. Lock 1 is used to lock the sun gear and the rotor of the electric motor to the vehicle frame. The transmission (gearbox) is driven by the ring gear of the planetary gear unit through a gear.

The operating characteristics of the planetary gear unit were discussed in detail in Chapter 10. They are repeated here for the reader's convenience.

The speeds, in rpm, of the sun gear, n_s, ring gear, n_r, and the yoke, n_y, have the relationship

$$n_y = \frac{1}{1+i_g} n_s + \frac{i_g}{1+i_g} n_r, \tag{12.5}$$

where i_g is the gear ratio defined as R_r/R_s as shown in Figure 12.14. The speeds n_s, n_r, and n_y are defined as positive in the direction shown in Figure 12.14. Defining $k_{ys} = (1 + i_g)$ and

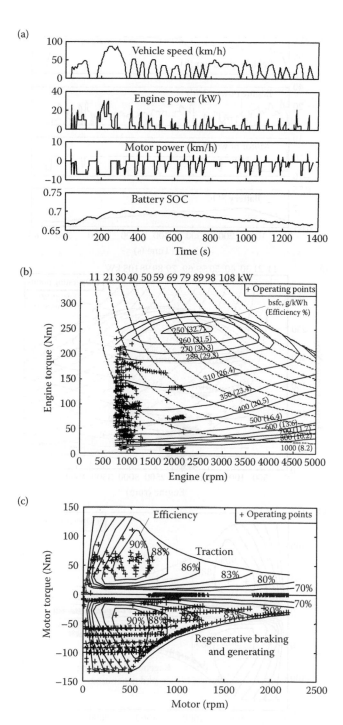

FIGURE 12.11
Simulation in FTP75 urban drive cycle: (a) vehicle speed, engine power, motor power, and battery SOC; (b) engine fuel consumption map and operating points; and (c) motor efficiency map and operating points.

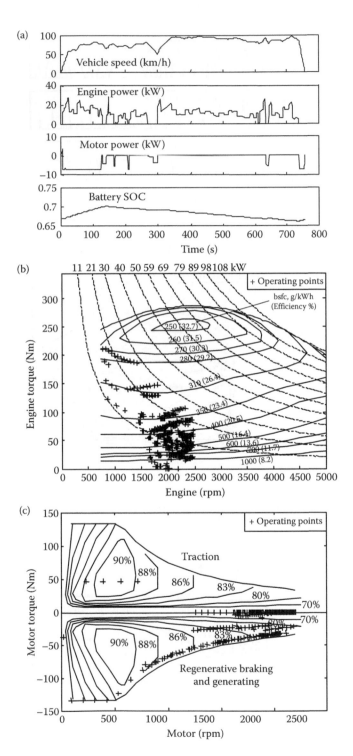

FIGURE 12.12
Simulation in FTP75 highway drive cycle: (a) vehicle speed, engine power, motor power, and battery SOC; (b) engine fuel consumption map and operating points; and (c) motor efficiency map and operating points.

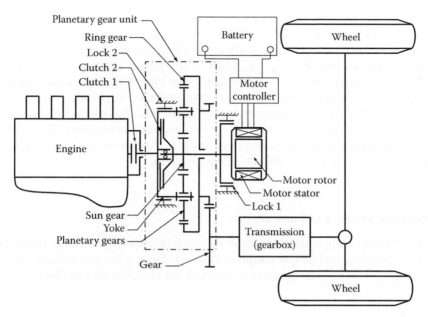

FIGURE 12.13
Series–parallel mild hybrid electric drivetrain with planetary gear unit.

$k_{yr} = (1 + i_g)/i_g$, Equation 12.5 can be further expressed as

$$n_y = \frac{1}{k_{ys}} n_s + \frac{1}{k_{yr}} n_r.$$ (12.6)

Ignoring the energy losses in the steady-state operation, the torques acting on the sun gear, the ring gear, and the yoke have the relationship

$$T_y = -k_{ys} T_s = -k_{yr} T_r.$$ (12.7)

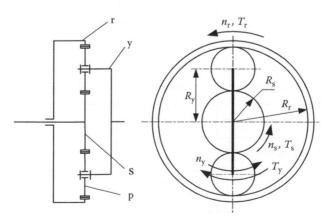

FIGURE 12.14
Planetary gear unit used as speed coupling.

Equation 12.7 indicates that the torques acting on the sun gear, T_s, and the ring gear, T_r, always have the same sign. In other words, they must always be in the same direction. However, the torque acting on the yoke, T_y, is always in the opposite direction of T_s and T_r. Equation 12.7 also indicates that with $i_g > 1$, which is the general case since $R_r > R_s$, T_s is the smallest, T_y is the largest, and T_r is in between. This means that the torque acting on the yoke is balanced by torques acting on the sun gear and the ring gear.

When one element among the sun gear, the ring gear, and the yoke is locked to the vehicle stationary frame, that is, one degree of freedom of the unit is constrained, the unit becomes a single-gear transmission (one input and one output). The speed and torque relationships, with different elements fixed, are shown in Figure 12.15.

12.3.2 Operating Modes and Control

As suggested by the configurations of the drivetrains, there are two distinct basic operating modes: speed coupling and torque coupling between the engine and the gearbox, depending on the engagement or disengagement states of the clutches and the lock.

12.3.2.1 Speed-Coupling Operating Mode

When the vehicle is starting from zero speed, and because the engine cannot run at zero speed and transmission has only a finite gear ratio, slip must exist between the input shaft and the output shaft of the transmission. The slip usually occurs in a clutch in manual transmissions or in a hydrodynamic torque converter in an automatic transmission. Thus, a certain amount of energy is lost in this slip. However, in the case of the drivetrain shown in Figure 12.13, this slip is performed between the engine and the electric motor (yoke and sun gear). In this case, clutch 1 connects the engine shaft to the yoke, clutch 2 releases the sun gear from the yoke, and locks 1 and 2 release the sun gear (motor) and the yoke (engine) from the vehicle frame at a given engine (yoke) speed and ring gear speed, which is proportional to the vehicle speed. According to Equation 12.6, the motor speed is

$$n_s = k_{ys}\left(n_y - \frac{n_r}{k_{yr}}\right). \tag{12.8}$$

Element fixed	Speed	Torque
Sun gear	$n_y = \dfrac{1}{k_{yr}}n_r$	$T_y = -k_{yr}T_r$
Ring gear	$n_y = \dfrac{1}{k_{ys}}n_s$	$T_y = -k_{ys}T_s$
Yoke	$n_s = -\dfrac{k_{ys}}{k_{ys}}n_r$	$T_s = \dfrac{k_{yr}}{k_{ys}}T_r$

FIGURE 12.15
Speed and torque relationships while one element is fixed.

When the first term on the right-hand side of Equation 12.8 is larger than the second term, that is, at low vehicle speed, the motor velocity is positive. However, from Equation 12.7, it is known that the motor torque must be negative, as expressed by

$$T_s = -\frac{T_y}{k_{ys}}. \tag{12.9}$$

Thus, the motor power is negative, and it operates as a generator. The generating power can be expressed as

$$P_{m/g} = \frac{2\pi}{60} T_s n_s = \frac{2\pi}{60}(-n_y T_y + n_t T_r) = -P_e + P_t, \tag{12.10}$$

where $P_{m/g}$ is the generating power of the motor (negative), P_e is the engine power, and P_t is the power going to the transmission for propelling the vehicle.

The vehicle speed (proportional to n_r) at which the value that makes n_s equal to zero is defined as synchronous speed. With further increase of vehicle speed, n_s becomes negative, and the electric motor goes into the motoring state.

In the speed-coupling operating mode, the engine speed is decoupled from the vehicle speed and can be controlled by the motor torque and engine throttle, as discussed previously.

12.3.2.2 Torque-Coupling Operating Mode

When clutch 1 is engaged, lock 2 releases the yoke, and clutch 2 engages the sun gear (motor) and the yoke (engine). The engine and motor speeds are forced to be the same. The ring gear speed and the yoke (engine) speed have the relationship (from Equation 12.6):

$$n_r = k_{yr}\left(1 - \frac{1}{k_{ys}}\right)n_y. \tag{12.11}$$

From the definitions of k_{yr} and k_{ys} by Equations 12.4 and 12.5, Equation 12.11 can be rewritten as

$$n_r = n_y. \tag{12.12}$$

Equation 12.12 implies that the gear ratio from the engine and the motor to the ring gear is 1.

The engine torque and the motor torque are added together by the planetary gear unit and then delivered to the transmission from the ring gear, which can be expressed as

$$T_r = T_e + T_{m/g}, \tag{12.13}$$

where T_e and T_m are the engine torque and motor torque, respectively.

12.3.2.3 Engine-Alone Traction Mode

The engine-alone traction mode can be realized in two ways. One is with the same operation of clutch 1, clutch 2, and lock 2 as in the torque-coupling mode (the previous section), but here, the motor is de-energized. In this case, the torque delivered to the transmission is expressed as

$$T_r = T_e. \tag{12.14}$$

The other is to lock the sun gear (motor shaft) to the vehicle frame by lock 1, and both clutch 2 and lock 2 are released. From Equation 12.6, the ring gear speed and the yoke (engine) speed have the relationship

$$n_r = k_{yr} n_y. \tag{12.15}$$

The torque delivered to the transmission can be expressed as

$$T_r = \frac{T_e}{k_{yr}}. \tag{12.16}$$

It is seen from the preceding discussion that the planetary gear unit plays the role of a two-gear transmission. The lower gear ratio is $1/k_{yr} < 1$, and the higher gear ratio is 1.

12.3.2.4 Motor-Alone Traction Mode

In this mode, the engine is shut down, and clutch 1 disengages the engine from the yoke. The motor-alone is used to propel the vehicle. There are two ways of doing this. One method is by coupling the sun gear to the yoke by clutch 2. In this way, the motor delivers its speed and torque to the ring gear as

$$n_m = n_r \tag{12.17}$$

and

$$T_m = T_r. \tag{12.18}$$

The gear ratio from the motor to the ring gear is 1.

The other method is by releasing the sun gear from the yoke by clutch 2 and locking the yoke to the vehicle frame by lock 2. In this manner, the speed and torque of the motor are related to the speed and torque of the ring gear by

$$n_{m/g} = -\frac{k_{ys}}{k_{yr}} n_r \tag{12.19}$$

and

$$T_{m/g} = -\frac{k_{yr}}{k_{ys}} T_r. \tag{12.20}$$

Equation 12.19 indicates that the motor turns in a direction opposite to that of the ring gear.

12.3.2.5 Regenerative Braking Mode

During braking, part of the braking energy can be recovered by the motor/generator. The operation of the drivetrain is the same as in electric traction mode, but the motor produces its torque in the direction opposite to that for traction.

12.3.2.6 Engine Starting

The engine can be started by the electric motor by engaging the sun gear to the yoke by clutch 2. The motor directly delivers its torque to the engine to start it.

12.3.3 Control Strategy

When the vehicle speed is lower than the synchronous speed, the speed-coupling operation mode is used. As explained in Section 12.3.2.1, the electric motor operates at a positive speed and a negative power. One part of the engine power is used to charge the batteries, and the other part is used to propel the vehicle.

When the vehicle speed is higher than the synchronous speed, the torque-coupling operation mode is used, and the drivetrain control strategy in this mode is as follows:

1. When the traction power demand is greater than the power that the engine can develop at full throttle, a hybrid traction mode is used. In this case, the engine is operated at full throttle, and the electric motor supplies extra power to meet the traction power demand.

2. When the traction power demand is less than the power that the engine can develop at full throttle, the operations of the engine and the electric motor are determined by the SOC of the batteries, as shown in Figure 12.16. In the battery-charging mode, the

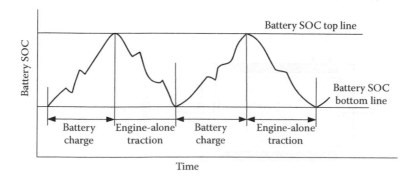

FIGURE 12.16
Battery charge and engine-alone traction, depending on battery SOC.

battery charging power may be determined by the maximum power of the electric motor or by the maximum engine power and demanded traction power.

12.3.4 Drivetrain with Floating-Stator Motor

An alternative mild hybrid electric drivetrain, which has characteristics similar to the drivetrain discussed previously, is shown in Figure 12.17.[6] This drivetrain uses an electric motor with a floating stator to replace the planetary gear unit and the electric motor.

As mentioned in Chapter 6, the angular velocity of the rotor is the summation of the angular velocities of the stator and the relative angular velocity between the stator and rotor, that is,

$$\omega_r = \omega_s + \omega_{rr}. \tag{12.21}$$

Due to the action and reaction effect, the torque acting on the stator and the rotor is always equal to the electromagnetic torque produced in the air gap (Figure 12.18), which is, in a

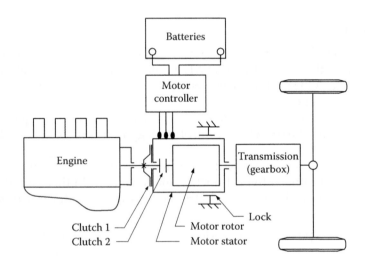

FIGURE 12.17
Series–parallel mild hybrid electric drivetrain with a floating-stator motor.

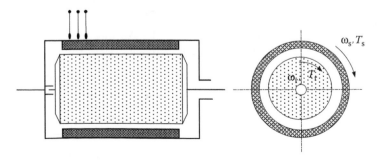

FIGURE 12.18
Electric motor with floating stator.

general sense, the electric motor torque. This relationship is described as

$$T_r = T_s = T_m,$$ (12.22)

where T_m is the electromagnetic torque in the air gap.

Comparing Equations 12.21 and 12.22 with Equations 12.6 and 12.7, it is known that both the planetary gear unit and the floating stator motor have the same operating characteristics. Therefore, the mild hybrid electric drivetrains shown in Figures 12.13 and 12.17 have the same operating principles and use the same control strategy. However, the design of the drivetrain with a planetary gear unit is more flexible since the gear ratio, i_g, is selectable.

Bibliography

1. Y. Gao, L. Chen, and M. Ehsani, Investigation of the effectiveness of regenerative braking for EV and HEV. *Society of Automotive Engineers (SAE) Journal*, SP-1466, Paper No. 1999-01-2901, 1999.
2. Y. Gao, K. M. Rahman, and M. Ehsani, The energy flow management and battery energy capacity determination for the drivetrain of electrically peaking hybrid. *Society of Automotive Engineers (SAE) Journal*, SP-1284, Paper No. 972647, 1997.
3. Y. Gao, K. M. Rahman, and M. Ehsani, Parametric design of the drivetrain of an electrically peaking hybrid (ELPH) vehicle. *Society of Automotive Engineers (SAE) Journal*, SP-1243, Paper No. 970294, 1997.
4. H. Gao, Y. Gao, and M. Ehsani, Design issues of the switched reluctance motor drive for propulsion and regenerative braking in EV and HEV. In *Society of Automotive Engineers (SAE) Future Transportation Technology Conference*, Costa Mesa, CA, Paper No. 2001-01-2526, August 2001.
5. Y. Gao and M. Ehsani, A mild hybrid drivetrain for 42-V automotive power system—design, control, and simulation. In *Society of Automotive Engineers (SAE) World Congress*, Detroit, MI, Paper No. 2002-02-1082, 2002.
6. Y. Gao and M. Ehsani, A mild hybrid vehicle drivetrain with a floating stator motor—configuration, control strategy, design, and simulation verification. In *Society of Automotive Engineers (SAE) Future Car Congress*, Crystal City, VA, Paper No. 2002-01-1878, June 2002.
7. Y. Gao and M. Ehsani, Investigation of battery technologies for the Army's hybrid vehicle application. In *Proceedings of the IEEE 56th Vehicular Technology Conference*, Vancouver, British Columbia, Canada, September 2002.
8. Y. Gao and M. Ehsani, Electronic braking system of EV and HEV—integration of regenerative braking, automatic braking force control and ABS. In *Society of Automotive Engineers (SAE) Future Transportation Technology Conference*, Costa Mesa, CA, Paper No. 2001-01-2478, August 2001.
9. C. Danzer, J. Liebold, E. Schreiterer, and J. Mueller, Low-cost powertrain platform for HEV and EV. SAE Technical Paper 2017-26-0088, 2017, doi: 10.4271/2017-26-0088.
10. Y. Jun, B. Jeon, and W. Youn, Equivalent consumption minimization strategy for mild hybrid electric vehicles with a belt driven motor. In SAE Technical Paper, 2017.
11. T. Q. Dinh, J. Marco, D. Greenwood, L. Harper, and D. Corrochano, Powertrain modelling for engine stop–start dynamics and control of micro/mild hybrid construction machines. *Proceedings of the Institution of Mechanical Engineers, Part K: Journal of Multi-body Dynamics*, 231(3), 2017: 439–456.
12. M. Awadallah, P. Tawadros, P. Walker, and N. Zhang, Dynamic modelling and simulation of a manual transmission based mild hybrid vehicle. *Mechanism and Machine Theory*, 112, 2017: 218–239.

13. M. Awadallah, P. Tawadros, P. Walker, and N. Zhang, Impact of low and high congestion traffic patterns on a mild-HEV performance. In *SAE International*, 2017.
14. S. Nüesch and A. G. Stefanopoulou, Multimode combustion in a mild hybrid electric vehicle. Part 1: Supervisory control. *Control Engineering Practice*, 57, 2016: 99–110.
15. S. Nüesch and A. G. Stefanopoulou, Multimode combustion in a mild hybrid electric vehicle. Part 2: Three-way catalyst considerations. *Control Engineering Practice*, 58, 2017: 107–116.
16. M. Dirnberger and H. G. Herzog, A verification approach for the optimization of mild hybrid electric vehicles. In *2015 IEEE International Electric Machines & Drives Conference (IEMDC)*, pp. 1494–1500, 2015.
17. M. Wüst, M. Krüger, D. Naber, L. Cross, A. Greis, S. Lachenmaier, and I. Stotz, Operating strategy for optimized Co2 and NOx emissions of diesel-engine mild-hybrid vehicles. In *15. Internationales Stuttgarter Symposium: Automobil- Und Motorentechnik*, 2015.
18. A. Babu and S. Ashok, Improved parallel mild hybrids for urban roads. *Applied Energy*, 144, 2015: 276–283.
19. M. Awadallah, P. Tawadros, P. Walker, and N. Zhang, Eliminating the torque hole: Using a mild hybrid EV architecture to deliver better driveability. In *Paper Presented at the 2016 IEEE Transportation Electrification Conference and Expo, Asia-Pacific (ITEC Asia-Pacific)*, 1–4 June 2016.

13

Peaking Power Sources and Energy Storage

"Energy storages" are defined in this book as devices that store energy, deliver energy outside (discharge), and accept energy from outside (charge). Several types of energy storage have been proposed for EV and HEV applications. These energy storages, so far, mainly include chemical batteries, ultracapacitors or supercapacitors, and ultra-high-speed flywheels. The fuel cell, which essentially is a type of energy converter, is discussed in Chapter 15.

There are many requirements for energy storages applied in an automotive application, such as specific energy, specific power, efficiency, maintenance requirement, management, cost, environmental adaptation and friendliness, and safety. For application on an EV, specific energy is the first consideration since it limits the vehicle range. On the other hand, for HEV applications, specific energy becomes less important, and specific power is the first consideration because all the energy is from the energy source (engine or fuel cell), and sufficient power is needed to ensure vehicle performance, particularly during acceleration, hill climbing, and regenerative braking. Of course, other requirements should be fully considered in the vehicle drivetrain development.

13.1 Electrochemical Batteries

Electrochemical batteries, more commonly referred to as "batteries," are electrochemical devices that convert electrical energy into potential chemical energy during charging and convert chemical energy into electric energy during discharging. A battery is composed of several cells stacked together. A cell is an independent and complete unit that possesses all the electrochemical properties. Basically, a battery cell consists of three primary elements: two electrodes (positive and negative) immersed into electrolyte, as shown in Figure 13.1.

Battery manufacturers usually specify a battery with coulometric capacity (ampere hours), which is defined as the number of ampere hours gained when discharging the battery from a fully charged state until the terminal voltage drops to its cut-off voltage, as shown in Figure 13.2. It should be noted that the same battery usually has a different number of ampere hours at different discharge rates. Generally, the capacity becomes smaller with a large discharge rate, as shown in Figure 13.3. Battery manufacturers usually specify a battery with a number of ampere hours along with a current rate. For example, a battery labeled as 100 Ah at a C/5 rate has a 100 Ah capacity at a 5-h discharge rate (discharging current = $100/5 = 20$ A).

Another important parameter of a battery is the state of charge (SOC). SOC is defined as the ratio of remaining capacity to fully charged capacity. With this definition, a fully charged battery has an SOC of 100%, and a fully discharged battery has an SOC of 0%. However, the term "fully discharged" sometimes causes confusion because of the different capacities at

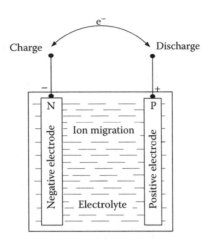

FIGURE 13.1
A typical electrochemical battery cell.

different discharge rates and different cut-off voltages (Figure 13.3). The change in SOC in a time interval, dt, with discharging or charging current i may be expressed as

$$\Delta SOC = \frac{i\,dt}{Q(i)}, \tag{13.1}$$

where $Q(i)$ is the ampere hour capacity of the battery at current rate i. For discharging, i is positive, and for charging, i is negative. Thus, the SOC of the battery can be expressed as

$$SOC = SOC_0 - \int \frac{i\,dt}{Q(i)}, \tag{13.2}$$

where SOC_0 is the initial value of the SOC.

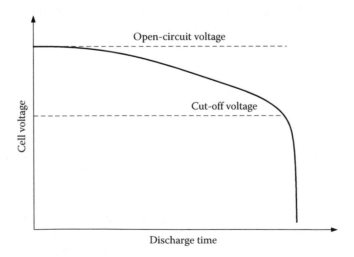

FIGURE 13.2
Cut-off voltage of a typical battery.

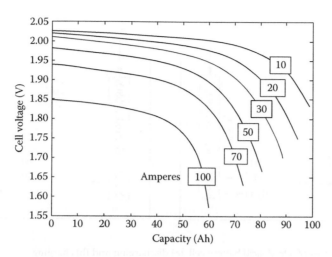

FIGURE 13.3
Discharge characteristics of a lead–acid battery.

For EVs and HEVs, the energy capacity is more important than the coulometric capacity (ampere hours) because it is directly associated with vehicle operation. The energy delivered from the battery can be expressed as

$$E_C = \int_0^t V(i, SOC)i(t)dt, \tag{13.3}$$

where $V(i, SOC)$ is the voltage at the battery terminals, which is a function of battery current and SOC.

13.1.1 Electrochemical Reactions

For simplicity, and because it is the most widespread battery technology in today's automotive applications, the lead–acid battery case is used as an example to explain the operating principle theory of electrochemical batteries. A lead–acid battery uses an aqueous solution of sulfuric acid $(2H^+ + SO_4^{2-})$ as the electrolyte. The electrodes are made of porous lead (Pb, anode, electrically negative) and porous lead oxide (PbO_2, cathode, electrically positive). The processes taking place during discharge are shown in Figure 13.4a, where the lead is consumed and lead sulfate is formed. The chemical reaction on the anode can be written

$$Pb + SO_4^{2-} \rightarrow PbSO_4 + 2e^-. \tag{13.4}$$

This reaction releases two electrons and, thereby, gives rise to an excess negative charge on the electrode that is relieved by a flow of electrons through the external circuit to the positive (cathode) electrode. At the positive electrode, the PbO_2 is also converted to $PbSO_4$ and, at the same time, water is formed. The reaction can be expressed as

$$PbO_2 + 4H^+ + SO_4^{2-} + 2e^- \rightarrow PbSO_4 + 2H_2O. \tag{13.5}$$

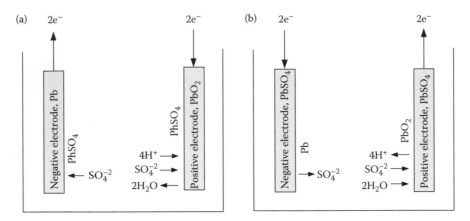

FIGURE 13.4
Electrochemical processes of a lead–acid battery cell: (a) discharging and (b) charging.

During charging, the reactions on the anode and the cathode are reversed, as shown in Figure 13.4a, and can be expressed as follows:

Anode:

$$PbSO_4 + 2e^- \rightarrow Pb + SO_4^{2-} \tag{13.6}$$

and

Cathode:

$$PbSO_4 + 2H_2O \rightarrow PbO_2 + 4H^+ + SO_4^{2-} + 2e^-. \tag{13.7}$$

The overall reaction in a lead–acid battery cell can be expressed as follows:

Overall:

$$Pb + PbO_2 + 2H_2SO_4 \underset{\text{Charge}}{\overset{\text{Discharge}}{\rightleftharpoons}} 2PbSO_4 + 2H_2O. \tag{13.8}$$

The lead–acid battery has a cell voltage of about 2.03 V in standard conditions, which is affected by the concentration of the electrolyte.

13.1.2 Thermodynamic Voltage

The thermodynamic voltage of a battery cell is closely associated with the energy released and the number of electrons transferred in the reaction. The energy released by a battery cell reaction is given by the change in Gibbs free energy, ΔG, usually expressed in per-mole quantities. The change in Gibbs free energy in a chemical reaction can be expressed as

$$\Delta G = \sum_{\text{Products}} G_i - \sum_{\text{Reactants}} G_j, \tag{13.9}$$

where G_i and G_j are the free energy in species i of products and species j of reactants. In a reversible process, ΔG is completely converted into electric energy, that is,

$$\Delta G = -nFV_r, \tag{13.10}$$

where n is the number of electrons transferred in the reaction, $F = 96,495$ is the Faraday constant in coulombs per mole, and V_r is the reversible voltage of the cell. In standard conditions (25°C temperature and 1 atm pressure), the open-circuit (reversible) voltage of a battery cell can be expressed as

$$V_r^0 = -\frac{\Delta G^0}{nF}, \tag{13.11}$$

where ΔG^0 is the change in Gibbs free energy in standard conditions.

The change of free energy, and thus the cell voltage, in a chemical reaction is a function of the activities of the solution species. From Equation 13.10 and the dependence of ΔG on reactant activities, the Nernst relationship is derived as

$$V_r = V_r^0 - \frac{RT}{nF} \ln\left[\frac{\prod (\text{activities of products})}{\prod (\text{activities of reactants})}\right], \tag{13.12}$$

where R is the universal gas constant, 8.31 J/mol K, and T is the absolute temperature in K.

13.1.3 Specific Energy

Specific energy is defined as the energy capacity per unit battery weight (Wh/kg). The theoretical specific energy is the maximum energy that can be generated per unit total mass of cell reactants. As discussed previously, the energy in a battery cell can be expressed by the Gibbs free energy ΔG. With respect to the theoretical specific energy, only the effective weights (molecular weights of reactants and products) are involved; then

$$E_{\text{spe,theo}} = -\frac{\Delta G}{3.6 \sum M_i} = \frac{nFV_r}{3.6 \sum M_i} (\text{Wh/kg}), \tag{13.13}$$

where $\sum M_i$ is the sum of the molecular weight of the individual species involved in the battery reaction. Taking the lead–acid battery as an example, $V_r = 2.03$ V, $n = 2$, and $\sum M_i = 642$ g, then $E_{\text{spe,theo}} = 170$ Wh/kg. From Equation 13.13, it is clear that the "ideal" couple would be derived from a highly electronegative element and a highly electropositive element, both of low atomic weight. Hydrogen, lithium, or sodium would be the best choice for the negative reactants, and the lighter halogens, oxygen, or sulfur would be the best choice for the positive reactants. To put such couples together in a battery requires electrode designs for effective utilization of the contained active materials, as well as electrolytes of high conductivity compatible with the materials in both electrodes. These constraints result in oxygen and sulfur being used in some systems as oxides and sulfides, rather than as the elements themselves. For operation at ambient temperature, aqueous electrolytes are advantageous because of their high conductivities. Here, alkali-group metals cannot be employed as electrodes, since these elements react with water. It is necessary to choose other metals that have a reasonable degree of electropositivity, such as zinc, iron, or aluminum. When considering electrode couples, it is preferable to exclude those elements that are in low abundance in the

Earth's crust, are expensive to produce, or are unacceptable from a health or environmental point of view.[1]

Examination of possible electrode couples has resulted in the study of more than 30 different battery systems with a view to developing a reliable, high-performance, inexpensive, high-power energy source for electric traction. The theoretical specific energies of the systems championed for EVs and HEVs are presented in Table 13.1.[1] Practical specific energies, however, are well below the theoretical maxima. Apart from electrode kinetic and other restrictions that serve to reduce cell voltage and prevent full utilization of reactants, there is a need for construction materials that add to the battery weight but are not involved in the energy-producing reaction.

To appreciate the extent to which the practical value of the specific energy is likely to differ from the theoretical values, it is instructive to consider the situation of the well-established lead–acid battery. A breakdown of the various components of a lead–acid battery designed to give a practical specific energy of 45 Wh/kg is shown in Figure 13.5.[1] It shows that only

TABLE 13.1

Theoretical Specific Energies of Candidate Batteries for EVs and HEVs

Battery		Cell Reaction		Specific Energy (Wh/kg)
(+)	(−)	Charge $\Leftarrow$	Discharge $\Rightarrow$	
Acidic Aqueous Solution				
PbO_2	Pb	$PbO_2 + 2H_2SO_4 + Pb$	$\Leftrightarrow 2PbSO_4 + 2H_2O$	170
Alkaline Aqueous Solution				
NiOOH	Cd	$2NiOOH + 2H_2O + Cd$	$\Leftrightarrow 2Ni(OH)_2 + Cd(OH)_2$	217
NiOOH	Fe	$2NiOOH + 2H_2O + Fe$	$\Leftrightarrow 2Ni(OH)_2 + Fe(OH)_2$	267
NiOOH	Zn	$2NiOOH + 2H_2O + Zn$	$\Leftrightarrow 2Ni(OH)_2 + Zn(OH)_2$	341
NiOOH	H_2	$2NiOOH + H_2$	$\Leftrightarrow 2Ni(OH)_2$	387
MnO_2	Zn	$2MnO_2 + H_2O + Zn$	$\Leftrightarrow 2MnOOH + ZnO$	317
O_2	Al	$4Al + 6H_2O + 3O_2$	$\Leftrightarrow 4Al(OH)_3$	2815
O_2	Fe	$2Fe + 2H_2O + O_2$	$\Leftrightarrow 2Fe(OH)_2$	764
O_2	Zn	$2Zn + 2H_2O + O_2$	$\Leftrightarrow 2Zn(OH)_2$	888
Flow				
Br_2	Zn	$Zn + Br_2$	$\Leftrightarrow ZnBr_2$	436
Cl_2	Zn	$Zn + Cl_2$	$\Leftrightarrow ZnCl_2$	833
$(VO_2)2SO_4$	VSO_4	$(VO_2)2SO_4 + 2HVSO_4 + 2H_2SO_4$	$2VOSO_4 + V_2(SO_4)_3 + 2H_2O$	114
Molten Salt				
S	Na	$2N3S\ a +$	$\Leftrightarrow Na_2S_3$	760
$NiCl_2$	Na	$2Na + NiCl_2$	$\Leftrightarrow 2NaCl$	790
FeS_2	LiAl	$4LiAl + FeS_2$	$\Leftrightarrow 2Li_2S + 4Al + Fe$	650
Organic Lithium				
$LiCoO_2$	Li − C	$Li(y + x)C6 +$ $Li(1-(y - x))CoO_2$	$\Leftrightarrow LiyC6 + Li(1 - y)CoO_2$	320^a

Source: A. J. Rand et al., *Batteries for Electric Vehicles*, Society of Automotive Engineers (SAE), Warrendale, PA, 1988.
[a] For maximum values of $x = 0.5$ and $y = 0$.

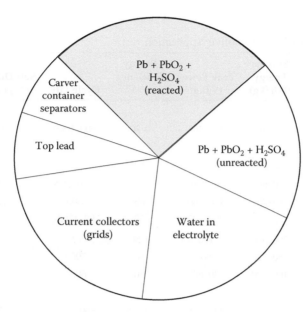

FIGURE 13.5
Weight distribution of the components of a lead–acid EV battery with a specific energy of 45 Wh/kg at the C5/5 rate. (D. A. J. Rand et al., *Batteries for Electric Vehicles,* Society of Automotive Engineers (SAE), Warrendale, PA, 1988.)

about 26% of the total weight of the battery is directly involved in producing electrical energy. The remainder is made up of (1) potential cell reactants that are not discharged at the rates required for EV operation, (2) water used as the solvent for the electrolyte (sulfuric acid alone is not suitable), (3) lead grids for current collection, (4) "top lead," that is, terminals, straps, and intercell connectors, and (5) cover, connector, and separators.

A similar ratio of practical-to-theoretical specific energy is expected for each of the candidate systems listed in Table 13.1. Present values realized by experimental cells and prototype batteries are listed in Table 13.2.[1] In recent years, some high-power batteries have been developed for application to HEVs.[2]

13.1.4 Specific Power

Specific power is defined as the maximum power of per-unit battery weight that the battery can produce in a short period. Specific power is important in the reduction of battery weight, especially in high-power-demand applications, such as HEVs. The specific power of a chemical battery depends mostly on the battery's internal resistance. With the battery model as shown in Figure 13.6, the maximum power that the battery can supply to the load is

$$P_{\text{peak}} = \frac{V_0^2}{4(R_c + R_{\text{int}})}, \tag{13.14}$$

where R_c is the conductor resistance (ohmic resistance), and R_{int} is the internal resistance caused by chemical reaction.

TABLE 13.2

Status of Battery System for Automotive Application

System	Specific Energy (Wh/kg)	Peak Power (V/kg)	Energy Efficiency (%)	Cycle Life	Self-Discharge (% per 48 h)	Cost (US$/kWh)
Acidic Aqueous Solution						
Lead/acid	35–50	150–400	>80	500–1000	0.6	120–150
Alkaline Aqueous Solution						
Nickel/cadmium	50–60	80–150	75	800	1	250–350
Nickel/iron	50–60	80–150	75	1500–2000	3	200–400
Nickel/zinc	55–75	170–260	65	300	1.6	100–300
Ni–MH	70–95	200–300	70	750–1200+	6	200–350
Aluminum/air	200–300	160	<50	?	?	?
Iron/air	80–120	90	60	500+	?	50
Zinc/air	100–220	30–80	60	600+	?	90–120
Flow						
Zinc/bromine	70–85	90–110	65–70	500–2000	?	200–250
Vanadium redox	20–30	110	75–85	–	–	400–450
Molten Salt						
Sodium/sulfur	150–240	230	80	800+	0[a]	250–450
Sodium/nickel chloride	90–120	130–160	80	1200+	0[a]	230–345
Lithium/iron sulfide (FeS)	100–130	150–250	80	1000+	?	110
Organic/Lithium						
Li–I	80–130	200–300	>95	1000+	0.7	200

[a] No self-discharge but some energy loss by cooling.

Internal resistance, R_{int}, represents the voltage drop, ΔV, which is associated with battery current. The voltage drop, ΔV, termed overpotential in battery terminology, includes two components: one caused by reaction activity ΔV_A and the other by electrolyte concentration ΔV_C. General expressions of ΔV_A and ΔV_C are[3]

$$\Delta V_A = a + b \log I \tag{13.15}$$

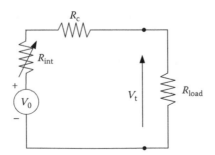

FIGURE 13.6
Battery circuit model.

and

$$\Delta V_C = -\frac{RT}{nF}\ln\left(1 - \frac{I}{I_L}\right),$$ (13.16)

where a and b are constants, R is the gas constant, 8.314 J/K mol, T is the absolute temperature, n is the number of electrons transferred in the reaction, F is the Faraday constant, 96,495 ampere-seconds per mole, and I_L is the limit current. Accurate determination of battery resistance or voltage drop by analysis is difficult and is usually obtained by measurement.[1] The voltage drop increases with an increase in discharging current, decreasing the stored energy in it (Figure 13.3).

Table 13.2 also shows the status of battery systems potentially available for EV. Although specific energies are high in advanced batteries, specific powers must improve. About 300 W/kg might be an optimistic estimate. However, SAFT has reported its Li-ion high-power for HEV application with a specific energy of 85 Wh/kg and a specific power of 1350 W/kg and its high-energy batteries for EV application with about 150 and 420 W/kg (at 80% state of discharge, 150 A current, and 30 s), respectively.[2]

13.1.5 Energy Efficiency

The energy or power losses during battery discharging and charging appear in the form of voltage loss. Thus, the efficiency of the battery during discharging and charging can be defined at any operating point as the ratio of the cell operating voltage to the thermodynamic voltage, that is:

During discharging

$$\eta = \frac{V}{V_0}$$ (13.17)

and

During charging

$$\eta = \frac{V_0}{V}.$$ (13.18)

The terminal voltage, as a function of battery current and energy stored in it or SOC, is lower in discharging and higher in charging than the electrical potential produced by chemical reaction. Figure 13.7 shows the efficiency of the lead–acid battery during discharging and charging. The battery has a high discharging efficiency with high SOC and a high charging efficiency with low SOC. The net cycle efficiency has a maximum in the middle range of the SOC. Therefore, the battery operation control unit of an HEV should control the battery SOC in its middle range to enhance the operating efficiency and depress the temperature rise caused by energy loss. High temperature would damage the battery.

13.1.6 Battery Technologies

The viable EV and HEV batteries consist of the lead–acid battery, nickel-based batteries, such as nickel/iron, nickel/cadmium, and nickel–metal hydride (Ni–MH) batteries, and

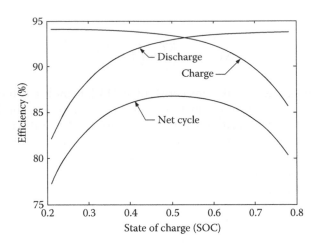

FIGURE 13.7
Typical battery charge and discharge efficiency.

lithium-based batteries such as lithium–polymer (Li–P) and lithium–ion (Li–I) batteries.[4] It seems that cadmium-based and lithium-based batteries would be the major candidates for EVs and HEVs.

13.1.6.1 Lead–Acid Battery

The lead–acid battery has been a successful commercial product for over a century and is still widely used as the electrical energy storage in the automotive field and other applications. Its advantages are its low cost, mature technology, and relatively high-power capability. These advantages are attractive for application in HEVs, where high-power is the first consideration. The materials involved (lead, lead oxide, and sulfuric acid) are rather low cost when compared with their more advanced counterparts. Lead–acid batteries also have several disadvantages. The energy density is low, mostly because of the high molecular weight of lead. The temperature characteristics are poor.[3] Below 10°C, specific power and specific energy are greatly reduced. This aspect severely limits the application of lead–acid batteries for the traction of vehicles operated in cold climates.

The presence of highly corrosive sulfuric acid is a potential safety hazard for vehicle occupants. The hydrogen released by the self-discharge reactions is another potential danger since this gas is extremely flammable even in tiny concentrations. The hydrogen emission is also a problem for hermetically sealed batteries. Indeed, to provide a good level of protection against acid spills, it is necessary to seal the battery, thereby trapping the parasitic gases in the casing. As a result, pressure may build up in the battery, causing swelling and mechanical constraints on the casing and the sealing. The lead in the electrodes is an environmental problem because of its toxicity. The emission of lead in connection with the use of lead–acid batteries may occur during the fabrication of the batteries, in the case of vehicle wreck (spill of electrolyte through cracks), or during their disposal at the end of battery life.

Different lead–acid batteries with improved performance are being developed for EVs and HEVs. Improvements of the sealed lead–acid batteries in specific energy over 40 Wh/kg, with the possibility of rapid charge, have been attained. One of these advanced sealed lead–acid batteries is Electrosource's Horizon battery. It uses a lead-wire-woven horizontal plate and hence offers competitive advantages of high specific energy (43 Wh/kg), high specific power (285 W/kg), long cycle life (over 600 cycles for on-road EV application), rapid recharge capability (50% capacity in 8 min and 100% in less than 30 min), low cost ($2000–$3000 per EV), mechanical ruggedness (robust structure of the horizontal plate), maintenance-free character (sealed battery technology), and environmental friendliness. Other advanced lead–acid battery technologies include bipolar designs and microtubular grid designs.

Advanced lead–acid batteries have been developed to remedy their disadvantages. The specific energy has been increased through the reduction of inactive materials such as the casing, current collector, separators, and so on. The lifetime has been increased by over 50%—at the expense of cost, however. The safety issue has been improved, with electrochemical processes designed to absorb the parasitic releases of hydrogen and oxygen.

13.1.6.2 Nickel-Based Batteries

Nickel is a lighter metal than lead and has very good electrochemical properties desirable for battery applications. There are four different nickel-based battery technologies: nickel–iron, nickel–zinc, nickel–cadmium, and Ni–MH.

13.1.6.2.1 Nickel–Iron Battery

The nickel/iron system was commercialized during the early years of the twentieth century. Applications included forklift trucks, mine locomotives, shuttle vehicles, railway locomotives, and motorized hand trucks.[1] The system comprises a nickel (III) oxyhydroxide (NIOOH) positive electrode and a metallic iron negative electrode. The electrolyte is a concentrated solution of potassium hydroxide (typically 240 g/L) containing lithium hydroxide (50 g/L). The cell reaction is given in Table 13.1, and its nominal open-circuit voltage is 1.37 V.

Nickel–iron batteries suffer from gassing, corrosion, and self-discharge problems. These problems have been partially or totally solved in prototypes that have yet to reach the market. These batteries are complex due to the need to maintain the water level and the safe disposal of the hydrogen and oxygen released during the discharge process. Nickel–iron battery performance also suffers at low temperatures, although less than lead–acid batteries. Finally, the cost of nickel is significantly higher than that of lead. Their greatest advantages are high-power density compared with lead–acid batteries, capable of withstanding 2000 deep discharges.

13.1.6.2.2 Nickel–Cadmium Battery

The nickel–cadmium system uses the same positive electrodes and electrolyte as the nickel–iron system, in combination with metallic cadmium negative electrode. The cell reaction is given in Table 13.1, and its nominal open-circuit voltage is 1.3 V. Historically, the development of the battery has coincided with that of nickel/iron, and they have similar performance.

Nickel–cadmium technology has seen enormous technical improvement because of the advantages of high specific power (over 220 W/kg), long cycle life (up to 2000 cycles), high tolerance of electric and mechanical abuse, a small voltage drop over a wide range of discharge currents, rapid charge capability (about 40%–80% in 18 min), wide operating temperature range (−40°C to −85°C), low self-discharge rate (<0.5% per day), excellent long-term storage due to negligible corrosion, and availability in a variety of size designs. However, the nickel–cadmium battery has some disadvantages, including high initial cost, relatively low cell voltage, and the carcinogenicity and environmental hazard of cadmium.

The nickel–cadmium battery can be generally divided into two major categories, vented and sealed types. The vented type consists of many alternatives. The vented sintered plate is a more recent development, which has a high specific energy but is more expensive. It is characterized by its flat discharge voltage profile, superior high current rate, and low-temperature performance. A sealed nickel–cadmium battery incorporates a specific cell design feature to prevent a build-up of pressure in the cell caused by gassing during overcharge. As a result, the battery requires no maintenance.

The major manufacturers of nickel–cadmium batteries for EV and HEV applications are SAFT and VARTA. Recent EVs powered by a nickel–cadmium battery have included the Chrysler TE Van, Citroën AX, Mazda Roadster, Mitsubishi EV, Peugeot 106, and Renault Clio.[4,5]

13.1.6.2.3 Ni–MH Battery

The Ni–MH battery has been on the market since 1992. Its characteristics are similar to those of the nickel–cadmium battery. The principal difference between them is the use of hydrogen, absorbed in a metal hydride, for the active negative electrode material in place of cadmium. Because of its specific energy, which is superior to that of Ni–Cd, and since it is free from toxicity or carcinogenicity, such as cadmium, the Ni–MH battery surpasses the Ni–Cd battery.

The overall reaction in an Ni–MH battery is

$$MH + NiOOH \leftrightarrow M + Ni(OH)_2. \tag{13.19}$$

When the battery is discharged, metal hydride in the negative electrode is oxidized to form metal alloy, and nickel oxyhydroxide in the positive electrode is reduced to nickel hydroxide. During charging, the reverse reaction occurs.

At present, Ni–MH battery technology has a nominal voltage of 1.2 V and attains a specific energy of 65 Wh/kg and a specific power of 200 W/kg. A key component of the Ni–MH battery is the hydrogen storage metal alloy, which should be formulated to obtain a material that is stable over many cycles. Two major types of metal alloys are being used. These are the rare-earth alloys based on lanthanum nickel, known as AB_5, and alloys consisting of titanium and zirconium, known as AB_2. The AB_2 alloys have a higher capacity than the AB_5 alloys. However, the trend is to use AB_5 alloys because of better charge retention and stability characteristics.

Because the Ni–MH battery is under continual development, its advantages based on present technology are summarized as follows: it has the highest specific energy (70–95 Wh/kg) and highest specific power (200–300 W/kg) of nickel-based batteries, environmental friendliness (cadmium free), a flat discharge profile (smaller voltage drop), and rapid recharge capability. However, this battery still suffers from its high initial cost. It may also have a memory effect and be exothermic during charging.

The Ni–MH battery has been considered an important near-term choice for EV and HEV applications. Several battery manufacturers, such as GM Ovonic, GP, GS, Panasonic, SAFT, VARTA, and YUASA, have actively engaged in the development of this battery technology, especially for powering EVs and HEVs. Since 1993, Ovonic has installed its Ni–MH battery in the Solectric GT Force EV for testing and demonstration. A 19-kWh battery has delivered over 65 Wh kg, 134 km/h, acceleration from 0 to 80 km/h in 14 s, and a city driving range of 206 km. Toyota and Honda have used the Ni–MH battery in their HEVs—Prius and Insight, respectively.[4,5]

13.1.6.3 *Lithium-Based Batteries*

Lithium is the lightest of all metals and possesses very interesting characteristics from an electrochemical point of view. Indeed, it allows a very high thermodynamic voltage, which results in a very high specific energy and specific power. There are two major technologies of lithium-based batteries: Li–P and Li–I.

13.1.6.3.1 *Li–P Battery*

Li–P batteries use lithium metal and a transition metal intercalation oxide (M_yO_z) for the negative and positive electrodes, respectively. This M_yO_z possesses a layered structure into which lithium ions can be inserted or from which they can be removed on discharge and charge, respectively. A thin solid polymer electrolyte (SPE) is used, which offers the merits of improved safety and flexibility in design. The general electrochemical reactions are:

$$xLi + M_yO_z \leftrightarrow Li_xM_yO_z. \tag{13.20}$$

On discharge, lithium ions formed at the negative electrode migrate through the SPE and are inserted into the crystal structure at the positive electrode. On charge, the process is reversed. By using a lithium foil negative electrode and a vanadium oxide (V_6O_{13}) positive electrode, the $Li/SPE/V_6O_{13}$ cell is the most attractive one within the family of Li–Ps. It operates at a nominal voltage of 3 V and has a specific energy of 155 Wh/kg and a specific power of 315 W/kg. The corresponding advantages are a very low self-discharge rate (about 0.5% per month), capability of fabrication in a variety of shapes and sizes, and safe design (reduced activity of lithium with solid electrolyte). However, it has the drawback of relatively weak low-temperature performance due to the temperature dependence of its ionic conductivity.[4]

13.1.6.3.2 *Li–I Battery*

Since the first announcement of the Li–I battery in 1991, the Li–I battery technology has seen an unprecedented rise to what is now considered to be the most promising rechargeable battery of the future. Although still in the stage of development, the Li–I battery has already gained acceptance for EV and HEV applications.

The Li–I battery uses a lithiated carbon intercalation material (Li_xC) for the negative electrode instead of metallic lithium, a lithiated transition metal intercalation oxide ($Li_{1-x}M_yO_z$) for the positive electrode, and a liquid organic solution or a solid polymer for the electrolyte. Lithium ions are swinging through the electrolyte between the positive and negative electrodes during discharge and charge. The general electrochemical reaction is described as

$$Li_xC + Li_{1-x}M_yO_z \leftrightarrow C + LiM_yO_z. \tag{13.21}$$

On discharge, lithium ions are released from the negative electrode, migrate via the electrolyte, and are taken up by the positive electrode. On charge, the process is reversed. Possible positive electrode materials include $Li_{1-x}CoO_2$, $Li_{1-x}NiO_2$, and $Li_{1-x}Mn_2O_4$, which have the advantages of stability in air, high voltage, and reversibility for the lithium intercalation reaction.

The $Li_xC/Li_{1-x}NiO_2$ type, loosely written as $C/LiNiO_2$ or simply called the nickel-based Li–I battery, has a nominal voltage of 4 V, a specific energy of 12 Wh/kg, an energy density of 200 Wh/L, and a specific power of 260 W/kg. The cobalt-based type has higher specific energy and energy density but with a higher cost and significant increase of the self-discharge rate. The manganese-based type has the lowest cost, and its specific energy and energy density lie between those of the cobalt-based and nickel-based types. It is anticipated that the development of the Li–I battery will ultimately move to the manganese-based type because of the low cost, abundance, and environmental friendliness of the manganese-based materials.

Many battery manufacturers, such as SAFT, GS Hitachi, Panasonic, SONY, and VARTA, have actively engaged in the development of Li–I batteries. Starting in 1993, SAFT focused on the nickel-based Li–I battery. Recently, SAFT reported the development of Li–I high-power batteries for HEV applications with a specific energy of 85 Wh/kg and a specific power of 1350 W/kg. SAFT also announced high-energy batteries for EV applications with about 150 Wh/kg and 420 W/kg (at 80% SOC, 150 A current, and 30 s), respectively.[2]

13.2 Ultracapacitors

Because of the frequent stop-and-go operation of EVs and HEVs, the discharging and charging profile of the energy storage is highly varied. The average power required from the energy storage is much lower than the peak power for acceleration and hill climbing in a relatively short time span. The ratio of peak power to average power can reach over 10:1 (Chapter 2). In HEV design, the peak power capacity of the energy storage is more important than its energy capacity and usually constrains its size reduction (Chapters 9 and 10). Based on present battery technology, battery design must make a trade-off among specific energy, specific power, and cycle life. The difficulty in simultaneously obtaining high values of specific energy, specific power, and cycle life has led to some suggestions that the energy storage system of EVs and HEVs should be a hybridization of an energy source and a power source. The energy source, mainly batteries and fuel cells, has high specific energy, whereas the power source has high specific power. Power sources can be recharged from the energy source during less demanding driving or regenerative braking. The power source that has received wide attention is the ultracapacitor.

13.2.1 Features of Ultracapacitors

The ultracapacitor is characterized by a much higher specific power but a much lower specific energy compared to batteries. Its specific energy is in the range of a few watt-hours per kilogram. However, its specific power can reach up to 3 kW/kg, much higher than any type of battery. Due to the low specific energy density and the dependence of terminal voltage on the SOC, it is difficult to use ultracapacitors alone as an energy storage for EVs and HEVs. Nevertheless, there are several advantages that can result from using an ultracapacitor as an auxiliary power source. One promising application is the so-called battery and ultracapacitor hybrid energy storage system for EVs and HEVs.[4,6] Specific energy and specific power requirements can be decoupled, affording an opportunity to design a battery that is optimized for specific energy and cycle life with little attention being paid to specific power. Due to the load-leveling effect of the ultracapacitor, the high current discharging from the battery and the high current charging to the battery by regenerative braking are minimized so that available energy, endurance, and life of the battery can be significantly increased.

13.2.2 Basic Principles of Ultracapacitors

Double-layer capacitor technology is the major approach to applying the ultracapacitor concept. The basic principle of a double-layer capacitor is illustrated in Figure 13.8. When two carbon rods are immersed in a thin sulfuric acid solution, separated from each other, and applied with increasing voltage from 0 to 1.5 V, almost nothing happens up to 1 V; then at slightly over 1.2 V, a small bubble appears on the surface of both electrodes. Bubbles at a voltage above 1 V indicate the electrical decomposition of water. Below the decomposition voltage, while the current does not flow, an "electric double layer" occurs at the boundary of electrode and electrolyte. The electrons are charged across the double layer and for a capacitor.

An electrical double layer works as an insulator only below the decomposing voltage. The stored energy, E_{cap}, is expressed as

$$E_{cap} = \frac{1}{2}CV^2,$$
(13.22)

where C is the capacitance in Faraday, and V is the usable voltage in volts. This equation indicates that the higher-rated voltage V is desirable for larger energy density capacitors. Up to now, a capacitor's rated voltage with aqueous electrolyte has been about 0.9 V per cell, and with a nonaqueous electrolyte, it is 2.3–3.3 V for each cell.

There is great merit in using an electric double layer in place of plastic or aluminum oxide films in a capacitor, since the double layer is very thin—as thin as one molecule with no pin holes—and the capacity per area is quite large, 2.5–5 $\mu F/cm^2$.

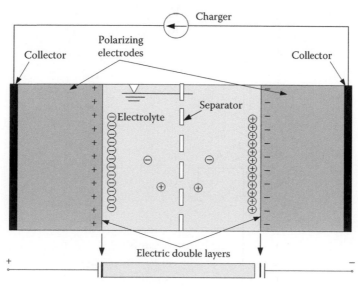

FIGURE 13.8
Basic principles of a typical electric double-layer capacitor.

Even if a few $\mu F/cm^2$ are obtainable, the energy density of capacitors is not large when using aluminum foil. For increasing the capacitance, electrodes are made from specific materials that have a very large area, such as activated carbons, which are well known for their surface areas of 1000–3000 m^2/g. Ions are adsorbed on those surfaces and result in 50 F/g ($1000\,m^2/g \times 5\,F/cm^2 \times 10,000\,cm^2/m^2 = 50\,F/g$). If the same weight of electrolyte is added, 25 F/g is quite a large capacity density. Nevertheless, the energy density of these capacitors is far smaller than that of batteries; the typical specific energy of ultracapacitors is at present about 2 Wh/kg, only 1/20 of 40 Wh/kg, which is the available value of typical lead–acid batteries.

13.2.3 Performance of Ultracapacitors

The performance of an ultracapacitor may be represented by terminal voltages during discharge and charge with different current rates. There are three parameters in a capacitor: the capacitance itself (its electric potential V_C), the series resistance R_S, and the dielectric leakage resistance, R_L, as shown in Figure 13.9. The terminal voltage of an ultracapacitor during discharge can be expressed as

$$V_t = V_C - iR_S. \tag{13.23}$$

The electric potential of the capacitor can be expressed as

$$\frac{dV_C}{dt} = -\left(\frac{i + i_L}{C}\right), \tag{13.24}$$

where C is the capacitance of the ultracapacitor.

On the other hand, the leakage current i_L can be expressed as

$$i_L = \frac{V_C}{R_L}. \tag{13.25}$$

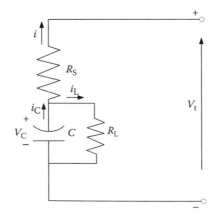

FIGURE 13.9
Ultracapacitor equivalent circuit.

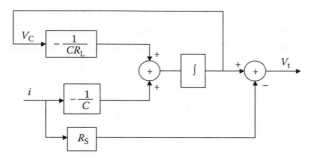

FIGURE 13.10
Block diagram of ultracapacitor model.

Substituting Equation 13.25 into Equation 13.24, one obtains

$$\frac{dV_C}{dt} = -\left(\frac{V_C}{CR_L} + \frac{i}{C}\right). \tag{13.26}$$

The terminal voltage of an ultracapacitor cell can be represented by the diagram shown in Figure 13.10. The analytical solution of Equation 13.26 is

$$V_C = \left[V_{C0} - \int_0^t \frac{i}{C}e^{t/CR_L}dt\right]e^{-(t/CR_L)}, \tag{13.27}$$

where i is the discharge current and is a function of time in real operation. The discharge characteristics of a Maxwell 2600F ultracapacitor are shown in Figure 13.11. At different discharge rates, the voltage linearly decreases with discharge time. At a large discharge rate the voltage decreases much faster than at a small rate.

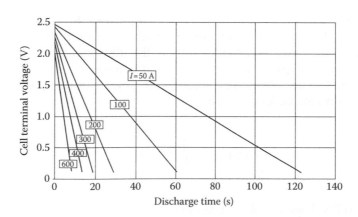

FIGURE 13.11
Discharge characteristics of Maxwell Technologies 2600F ultracapacitor.

A similar model can be used to describe the charging characteristics of an ultracapacitor, and readers who are interested may carry out their own analysis.

The operation efficiency in discharging and charging can be expressed as follows:

Discharging:

$$\eta_d = \frac{V_t I_t}{V_C I_C} = \frac{(V_C - I_t R_S)I_t}{V_C(I_t + I_L)} \tag{13.28}$$

and

Charging:

$$\eta_C = \frac{V_C I_C}{V_t I_t} = \frac{V_C(I_t - I_L)}{(V_C + I_t R_S)I_t}, \tag{13.29}$$

where V_t is the terminal voltage, and I_t is the current input to or output from the terminal. In actual operation, the leakage current I_L is usually very small (few mA) and can be ignored. Thus, Equations 13.28 and 13.29 can be rewritten as follows:

Discharging:

$$\eta_d = \frac{V_C - R_S I_t}{V_C} = \frac{V_t}{V_c} \tag{13.30}$$

and

Charging:

$$\eta_c = \frac{V_C}{V_C - R_S I_t} = \frac{V_C}{V_t}. \tag{13.31}$$

These equations indicate that the energy loss in an ultracapacitor is caused by the presence of series resistance. The efficiency decreases at a high current rate and low cell voltage, as shown in Figure 13.12. Thus, in actual operation, the ultracapacitor should be kept in its high-voltage region, for higher than 60% of its rated voltage.

The energy stored in an ultracapacitor can be obtained through the energy needed to charge it to a certain voltage level, that is,

$$E_C = \int_0^t V_C I_C dt = \int_0^V C V_C dV_C = \frac{1}{2}C V_C^2, \tag{13.32}$$

where V_C is the cell voltage in volts. At its rated voltage, the energy stored in the ultracapacitor reaches its maxima. Equation 13.32 indicates that increasing the rated voltage can significantly increase the stored energy, since the energy increases with the voltage squared.

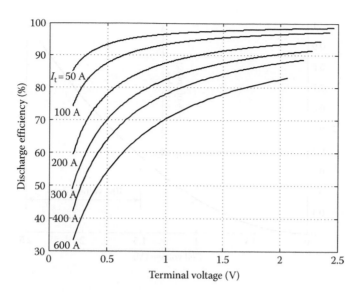

FIGURE 13.12
Discharge efficiency of Maxwell Technologies 2600F ultracapacitor.

In real operation, it is impossible to completely utilize the stored energy because of the low-power in the low state of energy (SOE). Thus, an ultracapacitor is usually given a bottom voltage, V_{Cb}, below which the ultracapacitor stops delivering energy. Consequently, the available or useful energy for use is less than its fully charged energy, which can be expressed as

$$E_u = \frac{1}{2}C\left(V_{CR}^2 - V_{Cb}^2\right), \tag{13.33}$$

where V_{CR} is the rated voltage of the ultracapacitor.

The usable energy in an ultracapacitor can also be expressed in SOE, which is defined as the ratio of the energy in the ultracapacitor at a voltage of V_C to the energy at full charged voltage, V_{CR}, as expressed by

$$\text{SOE} = \frac{0.5\,CV_C^2}{0.5\,CV_{CR}^2} = \left(\frac{V_C}{V_{CR}}\right)^2. \tag{13.34}$$

For example, 60% of the rated voltage is the bottom voltage, and 64% of the total energy is available for use, as shown in Figure 13.13.

13.2.4 Ultracapacitor Technologies

According to the goals set by the U.S. Department of Energy for the inclusion of ultracapacitors in EVs and HEVs, the near-term specific energy and specific power should exceed 5 Wh/kg and 500 W/kg, respectively, while the advanced performance values should be over 15 Wh/kg and 1600 W/kg. So far, none of the available ultracapacitors can fully satisfy

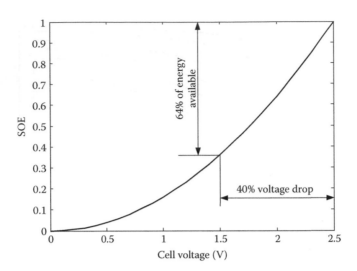

FIGURE 13.13
SOE versus cell voltage.

these goals. Nevertheless, some companies are actively engaged in the research and development of ultracapacitors for EV and HEV applications. Maxwell Technologies has claimed that its power BOOSTCAP® ultracapacitor cells (2600 F at 2.5 V) and integrated modules (145 F at 42 V and 435 F at 14 V) are in production. The technical specifications are listed in Table 13.3.

TABLE 13.3

Technical Specifications of Maxwell Technologies Ultracapacitor Cell and Integrated Modules

	BCA P0010 (Cell)	BMOD0115 (Module)	BMOD0117 (Module)
Capacitance (farads, −20%/+20%)	2600	145	435
Maximum series resistance ESR at 25°C (mΩ)	0.7	10	4
Voltage (V): continuous (peak)	2.5 (2.8)	42 (50)	14 (17)
Specific power at rated voltage (W/kg)	4300	2900	1900
Specific energy at rated voltage (Wh/kg)	4.3	2.22	1.82
Maximum current (A)	600	600	600
Dimensions (mm) (reference only)	60 × 172 (cylinder)	195 × 165 × 415 (box)	195 × 265 × 145 (box)
Weight (kg)	0.525	16	6.5
Volume (L)	0.42	22	7.5
Operating temperature (°C)	−35 to +65	−35 to +65	−35 to +65
Storage temperature (°C)	−35 to +65	−35 to +65	−35 to +65
Leakage current (mA) 12 h, 25°C	5	10	10

Source: Available at http://www.maxwell.com, Maxwell Technologies, 2007.

13.3 Ultra-High-Speed Flywheels

The use of flywheels for storing energy in mechanical form is not a new concept. More than 25 years ago, the Oerlikon Engineering Company in Switzerland made the first passenger bus solely powered by a massive flywheel. This flywheel, weighing 1500 kg and operating at 3000 rpm, was recharged by electricity at each bus stop. The traditional flywheel is a huge steel rotor with a mass of hundreds of kilograms that spins on the order of tens of hundreds of rpm. In contrast, the advanced flywheel is a lightweight composite rotor with a mass of tens of kilograms and rotating on the order of tens of thousands of rpm; it is a so-called ultra-high-speed flywheel.

The concept of ultra-high-speed flywheels appears to be a feasible means for fulfilling the stringent energy storage requirements for EV and HEV applications, namely, high specific energy, high specific power, long cycle life, high energy efficiency, quick recharge, maintenance-free characteristics, cost effectiveness, and environmental friendliness.

13.3.1 Operation Principles of Flywheels

A rotating flywheel stores energy in kinetic form as

$$E_f = \frac{1}{2}J_f\omega_f^2, \tag{13.35}$$

where J_f is the moment of inertia of the flywheel in kg m^2/s, and ω_f is the angular velocity of the flywheel in rad/s. Equation 13.35 indicates that enhancing the angular velocity of the flywheel is the key technique to increasing its energy capacity and reducing its weight and volume. At present, a speed of over 60,000 rpm has been achieved in some prototypes.

With current technology, it is difficult to use the mechanical energy stored in a flywheel directly to propel a vehicle due to the need for continuous variation in the transmission with a wide gear ratio range. The commonly used approach is to couple an electric machine to the flywheel directly or through a transmission to constitute a so-called mechanical battery. The electric machine, functioning as the energy input and output port, converts the mechanical energy into electric energy or vice versa, as shown in Figure 13.14.

Equation 13.35 indicates that the energy stored in a flywheel is proportional to the moment of inertia of the flywheel and flywheel rotating speed squared. A lightweight flywheel should be designed to achieve a large moment of inertia per unit mass and per unit volume by properly designing its geometric shape. The moment of inertia of a flywheel can be calculated by

$$J_f = 2\pi\rho \int_{R_1}^{R_2} W(r)r^3 dr, \tag{13.36}$$

where ρ is the material mass density, and $W(r)$ is the width of the flywheel corresponding to radius r, as shown in Figure 13.15. The mass of the flywheel can be calculated by

$$M_f = 2\pi\rho \int_{R_1}^{R_2} W(r)r\, dr. \tag{13.37}$$

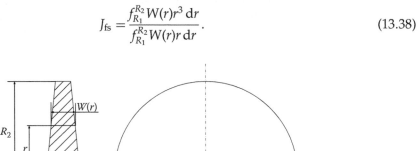

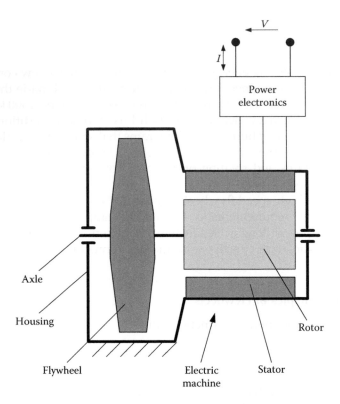

FIGURE 13.14
Basic structure of a typical flywheel system (mechanical battery).

Thus, the specific moment of inertia of a flywheel, defined as the moment of inertia per unit mass, can be expressed as

$$J_{fs} = \frac{\int_{R_1}^{R_2} W(r) r^3 \, dr}{\int_{R_1}^{R_2} W(r) r \, dr}. \tag{13.38}$$

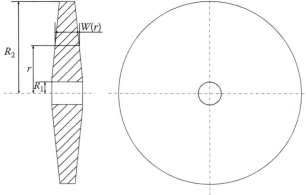

FIGURE 13.15
Geometry of a typical flywheel.

Equation 13.38 indicates that the specific moment of inertia of a flywheel is independent of its material mass density and dependent solely on its geometric shape $W(r)$.

For a flywheel of equal width, the moment of inertia is

$$J_f = 2\pi\rho(R_2^4 - R_1^4) = 2\pi\rho(R_2^2 + R_1^2)(R_2^2 - R_1^2).$$ (13.39)

The specific moment of inertia is

$$J_{fs} = R_2^2 + R_1^2.$$ (13.40)

The volume density of the moment of inertia, defined as the moment of inertia per unit volume, is, indeed, associated with the mass density of the material. The volume of the flywheel can be obtained by

$$V_f = 2\pi \int_{R_2}^{R_2} W(r)r\,dr.$$ (13.41)

The volume density of the moment of inertia can be expressed as

$$J_{fV} = \frac{\rho \int_{R_1}^{R_2} W(r)r^3\,dr}{\int_{R_1}^{R_2} W(r)r\,dr}.$$ (13.42)

For a flywheel of equal width, the volume density of the moment of inertia is

$$J_{fV} = \rho(R_2^2 + R_1^2).$$ (13.43)

Equations 13.42 and 13.43 indicate that heavy material can, indeed, reduce the volume of a flywheel with a given moment of inertia.

13.3.2 Power Capacity of Flywheel Systems

The power that a flywheel delivers or receives can be obtained by differentiating Equation 13.35 with respect to time, that is,

$$P_f = \frac{dE_f}{dt} = J_f\omega_f\frac{d\omega_f}{dt} = \omega_f T_f,$$ (13.44)

where T_f is the torque acting on the flywheel by the electric machine. When the flywheel discharges its energy, the electric machine acts as a generator and converts the mechanical energy of the flywheel into electrical energy. On the other hand, when the flywheel is charged, the electric machine acts as a motor and converts electrical energy into mechanical energy stored in the flywheel. Equation 13.44 indicates that the power capacity of a flywheel system depends completely on the power capacity of the electric machine.

An electric machine usually has the characteristics shown in Figure 13.16, which has two distinct operating regions—constant-torque and constant-power regions. In the constant-torque region, the voltage of the electric machine is proportional to its angular velocity, and the magnetic flux in the air gap is constant. However, in the constant-power region,

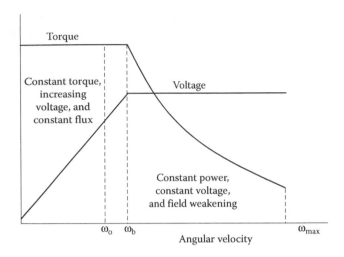

FIGURE 13.16
Typical torque and voltage profile versus rotational speed.

the voltage is constant, and the magnetic field is weakened with an increase in the machine angular velocity. During charging of the flywheel, that is, accelerating the flywheel from a low-speed, ω_0, to a high-speed, the maximum speed, ω_{max}, for example, the torque delivered from the electric machine is

$$T_m = J_f \frac{d\omega_f}{dt}, \tag{13.45}$$

where it is assumed that the electric machine is directly connected to the flywheel. The time, t, needed can be expressed as

$$t = \int_{\omega_0}^{\omega_{max}} \frac{J_f}{T_m} d\omega = \int_{\omega_0}^{\omega_b} \frac{J_f}{p_m/\omega_b} d\omega + \int_{\omega_b}^{\omega_{max}} \frac{J_f}{p_m/\omega} d\omega. \tag{13.46}$$

With the given accelerating time, t, the maximum power of the electric machine can be obtained from Equation 13.46 as

$$P_m = \frac{J_f}{2t} \left(\omega_b^2 - 2\omega_0\omega_b + \omega_{max}^2 \right). \tag{13.47}$$

Equation 13.47 indicates that the power of the electric machine can be minimized by designing its corner speed or base speed, ω_b, equal to the bottom speed of the flywheel, ω_0. This conclusion implies that the effective operating speed range of the flywheel should coincide with the constant-power region of the electric machine. The power of the electric machine can be minimized as

$$P_m = \frac{J_f}{2t} \left(\omega_{max}^2 - \omega_0^2 \right). \tag{13.48}$$

Another advantage achieved by matching the operating speed range of the flywheel with the constant-power speed range is that the voltage of the electric machine is always constant

TABLE 13.4

Composite Materials for Ultra-High-Speed Flywheel

	Tensile Strength σ (MPa)	Specific Energy ρ (kg/m³)	Ratio σ/ρ (Wh/kg)
E-glass	1379	1900	202
Graphite epoxy	1586	1500	294
S-glass	2069	1900	303
Kevlar epoxy	1930	1400	383

Source: C. C. Chan and K. T. Chau, *Modern Electric Vehicle Technology*, Oxford University Press, Oxford, 2001

(Figure 13.16), which significantly simplifies the power management system, such as DC/DC converters and their controls.

13.3.3 Flywheel Technologies

Although higher rotational speed can significantly increase stored energy (Equation 13.35), there is a limit beyond which the tensile strength σ of the material constituting the flywheel cannot withstand the stress resulting from the centrifugal force. The maximum stress acting on the flywheel depends on its geometry, specific density ρ, and rotational speed. Maximum benefit can be obtained by adopting flywheel materials that have a maximum ratio of σ/ρ. Note that if the speed of the flywheel is limited by material strength, the theoretical specific energy is proportional to the ratio of σ/ρ. Table 13.4 summarizes the characteristics of some composite materials for ultra-high-speed flywheels.

A constant-stress principle may be employed in the design of ultra-high-speed flywheels. To achieve maximum energy storage, every element in the rotor should be equally stressed to its maximum limit. This results in a shape of gradually decreasing thickness that theoretically approaches zero as the radius approaches infinity, as shown in Figure 13.17.[4]

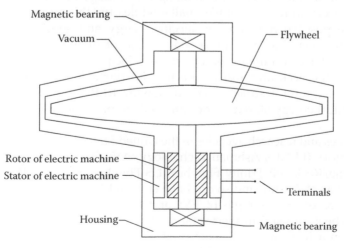

FIGURE 13.17
Basic structure of a typical flywheel system.

Due to the extremely high rotating speed and to reduce the aerodynamic loss and frictional loss, the housing inside which the flywheel is spinning is always highly vacuumed, and noncontact, magnetic bearings are employed.

The electric machine is one of the most important components in the flywheel system, since it has a critical impact on the performance of the system. At present, PM brushless DC motors are usually used in the flywheel system. Apart from possessing high-power density and high efficiency, the PM brushless DC motor has the unique advantage that no heat is generated inside the PM rotor, which is particularly essential for the rotor to work in a vacuum environment to minimize windage loss.

A switched reluctance machine (SRM) is also a very promising candidate for application in a flywheel system. SRM has a very simple structure and can operate efficiently at very high-speed. In addition, SRM possesses a large extended constant-power speed region, which allows more energy in the flywheel that can be delivered (Section 13.3.2). In this extended-speed region, only the machine excitation flux can be varied, which is easily realized. On the contrary, the PM brushless motor shows some difficulty in weakening the field flux induced by the PM.

In contrast to applying the ultra-high-speed flywheel for energy storage in stationary plants, its application to EVs and HEVs suffers from two specific problems. First, gyroscopic forces occur whenever a vehicle departs from its straight-line course, such as in turning and in pitching upward or downward from road grades. These forces essentially reduce the maneuverability of the vehicle. Second, if the flywheel is damaged, its energy stored in mechanical form is released in a very short period of time. The corresponding power released is very high, which can cause severe damage to the vehicle. For example, if a 1-kWh flywheel breaks apart in 1–5 s, it generates a huge amount of power, 720–3600 kW. Containment in the case of failure is presently the most significant obstacle to implementing the ultra-high-speed flywheel in EVs and HEVs.

The simplest way of alleviating gyroscopic forces is to use multiple smaller flywheels. By operating them in pairs (one half spinning in one direction and another half in the opposite direction), the net gyroscopic effect becomes theoretically zero. Practically, it still has some problems related to the distribution and coordination of these flywheels. Also, the overall specific energy and specific power of all flywheels may be smaller than a single one. Similarly, the simplest way of minimizing the damage due to breakage of the ultra-high-speed flywheel is to adopt multiple small modules, but this means vehicle performance suffers from the possible reduction of specific energy and specific power. Recently, a new failure containment has been proposed. Instead of diminishing the thickness of the rotor's rim to zero based on the maximum stress principle, the rim thickness is purposely enlarged. Hence, the neck area just before the rim (virtually a mechanical fuse) breaks first at the instant the rotor suffers a failure. Owing to the use of this mechanical fuse, only the mechanical energy stored in the rim needs to be released or dissipated in case of a failure.[4]

Many companies and research agencies in the United States (such as Lawrence Livermore National Laboratory (LLNL), Ashman Technology, AVCON, Northrop Grumman, Power R&D, Rocketdyne/Rockwell Trinity Flywheel US Flywheel Systems, Power Center at the University of Texas at Austin, and so on) have engaged in the development of ultra-high-speed flywheels as energy storages of EVs and HEVs. However, technologies of ultra-high-speed flywheels are still in their infancy. Typically, the whole ultra-high-speed flywheel system can achieve a specific energy of 10–150 Wh/kg and specific power of 2–10 kW/kg. LLNL has built a prototype (20-cm diameter and 30-cm height) that can achieve 60,000 rpm, 1 kWh, and 100 kW.

13.4 Hybridization of Energy Storages

13.4.1 Concept of Hybrid Energy Storage

The hybridization of energy storage involves combining two or more energy storages together so that the advantages of each can be brought out, and the disadvantages of each can be compensated by the advantages of the others. For instance, the hybridization of a chemical battery with an ultracapacitor can overcome problems such as the low specific power of chemical batteries and low specific energy of ultracapacitors, thereby achieving high specific energy and high specific power. Basically, hybridized energy storage consists of two basic energy storages, one with high specific energy and the other with high specific power. The basic operation of this system is illustrated in Figure 13.18. In high-power-demand operations, such as acceleration and hill climbing, both basic energy storages deliver their power to the load as shown in Figure 13.18a. On the other hand, in low-power-demand operations, such as constant-speed cruising, the high specific energy storage delivers its power to the load and charges the high specific power storage to recover the charge lost during high-power-demand operation, as shown in Figure 13.18b. In regenerative braking operations, the peak power is absorbed by the high specific power storage, and only a limited part is absorbed by the high specific energy storage. In this way, the whole system is much smaller in weight and size than if any one of them alone was the energy storage.

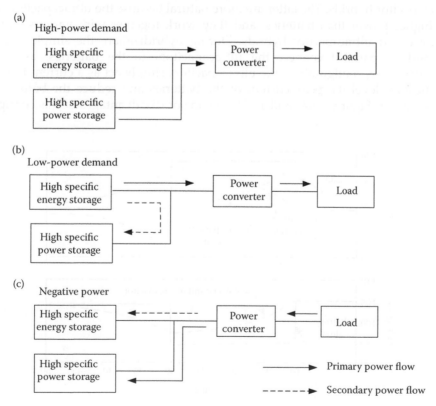

FIGURE 13.18
Concept of a hybrid energy storage operation. (a) Hybrid powering, (b) power split, and (c) hybrid charging.

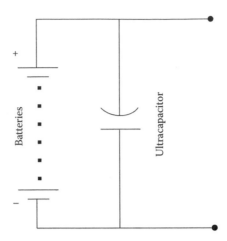

FIGURE 13.19
Direct and parallel connection of batteries and ultracapacitors.

13.4.2 Passive and Active Hybrid Energy Storage with Battery and Ultracapacitor

Based on the available technologies of various energy storages, there are several viable hybridization schemes for EVs and HEVs, typically, battery and battery hybrids and battery and ultracapacitor hybrids. The latter are more natural because the ultracapacitor can offer a much higher power than batteries, and they work together with various batteries to form battery and ultracapacitor hybrids. During hybridization, the simplest way is to directly and in parallell connect the ultracapacitors to the batteries, as shown in Figure 13.19. In this configuration, the ultracapacitors simply act as a current filter, which can significantly level the peak current of the batteries and reduce the battery voltage drop, as shown in Figures 13.20 and 13.21. The major disadvantages of this configuration

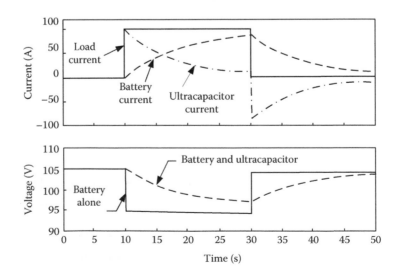

FIGURE 13.20
Variation of battery and ultracapacitor currents and voltages with a step current output change.

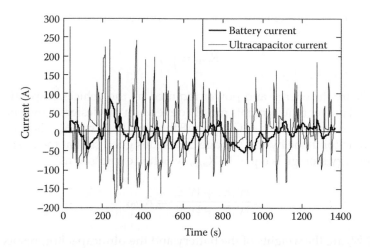

FIGURE 13.21
Battery and ultracapacitor currents during operation of HEV in FTP75 urban drive cycle.

are that the power flow cannot be actively controlled, and the ultracapacitor energy cannot be fully used.

Figure 13.22 shows a configuration in which a two-quadrant DC/DC converter is placed between the batteries and the ultracapacitors. This design allows the batteries and the ultracapacitors to have a different voltage; also, the power flow between them can be actively controlled, and the energy in the ultracapacitors can be fully used. In the long term, an ultra-high-speed flywheel replaces the batteries in hybrid energy storage to obtain a high-efficiency, compact, and long-life storage system for EVs and HEVs.

13.4.3 Battery and Ultracapacitor Size Design

The best design of a hybrid energy storage with a battery and an ultracapacitor is that the overall energy and power capacities just meet the energy and power requirements of the vehicle without much margin.[8] The energy and power requirements of a vehicle to its

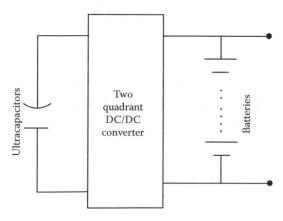

FIGURE 13.22
Actively controlled hybrid battery/ultracapacitor energy storage.

energy storage can be represented by the energy/power ratio, which is defined as

$$R_{e/p} = \frac{E_r}{P_r},$$ (13.49)

where E_r and P_r are the energy and power required by the vehicle, respectively.

The energy and power requirements mostly depend on the design of the vehicle drivetrain and control strategy, as discussed in Chapters 8 through 10. When $R_{e/p}$ is known, the battery and the ultracapacitor in the hybrid storage can be designed so that the energy/power ratio of the hybrid energy storage is equal to $R_{e/p}$, which is expressed as[8]

$$\frac{W_b E_b + W_c E_c}{W_b P_b + W_c P_c} = R_{e/p},$$ (13.50)

where W_b and W_c are the weights of the battery and the ultracapacitor, respectively; E_b and E_c are the specific energies of the battery and the ultracapacitor, respectively; and P_b and P_c are the specific powers of the battery and the ultracapacitor, respectively.

Equation 13.50 can be further written as

$$W_c = kW_b,$$ (13.51)

where

$$k = \frac{E_b - R_{e/p} P_b}{R_{e/p} P_c - E_c}.$$ (13.52)

Thus, the specific energy of the hybrid energy storage is

$$E_{spe} = \frac{W_b E_b + W_c E_c}{W_b + W_c} = \frac{E_b + kE_c}{1 + k}$$ (13.53)

and the specific power of the hybrid energy storage is

$$P_{spe} = \frac{W_b P_b + W_c P_c}{W_b + W_c} = \frac{P_b + kP_c}{1 + k}.$$ (13.54)

TABLE 13.5

Major Parameters of CHPS Battery Alternative at Standard Testing

CHPS Battery Alternative	Specific Energy (Wh/kg)	Specific Power (W/kg)	Energy/Power (h)
Lead–acid	28	75	0.373
NiCd	50	120	0.417
Ni–MH	64	140	0.457
Li–I (CHPS)[a]	100	1000[b]	0.1

Source: Y. Gao and M. Ehsani, *Power Electronics, IEEE Transactions,* 21(3), May 2006: 749–755.

[a] Combat hybrid power system sponsored by TACOM.
[b] Power capabilities depend on pulse length and temperature.

TABLE 13.6

Characteristic Data of a 42-V Ultracapacitor

Rated capacitance (DCC[a], 25°C)	(F)	145
Capacitance tolerance	(%)	±20
Rated voltage	(V)	42
Surge voltage	(V)	50
Max. series resis., ESR (DCC, 25°C)	(mΩ)	10
Specific power density (42 V)	(W/kg)	2900
Max. current	(A)	600
Max. stored energy	(J)	128,000
Specific energy density (42 V)	(Wh/kg)	2.3
Max. leakage current (12 h, 25°C)	(mA)	30
Weight	(kg)	15
Volume	(l)	22
Operating temperature	(°C)	−35 to +65
Storage temperature	(°C)	−35 to +65
Lifetime (25°C)	(year)	10, C < 20% of initial value, ESR < 200% of initial value
Cyclability (25°C, $I = 20$A)		500,000, C < 20% of initial value, ESR < 200% of initial value

Source: Available at http://www.maxwell.com, Maxwell Technologies, 2007.
[a] DCC: discharging at constant current.

TABLE 13.7

Characteristic Data of a 42-V Ultracapacitor

	Lead/Acid	Ni/Cd	Ni/MH	Li–I	Ultracap
Specific power (W/kg)	75	120	140	1000	2500
Specific energy (Wh/kg)	30	50	64	100	2
Total weight (kg)	667	417	357	50	1750

Source: Available at http://www.maxwell.com, Maxwell Technologies, 2007.

TABLE 13.8

Characteristic Data of a 42-V Ultracapacitor

	Lead/Acid	Ni/Cd	Ni/MH	Li–I
Specific power (W/kg)	378.5	581.4	703	1222
Specific energy (Wh/kg)	26.5	40.7	49.2	85.5
Battery weight (kg)	116	69	54	35
Ultracap. weight (kg)	16.5	16.7	16.9	6.05
Total weight (kg)	132	86	71	41

Source: Available at http://www.maxwell.com, Maxwell Technologies, 2007.

An example is shown in what follows.

Suppose that a vehicle needs a 50-kW energy storage; the desired energy/power ratio $R_{e/p} = 0.07$ h, that is, 3.5-kWh energy is required, and the battery and the ultracapacitor characteristics are shown in Tables 13.5 and 13.6. The weights needed for a single source and hybrid sources are listed in Tables 13.7 and 13.8. Comparing the total weights in Tables 13.7 and 13.8, it is obvious that the hybrid energy storage can save the weight significantly, especially with a battery that has low-power density.

Bibliography

1. D. A. J. Rand, R. Woods, and R. M. Dell, *Batteries for Electric Vehicles*, Society of Automotive Engineers (SAE), Warrendale, PA, 1988.
2. Available at http://www.saftbatteries.com, SAFT, The Battery Company, 2007.
3. T. R. Crompton, *Battery Reference Book*, Society of Automotive Engineers (SAE), Warrendale, PA, 1996.
4. C. C. Chan and K. T. Chau, *Modern Electric Vehicle Technology*, Oxford University Press, Oxford, 2001.
5. Y. Gao and M. Ehsani, Investigation of battery technologies for the army's hybrid vehicle application, In *Proceedings of the IEEE 56th Vehicular Technology Conference*, pp. 1505–1509, Fall 2002.
6. Y. Gao, H. Moghbelli, M. Ehsani, G. Frazier, J. Kajs, and S. Bayne, Investigation of high-energy and high-power hybrid energy storage systems for military vehicle application. *Society of Automotive Engineers (SAE) Journal*, Paper No. 2003-01-2287, Warrendale, PA, 2003.
7. Available at http://www.maxwell.com, Maxwell Technologies, 2007.
8. Y. Gao and M. Ehsnai, Parametric design of the traction motor and energy storage for series hybrid off-road and military vehicles. *Power Electronics, IEEE Transactions*, 21(3), May 2006: 749–755.
9. K. Itani et al. Regenerative braking modeling, control, and simulation of a hybrid energy storage system for an electric vehicle in extreme conditions. *IEEE Transactions on Transportation Electrification*, 2(4), 2016: 465–479.
10. J. Cao and A. Emadi, A new battery/ultracapacitor hybrid energy storage system for electric, hybrid, and plug-in hybrid electric vehicles. *IEEE Transactions on Power Electronics*, 27.1, 2012: 122–132.
11. R. Carter, A. Cruden, and P. J. Hall, Optimizing for efficiency or battery life in a battery/supercapacitor electric vehicle. *IEEE Transactions on Vehicular Technology* 61(4), 2012: 1526–1533.
12. M. Hedlund, J. Lundin, J. de Santiago, J. Abrahamsson, and H. Bernhoff, Flywheel energy storage for automotive applications. *Energies*, 8(10), 2015: 10636–10663.
13. M. M. Flynn, P. McMullen, and O. Solis, High-speed flywheel and motor drive operation for energy recovery in a mobile gantry crane. In *APEC 07-Twenty-Second Annual IEEE Applied Power Electronics Conference and Exposition*, pp. 1151–1157. IEEE.
14. J. Lundin, *Flywheel in an all-electric propulsion system*, Doctoral dissertation, PhD thesis, Uppsala University, 2011.
15. J. G. de Oliveira, *Power control systems in a flywheel based all-electric driveline*, Doctoral dissertation, PhD thesis, Uppsala University, 2011.
16. Eyer J., *Benefits from Flywheel Energy Storage for Area Regulation in California—Demonstration Results*. Sandia National Laboratories: Albuquerque, NM, USA, 2009.
17. M. Ahrens, L. Kucera, and R. Larsonneur, Performance of a magnetically suspended flywheel energy storage device. *IEEE Transactions on Control Systems Technology*, 4(5), 1996, 494–502.

18. M. T. Caprio, B. T. Murphy, and J. D. Herbst, *Spin Commissioning and Drop Tests of a 130 kW-hr Composite Flywheel*. CEM Publications, 2015.

19. K. Pullen and C. Ellis, Kinetic energy storage for vehicles, In *Hybrid Vehicle Conference*, IET The Institution of Engineering and Technology, 2006, pp. 91–108: IET.

20. O. Laldin, M. Moshirvaziri, and O. Trescases. Predictive algorithm for optimizing power flow in hybrid ultracapacitor/battery storage systems for light electric vehicles. *IEEE Transactions on Power Electronics* 28.8, 2013: 3882–3895.

14

Fundamentals of Regenerative Braking

One of the most important features of EVs, HEVs, and fuel cell vehicles (FCVs) is their ability to recover significant amounts of braking energy. The electric motors in EVs, HEVs, and FCVs can be controlled to operate as generators to convert the kinetic or potential energy of vehicle mass into electric energy that can be stored in the energy storage and then reused.

The braking performance of a vehicle is an important factor in vehicle safety. A successfully designed braking system for a vehicle must always meet the distinct demand of quickly reducing vehicle speed and maintaining vehicle direction controllable by the steering wheel. The former requires the braking system to be able to supply sufficient braking torque on all wheels. The latter requires proper braking force distribution on all wheels, as discussed in Chapter 2.

Generally, the braking torque required is much larger than the torque that an electric motor can produce, especially in heavy braking. In EVs, HEVs, and FCVs, mechanical friction braking systems must coexist with electrical regenerative braking. Therefore, this is a hybrid braking system. As in the hybrid propulsion system, there are many configurations and control strategies. However, the final goal of the design and control of such systems is to ensure the vehicle's braking performance and its ability to recover as much braking energy as possible.

14.1 Braking Energy Consumed in Urban Driving

A significant amount of energy is dissipated by braking.[1–3] Braking a 1500-kg vehicle from 100 km/h to zero speed dissipates about 0.16 kWh of energy [$(1/2)MV^2$] in a few tens of meters. If this amount of energy is dissipated by coasting and only by drag forces (rolling resistance and aerodynamic drag) without braking, the vehicle travels about 2 km, as shown in Figure 14.1.

When a vehicle is driving in a stop-and-go pattern in urban areas, a significant amount of energy is dissipated by frequent braking. Successful design of the hybrid braking system for recovering as much of the braking energy as possible requires a full understanding of braking behavior and its characteristics with respect to vehicle speed, braking power, deceleration rate, and so on during typical urban drive cycles.[2–4] The typical urban drive cycles that are used in this chapter are EPA FTP75, LA92, US06, New York City, and ECE-15.

While driving on a flat road, the driving power on the vehicle wheels can be calculated by

$$P_d = \frac{V}{1000}\left(Mgf_r + \frac{1}{2}\rho_a C_D A V^2 + M\delta\frac{dV}{dt}\right)(\text{kW}), \tag{14.1}$$

where M is the vehicle mass in kg, g is the gravitational acceleration, 9.81 m/s^2, f_r is the tire rolling resistance coefficient, ρ_a is the air mass density, 1.205 kg/m^3, C_D is the aerodynamic

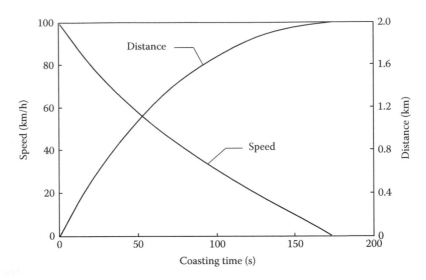

FIGURE 14.1
Coasting speed and distance.

drag coefficient, A is the frontal area of the vehicle in m^2, V is the vehicle speed in m/s, δ is the rotational inertia factor, and dV/dt is the vehicle acceleration in m/s^2 (and is negative for deceleration). For $P_d > 0$, the traction wheels receive power from the power plants and move the vehicle forward. In this case, the braking power is zero. In contrast, $P_d < 0$ when braking, and the kinetic energy of the vehicle mass is dissipated by the brake system. In this case, the driving power is zero.

Integrating Equation 14.1 over the driving time in a given driving cycle can give both the traction energy and the braking energy, as shown in Figure 14.2, for a typical passenger car with the parameters listed in Table 14.1 in the FTP75 urban drive cycle. The vehicle parameters used in this chapter are shown in Figure 14.3 and Table 14.1.

Figure 14.2 and Table 14.2 indicate that the braking energy in typical urban areas may reach up to more than 34% of the total traction energy. In large cities, such as New York City, it may reach up to 80%.

14.2 Braking Energy versus Vehicle Speed

Braking energy distribution over vehicle speed in typical urban drive cycles is useful information for the design and control of a regenerative brake system. In the speed range in which the braking energy is most dissipated, the operating efficiency of the electric motor, functioning as a generator, may be of most concern. In other speed ranges, regenerative braking may be abandoned with no significant compromise on energy recovered. Figure 14.4 shows such a diagram of braking energy distribution over vehicle speed while driving in the FTP75 urban drive cycle for a vehicle whose parameters are listed in Table 14.1. Figure 14.5 further shows the braking energy dissipated in a speed range that is less than a given speed. These two figures indicate that only 10% of the total braking energy is dissipated in the speed range below 15 km/h. Table 14.3 shows the braking energy dissipated in the speed range below 15 km/h while driving in other typical urban drive cycles.

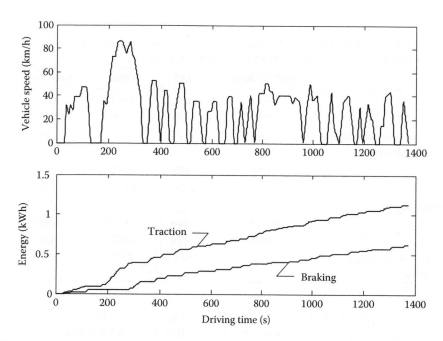

FIGURE 14.2
Traction and braking energy dissipation in an FTP75 urban drive cycle.

TABLE 14.1

Vehicle Parameters Used in This Paper

Item	Symbol	Unit	Value
Vehicle mass	M	kg	1500 (fully loaded), 1250 (unloaded)
Rolling resistance coefficient	f_r		0.01
Aerodynamic drag coefficient	C_D		0.3
Front area	A	m^2	2.2
Wheel base	L	m	2.7
Distance from gravity center to front wheel center	L_a	m	1.134 (fully loaded), 0.95 (unloaded)
Gravity center height	h_g	m	0.6 (fully loaded), 0.5 (unloaded)

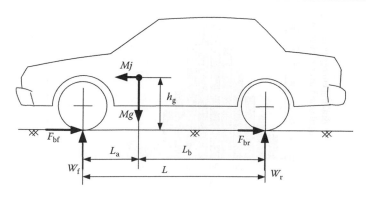

FIGURE 14.3
Forces acting on a vehicle in braking.

TABLE 14.2

Maximum Speed, Average Speed, Total Traction Energy, and Energies Dissipated by Drag and Braking per 100 km Traveling Distance in Different Drive Cycles

	FTP75 Urban	LA92	US06	New York	ECE15
Max. speed (km/h)	86.4	107.2	128.5	44.6	120
Ave. speed (km/h)	27.9	39.4	77.4	12.2	49.8
Traveling distance per cycle (km)	10.63	15.7	12.8	1.90	7.95
Traction energy (kWh)					
Per cycle	1.1288	2.3559	2.2655	0.2960	0.9691
Per km	1.1062	0.15	0.1769	0.1555	0.1219
Braking energy (kWh)					
Per cycle	0.6254	1.3666	0.9229	0.2425	0.3303
Per km	0.0589	0.0870	0.0721	0.1274	0.0416
Percentage of braking energy to traction energy	55.4	58.01	40.73	81.9	34.08

The braking energy dissipated in the low-speed range, such as below 15 km/h in all typical drive cycles, is insignificant. This result indicates that we need not attempt to obtain high operating efficiency at low speeds in the design and control of regenerative braking. In fact, it is difficult to regenerate at low speeds because of the low motor electromotive force (voltage) generated at low motor rotational speeds.

It should be noted that the rotational speed of a vehicle's drive wheels, which is proportional to the motor angular speed, is decoupled from the translational speed of the vehicle body when the vehicle's wheels are close to being locked. Thus, the operation of a hybrid brake system must be at speeds higher than a minimum threshold value. Electric regenerative braking should be applied primarily to recapture as much braking energy as possible. At speeds lower than this threshold, mechanical braking should be primarily applied to ensure the vehicle's braking performance.

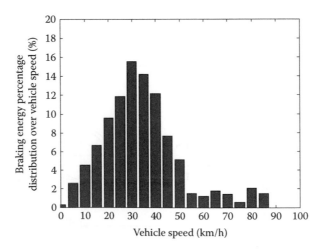

FIGURE 14.4

Braking energy distribution over vehicle speed in an FTP75 urban drive cycle.

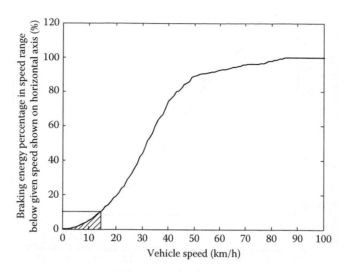

FIGURE 14.5
Braking energy dissipated over speed range below a given speed.

TABLE 14.3

Braking Energy Dissipated in Speed Range Below 15 km/h in Typical Urban Drive Cycles

	FTP75 Urban	LA92	US06	New York	ECE-15
Braking energy $\mid V < 15$ km/h (%)	10.93	5.51	3.27	21.32	4.25

14.3 Braking Energy versus Braking Power

Another important factor is braking energy versus braking power. Understanding braking energy versus braking power in a typical drive cycle is very helpful in the power capacity design of an electric motor drive and on-board energy storage, so that they can recover most of the braking energy without oversize design.

Figure 14.6 shows the braking simulation results for a vehicle whose parameters are listed in Table 14.1 while driving in an FTP75 urban drive cycle. This figure indicates that around 15% of the total braking energy is dissipated in a braking power range greater than 14.4 kW. This result implies that a 15-kW electric motor can recover about 85% of the total braking energy in this drive cycle. Table 14.4 shows the simulation results for other urban drive cycles, which also indicates the braking power range in which 85% of the total braking energy is dissipated. These data are good indicators of the design of the power capacity of the electric motor and the on-board energy storage from a braking point of view.

14.4 Braking Power versus Vehicle Speed

Another important consideration is the braking power characteristics versus vehicle speed in typical urban drive cycles. Understanding this is very helpful for proper design and control of the speed–power profile of electric motors to optimally match the driving application.

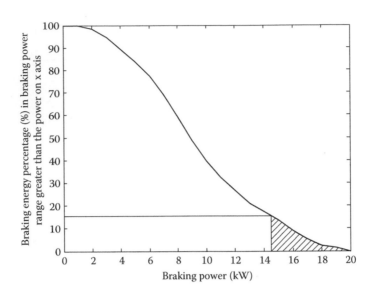

FIGURE 14.6
Braking energy percentage in range that is greater than the power shown on the horizontal axis.

TABLE 14.4

Braking Power Range in Which 85% of Braking Energy Is Dissipated in Typical Urban Drive Cycles

	FTP75 Urban	LA92	US06	New York	ECE-15
Power range in which 85% of total energy is consumed	0–14.4	0–44.5	0–46.5	0–18.5	0–33.5

Figure 14.7 shows the simulation results for the vehicle mentioned earlier. The bars in Figure 14.7 represent the maximum braking power in particular drive cycles at specified vehicle speeds. The solid lines represent the supposed motor speed–power profiles that can recover at least 85% of the braking energy, as indicated in Table 14.4.

The braking power versus vehicle speed profiles naturally match the power–speed characteristics of the motor, in that the power is proportional to the speed from zero speed to base speed (constant torque) and is constant beyond the base speed. Thus, electric motors do not need a special design and control for regenerative braking purposes.

14.5 Braking Energy versus Vehicle Deceleration Rate

Another important consideration is the braking energy distribution over the vehicle deceleration rate, which reflects the required braking force. Understanding this feature also helps the design and control of the hybrid braking system of EVs, HEVs, and FCVs. Figure 14.8 shows the braking energy consumed in the vehicle deceleration range of less than a value shown on the horizontal axis for the vehicle mentioned previously while driving in an FTP75 urban drive cycle. It can be seen in this figure that braking in this drive cycle is very gentle (the maximum deceleration rate is less than $0.15\,g$). Table 14.5 shows the

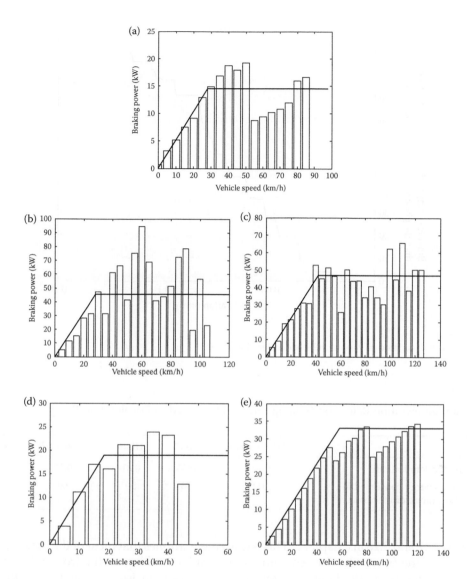

FIGURE 14.7
Braking power versus vehicle speed in typical urban drive cycles. (a) FTP75 urban, (b) LA92, (c) US06, (d) New York City, and (e) ECE-15.

maximum deceleration rates and braking energy in a deceleration rate of less than 0.15 g in other typical urban drive cycles.

14.6 Braking Energy on Front and Rear Axles

The braking performance for passenger cars requires the braking force distribution on the front and rear axles to be below the I curve but above the braking regulation curve, as shown

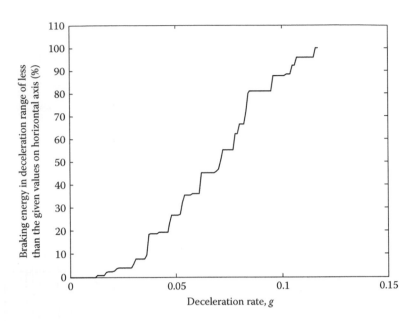

FIGURE 14.8
Braking energy dissipated in various vehicle deceleration rates.

TABLE 14.5

Maximum Deceleration Rates and Braking Energy Dissipated in Deceleration Range of Less Than 0.15 *g*

	FTP75 Urban	LA92	US06	New York	ECE-15
Maximum deceleration rate, *g*	0.12	0.40	0.31	0.27	0.14
% of braking energy consumed in deceleration range of less than 0.15 *g*	100	56	59	69	100

in Figures 2.27 and 2.28 in Chapter 2. This requirement implies that most of the braking force is applied on the front axle. Consequently, regenerative braking on the front axle is better than that on the rear axle. However, for other kinds of vehicles, such as trucks, the rear axle may be better.

14.7 Brake System of EV, HEV, and FCV

Regenerative braking in EVs, HEVs, and FCVs introduces slight complexity to the braking system design. Two basic questions arise: how to distribute the total required braking forces among regenerative braking and frictional braking to recover as much braking energy as possible and how to distribute the total braking forces on the front and rear axles to achieve stable braking performance. Usually, regenerative braking is effective only for the drive axle (the front axle for passenger cars). The electric motor must be controlled to produce the proper amount of braking force to recover as much braking energy as possible, and, at the same time, the total braking force must be sufficient to meet vehicle deceleration commanded by the driver. This chapter introduces two configurations of hybrid brake systems and their corresponding design and control principles. One is the parallel hybrid brake system, which has a simple structure and control and retains all the major components of conventional brakes. The other is a fully controllable hybrid brake system, which can fully control the braking force for each individual wheel, thus greatly enhancing the vehicle's braking performance on all types of roads.

The analysis in the following sections is based on the braking performance described in Section 2.9 of Chapter 2.

14.7.1 Parallel Hybrid Brake System

Perhaps the simplest system that is closest to conventional pure mechanical brakes (hydraulic or pneumatic) is the parallel hybrid brake system,[1,3] which retains all the major components of conventional mechanical brakes and adds electric braking directly on the front axle, as shown in Figure 14.9. The mechanical brake system consists of a master cylinder and booster. It may or may not have an ABS controller and actuator but has a brake caliper and brake disks. The electric motor directly applies its braking torque to the front axle and is controlled by the vehicle controller, based on vehicle speed and brake pedal position signals, which represent the desired braking strength and braking control strategy embedded in the vehicle controller. The feature of the parallel hybrid brake system is that only the electric braking force (torque) is electronically controlled, and the mechanical braking force (torque) is controlled by the driver through the brake pedal before the ABS starts its function. However, the mechanical braking force is controlled by the ABS when the wheels

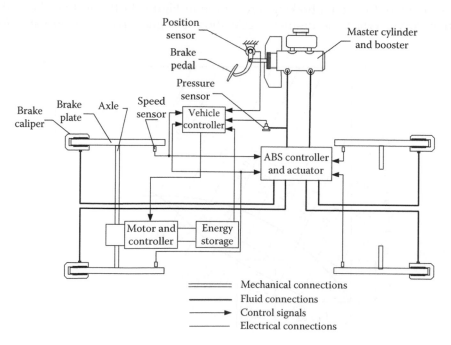

FIGURE 14.9
Schematic structure of parallel hybrid brake system.

are about to be locked, as in conventional brake systems. The key problem in the design and control of such a system is to properly control the electric braking force to recover as much braking energy as possible.

14.7.1.1 Design and Control Principles with Fixed Ratios between Electric and Mechanical Braking Forces

Figure 14.10 shows a braking force allocation strategy in which the mechanical brake has a fixed ratio of braking force distribution on the front and rear wheels represented by the β line. Curve I is the ideal braking force distribution of the vehicle. The ECE regulation that stipulates the minimum rear braking force is also plotted. The total braking force is the curve labeled mechanical + electrical. The braking force on the front wheels consists of mechanical braking, $F_{bf\text{-mech}}$, and electric braking, $F_{bf\text{-regen}}$, as shown in Figure 14.10. When the wheel speed is lower than a given threshold, 15 km/h for example, either very low vehicle speed or speed close to wheel lock-up, electric regenerative braking produces no braking force, and braking is performed solely by the mechanical system.

When the wheel speed is higher than the given threshold, and the desired vehicle deceleration is less than a given value (0.15 g in Figure 14.10), which is represented by the stroke of the brake pedal, all the braking force is produced by electric regenerative braking, and no mechanical braking force is applied to the front and rear wheels. As described in Section 12.7, most of the braking energy is in this deceleration range. Zero mechanical force may be used by employing larger clearances between the brake pads and plates or by carrying out small modifications to the conventional master cylinders. When the desired deceleration is greater than the given value (0.15 g in Figure 14.10), both mechanical and electric braking share the total front wheel braking force, as shown in Figure 14.10. The design of the electric braking portion is associated with the power capacity of the electric motor and on-board energy storage, but the total braking force curve in Figure 14.10 must be above the ECE regulation curve. When the desired deceleration is higher than a given value (0.6 g in

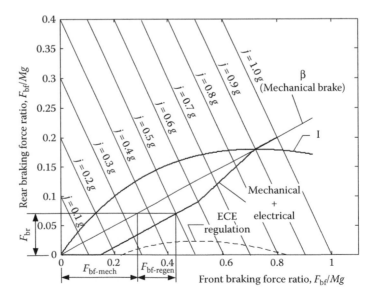

FIGURE 14.10
Braking forces varying with deceleration rate.

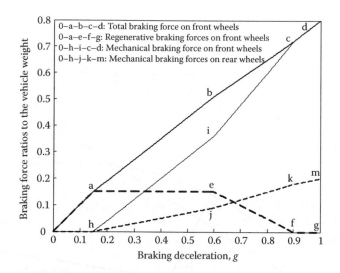

FIGURE 14.11
Braking force ratios relative to vehicle weight varying with deceleration rate.

Figure 14.10), the electric regenerative braking force is gradually reduced to zero with an increase of the desired deceleration (0.9 g in Figure 14.10). This design ensures that the actual front and rear wheel braking forces will be close to the ideal braking distribution curve, resulting in short braking distances and strong mechanical braking that may be more reliable in an emergency. Figure 14.11 shows the total braking force, regenerative braking force, and mechanical braking forces on the front and rear wheels, respectively, along with vehicle deceleration rate.

With the design principle described previously and the electric regenerative braking control rule with respect to wheel speed, the braking energy available for recovery can be computed in various typical drive cycles using a computer model. The simulation results are listed in Table 14.6. The data indicate that in normal urban driving, most of the braking energy can be recovered.

It should be noted that the maximum braking torque required of an electric motor is that which can produce 0.15 g of deceleration. Thus, a large-sized electric motor is not needed.

14.7.1.2 Design and Control Principles for Maximum Regenerative Braking

This design and control principle follows the rule that allocates the total braking force to the front wheels as much as possible, under the condition of meeting the braking regulations (the ECE regulation used here), that is, the total braking force distribution follows the maximum front braking force curve (minimum rear braking force) stipulated by the ECE regulation, which is represented by curve 0–a–b–c in Figure 14.12. More details are described later.

TABLE 14.6

Percentage of Total Braking Energy Available for Recovery

FTP75 Urban	LA92	US06	New York	ECE-15
89.69	82.92	86.55	76.16	95.75

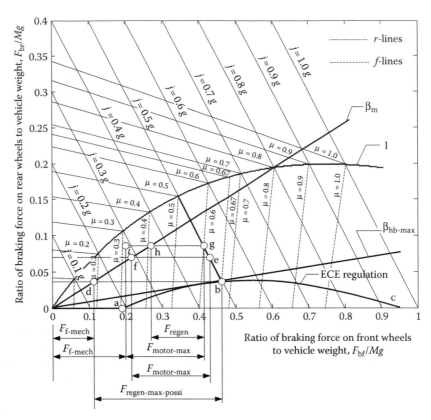

FIGURE 14.12
Schematic illustration of braking force distribution on front wheels (electrical + mechanical) and rear wheels.

When the braking strength is less than $0.2\,g$, all the braking force is allocated to the front wheels for regenerative braking, and no mechanical braking force is applied to the front or rear wheels. Motor torque may be controlled by sensing the braking pedal position. In this case, the master cylinder does not produce hydraulic pressure. When the braking strength is greater than $0.2\,g$, the mechanical system starts creating pressure, and the mechanical braking forces on the front and rear wheels start increasing, following the β_m line. At the same time, the electric motor adds its electric braking torque on the front wheels to make the total braking force follow the ECE regulation line. For example, when the braking strength required is $0.5\,g$, the total braking force is at point b, and the mechanical braking force on the front and rear wheels is at point d. The maximum possible braking force for regenerative braking is segment d–b, labeled $F_{\text{regen-max-possi}}$. However, to fully recover the maximum possible braking power, two conditions must be satisfied. One condition is that the electric motor should produce this braking force. If the maximum electric motor braking force is limited to that shown in Figure 14.12 by segment f–e, with a braking strength of $0.5\,g$, the operating point of the total braking force should be at point e and the mechanical braking force at point f. Assuming that the electric motor is powerful enough, another condition must be met to recover the maximum possible braking power limited by the ECE regulation. This condition is that the road adhesive coefficients must be larger than 0.67. Otherwise, the front wheels will lock. Figure 14.12 shows a case of braking strength $j = 0.5\,g$ and $\mu = 0.6$ and an operating point of total braking force at point g. In this case, the operation is different for brake systems with and without a mechanical ABS. For a system without ABS, to meet the braking force requirement on the rear wheels, the mechanical brake must operate at point h. Thus, the regenerative braking force takes segment h–g, which is smaller than the maximum braking force that the electric motor can produce. However, for a brake system with a mechanical ABS, when the front wheels are close to being locked, the ABS is activated, and the mechanical braking force on the front wheels is decoupled from the β line, instead following the f line of μ = 0.6; this limits further increases in the mechanical braking force on the front wheels. In this case, the electric motor can still produce its maximum braking force for maximum braking energy recovery, as shown in Figure 14.12 by the segment i–g, and the operating point of the mechanical braking is at point i.

The earlier design and control principle for maximum regenerative braking was based on the idea that the maximum possible braking force on the front wheels is limited by the ECE regulation. However, it can be seen from Figure 14.12 that the ECE regulation produces a nonlinear braking force distribution curve. This nonlinearity may lead to a complex design and control. A simple straight line can be used to replace the ECE regulation in the design and control of the hybrid brake system, as shown in Figure 14.12 by the line $\beta_{hb\text{-}max}$. The line $\beta_{hb\text{-}max}$ is generated as follows.

The minimum braking force on the rear wheels (maximum braking force on the front wheels) when the front wheels are locked, stipulated by the ECE regulation, is expressed as

$$\frac{F_{bf}}{W_f} \le \frac{q + 0.07}{0.85}, \tag{14.2}$$

where F_{bf} is the total braking force on the front wheels, W_f is the vertical loading on the front wheels, and q is the braking strength, $q = j/g$ (Figure 14.3). As in the definition of the β line in Section 2.9, the front wheel braking force can be expressed as

$$F_{bf} = \beta_{hb} F_b, \tag{14.3}$$

where F_b is the total braking force of the vehicle, which is related to the braking strength, q, as

$$F_b = M_j = Mgq. \tag{14.4}$$

The braking strength, q, and the vertical loading on the front wheels, W_f, have the relationship

$$W_f = \frac{Mg}{L}(L_b + qh_g), \tag{14.5}$$

where M is the mass of the vehicle, L is the wheel base, and L_b is the length from the vehicle's center of gravity to the rear axle, as shown in Figure 14.3.

Combining Equations 14.3 through 14.5, we obtain

$$\frac{F_{bf}}{W_f} = \frac{\beta_{bh}qL}{L_b + qh_g}. \tag{14.6}$$

From Equations 14.2 and 14.6 we obtain

$$\beta_{bh} \leq \frac{(q + 0.07)(L_b + qh_g)}{0.85qL}. \tag{14.7}$$

It can be seen from Equation 14.7 that the upper limit of β_{hb} to meet the ECE regulation is a function of braking strength q. The q at which the maximum β_{hb} is achieved can be obtained by

$$\left.\frac{d\beta_{bh}}{dq}\right|q = q^0 = 0, \tag{14.8}$$

with $q^0 = \sqrt{0.07L_b/h_g}$; then we obtain

$$\beta_{bh\text{-max}} = \frac{\sqrt[2]{0.07L_bh_g} + L_b + 0.07h_g}{0.85L}. \tag{14.9}$$

Equation 14.9 indicates that this braking force distribution ratio is only determined by the vehicle parameters. With $\beta_{bh\text{-max}}$, the braking force distribution on the front and rear wheels can be plotted as shown in Figure 14.12. This line can be used to replace the ECE regulation curve, and the design and control of the braking system may be simplified. The analysis for the braking process is similar to the preceding one and is left to the reader.

14.7.2 Fully Controllable Hybrid Brake System

In recent years, more advanced braking systems have emerged that allow one to control the braking force on each wheel independently.[1,2,5] Hydraulic electric brake systems (H-EBSs) and mechanical electric brake systems are two typical examples. Figure 14.13

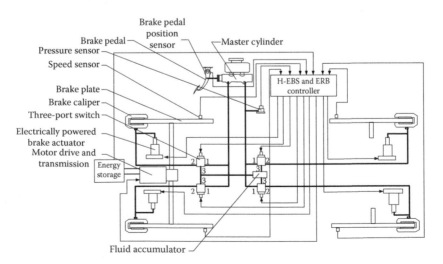

FIGURE 14.13
Fully controllable hybrid brake system with H-EBS and electric regenerative braking.

schematically shows a fully controllable hybrid brake system, which consists of a hydraulic electric brake and an electric regenerative brake.

A mechanical brake system consists mainly of a brake pedal and its position sensor, a master cylinder, an electrically operated and controlled brake actuator, electrically controlled three-port switches, a fluid accumulator, and a pressure sensor. In normal operation, ports 1 and 3 of the three-port switches are open, and port 2 is closed. The mechanical braking torque applied on each wheel is independently produced by the corresponding brake actuator, which is commanded by the H-EBS controller. The torque command to each wheel is generated in the H-EBSs, based on the pressure signal from the pressure sensor, the brake pedal stroke signal from the brake pedal position sensor, wheel speed signal from the wheel speed sensor, and the embedded control rule in the H-EBS controller. The brake fluid from the master cylinder flows into the fluid accumulator through the three-port switches to exert pressure and emulate the braking feeling of a conventional brake system. In the case of failure in any brake actuator, the corresponding three-port switch switches to the mode of ports 1 and 2 open and port 3 closed so the brake fluid from the master cylinder goes directly to the brake caliper cylinder and, thereby, maintains the braking torque.

An electric regenerative brake mainly includes an electric motor and its controller (drive) and an on-board energy storage. An ERB controller is used to control electric braking, based on wheel speed, brake pedal stroke, charge condition in the energy storage, and control rules embedded in the controller.

One of the key problems in this system is how to control the mechanical and electric braking torques to obtain acceptable braking performance and recover as much of the available regenerative braking energy as possible. In this chapter, two typical control strategies are introduced; one emphasizes the braking performance and the other the maximum regenerative braking energy recovery.

14.7.2.1 Control Strategy for Optimal Braking Performance

Due to the independent control of braking force on each wheel, a fully controllable hybrid brake system can be controlled to apply braking forces on the front and rear wheels by following the ideal braking force distribution curve. This control strategy can yield optimal brake performance.

Figure 14.14 illustrates the principle of this control strategy for a vehicle in which electric regenerative braking is available only on the front wheels. When the required total braking force on the front wheels is smaller than that produced by the electric motor, the electric motor produces the total braking force, and no mechanical braking force is applied. However, mechanical braking produces the total braking force for the rear wheels to follow the I curve, as shown by point a in Figure 14.14. When the required total braking force on the front wheels is greater than that produced by the electric motor, both electric braking and mechanical braking must be applied. For more braking energy recapture, the electric motor should be controlled to produce its maximum braking force, which is limited by the electric motor or energy storage. The remaining force is applied by the mechanical brake, as shown by point b in Figure 14.14.

It should be noted that at low front wheel speeds caused by actual low vehicle speed or almost locked wheels, it is hard to produce braking torque using the electric motor due to its low electromotive force (voltage) generated in the stator windings of the electric motor. Therefore, in this case, the mechanical brake must produce the total braking force as required.

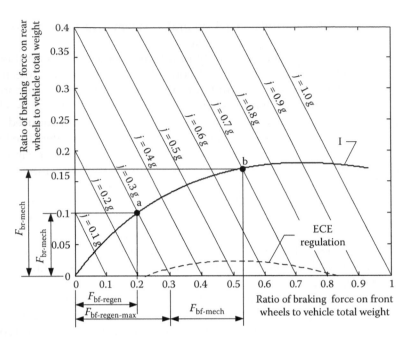

FIGURE 14.14
Control strategy for best braking performance.

As seen in Figure 14.14, a significant amount of braking energy is dissipated by the rear brakes, especially in weak braking (small deceleration). For example, at $j = 0.3\,g$, around 33% of the total braking energy is dissipated in the rear brakes. At $j = 0.1\,g$, this percentage reaches 37.8%. Unfortunately, this is just the case for most urban drive cycles. Considering no regenerative braking at low wheel speeds (<15 km/h), the braking energy available for recovery on the front wheels is considerably reduced. The simulation results, shown in Table 14.7, prove this conclusion.

14.7.2.2 Control Strategy for Optimal Energy Recovery

The principle of this control strategy aims to allocate more braking force to the front wheels under the condition of the front wheels never locking earlier than the rear wheels on a road

TABLE 14.7

Scenarios of Braking Energy in Typical Urban Drive Cycles

	FTP75 Urban	LA92	US06	New York	ECE-15
Percentage of braking energy on front wheels to total braking energy	61.52	63.16	62.98	62.57	61.92
Percentage of braking energy on rear wheels to total braking energy	38.48	36.84	37.02	37.43	38.08
Percentage of available regenerative braking energy on front wheels to total braking energy	55.16	59.85	60.89	50.26	59.27

with any adhesive coefficient. Thus, more braking energy is available for regenerative braking.

The details of this control strategy are explained below with the help of Figure 14.15.

When a vehicle brakes at an acceleration rate j on a road with an adhesive coefficient μ, and $j/g < \mu$, the braking forces on the front and rear wheels can be arbitrarily applied if the total braking force meets the requirements, that is, $F_{bf} + F_{br} = Mj$. However, braking performance requires that no wheel be locked and that the braking force on the rear wheels be above the ECE regulation curve, as shown in Figure 14.15. Thus, the braking forces on the front and rear wheels are variable in a certain range, which is dependent on the vehicle deceleration rate and road adhesive coefficient. Figure 14.15 shows the braking force ranges of a–b and c–d for deceleration rates of $j/g = 0.7$ and $j/g = 0.6$ (strong braking), respectively, on a road with adhesive coefficient $\mu = 0.9$ (concrete road). Obviously, for $j/g = 0.7$, the maximum braking force on the front wheels is determined by point b, which is dictated by the f line (front wheel locked) of $\mu = 0.9$. However, for $j/g = 0.6$, the maximum braking force on the front wheels is determined by point d, which is dictated by the ECE regulation. On this highly adhesive road, when $j/g < 0.7$, the braking force on the rear wheels can be very small, and almost all the braking force can be applied to the front wheels. However, when the road adhesive coefficient is smaller (slippery road), the braking force variable range is much smaller. Figure 14.15 shows a case with $\mu = 0.4$ (wet mud road), and $j/g = 0.3$ and $j/g = 0.2$. Obviously, point f determines the maximum braking force on the front wheels for $j/g = 0.3$ and point h for $j/g = 0.2$.

The preceding analysis provides only a control principle for a hybrid brake system to obtain maximum braking energy on the front wheels to make more braking energy

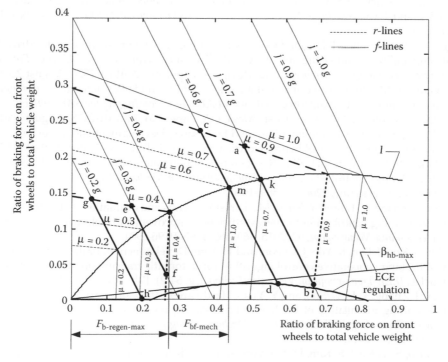

FIGURE 14.15
Control strategy for optimal energy recovery.

recoverable. However, the power capacities of the electric motor and energy storage are usually not big enough to handle the huge braking power when the braking is very strong. In this case, it is assumed that the electric motor will provide its maximum braking torque, and the mechanical brake provides the remaining torque. Figure 14.15 shows an electric motor that produces its maximum braking force on the front wheels to brake the vehicle at point n ($j/g = 0.4$). It is obvious that when the vehicle deceleration is lower, $j/g = 0.4$, the electric motor itself can handle it, and no mechanical braking is needed. However, when the braking deceleration is higher, $j/g = 0.6$ for example, the required braking force for the front wheels is more than the electric motor can handle. In this case, the mechanical brake must apply additional force to make the operation fall at any point in the range m–d. It is obvious that the best operation point is m.

With a simplified straight line βhb-max replacing the ECE regulation, very similar analysis can be done for this control strategy. This work is left to the reader.

Bibliography

1. S. R. Cikanek and K. E. Bailey, *Energy Recovery Comparison Between Series and Parallel Braking System for Electric Vehicles Using Various Drive Cycles*, Advanced Automotive Technologies, American Society of Mechanical Engineers (ASME), New York, DSC vol. 56/DE Vol. 86, pp. 17–31, 1995.
2. Y. Gao, L. Chu, and M. Ehsani, Design and control principle of hybrid braking system for EV, HEV and FCV, In *IEEE VPPC*, 2007.
3. Y. Gao, L. Chen, and M. Ehsani, Investigation of the effectiveness of regenerative braking for EV and HEV, *Society of Automotive Engineers (SAE) Journal*, SP-1466, Paper No. 1999-01-2901, 1999.
4. H. Gao, Y. Gao, and M. Ehsani, Design issues of the switched reluctance motor drive for propulsion and regenerative braking in EV and HEV, In *Proceedings of the SAE 2001 Future Transportation Technology Conference*, Costa Mesa, CA, Paper No. 2001-01-2526, August 2001.
5. Y. Gao and M. Ehsani, Electronic braking system of EV and HEV—integration of regenerative braking, automatic braking force control and ABS, In *Proceedings of the SAE 2001 Future Transportation Technology Conference*, Costa Mesa, CA, Paper No. 2001-01-2478, August 2001.
6. M. Panagiotidis, G. Delagrammatikas, and D. Assanis, Development and use of a regenerative braking model for a parallel hybrid electric vehicle, SAE Technical Paper 2000-01-0995, 2000.
7. Y. Hoon, S. Hwang, and H. Kim, Regenerative braking algorithm for a hybrid electric vehicle with CVT ratio control, *Proceedings of the Institution of Mechanical Engineers, Part D: Journal of Automobile Engineering*, 220(11): 1589–1600.
8. C. Lv, J. Zhang, Y. Li, and Y. Yuan, Mechanism analysis and evaluation methodology of regenerative braking contribution to energy efficiency improvement of electrified vehicles, *Energy Conversion and Management*, 92, 2015: 469–482, ISSN 0196-8904.
9. J. Ko, S. Ko, H. Son, B. Yoo, J. Cheon, and H. Kim, Development of brake system and regenerative braking cooperative control algorithm for automatic-transmission-based hybrid electric vehicles, *IEEE Transactions on Vehicular Technology*, 64(2), Feb. 2015: 431–440.
10. C. Lv, J. Zhang, Y. Li, D. Sun, and Y. Yuan, Hardware-in-the-loop simulation of pressure-difference-limiting modulation of the hydraulic brake for regenerative braking control of electric vehicles, *Proceedings of the Institution of Mechanical Engineers, Part D: Journal of Automobile Engineering*, 228(6), 2014: 649–662.
11. J. W. Ko, S. Y. Ko, I. S. Kim, D. Y. Hyun, and H. S. Kim, Co-operative control for regenerative braking and friction braking to increase energy recovery without wheel lock, *International Journal of Automotive Technology*, 15(2), 2014: 253–262.

12. L. Li, Y. Zhang, C. Yang, B. Yan, and C. M. Martinez, Model predictive control-based efficient energy recovery control strategy for regenerative braking system of hybrid electric bus, *Energy Conversion and Management*, 111, 2016: 299–314, ISSN 0196-8904.

13. P. Gyan, I. Baghdadi, O. Briat, and J. Del, Lithium battery aging model based on Dakin's degradation approach. *Journal of Power Sources*, 325, 2016: 273–285. http://dx.doi.org/10.1016/j.jpowsour.2016.06.036.

14. S. F. Schuster, T. Bach, E. Fleder, J. Müller, M. Brand, G. Sextl et al., Nonlinear aging characteristics of lithium-ion cells under different operational conditions. *Journal of Energy Storage*, 1, 2015: 44–53. http://dx.doi.org/10.1016/j.est.2015.05.003.

15. I. Gandiaga and I. Villarreal, Cycle ageing analysis of a LiFePO4/graphite cell with dynamic model validations: towards realistic lifetime predictions. *Journal of Power Sources*, 275, 2015: 573–587. http://dx.doi.org/10.1016/j.jpowsour.2014.10.153.

16. A. Castaings, W. Lhomme, R. Trigui, and A. Bouscayrol, Comparison of energy management strategies of a battery/supercapacitors system for electric vehicle under real-time constraints. *Applied Energy*, 163, 2016: 190–200. http://dx.doi.org/10.1016/j.apenergy.2015.11.020.

17. R. E. Arajo, R. De Castro, C. Pinto, P. Melo, and D. Freitas, Combined sizing and energy management in EVs with batteries and supercapacitors. *IEEE Transactions on Vehicular Technology*, 63, 2014: 3062–3076. http://dx.doi.org/10.1109/TVT.2014.2318275.

18. A. Chauvin, A. Hijazi, E. Bideaux, and A. Sari, Combinatorial approach for sizing and optimal energy management of HEV including durability constraints. In *IEEE International Symposium on Industrial Electronics*, pp. 1236–1241, 2015. http://dx.doi.org/10.1109/ISIE.2015.7281649.

19. R. Hemmati and H. Saboori, Emergence of hybrid energy storage systems in renewable energy and transport applications, A review. *Renewable and Sustainable Energy Reviews*, 65, 2016: 11–23. http://dx.doi.org/10.1016/j.rser.2016.06.029.

20. Z. Song, H. Hofmann, J. Li, J. Hou, X. Han, and M. Ouyang, Energy management strategies comparison for electric vehicles with hybrid energy storage system. *Applied Energy*, 134, 2014: 321–331. http://dx.doi.org/10.1016/j.

21. V. Herrera, A. Milo, H. Gaztañaga, I. Etxeberria-Otadui, I. Villarreal, and H. Camblong, Adaptive energy management strategy and optimal sizing applied on a battery-supercapacitor based tramway. *Applied Energy*, 169, 2016: 831–845. http://dx.doi.org/10.1016/j.apenergy.2016.02.079.

22. N. Devillers, S. Jemei, M.-C. Péra, D. Bienaimé, and F. Gustin, Review of characterization methods for supercapacitor modelling. *Journal of Power Sources*, 246, 2014: 596–608. http://dx.doi.org/10.1016/j.jpowsour.2013.07.116.

15

Fuel Cells

In recent decades, the application of fuel cells in vehicles has been the focus of increased attention. In contrast to a chemical battery, the fuel cell generates electrical energy rather than storing it and continues to do so as long as the fuel supply is maintained. Compared with battery-powered EVs, a fuel-cell-powered vehicle has the advantages of a longer driving range without a long battery charging time. Compared with internal combustion engine vehicles, it has the advantages of high energy efficiency and much lower emissions due to the direct conversion of free energy in the fuel into electrical energy, without undergoing combustion.

15.1 Operation Principles of Fuel Cells

A fuel cell is a galvanic cell in which the chemical energy of a fuel is converted directly into electrical energy by means of electrochemical processes. The fuel and oxidizing agent are continuously and separately supplied to the two electrodes of the cell, where they undergo a reaction. An electrolyte is necessary to conduct the ions from one electrode to the other, as shown in Figure 15.1. Fuel is supplied to the anode or positive electrode, where electrons are released from the fuel under catalyst. The electrons, under the potential difference between these two electrodes, flow through the external circuit to the cathode electrode or negative electrode where combining positive ions and oxygen, reaction products, or exhaust are produced.

The chemical reaction in a fuel cell is similar to that in a chemical battery. The thermodynamic voltage of a fuel cell is closely associated with the energy released and the number of electrons transferred in the reaction.[1,2] The energy released by the cell reaction is given by the change in Gibbs free energy, ΔG, usually expressed in per-mole quantities. The change in Gibbs free energy in a chemical reaction can be expressed as

$$\Delta G = \sum_{\text{Products}} G_i - \sum_{\text{Reactants}} G_j, \tag{15.1}$$

where G_i and G_j are the free energies in species i of products and species j of reactants. In a reversible process, ΔG is completely converted into electric energy, that is,

$$\Delta G = -nFV_r, \tag{15.2}$$

where n is the number of electrons transferred in the reaction, $F = 96{,}495$ is the Faraday constant in coulombs per mole, and V_r is the reversible voltage of the cell. At standard conditions (25°C temperature and 1 atm. pressure), the open-circuit (reversible) voltage of a cell

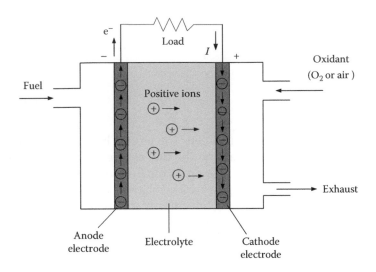

FIGURE 15.1
Basic operation of a fuel cell.

can be expressed as

$$V_r^0 = -\frac{\Delta G^0}{nF},$$ (15.3)

where ΔG^0 is the change in Gibbs free energy under standard conditions. ΔG is expressed as

$$\Delta G = \Delta H - T\,\Delta S,$$ (15.4)

where ΔH and ΔS are the enthalpy and entropy changes, respectively, in the reaction at absolute temperature T. Table 15.1 shows the values of standard enthalpy, entropy, and Gibbs

TABLE 15.1

Standard Enthalpy of Formation and Gibbs Free Energy for Typical Fuels

Substance	Formula	ΔH_{298}^0 (kJ/mol)	ΔS_{298}^0 (kJ/mol K)	ΔG_{298}^0 (kJ/mol)
Oxygen	O (g)	0	0	0
Hydrogen	H (g)	0	0	0
Carbon	C (s)	0	0	0
Water	H_2O (l)	−286.2	−0.1641	−237.3
Water	H_2O (g)	−242	−0.045	−228.7
Methane	CH_4 (g)	−74.9	−0.081	−50.8
Methanol	CH_3OH (l)	−238.7	−0.243	−166.3
Ethanol	C_2H_5OH (l)	−277.7	−0.345	−174.8
Carbon monoxide	CO (g)	−111.6	0.087	−137.4
Carbon dioxide	CO_2	−393.8	0.003	−394.6
Ammonia	NH_3 (g)	−46.05	−0.099	−16.7

TABLE 15.2

Thermodynamic Data for Different Reactions at 25°C and 1 atm. Pressure

	ΔH^0_{298} (kJ/mol)	ΔS^0_{298} (kJ/mol K)	ΔG^0_{298} (kJ/mol)	n	E (V)	η_{id} (%)
$H_2 + \frac{1}{2}O_2 \rightarrow H_2O\,(l)$	−286.2	−0.1641	−237.3	2	1.23	83
$H_2 + \frac{1}{2}O_2 \rightarrow H_2O\,(g)$	−242	−0.045	−228.7	2	1.19	94
$C + \frac{1}{2}O_2 \rightarrow CO\,(g)$	−116.6	0.087	−137.4	2	0.71	124
$C + O_2 \rightarrow CO_2\,(g)$	−393.8	0.003	−394.6	4	1.02	100
$C + \frac{1}{2}O_2 \rightarrow CO_2(g)$	−279.2	−0.087	−253.3	2	1.33	91

free energy of some typical substances.[3] Table 15.2 shows the thermodynamic data for some reactions in a fuel cell at 25° and 1 atm. pressure.[3]

The "ideal" efficiency of a reversible galvanic cell is related to the enthalpy for the cell reaction by

$$\eta_{id} = \frac{\Delta G}{\Delta H} = 1 - \frac{\Delta S}{\Delta H} T. \tag{15.5}$$

η_{id} will be 100% if the electrochemical reaction involves no change in the number of gas moles, that is, when ΔS is zero. This is the case for reactions $C + O_2 = CO_2$. However, if the entropy change, ΔS, of a reaction is positive, then the cell—in which this reaction proceeds isothermally and reversibly—has at its disposal not only the chemical energy, ΔH, but also (in analogy to a heat pump) a quantity of heat, $T\Delta S$, absorbed from the surroundings for conversion into electrical energy (Table 15.2) (Figure 15.2).

The change of free energy, and thus cell voltage, in a chemical reaction is a function of the activities of the solution species. The dependence of cell voltage on the reactant activities is expressed as

$$V_r = V_r^0 - \frac{RT}{nF} \ln\left[\frac{\prod (\text{activities of products})}{\prod (\text{activities of reactants})}\right], \tag{15.6}$$

where R is the universal gas constant, 8.31 J/mol K, and T is the absolute temperature in K. For gaseous reactants and products, Equation 15.6 can be expressed as

$$V_r = V_r^0 - \frac{RT}{nF} \sum_i v_i \ln\left(\frac{p_i}{p_i^0}\right), \tag{15.7}$$

where V_r is the voltage of the cell in which the reaction proceeds with gaseous participants at nonstandard pressure p_i, V_r^0 is the corresponding cell voltage with all gases at the standard pressure p_i^0 (normally 1 atm.), and v_i is the number of moles of species i accounted as positive for products and negative for reactants.

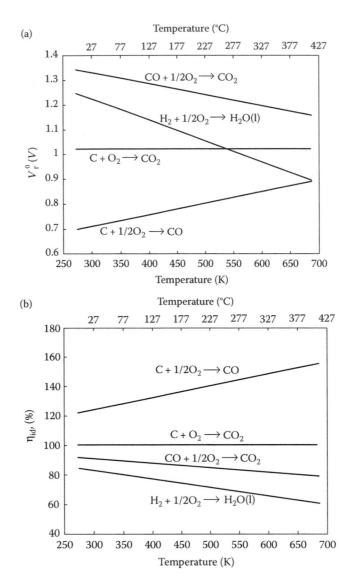

FIGURE 15.2
(a) Temperature dependence of cell voltage and (b) reversible efficiency.

15.2 Electrode Potential and Current–Voltage Curve

Experiments have shown that the rest voltage, V, is usually lower than the reversible voltage, V_r^0, calculated from the ΔG value. The voltage drop is called the rest voltage drop, ΔV_0. The reason may be the existence of a significant kinetic hindrance to the electrode process or because the process does not take place in the manner assumed in the thermodynamic calculation of V_r^0. This rest voltage drop depends, in general, on the electrode materials and the kind of electrolyte.

When current is drawn from a cell, a voltage drop is caused by the existence of ohmic resistance in the electrode and electrolyte, which increases in direct proportion to the current density, that is,

$$\Delta V_\Omega = R_e i, \tag{15.8}$$

where R_e is the equivalent ohmic resistance per area, and i is the current density.

In a fuel cell, part of the generated energy is lost in pushing the species to react because extra energy is required to overcome the activation barriers. These losses are called activation losses and are represented by an activation voltage drop, ΔV_a. This voltage drop is closely related to the materials of electrodes and the catalysts. The Tafel equation is most commonly used to describe this behavior, by which the voltage drop is expressed as[3]

$$\Delta V_a = \frac{RT}{\beta nF} \ln\left(\frac{i}{i_0}\right). \tag{15.9}$$

More conveniently, it is written

$$\Delta V_a = a + b \ln(i), \tag{15.10}$$

where $a = -(RT/\beta nF) \ln(i_0)$ and $b = RT/\beta nF$; i_0 is the exchange current at the equilibrium state, and b is constant depending on the process. For a more detailed theoretical description, refer to pp. 230–236 of Messerle.[3]

When current flows, ions are discharged near the negative electrode, and as a result, the concentration of ions in this region tends to decrease. If the current is to be maintained, ions must be transported to the electrode. This takes place naturally by diffusion of ions from the bulk electrolyte and by direct transport due to fields caused by concentration gradients. Bulk movement of the electrolyte by convection or stirring also helps to bring the ions up.

The voltage drop caused by the lack of ions is called a concentration voltage drop since it is associated with a decrease in the concentration of the electrolyte in the immediate vicinity of the electrode. For small current densities, the concentration voltage drop is generally small. However, as the current density increases, it reaches a limit when the maximum possible rate of transport of ions to the electrode is approached as the concentration at the electrode surface falls to zero.

The voltage drop caused by the concentration at the electrode where ions are removed (cathode electrode in fuel cell) can be expressed as[3]

$$\Delta V_{c1} = \frac{RT}{nF} \ln\left(\frac{i_L}{i_L - i}\right) \tag{15.11}$$

and at the electrode where ions are formed (anode electrode in fuel cell) as

$$\Delta V_{c2} = \frac{RT}{nF} \ln\left(\frac{i_L + i}{i_L}\right), \tag{15.12}$$

where i_L is the limiting current density.

The voltage drop caused by concentration is not only restricted to the electrolyte. When either the reactant or the product is gaseous, a change in partial pressure in the reacting zones also represents a change in concentration. For example, in a hydrogen–oxygen fuel cell, oxygen may be introduced into the air. When the reaction takes place, oxygen is removed near the electrode surface in the pores of the electrode, and the partial pressure of oxygen drops there when compared with that in bulk air. The change in partial pressure causes a voltage drop, which is determined by

$$\Delta V_{cg} = \frac{RT}{nF} \ln\left(\frac{p_s}{p_0}\right), \tag{15.13}$$

where p_s is the partial pressure at the surface, and p_0 is the partial pressure in the bulk feed. For more details, see pp. 236–238 of Messerle.[3]

Figure 15.3 shows the voltage–current curves of a hydrogen–oxygen fuel cell with a temperature of 80°C. The drop caused by the chemical reaction, including activation and concentration, is the source of the voltage drop. This figure also indicates that improving the electrode materials and manufacturing using advanced technology, such as nanotechnology, and advanced catalysts significantly reduces the voltage drop and consequently improves the efficiency of the fuel cell.

The energy loss in a fuel cell is represented by the voltage drop. Thus, the efficiency of the fuel cell can be written

$$\eta_{fc} = \frac{V}{V_r^0}, \tag{15.14}$$

where V_r^0 is the cell reversible voltage at standard conditions ($T = 298\,\text{K}$ and $p = 1\,\text{atm.}$). The efficiency curve is strictly homothetic to the voltage curve. An efficiency–current

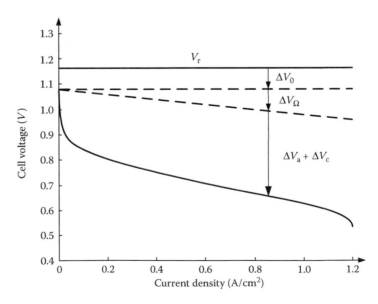

FIGURE 15.3
Current–voltage curves for hydrogen–oxygen fuel cell at $T = 80°C$.

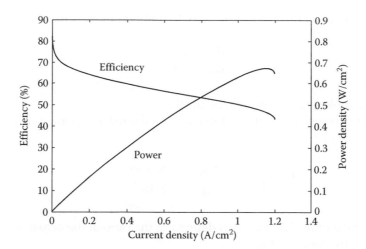

FIGURE 15.4
Operating efficiency and power density along with current density in hydrogen–oxygen fuel cell.

curve for a hydrogen–oxygen fuel cell (Figure 15.3) is shown in Figure 15.4. Figure 15.4 indicates that the efficiency decreases and power increases with the increase of current. Therefore, operating a fuel cell at its low current, and then at low-power, results in high operating efficiency. However, taking account of the energy consumed by its auxiliaries, such as the air circulating pump, the cooling water circulating pump, and so on, very low-power (<10% of its maximum power) results in low operating efficiency due to the larger percentage of power consumption in the auxiliary. This is discussed in more detail later.

15.3 Fuel and Oxidant Consumption

The fuel and oxidant consumptions in a fuel cell are proportional to the current drawn from the fuel cell. The chemical reaction in a fuel cell can be generally described by Equation 15.15, where A is the fuel, B is the oxidant, and C and D are the products:

$$A + x_B B \rightarrow x_C C + x_D D. \tag{15.15}$$

The mass flow of the fuel, associated with the current drawn from the fuel cell, can be expressed as

$$\dot{m}_A = \frac{W_A I}{1000 nF} \text{ (kg/s)}, \tag{15.16}$$

where W_A is the molecular weight of fuel A, I is the fuel cell current, n is the electrons transferred in the reaction of Equation 15.15, and $F = 96{,}495$ C/mol is the Faraday constant.

The stoichiometric ratio of oxidant mass flow to fuel mass flow can be expressed as

$$\frac{\dot{m}_B}{\dot{m}_A} = \frac{x_B W_B}{W_A}, \tag{15.17}$$

where W_B is the molecular weight of oxidant B.

For a hydrogen–oxygen fuel cell (see Table 15.2 for the reaction), the stoichiometric ratio of hydrogen to oxygen is

$$\left(\frac{\dot{m}_H}{\dot{m}_O}\right)_{stoi} = \frac{0.5 W_O}{W_H} = \frac{0.5 \times 32}{2.016} = 7.937. \tag{15.18}$$

The equivalent ratio of oxidant to fuel is defined as the ratio of the actual oxidant/fuel ratio to the stoichiometric ratio, that is,

$$\lambda = \frac{(\dot{m}_B / \dot{m}_A)_{actual}}{(\dot{m}_B / \dot{m}_A)_{stoi}}. \tag{15.19}$$

When $\lambda < 1$, the reaction is fuel rich; when $\lambda = 1$, the reaction is stoichiometric; and when $\lambda > 1$, the reaction is fuel lean. In practice, fuel cells are always operated at $\lambda > 1$, that is, excessive air over the stoichiometric value is supplied to reduce the voltage drop caused by concentration. For fuel cells using O_2 as oxidant, air is usually used rather than pure oxygen. In this case, the stoichiometric ratio of fuel to air can be expressed as

$$\frac{\dot{m}_{air}}{\dot{m}_A} = \frac{(x_O W_O)/0.232}{W_A}, \tag{15.20}$$

where it is assumed that the oxygen mass takes 23.2% of the air mass. For hydrogen–air fuel cells, Equation 15.19 becomes

$$\left(\frac{\dot{m}_{air}}{\dot{m}_H}\right)_{stoi} = \frac{(0.5 W_O)/0.232}{W_H} = \frac{(0.5 \times 32)/0.232}{2.016} = 34.21. \tag{15.21}$$

15.4 Fuel Cell System Characteristics

In practice, fuel cells need auxiliaries to support their operation. The auxiliaries mainly include an air circulating pump, a coolant circulating pump, a ventilation fan, a fuel supply pump, and electrical control devices, as shown in Figure 15.5. Among the auxiliaries, the air circulating pump is the largest energy consumer. The power consumed by the air circulating pump (including its drive motor) may take about 10% of the total power output of the fuel cell stack. The other auxiliaries consume much less energy compared with the air circulating pump.

In a fuel cell, the air pressure on the electrode surface, p, is usually higher than the atmospheric pressure, p_0, in order to reduce the voltage drop (Equation 15.13). According to

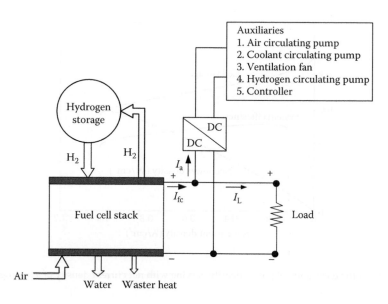

FIGURE 15.5
Hydrogen–air fuel cell system.

thermodynamics, the power needed to compress air from low-pressure p_0 to high-pressure p with a mass flow $\dot{m}_{air}$ can be calculated by[1,2]

$$P_{air-comp} = \frac{\gamma}{\gamma - 1} \dot{m}_{air} RT \left[\left(\frac{p}{p_0} \right)^{\gamma - 1/\gamma} - 1 \right] (W), \qquad (15.22)$$

where γ is the ratio of specific heats of air ($\gamma = 1.4$), R is the gas constant of air ($R = 287.1$ J/kg K), and T is the temperature at the inlet of the compressor in K. When calculating the power consumed by the air circulating pump, the energy losses in the air pump and motor drive must be taken into account. Thus, the total power consumed is

$$P_{air-cir} = \frac{P_{air-comp}}{\eta_{ap}}, \qquad (15.23)$$

where η_{ap} is the efficiency of the air pump plus motor drive.

Figure 15.6 shows an example of the operation characteristics of the hydrogen–air fuel cell system, where $\lambda = 2$, $p/p_0 = 3$, and $\eta_{ap} = 0.8$, and the net current and net power are the current and power that flow to the load (Figure 15.5). This figure indicates that the optimal operation region of the fuel cell system is in the middle region of the current range, say, 7%–50% of the maximum current. A large current leads to low efficiency due to the large voltage drop in the fuel cell stack, and, conversely, a very small current leads to low efficiency due to the increase in the percentage of the auxiliaries' energy consumption.

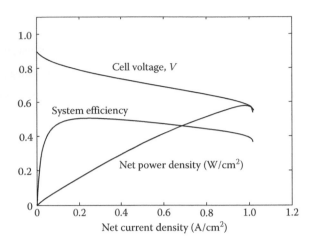

FIGURE 15.6
Cell voltage, system efficiency, and net power density varying with net current density of a hydrogen–air fuel cell.

15.5 Fuel Cell Technologies

It is possible to distinguish six major types of fuel cells, depending on the type of their electrolyte.[4] They are proton exchange membrane (PEM) or polymer exchange membrane fuel cells (PEMFCs), alkaline fuel cells (AFCs), phosphoric acid fuel cells (PAFCs), molten carbonate fuel cells (MCFCs), solid oxide fuel cells (SOFCs), and direct methanol fuel cells (DMFCs). Table 15.3 lists their normal operation temperature and the state of electrolyte.

15.5.1 Proton Exchange Membrane Fuel Cells

PEMFCs use solid polymer membranes as the electrolyte. The polymer membrane is perfluorosulfonic acid, which is also referred to as Nafion (Dupont®). This polymer membrane is acidic; therefore, the ions transported are hydrogen ions (H^+) or protons. The PEMFC is fueled with pure hydrogen and oxygen or air as the oxidant.

TABLE 15.3

Operating Data of Various Fuel Cell Systems

Cell System	Operating Temperature (°C)	Electrolyte
PEMFCs	60–100	Solid
AFCs	100	Liquid
PAFCs	60–200	Liquid
MCFCs	500–800	Liquid
SOFCs	1000–1200	Solid
DMFCs	100	Solid

Source: P. J. Berlowitz and C. P. Darnell, *Society of Automotive Engineers (SAE) Journal*, Paper No. 2001-01-0003, Warrendale, PA, 2002.

The polymer electrolyte membrane is coated with a carbon-supported catalyst. The catalyst is in direct contact with both the diffusion layer and the electrolyte for a maximized interface. The catalyst constitutes the electrode. Directly on the catalyst layer is the diffusion layer. The assembly of the electrolyte, catalyst layers, and gas diffusion layers is referred to as the membrane–electrode assembly.

The catalyst is a critical issue in PEMFCs. In early realizations, very high loadings of platinum were required for the fuel cell to operate properly. Tremendous improvements in catalyst technology have made it possible to reduce the loading from 28 mg to $0.2 \, mg/cm^2$. Because of the low operating temperature of the fuel cell and the acidic nature of the electrolyte, noble metals are required for the catalyst layer. The cathode is the most critical electrode because the catalytic reduction of oxygen is more difficult than the catalytic oxidation of hydrogen.

Another critical issue in PEMFCs is water management. To operate properly, the polymer membrane needs to be kept humid. Indeed, the conduction of ions in polymer membranes requires humidity. If the membrane is too dry, there will not be enough acid ions to carry the protons. If it is too wet (flooded), the pores of the diffusion layer will be blocked, and reactant gases will not be able to reach the catalyst.

In PEMFCs, water is formed on the cathode. It can be removed by keeping the fuel cell at a certain temperature and flowing enough to evaporate the water and carry it out of the fuel cell as a vapor. However, this approach is difficult because the margin of error is narrow. Some fuel cell stacks run on a large excess of air that would normally dry the fuel cell and use an external humidifier to supply water by the anode.

The last major critical issue in PEMFCs is poisoning. The platinum catalyst is extremely active and thus provides great performance. The trade-off of this great activity is a greater affinity for carbon monoxide (CO) and sulfur products than oxygen. The poisons bind strongly to the catalyst and prevent hydrogen or oxygen from reaching it. The electrode reactions cannot take place on the poisoned sites, and the fuel cell performance is diminished. If hydrogen is fed from a reformer, the stream will contain some carbon monoxide. The carbon monoxide may also enter the fuel cell in the air stream if the air is pumped from the atmosphere of a polluted city. Poisoning by carbon monoxide is reversible, but it comes at a cost and requires the individual treatment of each cell.

The first PEMFCs were developed in the 1960s for the needs of the U.S.-manned space program. It is now the most investigated fuel cell technology for automotive applications by manufacturers such as Ballard. It is operated at 60–100°C and can offer a power density of $0.35–0.6 \, W/cm^2$. The PEMFC has some definite advantages in its favor for EV and HEV applications.[5] First, its low-temperature operation and hence its fast startup are desirable for an EV and HEV. Second, the power density is the highest among all the available types of fuel cells. The higher the power density, the smaller the size of the fuel cell that needs to be installed for the desired power demand. Third, its solid electrolyte does not change, move, or vaporize from the cell. Finally, since the only liquid in the cell is water, the possibility of any corrosion is essentially delimited. However, it also has some disadvantages, such as the expensive noble metal needed, expensive membrane, and easily poisoned catalyst and membrane.[6]

15.5.2 Alkaline Fuel Cells

AFCs use an aqueous solution of potassium hydroxide (KOH) as the electrolyte to conduct ions between electrodes. Potassium hydroxide is alkaline. Because the electrolyte is alkaline, the ion conduction mechanism is different from that of PEMFCs. The ion carried by the alkaline electrolyte is a hydroxide ion (OH^-). This affects several other aspects of the fuel cell. The half reactions are as follows:

$$Anode: 2H_2 + 4OH^- \rightarrow 4H_2O + 4e^-.$$

$$Cathode: O_2 + 4e^- + 2H_2O \rightarrow 4OH^-.$$

Unlike in acidic fuel cells, water is formed on the hydrogen electrode. In addition, water is needed at the cathode by oxygen reduction. Water management becomes an issue that is sometimes resolved by making the electrodes waterproof and keeping the water in the electrolyte. The cathode reaction consumes water from the electrolyte, whereas the anode reaction rejects its product water. The excess water (2 mol per reaction) is evaporated outside the stack.

AFCs can operate over a wide range of temperatures (from 80 to 230°C) pressures[3] (from 2.2 to 45 atm.). High-temperature AFCs also make use of a highly concentrated electrolyte, so highly concentrated that the ion transport mechanism changes from aqueous solution to molten salt.

AFCs can achieve very high efficiencies because of the fast kinetics allowed by the hydroxide electrolyte. The oxygen reaction ($O_2 \rightarrow OH^-$) in particular is easier than the oxygen reduction in acidic fuel cells. As a result, the activation losses are very low. The fast kinetics in AFCs allows using silver or nickel as catalysts instead of platinum. The cost of the fuel cell stack is thus greatly reduced.

AFC kinetics is further improved by the eventual circulation of the electrolyte. When the electrolyte is circulated, the fuel cell is said to be a "mobile electrolyte fuel cell." The advantages of such an architecture are as follows: easy thermal management because the electrolyte is used as coolant; more homogeneous electrolyte concentration, which solves problems of concentration around the cathode; the possibility of using the electrolyte for water management; the possibility of replacing the electrolyte if it has been too polluted by carbon dioxide; and, finally, the possibility of removing the electrolyte from the fuel cell when it is turned off, which has the potential to greatly lengthen the lifetime of the stack.

The use of a circulated electrolyte, however, poses some problems. The greatest problem is the increased risk of leakage: potassium hydroxide is highly corrosive and has a natural tendency to leak, even through the tightest seals. The construction of the circulation pump, the heat exchanger, and the eventual evaporator is further complicated. Another problem is the risk of internal electrolytic short circuit between two cells if the electrolyte is circulated too violently or if the cells are not isolated enough. A circulating electrolyte AFC is pictured in Figure 15.7.[7]

The greatest problem with AFCs is the poisoning by carbon dioxide. The alkaline electrolyte has great affinity for carbon dioxide, and together they form carbonate ions (CO_3^{2-}). These ions do not participate in the fuel cell reaction and diminish its performance. There is also a risk that the carbonate will precipitate and obstruct the electrodes. This last issue may be taken care of by circulating the electrolyte. The solution, which adds to the cost and complexity, is to use a carbon dioxide scrubber that removes the gas from the air stream.

The advantages of AFCs are that they require a cheap catalyst, a cheap electrolyte, and high-efficiency and low-temperature operation. However, they also have some disadvantages such as impaired durability due to corrosive electrolyte, water produced on the fuel electrode, and poisoning by carbon dioxide.

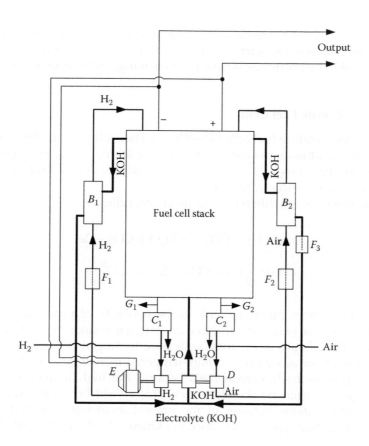

FIGURE 15.7

Circulating electrolyte and supplies of hydrogen and air in an AFC; B_1, B_2, heater exchangers; C_1, C_2, condensers; D, pampers; E, motor; F_1, F_2, F_3, controls; and G_1, G_2, outlets. (W. Vielstich, *Fuel Cells—Modern Processes for Electrochemical Production of Energy*, John Wiley & Sons, New York, ISBN 0-471-90695-6, 1970.)

15.5.3 Phosphoric Acid Fuel Cells[7]

Like PEMFCs, PAFCs rely on an acidic electrolyte to conduct hydrogen ions. The anode and cathode reactions are the same as the PEMFC reactions. Phosphoric acid (H_3PO_4) is a viscous liquid that is contained by capillarity in the fuel cell in a porous silicon carbide matrix.

PAFC was the first fuel cell technology to be marketed. Many hospitals, hotels, and military bases make use of a PAFC to cover part or the totality of their electricity and heating requirements. Very little work has been done to apply this technology to vehicles, probably because of temperature problems.

The phosphoric acid electrolyte temperature must be kept above 42°C, which is its freezing point. Freezing and rethawing the acid unacceptably stresses the stack. Keeping the stack above this temperature requires extra hardware, which adds to the cost, complexity, weight, and volume. Most of these issues are minor in the case of a stationary application but are incompatible with a vehicular application. Another problem arising from the high operating temperature (above 150°C) is the energy consumption associated with warming up the stack. Every time the fuel cell is started, some energy (i.e., fuel) must be spent to heat it up to operating temperature, and every time the fuel cell is turned off, heat (i.e., energy) is wasted. The loss is significant for short travel times, which usually occurs during city driving. However, this issue seems to be minor in the case of mass transportation such as buses.

The advantages of PAFCs are that they require a cheap electrolyte, a low operating temperature, and a reasonable startup time. The disadvantages are the expensive catalyst (platinum), corrosion by acidic electrolyte, CO poisoning, and low efficiency.

15.5.4 Molten Carbonate Fuel Cells

MCFCs are high-temperature fuel cells (500–800°C). They rely on a molten carbonate salt to conduct ions, usually lithium–potassium carbonate or lithium–sodium carbonate. The ions conducted are carbonate ions (CO_3^{2-}). The ion conduction mechanism is that of a molten salt like in PAFCs or highly concentrated AFCs.

The electrode reactions are different from other fuel cells:

$$Anode:\ H_2 + CO_3^{2-} \rightarrow H_2O + CO_2 + 2e^-.$$

$$Cathode:\ \frac{1}{2}O_2 + CO_2 + 2e^- \rightarrow CO_3^{2-}.$$

The major difference from other fuel cells is the necessity of providing carbon dioxide at the cathode. It is not necessary to have an external source since it can be recycled from the anode. MCFCs are never used with pure hydrogen but rather with hydrocarbons. Indeed, the major advantage of high-temperature fuel cells is their capability to almost directly process hydrocarbon fuels because the high temperature allows their decomposition to hydrogen on the electrodes. This would be a tremendous advantage for automotive applications because of the present availability of hydrocarbon fuels. In addition, the high temperatures enhance the kinetics to the point that cheap catalysts may be used.

MCFCs, however, pose many problems because of the nature of the electrolyte and the operating temperature required. The carbonate is an alkali and is extremely corrosive, especially at high temperature. Not only is this unsafe, but there is also the problem of corrosion on the electrodes. It is unsafe to have a large device at 500–800°C under the hood of a vehicle. While it is true that temperatures in internal combustion engines do reach above 1000°C, these temperatures are restricted to the gases themselves, and most parts of the engine are kept cool (around 100°C) by the cooling system. The fuel consumption associated with heating up the fuel cell is also a problem, worsened by the very high operating temperature and the latent heat necessary to melt the electrolyte. These problems are likely to confine MCFCs to stationary or steady power applications such as ships.

The major advantages of MCFCs are that they are fueled with hydrocarbon fuels, require a low-cost catalyst, have improved efficiency due to fast kinetics, and have a lower sensitivity to poisoning. The major disadvantages are slow startup and reduced material choice due to high temperature, a complex fuel cell system due to CO_2 cycling, corrosive electrolyte, and slow power response.

15.5.5 Solid Oxide Fuel Cells

SOFCs conduct ions in a ceramic membrane at high temperature (1000–1200°C). Usually, the ceramic is an yttrium-stabilized zirconia (YSZ) that conducts oxygen ions (O^{2-}), but other ceramics conduct hydrogen ions. The conduction mechanism is similar to that observed in semiconductors, often called solid-state devices. The name of the fuel cell is

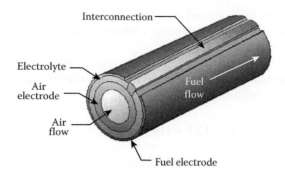

FIGURE 15.8
Tubular SOFC.

derived from this similarity. The half reactions are as follows:

$$Anode: H_2 + O^{2-} \rightarrow H_2O + 2e^-.$$

$$Cathode: \frac{1}{2}O_2 + 2e^- \rightarrow O^{2-}.$$

Water is produced at the fuel electrode. The greatest advantage of SOFCs is this static electrolyte. There is no moving part, except perhaps in the ancillaries. The very high operating temperature allows the use of hydrocarbon fuels as in MCFCs. It should also be noted that SOFCs are not poisoned by carbon monoxide and that they process it about as efficiently as hydrogen. The anode reaction is then

$$CO + O^{2-} \rightarrow CO_2 + 2e^-.$$

SOFCs also benefit from reduced activation losses due to their high operating temperature. The losses are dominated by the ohmic component. SOFCs may be of two kinds: planar and tubular. The planar type is a bipolar stack similar to other fuel cell technologies. A tubular SOFC is described in Figure 15.8. The major advantages of tubular technologies include easier sealing and reduced constraints on the ceramics. Disadvantages include lower efficiency and power density.

Like MCFCs, the disadvantages of SOFCs are mostly associated with their high operating temperature (safety and fuel economy). Supplementary problems arise because the ceramic electrolyte and electrodes are extremely brittle. This is a major disadvantage for vehicular applications where vibrations are a common occurrence. Thermal cycling further stresses the ceramics and is a major concern for planar fuel cells.

15.5.6 Direct Methanol Fuel Cells

Instead of using hydrogen, methanol can be directly used as the fuel for a fuel cell; this is the so-called DMFC. There are some definite motivations for applying DMFCs to vehicles. First, methanol is a liquid fuel that can be easily stored, distributed, and marketed for vehicle application; hence, the current infrastructure of fuel supply can be used without much further investment. Second, methanol is the simplest organic fuel that can be most economically and efficiently produced on a large scale from relatively abundant fossil fuel, namely coal and natural gas. Furthermore, methanol can be produced from agriculture products, such as sugar cane.[8]

In the DMFCs, both the anode and cathode adopt platinum or platinum alloys as an electrocatalyst. The electrolyte can be trifluoromethanesulfonic acid or PEM. The chemical reaction in a DMFC is as follows:

$$\text{\textit{Anode}: CH}_3\text{OH} + \text{H}_2\text{O} \rightarrow \text{CO}_2 + 6\text{H}^+ + 6\text{e}^-.$$

$$\text{\textit{Cathode}: } \frac{3}{2}\text{O}_2 + 6\text{H}^+ + 6\text{e}^- \rightarrow 3\text{H}_2\text{O}.$$

$$\text{\textit{Overall}: CH}_3\text{OH} + \frac{3}{2}\text{O}_2 \rightarrow \text{CO}_2 + 2\text{H}_2\text{O}.$$

The DMFC is relatively immature among the aforementioned fuel cells. At the present status of DMFC technology, it generally operates at 50–100°C. Compared with direct hydrogen fuel cells, DMFCs have low-power density, slow power response, and low efficiency.[8–10]

15.6 Fuel Supply

Fuel supply to on-board fuel cells is the major challenge for FCV applications. As mentioned previously, hydrogen is the ideal fuel for fuel-cell-powered vehicles.[4] Hence, hydrogen production and storage on board are the major concern. Generally, there are two ways of supplying hydrogen to fuel cells. One is to produce hydrogen in ground stations and store pure hydrogen on board. The other is to produce hydrogen on board from an easy-carry hydrogen carrier and directly feed the fuel cells.

15.6.1 Hydrogen Storage

So far, there are three methods of storing of hydrogen on board: compressed hydrogen in a container at ambient temperature, cryogenic liquid hydrogen at low temperature, and the metal hydride method. All these methods have their advantages and disadvantages.

15.6.1.1 Compressed Hydrogen

Pure hydrogen may be stored on board a vehicle under pressure in a tank. The ideal gas equation can be used to calculate the mass of hydrogen stored in a container with volume V and pressure p, that is,

$$m_\text{H} = \frac{pV}{RT} W_\text{H}, \tag{15.24}$$

where p and V are the pressure and volume of the container, R is the gas constant (8.31 J/mol K), T is the absolute temperature, and W_H is the molecular weight of hydrogen

(2.016 g/mol). The energy stored in hydrogen can be calculated as

$$E_H = m_H\,HV,\qquad(15.25)$$

where HV is the heating value of hydrogen. The heating value is either the high heating value ($HHV_H = 144$ mJ/kg) or the low heating value ($LHV_H = 120$ mJ/kg), depending on the condensation energy of produced water. For a convenient comparison with internal combustion engines, LHVH is most often used.

Figure 15.9 shows the mass and energy in 1 L of hydrogen and the equivalent liters of gasoline under different pressure and at room temperature (25°C). The equivalent liters of gasoline are defined as the number of liters of gasoline in which the same amount of energy is contained as that in 1 L of hydrogen. Figure 15.9 also indicates that at a pressure of 350 bar, the energy per liter of hydrogen is less than 1 kWh and is equivalent to about 0.1 L of gasoline. Even if the pressure is increased to 700 bar, which is believed to be the maximum pressure that can be reached, the energy per liter of hydrogen is still less than 2.0 kWh and is equivalent to about 0.2 L of gasoline.

In addition, a certain amount of energy is needed to compress hydrogen from low pressure to high pressure. The process in hydrogen compression may be assumed to be an adiabatic process, that is, no heat exchange occurs during the process. The energy consumed can be expressed as

$$E_{comp} = \frac{\gamma}{\gamma-1}\frac{m}{W_H}RT\left[\left(\frac{p}{p_0}\right)^{\gamma-1/\gamma} - 1\right],\qquad(15.26)$$

where m is the mass of hydrogen, W_H is the molecular weight of hydrogen, γ is the ratio of the specific heat ($\gamma = 1.4$), p is the pressure of hydrogen, and p_0 is atmospheric pressure. This energy consumption is also drawn in Figure 15.9. It shows that about 20% of hydrogen

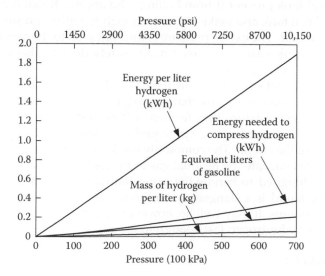

FIGURE 15.9
Energy per liter of hydrogen and equivalent liters of gasoline versus pressure.

energy must be consumed to compress it to high pressure. Considering the inefficiency of the compressor and the electric motor, it is estimated that about 25% of hydrogen energy is consumed.

To contain a gas at several hundred atmospheres requires a very strong tank. To keep the weight as low as possible and the volume reasonable, today's hydrogen tanks for automotive applications use composite materials such as carbon fiber. The cost of a compressed hydrogen tank is thus likely to be high.

The hazards of compressed hydrogen on board a vehicle must be considered. Besides the risk of leakages through cracks in the tank walls, seals, and so on, there is the problem of permeation of hydrogen through the material of the wall. The hydrogen molecule is so small that it can diffuse through some materials.

In addition, a compressed gas tank is a potential bomb in case of a crash. The dangers are even greater in the case of hydrogen, which has a very wide explosive range in air from 4% to 77%[11] and is capable of mixing very quickly with air. This is to be compared with gasoline, which has an explosive range from only 1% to 6% and is a liquid. It should be noted that hydrogen has a high autoignition temperature of 571°C, whereas gasoline autoignites at around 220°C but must be vaporized first.

So far, the technology of compressed hydrogen storage on board is still a huge challenge for vehicle application.

15.6.1.2 Cryogenic Liquid Hydrogen

Another alternative solution to storing hydrogen on board a vehicle is to liquefy the gas at cryogenic temperatures (−259.2°C). The thus stored hydrogen is commonly referred to as "LH_2." LH_2 storage is affected by the same density problems that affect compressed hydrogen. Indeed, the density of liquid hydrogen is very low, and 1 L of liquid hydrogen only weighs 71×10^{-3} kg. This low density results in an energy content of about 8.52 mJ/L of liquid hydrogen.

Containing a liquid at such a low temperature as −259.2°C is technically challenging. It requires a heavily insulated tank to minimize the heat transfer from the ambient air to the cryogenic liquid and thus prevent it from boiling. The approach usually taken is to build a significantly insulated tank and make it strong enough to withstand some of the pressure resulting from the boil-off. The excess pressure is then released to the atmosphere by means of a safety valve. The tank insulation, strength, and safety devices also add significantly to the weight and cost of LH_2 storage.

The boil-off is a problematic phenomenon: if the vehicle is parked in a closed area (garage, underground parking), there is the risk that hydrogen will build up in the confined atmosphere, and the explosive mixture thus formed will explode at the first spark (e.g., light switch, lighter). The refueling of a tank with liquid hydrogen requires specific precautions: air must be kept out of the circuit. The commonly used method is to fill the tank with nitrogen prior to fueling to evacuate the residual gas in the tank. It is also necessary to use specialized equipment designed to handle the explosion hazard and the cryogenic hazards. Indeed, a cryogenic liquid is a dangerous compound for living beings because it burn-freezes the skin and organs. It may well be, however, that the ambient temperature would evaporate the cryogenic hydrogen fast enough to limit or eliminate this risk.

15.6.1.3 Metal Hydrides

Some metals can combine with hydrogen to form stable compounds that can later be decomposed under certain pressure and temperature conditions. These metals may be iron,

TABLE 15.4

Theoretical Hydrogen Storage Densities in Compressed, Liquid, and Metal Hydride Approaches

Material	H Atoms per cm^3 ($\times 10^{22}$)	% of Weight That is Hydrogen
H$_2$ gas, 200 bar (2900 psi)	0.99	100
H$_2$ liquid, 20°K (-253°C)	4.2	100
H$_2$ liquid, 4.2°K (-269°C)	5.3	100
Mg–H$_2$	6.5	7.6
Mg$_2$Ni–H$_2$	5.9	3.6
FeTi–H$_2$	6.0	1.89
LaNi–H$_6$	5.5	1.37

Source: S. E. Gay et al. *Society of Automotive Engineers (SAE) Journal,* Paper No. 2002-01-0097, Warrendale, PA, 2002.

titanium, manganese, nickel, lithium, and some alloys of these metals. Metal hydrides are stable under normal temperature and pressure conditions and are capable of releasing hydrogen only when required.

The hydrogen storage metals and metal alloys are Mg, Mg$_2$Ni, FeTi, and LaNi$_5$. These metals and metal alloys absorb hydrogen to form Mg–H$_2$, Mg$_2$Ni–H$_4$, FeTi–H$_2$, and LaNi$_5$–H$_6$. Theoretically, metal and metal alloys store hydrogen at a higher density than pure hydrogen, as shown in Table 15.4. In practice, the hydrogen storage capacity depends heavily on the surface area of the material on which the hydrogen molecules are absorbed. A large surface area per unit weight of material can be obtained by fine porous modules made of finely ground powder of the metals or metal alloys. Figure 15.10 shows the practical mass and volume needed to store 6 kg of hydrogen (22 L of gasoline equivalent). This figure indicates that Mg–H$_2$ is the promising technology.

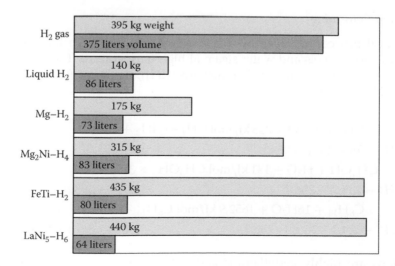

FIGURE 15.10
Current mass and volume needed to store 6 kg of hydrogen (22 L of gasoline equivalent) in various hydrogen storage devices. (C. E. Thomas et al. *Society of Automotive Engineers (SAE) Journal,* Paper No. 982496, Warrendale, PA, 2002.)

Alkaline metal hydrides are possible alternatives to metal hydride absorption. These hydrides react violently with water to release hydrogen and a hydroxide. The example of sodium hydride is shown here:

$$NaH + H_2O \rightarrow NaOH + H_2.$$

The major disadvantage is the necessity of carrying a highly reactive hydride and a corrosive solution of hydroxide in the same vehicle. The storage density is good enough in comparison to many other hydrogen storage techniques, but it falls short in comparison to gasoline. The manufacture of these hydrides and their recycling are also challenging.

Carbon nanotubes, discovered in 1991, would be a prospective method for hydrogen storage systems because of their potential high hydrogen-absorbing capability and light weight. However, carbon nanotube technology is still in its infancy and has a long way to go before its practical utility can be assessed.

15.6.2 Hydrogen Production

At present, hydrogen is mostly produced from hydrocarbon fuels through reforming. Reforming is a chemical reaction that extracts hydrogen from hydrocarbons. During this reaction, the energy content of the fuel is transferred from the carbon–hydrogen bonds to the hydrogen gas. Hydrocarbons such as gasoline, methane, or methanol are the most likely candidates because of their ease of reforming.

There are three major methods of reforming: steam reforming (SR), autothermal reforming (ATR), and partial oxidation (POX) reforming. SR may be used with methanol, methane, or gasoline, whereas ATR and POX reforming are the most commonly used for processing gasoline.

15.6.2.1 Steam Reforming

SR is a chemical process in which hydrogen is produced through the chemical reaction between hydrocarbon fuels and water steam at high temperature. The following chemical equations describe the reforming, using methane (CH_4), methanol (CH_3OH), and gasoline (iso-octane C_8H_{18}) as the fuels:

$$CH_4 + 2H_2O + 258\,kJ/mol\ CH_4 \rightarrow 4H_2 + CO_2$$
$$\Delta H^\circ - 79.4\ 2 \times (-286.2) \qquad\qquad 0 \quad -393.8$$
$$CH_3OH + H_2O + 131\,kJ/mol\ CH_3OH \rightarrow 3H_2 + CO_2$$
$$\Delta H^\circ - 238.7 - 286.2 \qquad\qquad 0 \quad -393.8$$
$$C_8H_{18} + 16H_2O + 1652.9\,kJ/mol\ C_8H_{18} \rightarrow 25H_2 + 8CO_2$$
$$\Delta H^\circ - 224.1 \quad 16 \times (-286.2) \qquad\qquad 0 \quad 8 \times (-393.8)$$

These reactions are highly endothermic and need to be powered by some burning of fuels. Also, these reactions yield some carbon monoxide (CO) in the products, which is a poison to some electrolytes such as PEMFCs, AFCs, and PAFCs. The carbon monoxide can be further converted into hydrogen and carbon dioxide by means of a water–gas shift reaction:

$$CO + H_2O + 4\,kJ/mol\,CO \rightarrow H_2 + CO_2$$
$$\Delta H° \quad -111.6 \quad -286.2 \qquad\qquad 0 \quad -393.8$$

In SR, it is particularly preferred to use methanol as the fuel, since there is no theoretical need for a water–gas shift reaction and since the processing temperature is low (250°C). The hydrogen yield is also particularly high. Among its disadvantages, the most significant are the poisoning of the reformer catalysts by impurities in methanol and the need for external heat input to the endothermic reaction. The heat requirements slow the reaction down and impose a slow startup time between 30 and 45 min.[12] The methanol steam reformer also has slow output dynamics. Although feasible, the SR of gasoline is not commonly used.

15.6.2.2 POX Reforming

POX reforming combines fuel with oxygen to produce hydrogen and carbon monoxide. This approach generally uses air as the oxidant and results in a reformate that is diluted with nitrogen. Then the carbon monoxide further reacts with water steam to yield hydrogen and carbon dioxide (CO_2), as mentioned previously. POX reforming usually uses gasoline (iso-octane) as its fuel. The reaction is expressed as

$$C_8H_{18} + 4O_2 + 16N_2 \rightarrow 8CO + 9H_2 + 16N_2 + 668.7\,kJ/mol\,C_8H_{18}$$
$$\Delta H° \quad -224.1 \quad 0 \qquad 0 \qquad 8 \times (-111.6) \quad 0 \qquad 0$$
$$8CO + 8H_2O + 32\,KJ/mol\,8CO \rightarrow 8H_2 + 8CO_2$$
$$\Delta H° 8 \times (-111.6)\ 8 \times (-286.2) \qquad\qquad 0 \qquad 8 \times (-393.8)$$

POX reforming is highly exothermic, which thus has the advantages of a very fast response to transients and a capability for very fast startups. POX reformers are also fuel flexible and can treat a wide variety of fuels. The disadvantages include a high operating temperature (800–1000°C) and a difficult construction due to heat-integration problems between the different steps of the reaction.[12] In addition, it can be seen from the previous chemical equation that the heat produced from the first reaction is much more than that absorbed in the second reaction, and so POX reforming is somewhat less efficient than the SR of methanol.

Figure 15.11 shows a fuel processing system developed by Epyx Cooperation.[12]

15.6.2.3 Autothermal Reforming

ATR combines fuel with both water and steam so that the exothermic heat from the POX reaction is balanced by the endothermic heat of the SR reaction. The chemical equation in this reaction is

$$C_8H_{18} + nO_2 + (8 - 2n)H_2O \rightarrow 8CO + (17 - 2n)H_2$$
$$\Delta H° \quad -224.1 \qquad\qquad (8 - 2n) \times (-286.2) \qquad 8 \times (-111.6)$$

Zero heat produced in this equation yields $n = 2.83$. The CO produced in this reaction can further react with water steam to produce hydrogen by the water-shift reaction mentioned earlier.

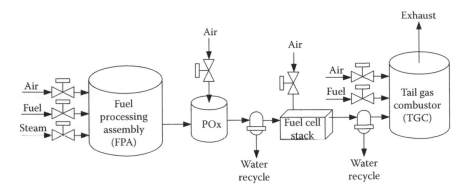

FIGURE 15.11
Fuel processing diagram. (Hydrogen at GKSS: Storage Alternative. Available at http://www.gkss.de/, last visited in May 2003.)

ATR yields a more concentrated hydrogen stream than POX reforming but less than SR. The heat integration is easier than for POX reforming, but a catalyst is required. ATR is potentially more efficient than POX reforming.

15.6.3 Ammonia as Hydrogen Carrier

Ammonia is a non-carbon-based chemical that possesses interesting characteristics as a hydrogen source. The extraction of hydrogen from ammonia, called "cracking," is shown here:

$$2NH_3 \rightarrow N_2 + 3H_2$$

This reaction is easily achieved by heating ammonia, either alone or over a catalyst bed, which has the advantage of lowering the reaction temperature. The energy requirement for this reaction is minimal because it is reversible. Ammonia possesses great advantages in terms of storage as it is easily liquefied at low pressure (about 10 atm.) or at mildly low temperatures ($-33°C$). Other advantages include a very high autoignition temperature ($651°C$) and a limited explosive range in air (15%–28%).

Despite its many advantages, ammonia has a major disadvantage: it is toxic. Ammonia is an alkali that has an extreme affinity for water; hence, it strongly attacks the eyes and lungs, and it causes severe burns. This causticity makes it challenging to conceive of ammonia as a fuel for fuel-cell-powered automobiles.

15.7 Non-Hydrogen Fuel Cells

Some fuel cell technologies can directly process fuels other than hydrogen.[4] Some likely couples are listed here:

- Direct methanol PEMFCs
- Ammonia AFCs
- Direct hydrocarbon MCFCs or SOFCs

Like their hydrogen counterparts, direct methanol PEMFCs are actively studied and have many advantages such as the absence of a reformer, the handling of a liquid fuel, and the absence of high temperatures in the system. The major disadvantages are the necessity of diluting the methanol in liquid water to feed the fuel electrode and a strong crossover of methanol—due to its absorption in the polymer membrane, but due mostly to its slow kinetics.

Ammonia AFCs[13] are possible alternatives to the thermal cracking of ammonia. Ammonia gas is directly fed to the fuel cell and is catalytically cracked on the anode. The ammonia fuel cell reaction yields a slightly lower thermodynamic voltage and higher activation losses than hydrogen AFCs. The activation losses may be reduced by improving the catalyst layer. Interestingly, it would be possible to use ammonia directly with other fuel cell technologies if it were not for the fact that the acidic nature of their electrolyte would be destroyed by the alkaline ammonia.

MCFCs and SOFCs have the capability of directly cracking hydrocarbons because of their high operating temperature. Therefore, they do not directly consume the hydrocarbons but internally extract the hydrogen from them. This option obviously has all the disadvantages of high-temperature fuel cells as discussed in the section on fuel cell technologies.

Bibliography

1. J. Bevan Ott and J. Boerio-Goates, *Chemical Thermodynamics—Advanced Applications*, Academic Press, New York, ISBN 0-12-530985-6, 2000.
2. S. I. Sandler, *Chemical and Engineering Thermodynamics*, Third Edition, John Wiley & Sons, New York, ISBN 0-471-18210-9, 1999.
3. H. K. Messerle, *Energy Conversion Statics*, Academic Press, New York, 1969.
4. P. J. Berlowitz and C. P. Darnell, Fuel choices for fuel cell powered vehicles. *Society of Automotive Engineers (SAE) Journal*, Paper No. 2000-01-0003, Warrendale, PA, 2002.
5. F. Michalak, J. Beretta, and J.-P. Lisse, Second generation proton exchange membrane fuel cell working with hydrogen stored at high pressure for fuel cell electric vehicle. *Society of Automotive Engineers (SAE) Journal*, Paper No. 2002-01-0408, Warrendale, PA, 2002.
6. J. Larminie and A. Dicks, *Fuel Cell Systems Explained*, John Wiley & Sons, New York, 2000.
7. W. Vielstich, *Fuel Cells—Modern Processes for Electrochemical Production of Energy*, John Wiley & Sons, New York, ISBN 0-471-90695-6, 1970.
8. R. M. Moore, Direct methanol fuel cells for automotive power system. In *Fuel Cell Technology for Vehicles*, R. Stobart (ed.), Society of Automotive Engineers (SAE), ISBN: 0-7680-0784-4, Warrendale, PA, 2001.
9. N. Q. Minh and T. Takahashi, *Science and Technology of Ceramic Fuel Cells*, Elsevier, Amsterdam, 1995.
10. M. Baldauf and W. Preidel, Status of the development of a direct methanol fuel cell. In *Fuel Cell Technology for Vehicles*, R. Stobart (ed.), Society of Automotive Engineers (SAE), ISBN: 0-7680-0784-4, Warrendale, PA, 2001.
11. S. E. Gay, J. Y. Routex, M. Ehsani, and M. Holtzapple, Investigation of hydrogen carriers for fuel cell based transportation. *Society of Automotive Engineers (SAE) Journal*, Paper No. 2002-01-0097, Warrendale, PA, 2002.
12. Hydrogen at GKSS: Storage Alternative. Available at http://www.gkss.de/, last visited in May 2003.

13. C. E. Thomas, B. D. James, F. D., Lomax Jr, and I. F., Kuhn Jr, Societal impacts of fuel options for fuel cell vehicles. *Society of Automotive Engineers (SAE) Journal*, Paper No. 982496, Warrendale, PA, 2002.

14. R. O'Hayre et al. *Fuel Cell Fundamentals*. John Wiley & Sons, 2016.

15. V. M. Ortiz-Martínez, M. J. Salar-García, A.P. de los Ríos, F. J. Hernández-Fernández, J. A. Egea, and L. J. Lozano, Developments in microbial fuel cell modeling. *Chemical Engineering Journal*, 271, 2015: 50–60, ISSN 1385-8947.

16. H. El Fadil, F. Giri, J. M. Guerrero, and A. Tahri, Modeling and nonlinear control of a fuel cell/supercapacitor hybrid energy storage system for electric vehicles. *IEEE Transactions on Vehicular Technology*, 63(7), September 2014: 3011–3018.

17. H. Xiaosong, L. Johannesson, N. Murgovski, and B. Egardt, Longevity-conscious dimensioning and power management of the hybrid energy storage system in a fuel cell hybrid electric bus. *Applied Energy*, 137, 2015: 913–924, ISSN 0306-2619.

18. C. Bao and W. G. Bessler, Two-dimensional modeling of a polymer electrolyte membrane fuel cell with long flow channel. Part II. Physics-based electrochemical impedance analysis. *Journal of Power Sources*, 278, 2015: 675–682, ISSN 0378-7753.

19. P. Hong, L. Xu, J. Li, and M. Ouyang, Modeling of membrane electrode assembly of PEM fuel cell to analyze voltage losses inside. *Energy*, 2017, ISSN 0360-5442.

20. E. Miller, K. Randolph, D. Peterson, N. Rustagi, K. Cierpik-Gold, B. Klahr, and J. Gomez, Innovative approaches to addressing the fundamental materials challenges in hydrogen and fuel cell technologies. *MRS Advances*, 1(46), 2016: 3107–3119. doi: 10.1557/adv.2016.271.

21. A. A. Fardoun, H. A. N. Hejase, A. Al-Marzouqi, and M. Nabag, Electric circuit modeling of fuel cell system including compressor effect and current ripples. *International Journal of Hydrogen Energy*, 42(2), 2017: 1558–1564, ISSN 0360-3199.

22. R. Cozzolino and L. Tribioli, On-board diesel autothermal reforming for PEM fuel cells: Simulation and optimization. *American Institute of Physics*, 1648, 2015.

23. L. Tribioli, P. Iora, R. Cozzolino, and D. Chiappini, Influence of fuel type on the performance of a plug-in fuel cell/battery hybrid vehicle with on-board fuel processing. *SAE Technical Paper 2017-24-0174*, 2017, doi:10.4271/2017-24-0174.

24. Y. Nonobe, Development of the fuel cell vehicle Mirai. *IEEJ Transactions on Elec Electronics Engineering*, 12, 2017: 5–9, doi: 10.1002/tee.22328.

25. I. Gandiaga and I. Villarreal, Cycle ageing analysis of a LiFePO4/graphite cell with dynamic model validations: Towards realistic lifetime predictions. *Journal of Power Sources*, 275, 2015: 573–587. http://dx.doi.org/10.1016/j.jpowsour.2014.10.153.

16

Fuel Cell Hybrid Electric Drivetrain Design

Fuel cells, as discussed in Chapter 15, are one of the advanced power sources for applications in transportation. Compared with internal combustion (IC) engines, fuel cells have the advantages of high energy efficiency and much lower emissions. This is because they directly convert the free energy in fuel into electrical energy without undergoing combustion. However, vehicles powered solely by fuel cells have some disadvantages, such as a heavy and bulky power unit caused by the low power density of the fuel cell system, long startup time, and slow power response. Furthermore, in propulsion applications, the extremely large power output at sharp acceleration and the extremely low power output at low-speed driving lead to low efficiency, as shown in Figure 16.1.

Hybridization of a fuel cell system with a peaking power source is an effective technique to overcome the disadvantages of fuel-cell-alone-powered vehicles. The fuel cell HEV is totally different from the conventional IC-engine-powered vehicles and IC-engine-based hybrid drivetrains. Therefore, a totally new design methodology is necessary.[1] In this chapter, a general systematic design methodology and a control strategy for fuel cell hybrid electric drivetrains are discussed. Along with the discussion, a design example for a passenger car drivetrain is introduced.

16.1 Configuration

A fuel-cell-powered hybrid drivetrain has the construction as shown in Figure 16.2.[1,2] It mainly consists of a fuel cell system as the primary power source (PPS), an electric motor drive (motor and its controller), a vehicle controller, and an electronic interface between the fuel cell system and the PPS.[1] According to the power or torque command received from the accelerator or the brake pedal and other operating signals, the vehicle controller controls the motor power (torque) output, and the energy flows between the fuel cell system, the PPS, and the drivetrain. For peak power demand, for instance, in sharp acceleration, both the fuel cell system and the PPS supply propulsion power to the electric motor drive. In braking, the electric motor, working as a generator, converts part of the braking energy into electric energy and stores it in the PPS. The PPS can also restore the energy coming from the fuel cell system when the load power is less than the rated power of the fuel cell system. Thus, with proper design and control strategy, the PPS will never need to be charged from outside of the vehicle.

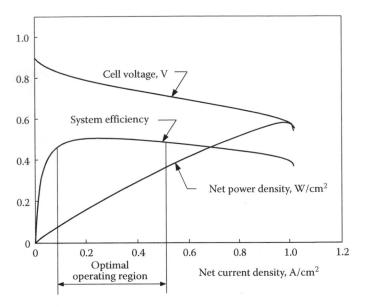

FIGURE 16.1
Typical operating characteristics of a fuel cell system.

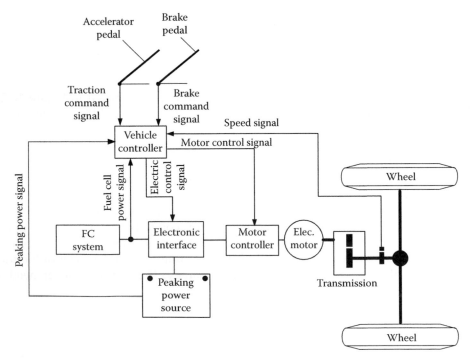

FIGURE 16.2
Configuration of a typical fuel cell hybrid drivetrain.

16.2 Control Strategy

The control strategy that is preset in the vehicle controller is to control the power flow between the fuel cell system, the PPS, and the drivetrain. The control strategy should ensure the following:

1. The power output of the electric motor always meets the power demand.
2. The energy level in the PPS is always maintained within its optimal region.
3. The fuel cell system operates within its optimal operating region.

The driver gives a traction command or a brake command through the accelerator pedal or the brake pedal (Figure 16.3), which is represented by a power command, P_{comm}, that the motor is expected to produce. Thus, in traction mode, the electric power input to the motor drive can be expressed as

$$P_{\text{m-in}} = \frac{P_{comm}}{\eta_m}, \tag{16.1}$$

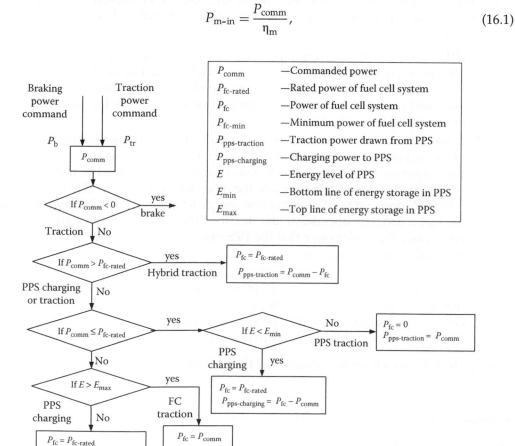

FIGURE 16.3
Flowchart of control strategy.

where η_m is the efficiency of the motor drive. However, in braking, the motor drive functions as a generator, and the electric power output from the motor is expressed as

$$P_{\text{m-out}} = P_{\text{mb-comm}}\eta_m, \tag{16.2}$$

where $P_{\text{mb-comm}}$ is the braking power command to the motor, which may be different from the power command, P_{comm}, coming from the brake pedal, since not all the braking power, P_{comm}, may be supplied by the regenerative braking, as discussed in Chapter 14.

According to the motor power command and other vehicle information, such as the energy level in the PPS and minimum operating power of the fuel cell system—below which the efficiency of the fuel cell decreases significantly (Figure 16.1)—the fuel cell system and the PPS are controlled to produce the corresponding power. Various operating modes of the drivetrain and the corresponding power control strategy are described in detail.

Standstill mode: Neither the fuel cell system nor the PPS supplies power to the drivetrain. The fuel cell system may operate at idle.

Braking mode: The fuel cell system operates at idle, and the PPS absorbs the regenerative braking energy, according to the brake system operating characteristics.

Traction mode:

1. If the commanded motor input power is greater than the rated power of the fuel cell system, hybrid traction mode is used, in which the fuel cell system operates at its rated power, and the remaining power demanded is supplied by the PPS. The rated power of the fuel cell system may be set as the top line of the optimal operating region of the fuel cell.

2. If the commanded motor input power is smaller than the preset minimum power of the fuel cell system, and the PPS needs charging (the energy level is less than the minimum value), the fuel cell system operates at its rated power—part of which goes to the drivetrain, while the other part goes to the PPS. Otherwise, if the PPS does not need charging (the energy level is near its maximum value), the fuel cell system operates at idle, and the PPS alone drives the vehicle. In the latter case, the peak power that the PPS can produce is greater than the commanded motor input power.

3. If the load power is greater than the preset minimum power and less than the rated power of the fuel cell, and the PPS does not need charging, the fuel cell system alone drives the vehicle. Otherwise, if the PPS does need charging, the fuel cell system operates at its rated power—part of which goes to the drivetrain to drive the vehicle, while the other part is used to charge the PPS.

Figure 16.3 shows a flowchart diagram of this control strategy.

16.3 Parametric Design

Similar to the design of the engine-based hybrid drivetrain, the parametric design of a fuel-cell-powered hybrid drivetrain includes the design of the traction motor power, the fuel cell system power, and the PPS power and energy capacity.

16.3.1 Motor Power Design

The motor power is required to match the acceleration performance of the vehicle, as discussed in previous chapters. Figure 16.4 shows the motor power for a 1500-kg passenger car, with respect to the acceleration time from 0 to 100 km/h and a constant speed on a flat road and a 5% grade road. The parameters used in this example are vehicle mass 1500 kg, rolling resistance coefficient 0.01, aerodynamic drag coefficient 0.3, and front area 2 m². Accelerating the vehicle from zero speed to 100 km/h in 12 s needs about 70 kW of motor power. Figure 16.4 also shows the required power while driving at a constant speed on a flat road and a 5% grade road. It can be seen that 33 kW of motor power can support the vehicle driving at about 150 and 100 km/h on a flat road and a 5% grade road, respectively. In other words, the motor power required for acceleration performance is much greater than that required for constant speed driving. Thus, a 70-kW traction motor is the proper design for this vehicle example. Figure 16.5 shows the acceleration time and distance covered by the vehicle during acceleration driving.

16.3.2 Power Design of Fuel Cell System

The PPS, as discussed in previous chapters, is only used to supply peak power in short time periods and has a limited amount of energy in it. Thus, the fuel cell system must be able to supply sufficient power to support the vehicle while it drives at high constant speeds on a long trip (e.g., highway driving between cities), and to support the vehicle to overcome a mild grade at a specified speed without the help of the PPS.

For the 1500-kg example passenger car, as indicated in Figure 16.4, 33 kW of motor power is sufficient to meet the power demand at about 150 km/h of constant speed on a flat road and 100 km/h on a 5% grade road. Considering the inefficiency of the motor drive, a fuel cell system of about 40 kW power is needed to support long trip driving (in the fuel cell system design, the maximum power may be designed slightly larger than that dictated by the constant-speed driving).[1]

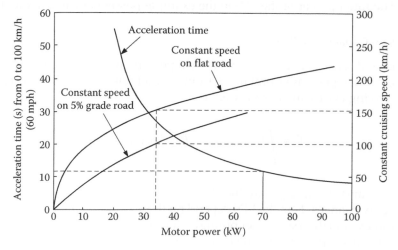

FIGURE 16.4
Motor power versus acceleration time and vehicle cruising speed.

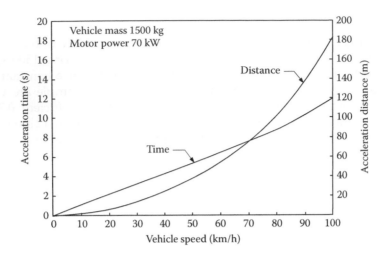

FIGURE 16.5
Acceleration time and distance versus vehicle speed of passenger car example.

16.3.3 Design of Power and Energy Capacity of PPS

16.3.3.1 Power Capacity of PPS

Based on the maximum power of the motor determined by the specified acceleration performance and the rated power of the fuel cell system determined by constant-speed driving, the rated power of the peaking power source can be determined by

$$P_{pps} = \frac{P_{motor}}{\eta_{motor}} - P_{fc},$$ (16.3)

where P_{pps} is the rated power of the peaking power source, P_{motor} is the maximum motor power, η_{motor} is the efficiency of the motor drive, and P_{fc} is the rated power of the fuel cell system. The rated power of the PPS in the example passenger car is about 43 kW.

16.3.3.2 Energy Capacity of PPS

The PPS supplies its energy to the drivetrain when peaking power is needed and restores its energy storage from regenerative braking or from the fuel cell system. The energy changes in the PPS in a driving cycle can be expressed as

$$E = \int_t (P_{pps\_charge} - P_{pps\_discharge})\, dt$$ (16.4)

where $P_{pps\text{-}charge}$ and $P_{pps\text{-}discharge}$ are the charge and discharge power of the PPS, respectively. The energy changes, E, in the PPS depend on the size of the fuel cell system, the vehicle control strategy, and the load power profile over time. Figure 16.6 shows the time profiles of the vehicle speed, the power of the fuel cell system, the PPS power, and the energy change in the PPS for a 1500-kg passenger car with a 40-kW rated power fuel cell system, driving in an FTP75 urban drive cycle and using the control strategy mentioned earlier. Figure 16.6

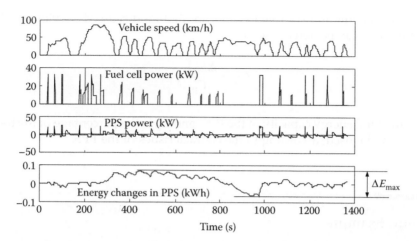

FIGURE 16.6
Vehicle speed, fuel cell power, power of PPS, and energy changes in PPS.

indicates that the maximum energy change, ΔEmax, in the PPS is quite small (about 0.1 kWh). This result implies that the PPS does not need much stored energy to support a vehicle driving in this drive cycle.

It should be noted that the power-producing capability of a fuel cell system is limited before the fuel cell system is warmed up, and the propulsion of the vehicle relies on the PPS. In this case, the energy in the PPS is delivered quickly. Figure 16.7 shows the energy changes in the PPS in an FTP75 urban drive cycle for a 1500-kg passenger car, while the PPS alone propels the vehicle. It indicates that about 1 kWh of energy in the PPS is needed to complete the drive cycle (approximately 10.62 km (6.64 miles) in 23 min), and about 43.5 Wh of energy from the PPS is discharged each minute. Assuming that 10 min is needed to warm up the fuel cell system,[3] about 435 Wh of energy in the PPS is discharged.

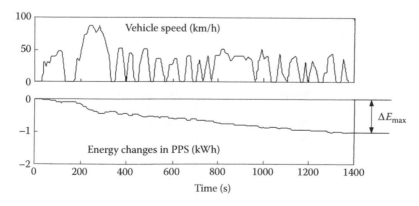

FIGURE 16.7
Energy changes in PPS while powered by PPS alone in FTP75 urban drive cycle.

Based on the maximum discharged energy in the PPS discussed previously, the energy capacity of the PPS can be determined by

$$C_E = \frac{\Delta E_{max}}{C_p},$$ (16.5)

where C_E is the total energy capacity of the PPS, and C_p is the percentage of the total energy capacity that can be used, according to the characteristics of the PPS.

16.4 Design Example

Using the design methodology developed in previous sections, a fuel-cell-powered hybrid drivetrain for a passenger car was designed.[1] For comparison, a fuel-cell-system-alone-powered passenger car of the same size was also simulated. The simulation results are shown in Table 16.1 and Figures 16.8 and 16.9. The design and simulation results indicate that the hybrid vehicle has much higher fuel efficiency and the same performance when compared with the fuel-cell-system-alone-powered vehicle.

TABLE 16.1

Simulation Results for 1500-kg Hybrid and Fuel-Cell-Alone-Powered Passenger Cars

		Hybrid	**Fuel Cell**
Vehicle mass (kg)		1500	1500
Rated motor power (kW)		70	70
Rated power of fuel cell system (kW)		40	83
Maximum power of PPS (kW)		43	–
Maximum energy storage in PPS (kWh)		1.5	–
Acceleration time (0–100 km/h or 60 mph) (s)		12	12
Gradeability (at 100 km/h or 60 mph) (%)		5	5
Fuel economy	Constant speed, at 100 km/h or 60 mph	1.81 L/100 km or 130 mpg (gas. equi.) 0.475 kg H_2/100 km or 131 mile/kg H_2	1.91 L/100 km or 123 mpg (gas. equi.) 0.512 kg H_2/100 km or 124 mile/kg H_2
	FTP75 urban drive cycle	2.93 L/100 km or 80 mpg (gas. equi.) 0.769 kg H_2/100 km or 80.4 mile/kg H_2	4.4 L/100 km or 53.4 mpg (gas. equi.) 1.155 kg H_2/100 km or 53.7 mile/kg H_2
	FTP75 highway drive cycle	2.65 L/100 km or 88.7 mpg (gas. equi.) 0.695 kg H_2/100 km or 89.1 mile/kg H_2	2/9 L/100 km or 81 mpg (gas. equi.) 0.762 kg H_2/100 km or 81.4 mile/kg H_2

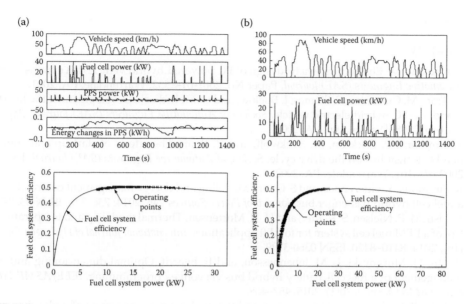

FIGURE 16.8
Operating simulation of fuel cell hybrid and fuel-cell-alone-powered passenger car in FTP75 urban drive cycle: (a) hybrid drivetrain and (b) fuel-cell-alone-powered drivetrain.

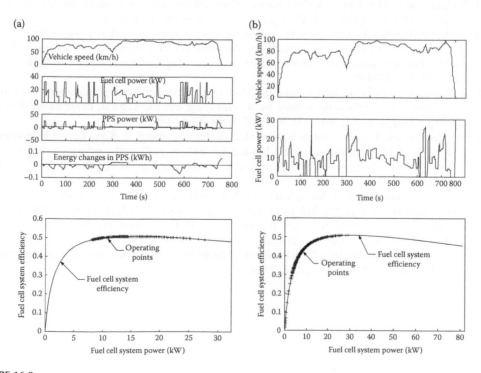

FIGURE 16.9
Operating simulation of fuel cell hybrid and fuel-cell-alone-powered passenger car in FTP75 highway drive cycle: (a) hybrid drivetrain and (b) fuel-cell-alone-powered drivetrain.

Bibliography

1. Y. Gao and M. Ehsani, Systematic design of FC powered hybrid vehicle drive trains. *Society of Automotive Engineers (SAE) Journal*, Paper No. 2001-01-2532, Warrendale, PA, 2001.
2. D. Tran, M. Cummins, E. Stamos, J. Buelow, and C. Mohrdieck, Development of the Jeep Commander 2 FC hybrid electric vehicle. *Society of Automotive Engineers (SAE) Journal*, Paper No. 2001-01-2508, Warrendale, PA, 2001.
3. T. Simmons, P. Erickson, M. Heckwolf, and V. Roan, The effects of start-up and shutdown of a FC transit bus on the drive cycle. *Society of Automotive Engineers (SAE) Journal*, Paper No. 2002-01-0101, Warrendale, PA, 2002.
4. K. Simmons, Y. Guezennec, and S. Onori, Modeling and energy management control design for a fuel cell hybrid passenger bus. *Journal of Power Sources* 246, 2014: 736–746, ISSN 0378-7753.
5. V. Liso, M. P. Nielsen, S. K. Kær, and H. H. Mortensen, Thermal modeling and temperature control of a PEM fuel cell system for forklift applications. *International Journal of Hydrogen Energy* 39 (16), 2014: 8410–8420, ISSN 0360-3199.
6. X. Hu, N. Murgovski, L. M. Johannesson, and B. Egardt, Optimal dimensioning and power management of a fuel cell/battery hybrid bus via convex programming. *IEEE/ASME Transactions on Mechatronics* 20(1), 2015: 457–468.
7. L. Tribioli, R. Cozzolino, D. Chiappini, and P. Iora, Energy management of a plug-in fuel cell/battery hybrid vehicle with on-board fuel processing. *Applied Energy* 184, 2016: 140–154, ISSN 0306-2619.
8. P. Thounthong, P. Tricoli, and B. Davat, Performance investigation of linear and nonlinear controls for a fuel cell/supercapacitor hybrid power plant. *International Journal of Electrical Power & Energy Systems* 54, 2014: 454–464, ISSN 0142-0615.
9. S. Kang and K. Min, Dynamic simulation of a fuel cell hybrid vehicle during the federal test procedure-75 driving cycle. *Applied Energy* 161, 2016: 181–196, ISSN 0306-2619.
10. H. Aouzellag, K. Ghedamsi, and D. Aouzellag, Energy management and fault tolerant control strategies for fuel cell/ultra-capacitor hybrid electric vehicles to enhance autonomy, efficiency and life time of the fuel cell system. *International Journal of Hydrogen Energy* 40(22), 2015: 7204–7213, ISSN 0360-3199.
11. Z. Mokrani, D. Rekioua, and T. Rekioua, Modeling, control and power management of hybrid photovoltaic fuel cells with battery bank supplying electric vehicle. *International Journal of Hydrogen Energy* 39(27), 2014: 15178–15187, ISSN 0360-3199.
12. S. J. Andreasen, L. Ashworth, S. Sahlin, H.-C. B. Jensen, and S. K. Kær, Test of hybrid power system for electrical vehicles using a lithium-ion battery pack and a reformed methanol fuel cell range extender. *International Journal of Hydrogen Energy* 39(4), 2014: 1856–1863, ISSN 0360-3199.
13. X. Hu, L. Johannesson, N. Murgovski, and B. Egardt, Longevity-conscious dimensioning and power management of the hybrid energy storage system in a fuel cell hybrid electric bus. *Applied Energy* 137, 2015: 913–24, http://dx.doi.org/10.1016/j.apenergy.2014.05.013.
14. N. Marx, D. Hissel, F. Gustin, L. Boulon, and K. Agbossou, On the sizing and energy management of an hybrid multistack fuel cell—Battery system for automotive applications. *International Journal of Hydrogen Energy* 42, 2016: 1–9.

17

Design of Series Hybrid Drivetrain for Off-Road Vehicles

Off-road vehicles (military, agricultural, and construction) usually operate on unprepared ground and need to overcome very complex and difficult ground obstacles, such as a steep grade and very soft ground. Depending on the functional requirements, different criteria are used to evaluate the performance of various types of off-road vehicles. For tractors, their main function is to provide adequate draft to pull various types of implements and machinery. Drawbar performance is of primary interest; this may be characterized by the ratio of drawbar pull to vehicle weight, drawbar power, and drawbar efficiency. For off-road transport vehicles, the transport productivity and efficiency are often used as basic criteria for evaluating their performance. For military vehicles, the maximum feasible operating speed at two specific points in a given area may be employed as a criterion for the evaluation of their agility.[1]

Although different criteria are used to evaluate the performance of different types of off-road vehicles, there is a common requirement for all: mobility over unprepared terrain. Mobility, in the broad sense, is concerned with performance of the vehicle in relation to soft terrain, obstacle negotiation and avoidance, ride quality over rough terrain, water crossing, and so on.

This chapter discusses the design principle of an off-road hybrid electric tracked vehicle, focusing on power ratings of the traction motor, engine/generator, and energy storage for satisfying specified vehicle performance indices, such as gradeability, acceleration, and steering on various types of ground. For drivetrain control, refer to Chapter 8.

17.1 Motion Resistance

In addition to aerodynamic drag, rolling distance caused by tire deformation, and track friction as discussed in Chapter 2, the motion resistance of an off-road vehicle mostly stems from significant deformations of the ground while the vehicle is moving on it. In off-road operations, various types of terrain with different characteristics, ranging from desert sand through deep mud to snow, may be encountered. The mechanical properties of the terrain quite often impose severe limitations on the mobility and performance of off-road vehicles. The study of the relationship between the performance of an off-road vehicle and its physical environment (terrain) has now become known as terramechanics.[1]

Although a study of terramechanics is beyond the scope of this book, for the proper design of a vehicle power train it is necessary to briefly introduce some terramechanics concepts, especially the motion resistance and thrust that a terrain can support for off-road vehicle normal operation.

This section briefly explains the calculation method of a tracked vehicle's motion resistance and thrust, accompanied by an example tracked vehicle. For more details on wheeled vehicles, see Wong[1] and Bekker.[2-4]

17.1.1 Motion Resistance Caused by Terrain Compaction

Motion resistance, supported by consuming vehicle energy, to compact terrain is studied using penetration tests as shown in Figure 17.1. A plate is used to simulate the contact area of a track. A vertical load, P, is placed on the plate, resulting in sinkage, z, and terrain reaction pressure, p.

The work done by the load P can be expressed as

$$W = bl \int_0^{z_0} p\,\mathrm{d}z, \tag{17.1}$$

where b and l are the plate dimensions of the short and long sides, as shown in Figure 17.1. The relationship between pressure, p, and sinkage, z, depends on the terrain characteristic, which is determined by experiment and has the expression[1]

$$z = \left(\frac{p}{k_c/b + k_\phi}\right)^{1/n}, \tag{17.2}$$

where k_c, $k_{\phi,g}$, and n are terrain parameters, and b is the dimension of the shorter side of the plate. Parameter k_c reflects the cohesion characteristic and k_ϕ the internal friction characteristic of the terrain. Parameter n reflects the "hardness" of the terrain. The terrain has a linear characteristic when $n = 1$, a hard characteristic when $n < 1$, and a soft characteristic when $n > 1$. Typical terrain parameters are listed in Table 17.1.

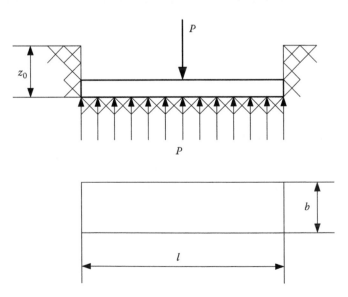

FIGURE 17.1
Terrain penetration test.

TABLE 17.1

Terrain Values

Terrain	Moisture Content (%)	n	k_c		k_p		c		
			lb/in.$^{n+1}$	kN/m$^{(n+1)}$	lb/in.$^{(n+2)}$	kN/m$^{(n+2)}$	lb/in.^{2}	kPa	ϕ (deg)
Dry sand (Land Locomotion Lab. [LLL])	0	1.1	0.1	0.95	3.9	1528.43	0.15	1.04	28
Sand loam (LLL)	15	0.7	2.3	5.27	16.8	1515.04	0.25	1.72	29
	22	0.2	7	2.56	3	43.12	0.2	1.38	38
Sand loam Michigan (Strong, Buchele)	11	0.9	11	52.53	6	1127.97	0.7	4.83	20
	23	0.1	15	11.42	27	808.96	1.4	9.65	35
Sandy loam (Hanamoto)	26	0.3	5.3	2.79	6.8	141.11	2.0	13.79	22
	32	0.5	0.7	0.77	1.2	51.91	0.75	5.17	11
Clayey soil (Thailand)	38	0.5	12	13.91	16	692.15	0.6	4.14	13
	55	0.7	7	16.03	14	1262.53	0.3	2.07	10
Heavy clay (Waterways Experiment Station [WEBS])	25	0.13	45	12.70	140	1555.95	10	68.95	31
	10	0.11	7	1.81	10	103.27	3	20.69	6
Lean clay (WES)	22	0.2	45	16.43	120	1724.69	10	68.95	20
	32	0.15	5	1.52	10	119.61	2	13.79	11
Snow (Harrison)		1.6	0.07	4.37	0.08	196.72	0.15	1.03	19.7
		1.6	0.04	2.49	0.10	245.90	0.09	0.62	23.2

Substituting for p from Equation 17.2 into Equation 17.1 yields

$$W_c = bl\left(\frac{k_c}{b} + k_\phi\right)\left(\frac{z_0^{n+1}}{n+1}\right). \tag{17.3}$$

The interaction between the track of a tracked vehicle and the terrain is similar to that between a plate and the terrain as shown in Figure 17.1. Using Equation 17.3, a vehicle's motor resistance caused by terrain compaction can be expressed as

$$R_c = \frac{W_c}{l} = b\left(\frac{k_c}{b} + k_\phi\right)\left(\frac{z_0^{n+1}}{n+1}\right). \tag{17.4}$$

Example

A tracked vehicle with 196 kN gross weight and track dimensions of $l = 3.6$ m and $b = 1.0$ m is operating on terrain with parameters $n = 1.6$, $k_c = 4.37$ kN/m$^{2.6}$, and $k_\phi = 196.73$ kN/m$^{3.6}$. The vehicle resistance caused by terrain compaction can be calculated as follows:

Pressure:

$$p = \frac{P}{bl} = \frac{196/2}{3.6 \times 1.0} = 27.2 \, \text{kN/m}^2.$$

Sinkage:

$$z_0 = \left(\frac{p}{k_c/b + k_\phi}\right)^{1/n} = \left(\frac{27.2}{4.37/1.0 + 196.73}\right)^{1/1.6} = 0.2864 \, \text{m}.$$

Motion resistance:

$$R_c = \frac{W_c}{l} = 2b\left(\frac{k_c}{b} + k_\phi\right)\left(\frac{z_0^{n+1}}{n+1}\right)$$

$$= 2 \times 1.0\left(\frac{4.37}{1.0} + 196.73\right)\left(\frac{0.2864^{2.6}}{2.6}\right) = 5.99 \, \text{kN}.$$

17.1.2 Motion Resistance Caused by Terrain Bulldozing

Another type of motion resistance may exist, caused by bulldozing the soil in front of the vehicle track. This motion resistance is called bulldozing resistance. In this section, equations and diagrams are introduced for the purpose of calculating the bulldozing resistance.[1]

In predicting the bulldozing resistance, Bekker[2,3] proposed equations by assuming that it was equivalent to the horizontal force acting on a vertical blade:

$$R_b = b(cz_0 k_{pc} + 0.5 \, z_0^2 \gamma_s K_{p\gamma}), \tag{17.5}$$

where b is the width of the track, c is the cohesion of the terrain (Table 17.1), γ_s is the specific weight of the terrain, z_0 is sinkage, and

$$K_{pc} = (N_c - \tan\phi)\cos^2\phi$$

and

$$K_{p\gamma} = \left(\frac{2N_\gamma}{\tan\phi} + 1\right)\cos^2\phi,$$

where N_c and N_γ are the terrain's bearing capacity factors, as shown in Figure 17.2, and ϕ is the angle of internal shearing resistance of the terrain. In soft terrain or loose soils, local

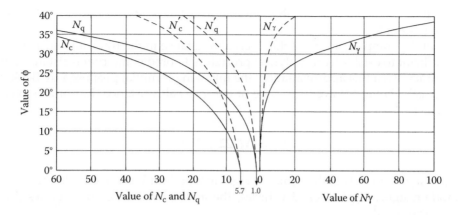

FIGURE 17.2
Variation of terrain bearing capacity factors with angle of internal shearing resistance. (Adapted from J. W. Wong, *Theory of Ground Vehicles*, John Wiley & Sons, New York, 1978.)

failure in front of the wheel or track may be assumed, and the bulldozing resistance may be estimated using the following equation[1]:

$$R_b = b(0.67cz_0 K'_{pc} + 0.5z_0^2 \gamma_s K'_{p\gamma}), \tag{17.6}$$

where

$$K'_{pc} = (N'_c - \tan\phi') \cos^2\phi'$$

and

$$K'_{p\gamma} = \left(\frac{2N'_\gamma}{\tan\phi'} + 1\right) \cos^2\phi'.$$

N'_c and N'_γ are bearing capacity factors for local shear failure shown in Figure 17.2, and $\tan\phi' = 2/3 \tan\phi$.

Example
The same vehicle as above is operating on the same terrain as shown in the example in Section 17.1.1, with $c = 1.0$ kPa, internal shear angle $\phi = 19.7$, and $\gamma_s = 2646$ N/m³.
From Figure 17.2, for $\phi = 19.7$, $N'_c = 11.37$, and $N'_\gamma = 1.98$, then $K'_{pc} = 10.53$ and $K'_{p\gamma} = 16.64$. By Equation 17.6, $R_b = 7.79$ kN is obtained.
The total motion resistance, compacting and bulldozing terrain, is $R_{terr} = 5.99 + 7.79 = 13.78$ kN. The resistance coefficient defined as the motion resistance per unit vehicle weight is $f_{terr} = 13.78/196 = 0.07$.

17.1.3 Internal Resistance of Running Gear

For wheeled vehicles, the internal resistance of the running gear is mainly caused by the hysteresis of tire materials, as discussed in Chapter 2. For tracked vehicles, the internal resistance of the track and the associated suspension system may be substantial. Frictional losses in track pins, driving sprockets and sprocket hubs, and road wheel bearing constitute the major portion of the internal resistance of the track and associated suspension system.[1]

Because of the complex nature of the internal resistance in the track and suspension system, it is difficult to establish an analytical procedure to predict internal resistance with sufficient accuracy. As a first approximation, the following formula, proposed by Bekker, may be used for calculating the average value of the internal resistance, R_{in}, of a conventional tracked vehicle[1,4]:

$$R_{in} = W(222 + 3V),$$ (17.7)

where R_{in} is in newtons, W is the vehicle weight in tons, and V is the vehicle speed in km/h.

For modern lightweight tracked vehicles, the internal resistance may be less, and the empirical formula is[1,4]

$$R_{in} = W(133 + 2.5V).$$ (17.8)

17.1.4 Tractive Effort of Terrain

The tractive effort of a track is produced by the shearing of the terrain, as shown in Figure 17.3. The maximum tractive effort, $F_{t,max}$, that can be developed by a track is determined by the shear strength of the terrain, which can be expressed by[1]

$$F_{t,max} = Ac + W \tan \phi,$$ (17.9)

where W and A are the vertical load and contact area of the track to the ground, respectively, and c and ϕ are the apparent cohesion and angle of internal shearing resistance of the terrain, respectively. Their typical values are shown in Table 17.1.

When the vehicle motion resistance is greater than the maximum tractive effort that the terrain can develop, complete skidding occurs if the torque developed by the mover is large enough, and the vehicle cannot move.

For the example vehicle in Sections 17.1.2 and 17.1.3, the terrain parameters are $c = 1.0$ kPa, internal shear angle $\phi = 19.7$, $W = 196$ kN, and $A = 2 \times l \times B = 7.2$ m^2. The maximum tractive effort is:

$$F_{t,max} = Ac + W \tan \phi = 7.2 + 196 \times \tan 19.7° = 77.4 \text{ kN}.$$

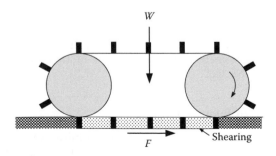

FIGURE 17.3
Shearing action of a track.

17.1.5 Drawbar Pull

For off-road vehicles designed for traction (i.e., tractors), drawbar performance is of prime importance because it denotes the ability of a vehicle to pull or push various types of working machinery, including implements and construction and earthmoving equipment. Drawbar pull, F_d, is the force available at the drawbar and is equal to the difference between the tractive effort and the total resistance $\sum R$. That is,

$$F_d = F_t - \sum R. \tag{17.10}$$

The tractive effort, F_t, may be determined by the power plant and the transmission of the vehicle on hard terrain, for example, a paved road, or by the maximum thrust that the terrain can support on soft terrain, as discussed in Section 17.1.4. Because this book focuses on transportation vehicles, drawbar performance is not discussed further. However, the principles developed in this book can be directly applied to the analysis of vehicle drawbar performance.

17.2 Tracked Series Hybrid Vehicle Drivetrain Architecture

A tracked series hybrid off-road-vehicle drivetrain mainly contains subsystems of a primary power source, a secondary power source, traction motors and their controllers, a power converter, and a vehicle controller, as shown in Figure 17.4.

Primary power source: The primary power source, generally a diesel engine and a generator, is used to provide power to meet the average power demand of the load. The diesel engine drives the generator to generate electricity to charge the secondary power source and the batteries/ultracapacitors, or it provides power directly to the electric motor drives. The generator generates AC power. The nontraction AC power devices may be directly connected to the output of the generator through transformers if the voltages of the AC devices are different from the output voltage of the generator. An AC/DC converter (rectifier) is used to convert the AC power into DC power.

Secondary power source: The secondary power source, generally a battery/ultracapacitor, is used to supply peaking power to the traction motor drives to meet the peak power demand of the load. The secondary power source can be charged from the primary power source and/or the regenerative braking in which the motor functions as a generator to convert all or part of the braking power of the vehicle into electrical power to charge the secondary power source. In normal operation, the total output energy from the secondary power source should be equal to the total charging energy over the mission period. Furthermore, the charging rate should be controlled to be in an acceptable range.

Traction motors and controllers: The traction motors deliver their torque to the sprockets through transmissions to propel the vehicle. The motor drives are powered by the primary and/or secondary power sources and controlled by their controllers to provide correct torques and speeds that meet the maneuver requirements, according to the commands of the driver. These maneuvers include acceleration, deceleration, forward and reverse movement, and steering.

Power converter: The power converter is an assembly of controllable power electronics. It is used to control the power flow between the primary power source, the secondary power

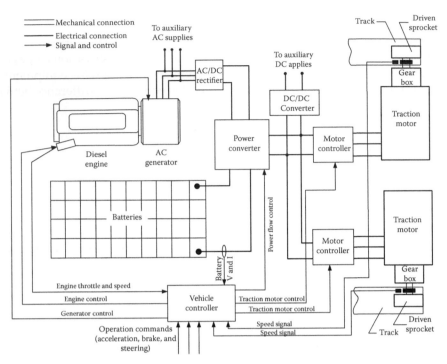

FIGURE 17.4
Tracked series hybrid off-road vehicle architecture.

source, and the motor drives. All operation modes of the drivetrain are implemented by controlling the power converter. The operation modes of the drivetrain mainly include engine/generator-alone-powered operation, battery/ultracapacitor-alone-powered operation, both engine/generator and battery/ultracapacitor-powered operation (hybrid traction), regenerative braking operation, and battery/ultracapacitor charging operation. The various architectures of the power converter were discussed in Chapter 8.

Vehicle controller: The vehicle controller is the highest-level microprocessor-based system controller. The vehicle controller receives operation commands from the driver (e.g., acceleration, deceleration, forward or reverse moving, steering) and drivetrain real-time operating information, such as vehicle speed, and component real-time operating information, such as the voltages and currents of batteries/ultracapacitors, engine throttle position, speed. Based on all the information received and the control strategy (software code stored in the vehicle controller), the vehicle controller generates the necessary control signals and sends them to the components (e.g., engine/generator, power converter, motor drives). The components follow the control commands of the vehicle controller.

17.3 Parametric Design of Drivetrain

The parametric design of the drivetrain mainly includes traction motor power design, engine/generator power design, and energy storage (battery/ultracapacitor) power and energy design.

17.3.1 Traction Motor Power Design

In motor power design, the acceleration performance, maximum gradeability, and steering are the highest considerations.

17.3.1.1 Vehicle Thrust versus Speed

A well-controlled traction motor usually has the torque and power characteristics of constant torque in the low-speed range and constant power in the high-speed range, as shown in Figure 17.5. The corner speed is referred to as the base speed. In traction motor design, two important parameters, maximum power and extended speed ratio x, which is defined as the ratio of its maximum speed to its base speed, must be determined first. For a given power rating, that is, power in a constant power range, motor torque is expressed as

$$
\begin{aligned}
T_{\mathrm{m}} &= \frac{30 P_{\mathrm{m}}}{\pi n_{\mathrm{mb}}}, \quad n_{\mathrm{m}} \le n_{\mathrm{mb}}, \\
&= \frac{30 P_{\mathrm{m}}}{\pi n_{\mathrm{m}}}, \quad n_{\mathrm{m}} > n_{\mathrm{mb}},
\end{aligned}
\tag{17.11}
$$

where P_{m} is the motor power rating defined previously, n_{mb} is the motor base speed (rpm) as shown in Figure 17.5, and n_{m} is the motor speed (rpm) varying from zero to its maximum.

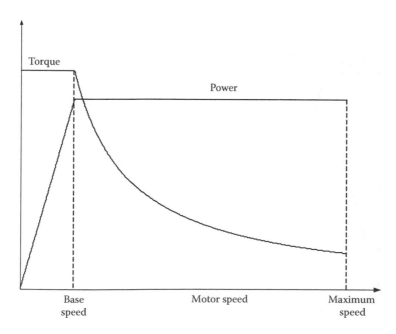

FIGURE 17.5
Typical torque and power profiles of traction motor versus motor speed.

The motor torque and power profiles can be translated into the thrust of the vehicle versus vehicle speed by

$$F_t = \frac{T_m \eta_t i_g}{r} \qquad (17.12)$$

and

$$V = \frac{\pi n_m r}{30 i_g}, \qquad (17.13)$$

where ηt and i_g are the transmission efficiency and gear ratio from the traction motor to the driving sprockets, and r is the radius of driving sprockets. The transmission may be single gear or multigear (for more details, see Chapter 4). The typical thrust and power profiles are illustrated in Figure 17.6. The motor power design given below is based on these profiles.

17.3.1.2 Motor Power and Acceleration Performance

The motor power required by the acceleration performance includes the power for overcoming various mechanical resistances (losses in the locomotive mechanism as described by Equations 17.7 and 17.8, losses caused by road deformation as discussed in Section 17.1, and aerodynamic drag) and the power for accelerating vehicle mass (inertial resistance). On a hard-surface road, the loss due to road deformation is negligible,

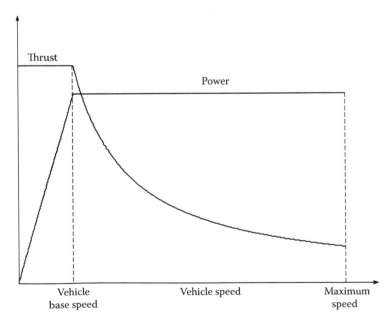

FIGURE 17.6
Typical thrust and power profiles versus vehicle speed with an electric motor as the mover.

TABLE 17.2

Vehicle Parameter Values

M	20,000 kg
t^a	8 s
V_f	48 km/h (30 mph)
C	0.0138[a]
D	0.000918[a]
CD	1.17
Af	6.91 m²

[a] c and d correspond to the constants of 133 and 2.5 with vehicle weight in newtons and V in m/s.

and the motor power for acceleration can be expressed as

$$P_{\mathrm{acc}} = \frac{M}{2t_a}\left(V_b^2 + V_f^2\right) + MV_f\left(\frac{2}{3}c + \frac{1}{2}dV_f\right) + \frac{1}{5}\rho_a C_D A_f V_f^3 (W), \tag{17.14}$$

where M is the vehicle mass in kg; t_a is the time in seconds for accelerating the vehicle from zero speed to a specified final speed; V_f, V_b is the base speed in m/s, as shown in Figure 17.6; ρ_a is the air mass density in kg/m³; CD is the coefficient of aerodynamic drag; A_f is the front area of the vehicle in m²; and c and d are constants, representing the constant term and the term proportional to the vehicle speed of the resistance of the locomotive mechanism as described in Equation 17.8. The resistance coefficient is expressed as $f_r = c + dV$. The vehicle parameters used in the motor power design are shown in Table 17.2.

The motor powers required for acceleration performance (8 s from 0 to 48 km/h [30 mph]) with different extended speed ratios, x, of the thrust are shown in Figure 17.7. The motor power decreases with increasing extended speed ratio, x. However, when $x \geq 6$, further decrease in the motor power is not significant with further increase in the extended speed ratio, x.

17.3.1.3 Motor Power and Gradeability

The motor power requirement in an uphill operation can be expressed as

$$P_{\mathrm{grade}} = \left(Mgf_r + \frac{1}{2}\rho_a C_D A_f V^2 + Mg \sin \alpha\right)V(W), \tag{17.15}$$

where α is the slope angle of the road, and V is the vehicle speed specified by the gradeability. When the vehicle is climbing its maximum slope (60%) in real operation, the ground surface is usually unprepared, and thus, resistance is much larger than on prepared roads. Therefore, in the calculation of motor power required by gradeability, additional resistance should be included to reflect this situation. The resistance due to terrain deformation can be obtained as discussed in Section 17.1. In the following sections, 0.06 of additional resistance coefficient is introduced (snow or sandy loam with about 21% moisture content).[5,6]

Based on the specified gradeability, for example 60%, the tractive effort profiles versus vehicle speed with different extended speed ratios and motor powers are shown in

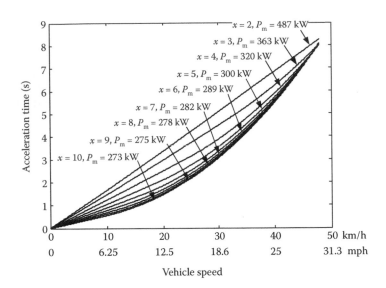

FIGURE 17.7
Motor power required by acceleration performance with different extended speed ratios, x.

Figure 17.8. With the same gradeability, a larger extended speed ratio, x, will result in smaller motor power demand.

Figure 17.9 summarizes the motor power demand by acceleration and gradeability performances along with the extended speed ratio, x, of the motor drive. It can be clearly seen that the motor power required by gradeability is larger than the motor power required by acceleration, especially with a small x.

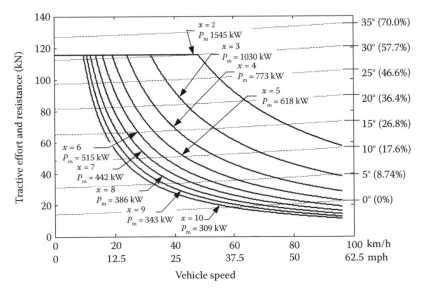

FIGURE 17.8
Tractive effort versus vehicle speed with different x and motor power.

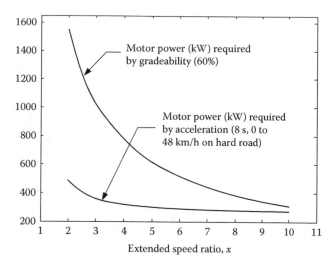

FIGURE 17.9
Motor power required by acceleration and gradeability along with extended speed ratios, x.

The previous analysis indicates that one of the effective approaches for reducing the motor power rating is to increase the extended speed ratio. However, the extended speed ratio of the motor drive is closely and naturally related to motor type. PM motor drives have very small x, usually less than 2, due to their rather limited field weakening capability.[7] To keep the power of the motor drive in a reasonable range and meet the gradeability requirement, a multigear transmission with three or four gears is required. A common induction motor with a speed adjustment control usually has an extended speed ratio of 2. Nevertheless, a properly designed induction motor, for example a spindle motor, with field orientation control can achieve field weakening in a range of about three to five times its base speed.[8–11] Even with this special design, a double-gear transmission is still needed. An SRM drive can inherently operate in an extremely long constant power range. Both 6–4 and 8–6 SRMs can reach 6–8[12]; thus, a single-gear transmission would serve this requirement.

17.3.1.4 Steering Maneuver of a Tracked Vehicle

The steering maneuver of a tracked vehicle is quite different from that of wheeled vehicles. There are several methods can be used to accomplish the steering of a tracked vehicle. These include skid-steering, steering by articulation in multibody vehicles, and curved track steering. For single-body vehicles, skid steering is the common method. In this book, only skid steering, which is closely related to the thrusts on both side tracks, is discussed. For other steering methods, readers are referred to Wong.[1]

In skid steering, the thrust of one track is increased and that of the other is reduced to create a turning moment to overcome the moment of resistance due to the skidding of the track on the ground and the rotational inertia of the vehicle. Since the moment of the turning resistance is usually considerable, significantly more power may be required during a turn than in straight-line motion.[1]

The turning behavior of a tracked vehicle using skid steering depends on the thrusts of the outside and inside tracks, F_{to} and F_{ti}, the resultant resisting force, R_{tot}, the moment of turning

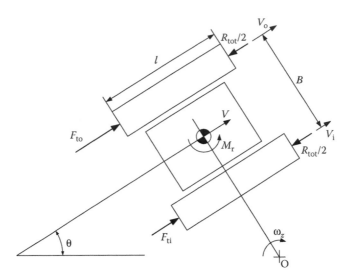

FIGURE 17.10
Skid steering behavior.

resistance, M_r, exerted on the track by the ground, and vehicle parameters as shown in Figure 17.10. At low speeds on level ground, the centrifugal force may be neglected, and the behavior of the vehicle can be described by the following motion equations:

$$M\frac{dV}{dt} = F_{to} + F_{ti} - R_{tot},\qquad(17.16)$$

$$I_z\frac{d\omega_z}{dt} = \frac{B}{2}(F_{to} - F_{ti}) - M_r,\qquad(17.17)$$

where I_z is the moment of inertia of the vehicle about the vertical axis passing through its center of gravity, and ω_z is the turning angle velocity of the vehicle.

At low speeds and under steady-state conditions with zero linear and angular acceleration, that is, $dV/dt = 0$ and $d\omega_z/dt = 0$, the thrusts of the outside and inside tracks can be expressed as

$$F_{to} = \frac{R_{tot}}{2} + \frac{M_r}{B} = \frac{Mgf_r}{2} + \frac{M_r}{B},\qquad(17.18)$$

$$F_{ti} = \frac{R_{tot}}{2} - \frac{M_r}{B} = \frac{Mgf_r}{2} - \frac{M_r}{B},\qquad(17.19)$$

where M and g are the vehicle mass and the acceleration of gravity, respectively, and f_r is the resistance coefficient due to ground deformation.

The moment of turning resistance, M_r, can be determined experimentally or analytically. If the normal pressure is assumed to be uniformly distributed along the track, the lateral resistance per unit length of the track, R_l, can be expressed by

$$R_l = \frac{\mu_t Mg}{2l},$$ (17.20)

where μ_t is the coefficient of lateral resistance, and l is the contact length of the track as shown in Figure 17.11. The value of μ_t depends not only on the terrain but also on the design of the track. Over soft terrain, the vehicle sinks into the ground, and the tracks together with the grousers slide on the surface and displace the soil laterally during steering. The lateral force acting on the track and grousers due to displacing the soil laterally forms part of the lateral resistance. Table 17.3 shows the values of μ_t for steel and rubber tracks over various types of ground.[1]

By referring to Figure 17.11, the resultant moment of the lateral resistance about the center of the two tracks (i.e., moment of turning resistance) can be determined by

$$M_r = 4 \int_0^{1/2} R_l x \, dx = 4 \frac{Mg\mu_t}{2l} \int_0^{1/2} x \, dx = \frac{Mgl\mu_t}{4}.$$ (17.21)

Accordingly, Equations 17.18 and 17.19 can be rewritten as

$$F_{to} = \frac{Mg}{2}\left(f_r + \frac{l\mu_t}{2B}\right),$$ (17.22)

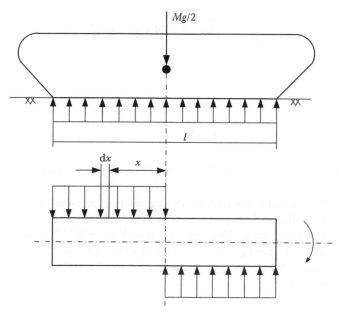

FIGURE 17.11
Moment of turning resistance of a track.

TABLE 17.3

Values of Lateral Resistance of Tracks Over Various Surfaces

Track Materials	Coefficient of Lateral Resistance, μ_t		
	Concrete	Hard Ground (not Paved)	Grass
Steel	0.50–0.51	0.55–0.85	0.87–1.11
Rubber	0.09–0.91	0.65–0.66	0.67–1.14

Source: J. W. Wong, *Theory of Ground Vehicles*, John Wiley & Sons, New York, 1978.

$$F_{ti} = \frac{Mg}{2}\left(f_r - \frac{l\mu_t}{2B}\right). \tag{17.23}$$

From Equation 17.23, if $l\mu_t/2B > f_r$, the thrust of the inside track, F_{ti}, will be negative. This implies that to achieve a steady state, a conventional vehicle with a diesel engine as its mover must apply a braking force on the inside track. In a series hybrid drivetrain, as shown in Figure 17.3, the inside track motor can apply a negative torque (regenerative braking) to this track.

Referring to the maximum terrain tractive effort and for normal operation and steerability, the thrust of the outside track should be smaller than the maximum terrain tractive effort, that is,

$$F_{to} \leq cbl + \frac{Mg\tan\phi}{2}, \tag{17.24}$$

$$\frac{l}{B} \leq \frac{1}{\mu_r}\left(\frac{4cA}{Mg} + 2\tan\phi - 2f_r\right), \tag{17.25}$$

or

$$\frac{l}{B} \leq \frac{2}{\mu_r}\left(\frac{c}{p} + \tan\phi - f_r\right), \tag{17.26}$$

where A is the contact area of the track ($A = b \times l$), and p is the normal pressure, which is equal to $Mg/2A$.

From Equation 17.22, in addition to overcoming motion resistance, the motor drive of the outside track has to produce additional thrust to overcome the turning resistance. The most difficult situation is steering on a slope, as shown in Figure 17.12. In this case, the traction motor of the outside track must produce a large traction torque to overcome terrain resistance, grade resistance, and steering resistance. The total resistance of the vehicle, in this case, can be expressed by

$$R_o = Mg\left(f_r + \frac{l\mu_t}{2B}\right) + \frac{1}{2}\rho_a C_D A_f V^2 + Mg\sin a, \tag{17.27}$$

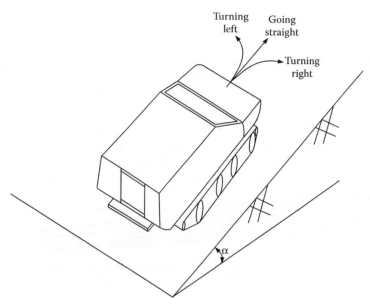

FIGURE 17.12
A vehicle going straight and steering on sloped ground.

where f_r is the resistance coefficient that includes the internal resistance of the track and resistance caused by terrain deformation.

Figure 17.13 shows vehicle resistance that includes motion resistance, aerodynamic drag, grade resistance, and steering resistance as expressed in Equation 17.27. The thrusts produced by the electric motors have the same powers as shown in Figure 17.8 (e.g., straight line on 60% slope ground) and different extended speed ratios, x. Since significant steering resistance is involved, the ground grade on which the vehicle can operate is greatly reduced. In motor and transmission design, this situation should be taken into consideration.

17.4 Engine/Generator Power Design

The engine/generator power should be designed to meet the requirements of constant-speed operation at high speed (near its maximum speed) on hard-surface roads, medium speed on soft-surface roads for long-distance trips, and also be larger than the average power at variable speed (driving cycle) in order to prevent the energy storage from being fully discharged during the mission. The engine/generator unit also needs to produce additional power to support the nontraction, continuous power, such as communication, lights, hotel loads, reconnaissance, and the auxiliaries (e.g., coolant circulation, cooling fans). The peaking powers required by acceleration, hill climbing, steering, and high-magnitude pulsed power needed by nontraction devices are provided by the primary power source (PPS): batteries or a combination of batteries and ultracapacitors.

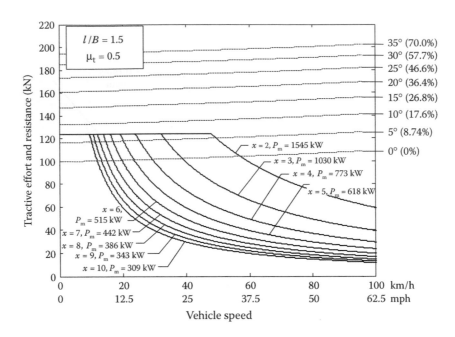

FIGURE 17.13
Vehicle gradeability with steering versus vehicle speed with respect to various x and motor powers.

Due to the absence of sufficient information about variable-speed operation (drive cycle) for off-road vehicles, the engine/generator power design is based on constant-speed operation on hard roads and soft grounds and can be expressed by

$$P_{m/g} = \eta_t \eta_m \left(M g f_r + \frac{1}{2} \rho_a C_D A_f V^2 \right) V, \qquad (17.28)$$

where η_t and η_m are transmission efficiency and motor drive efficiency, respectively. On hard roads, the resistance coefficient, f_r, includes only the internal resistance of run gear, which is described by $c + dV$, as shown in Table 17.2, and no road deformation loss. However, on soft roads, an additional resistance coefficient of 0.06 caused by ground deformation is added. Using the parameters shown in Table 17.2 and assuming the total efficiency of converters and electric motor drives as 0.85, the traction powers of the engine/generator versus vehicle speeds on soft and hard surfaces are shown in Figure 17.14.

The engine/generator unit needs to produce additional power to support nontraction continuous loads, such as communications, lights, hotel loads, reconnaissance, and the auxiliaries (e.g., coolant circulation, cooling fans). Due to the lack of accurate data from the nontraction loads for off-road vehicles, we take civilian vehicles as reference. The nontraction continuous power may be estimated as 40–50 kW. Thus, the total engine/generator power is designed around 350 kW.

It should be noted that the power of the engine/generator, designed previously, is the maximum power. In real-time operation, this power may be smaller, depending on the real operation conditions and the overall drive train control strategy.

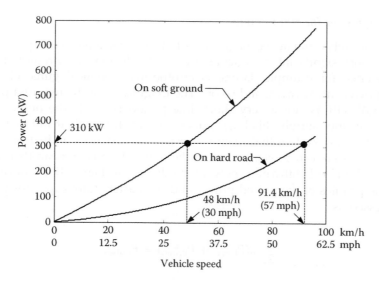

FIGURE 17.14
Traction power on hard road and soft ground at constant speeds.

17.5 Power and Energy Design of Energy Storage

Batteries or a combination of batteries and ultracapacitors is commonly employed as the energy storage for providing peaking power to the drivetrain. Peaking power can be divided into traction power and nontraction power. Traction power for peaking operation mainly includes the power in acceleration, hill climbing, and steering, and nontraction power mainly includes the power needed, for example, by high-power detection devices and electric weapon systems in military vehicles.[5,6]

17.5.1 Peaking Power for Traction

In high-power traction, the drivetrain is powered by both the engine/generator and the batteries/ultracapacitors, and the maximum power is constrained by the designed maximum power of the traction motors as previously specified. Thus, the traction peaking power for batteries/ultracapacitors is

$$P_b = \frac{P_{m,max}}{\eta_m} - P_{e/g},$$
(17.29)

where $P_{m,maxr}$ and η_m are the maximum output power and efficiency of the motor drives, respectively, and $P_{e/g}$ is the engine/generator output power. It should be noted that in acceleration and hill-climbing operations, the engine is not always called upon to operate at its maximum power due to efficiency concerns. Suppose, in the example vehicle, the traction motor drives have an efficiency of 0.85; the batteries'/ultracapacitors' power for traction is then around 375 kW (515/0.85–230), where the engine/generator delivers 230 kW of power to the traction motor (75% of maximum traction power of the engine/generator [310 kW]).

17.5.2 Peaking Power for Nontraction

It is hard to accurately estimate the magnitude of the nontraction peaking power. In military vehicles, the most significant nontraction pulsed loads may be presented by "electric weapon" systems, for example, lasers, electrothermal chemical guns, electromagnetic armor, high-power microwaves, and so on. The magnitude of the required pulsed power may reach 1 GW (10^9 W) for a very short time period (10^{-3} s). The on-board batteries/ultracapacitors cannot supply this huge pulsed power due to their internal impedances. Thus, a pulse-forming system that mainly consists of capacitors, inductors, and resistors is needed. This system can be charged from the main DC bus and can then discharge its energy to the pulsed load with a huge power for a short time. Figure 17.15 conceptually illustrates the time profiles of the pulsed power and the battery/ultracapacitor power, which may be expressed as

$$\frac{1}{2\varepsilon} t_p (P_{b,max} + P_{b,min}) = E_{pulse}. \tag{17.30}$$

Thus,

$$P_{b,max} = \frac{2E_{pulse}}{t_p(1 + D)}, \tag{17.31}$$

where $D = P_{b,max}/P_{b,min}$ is the charging power ratio and T_p is the period of the pulsed power load.

Figure 17.16 illustrates the battery/ultracapacitor power for the pulsed power load, varying with the charging power ratio, D, and the period of the pulsed power load, T_p. In this design, $T_p = 4$ s and $D = 0.6$ would be a good estimate. Thus, the maximum battery/ultracapacitor power is around 300 kW. Adding this to the traction power, the total power requirement is estimated to be around 675 kW.

It should be noted that the battery/ultracapacitor power capacity must be maintained above the designed value for a certain period of time to support the peaking power operation. For traction power demand, this time period is required by acceleration, hill climbing,

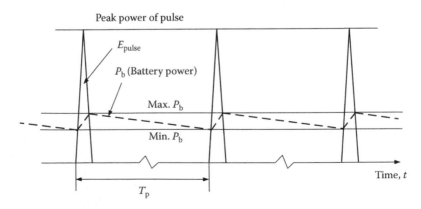

FIGURE 17.15
Conceptual illustration of time profiles of pulsed power and battery power.

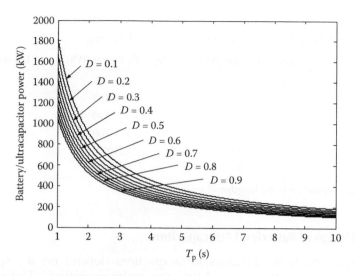

FIGURE 17.16
Peaking power of batteries for pulsed power load varying with D and Tp.

obstacle negotiation, and steering and may be over 20–30 s. For nontraction peaking power demand, it is dependent on the mission requirements.

Figure 17.17 shows the discharging power characteristics, in 18 s, of the Li-ion batteries provided by SAFT America and tested by CHPS (Combat Hybrid Power System sponsored by TACOM). It indicates that the battery power is very dependent on temperature and depth of discharge (DoD). Table 17.4 gives the specific power, specific energy, and energy density of the CHPS battery alternatives at standard testing. In this design, 1000 W/kg of specific power and 100 Wh/kg of specific energy would be a good estimate.

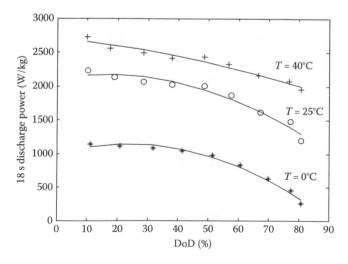

FIGURE 17.17
Eight-second discharge power of SAFT Li-ion batteries at different operation temperatures and DoD.

TABLE 17.4

Major Parameters of CHPS Battery Alternatives at Standard Testing

CHPS Battery Alternative	Specific Energy (Wh/kg)	Specific Power (Wh/kg)	Energy Density (Wh/L)
Lead acid	28	75	73
Ni–Cd	50	120	80
Ni–MH	64	140	135
Li-ion (high energy)	144	700	308
Li-ion (CHPS)	100	1000[a]	214
Li-ion (high power)	80	1400[a]	150

[a] Power capabilities depend on pulse length and temperature.

17.5.3 Energy Design of Batteries/Ultracapacitors

The energy requirements for batteries/ultracapacitors depend on the specific mission requirements, for example, the required time for stealth operation, silent watch, "electric weapon" operation, and so on. However, when the power capacity is determined, the energy capacity of the batteries can be obtained from the energy/power ratio of the selected batteries.

As mentioned previously, the battery power demand is around 675 kW, and the energy/power ratio of Li-ion battery is 0.1 h (specific energy/specific power). Thus, 67.5 kWh of energy capacity is obtained. The battery weight is around 675 kg.

17.5.4 Combination of Batteries and Ultracapacitors

In addition to batteries, ultracapacitors are another possible PPS. Compared with batteries, ultracapacitors have some advantages, such as two to three times the specific power density of Li-ion batteries (Tables 17.4 and 17.5), wide temperature adaptability, high efficiency (low resistance), and fast response to charging and discharging. Hence, it may be a good selection

TABLE 17.5

.Technical Specifications of Maxwell MBOD 0115 Ultracapacitor Module

Capacitance		145 Faradays (−20%/+20%)
Max. series resistance ESR	25C	10 mohm
Specific power density	42V	2900 W/kg
Voltage	Continuous	42 V
	Peak	50 V
Max. current		600 A
Dimensions	(Reference only)	$195 \times 265 \times 415$ mm
Weight		16 kg
Volume		22 L
Temperature[a]	Operating	−35°C to 65°C
	Storage	−35°C to 65°C
Leakage current	12 h, 25°C	10 mA

Source: Available at http://www.maxwell.com, Maxwell Technologies.
[a] Steady-state temperature.

as a pulsed power source. However, the ultracapacitor has the major disadvantage of a low specific energy density of less than 5 W/kg. It cannot sustain its power for more than a couple of minutes. Thus, it is difficult for ultracapacitors alone to supply the peaking power for a vehicle.

A good design for the PPS of a hybrid vehicle may be to combine Li-ion batteries and ultracapacitors to constitute a hybrid energy storage, in which the batteries supply the energy and the ultracapacitors supply the power.[6] The combination of batteries and ultracapacitors meets the power and energy requirements, that is,

$$P_{tot} = W_b P_b + W_c P_c, \tag{17.32}$$

$$E_{tot} = W_b E_b + W_c E_c, \tag{17.33}$$

where P_{tot} and E_{tot} are the total power and energy required, W_b and W_c are the weights of the battery and the ultracapacitor, P_b and P_c are the specific powers of the battery and the ultracapacitor, and E_b and E_c are the specific energies of the battery and the ultracapacitor. For a given P_{tot}, E_{tot}, P_b, P_c, E_b, and E_c, the battery and ultracapacitor weights can be obtained as

$$W_b = \frac{P_c E_{tot} - P_{tot} E_c}{P_c E_b - P_b E_c}, \tag{17.34}$$

$$W_c = \frac{P_{tot} E_b - P_b E_{tot}}{P_c E_b - P_b E_c}. \tag{17.35}$$

Figure 17.18 shows the weights of batteries, ultracapacitors, and the hybrid energy storage that can supply 675 kW of total power. When the total energy requirement is less

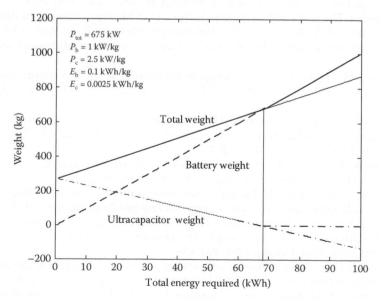

FIGURE 17.18
Battery weight, ultracapacitor weight, and total weight of hybrid energy storage versus total energy.

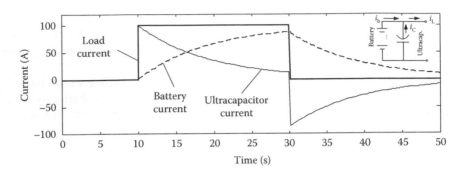

FIGURE 17.19
Current profiles of the battery/ultracapacitor energy storage.

than 67.5 kWh, the hybrid energy storage has less weight than the battery-alone energy storage. When the total energy requirement is greater than 67.5 kWh, a battery alone should be used.

Another advantage of hybrid energy storage is the leveling of the battery peak current, as shown in Figure 17.19. This simplifies thermal management of the batteries, extends the battery life cycle, and offers fast power response due to the very low resistance in the ultracapacitors. Other advanced configurations of hybrid energy storage may be used for better performance (refer to Chapter 13 for more details).

Bibliography

1. J. W. Wong, *Theory of Ground Vehicles*, John Wiley & Sons, New York, 1978.
2. M. G. Bekker, *Theory of Land Locomotion*, University of Michigan Press, Ann Arbor, 1956.
3. M. G. Bekker, *Off-the-Road Locomotion*, University of Michigan Press, Ann Arbor, 1960.
4. M. G. Bekker, *Introduction of Terrain-Vehicle Systems*, University of Michigan Press, Ann Arbor, 1969.
5. Y. Gao and M. Ehsani, Parametric design of the traction motor and energy storage for series hybrid off-road and military vehicles, *IEEE Transactions on Power Electronics*, 21(3), May 2006: 749–755.
6. Y. Gao and M. Ehsani, Investigation of battery technologies for the army's hybrid vehicle application, In *Vehicular Technology Conference, 2002. Proceedings, VTC 2002-Fall, 2002 IEEE 56th*, Vol. 3, pp. 1505–1509. September 24–28, 2002.
7. M. Ehsani, K. Rahman, and A. Toliyat, Propulsion system design of electric and hybrid vehicles, *IEEE Transactions on Industrial Electronics*, 44(1), February 1997: 19–27.
8. A. Boglietti, P. Ferraris, and M. Lazzari, A new design criteria for spindle induction motors controlled by field orientated technique, *Electric Machine and Power Systems*, 21, 1993: 171–182.
9. T. Kume, T. Iwakane, T. Yoshida, and I. Nagai, A wide constant power range vector-controlled AC motor drive using winding changeover technique, *IEEE Transactions on Industry Applications*, 27(5), September/October 1991: 934–939.
10. M. Osama and T. A. Lipo, A new inverter control scheme for induction motor drives requiring speed range, In *Proceedings of the IEEE-IAS Annual Meeting*, pp. 350–355. Orlando, FL, 1995.
11. R. J. Kerkman, T. M. Rowan, and D. Leggate, Indirect field-oriented control of an induction motor in the field weakened region, *IEEE Transactions on Industry Applications*, 28(4), 1992: 850–857.

12. K. Rahman, B. Fahimi, G. Suresh, A. Rajarathnam, and M. Ehsani, Advanced of switched reluctance motor applications to EV and HEV: Design and control issues, *IEEE Transactions on Industry Applications*, 36(1), January/February 2000: 11.
13. Available at http://www.maxwell.com, Maxwell Technologies.
14. S. De Breucker and P. Coenen, Series-hybrid drivetrain with FOC salient-pole synchronous generator and supercapacitor storage for large off-road vehicle, In *2015 IEEE Vehicle Power and Propulsion Conference (VPPC)*, pp. 1–5. Montreal, QC, 2015.
15. O. Hegazy, R. Barrero, P. Van den Bossche, M. E. Baghdadi, J. Smekens, J. Van Mierlo, W. Vriens, and B. Bogaerts, Modeling, analysis and feasibility study of new drivetrain architectures for off-highway vehicles, *Energy*, 109, 2016: 1056–1074, ISSN 0360-5442.
16. F. Herrmann and F. Rothfuss, Introduction to hybrid electric vehicles, battery electric vehicles, and off-road electric vehicles, *Advances in Battery Technologies for Electric Vehicles* 2015: 3–16.
17. R. V. Wagh and N. Sane, Electrification of heavy-duty and off-road vehicles, In *2015 IEEE International Transportation Electrification Conference (ITEC)*, pp. 1–3. Chennai, 2015.
18. H. Ragheb, M. El-Gindy, and H. A. Kishawy, On the multi-wheeled off-road vehicle performance and control, *International Journal of Vehicle Systems Modelling and Testing*, 8.3, 2013: 260–281.
19. H. Zhang, Y. Zhang, and C. Yin, Hardware-in-the-loop simulation of robust mode transition control for a series–parallel hybrid electric vehicle, *IEEE Transactions on Vehicular Technology*, 65.3, 2016: 1059–1069.
20. M. F. M. Sabri, K. A. Danapalasingam, and M. F. Rahmat, A review on hybrid electric vehicles architecture and energy management strategies, *Renewable and Sustainable Energy Reviews*, 53, 2016: 1433–1442.

18

Design of Full-Size-Engine HEV with Optimal Hybridization Ratio

Commercially available HEVs have been around for more many years. However, their market share remains small. Focusing only on the improvement of fuel economy, the design tends to reduce the size of the internal combustion (IC) engine in the HEV and uses the electrical drive to compensate for the power gap between the load demand and the engine capacity. Unfortunately, the low power density and the high cost of the combined electric motor drive and battery packs dictate that HEVs have either worse performance or a much higher price than conventional vehicles. In this chapter, a new design philosophy for a parallel HEV is proposed that uses a full-size engine to guarantee that vehicle performance will be at least as good as that of conventional vehicles and hybridize with an electrical drive in parallel to improve the fuel economy and performance beyond the conventional vehicle.

For a full-size-engine vehicle, any hybridization will increase the cost and weight of the vehicle. The additional cost will erode the benefits for the vehicle owner, while the additional weight of the electric motor and battery pack will have negative impacts on the vehicle's fuel economy. We implemented a dynamic programming algorithm for an analysis of the sensitivity of fuel consumption to increasing hybridized electrical drive power to determine the optimal hybridization ratio for a full-size-engine HEV. This will optimize the power of hybridized electrical drives for fuel economy, performance, and cost.

To show the advantages of the full-size-engine HEV over conventional vehicles and HEVs, comparisons of fuel economy, vehicle performance, and payback time are made with a typical conventional car, the 2011 Toyota Corolla, and the best-selling hybrid car in the United States, the 2011 Toyota Prius.

18.1 Design Philosophy of Full-Size-Engine HEV

A next-generation HEV should never be worse than a conventional vehicle. This means better fuel economy and better performance at cost parity with the conventional vehicle, or equal fuel economy and equal performance to the conventional vehicle at lower cost. The consumer can accept a slightly higher vehicle cost for achieving real improvements in the fuel economy and vehicle performance. This is illustrated by the following example. During the past 60 years, the average price (adjusted for inflation) of new cars has been increasing from \$12,000 to \$24,000, as shown in Figure 18.1, while the number of vehicles increased from 240 per 1000 people to 840 per 1000 people in the United States.[1] This means that the consumer is willing to pay more for arguably better cars.

Based on the previous discussion, we propose a new design philosophy for a parallel HEV, which we call a full-size-engine HEV. It starts from a conventional vehicle with a full-size IC engine as its primary power source, and it hybridizes the IC engine with an

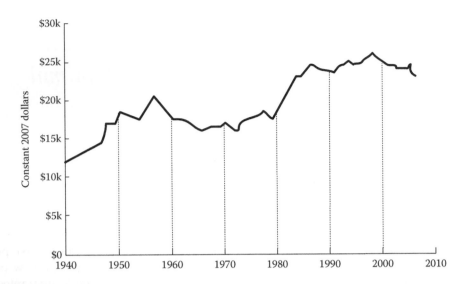

FIGURE 18.1
New car prices: over the past 70 years. (Adapted from Energy efficiency and renewable energy. Vehicle technologies program, fact of the week [Online]. Available: http://www1.eere.energy.gov/vehiclesandfuels/index.html.)

electrical drive in parallel. The full-size engine means that the IC engine has the capacity to propel the vehicle alone under typical driving conditions and guarantees travel range and performance at least as good as the conventional vehicle. In this case, connecting an electrical drive in parallel can improve the fuel economy and the performance beyond the conventional vehicle.

Figure 18.2 shows the configuration of a full-size-engine parallel HEV. A full-size IC engine connects to a torque coupler via a transmission (Trans. 1 in Figure 18.2). Since a

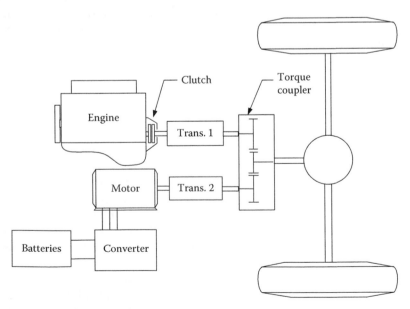

FIGURE 18.2
Configuration of full-size-engine HEV.

well-controlled electric motor has good characteristics for vehicle traction application, the transmission (Trans. 2 in Figure 18.2) has a single gear that functions as a speed reducer with a fixed-gear ratio.

18.2 Optimal Hybridization Ratio

Given a full-size-engine vehicle, any hybridization will increase the cost and weight of the vehicle. The additional cost will erode the benefits for the vehicle owner, while the additional weight of the electric motor and the battery pack will negative impacts on the vehicle's fuel economy. However, it is possible to find an optimal hybridization ratio or range of ratios in which the electrical drive can cause significant fuel economy improvements in typical drive cycles.[2,3] The optimal hybridization ratio can be achieved by fuel consumption sensitivity analysis using dynamic programming optimization.[4,5]

The following is an example for the design of a 1200-kg passenger car drivetrain. The major parameters of the vehicle are listed in Table 18.1. The engine is designed to have a peak power of 80 kW. The engine specific fuel consumption (sfc) map is shown in Figure 18.3. The operating points with low fuel consumption are usually in the region of middle speed and high power output.

To determine the optimal hybridization ratio for the full-size-engine HEV, the sensitivity of the minimal fuel consumption (maximum miles per gallon [mpg]) to the increasing of the hybridized electrical drive power is analyzed. The additional weight of the electrical drive, which affects the fuel economy of the HEV, should be considered during this sensitivity analysis. The weight for each electrical drive is determined and listed in Table 18.2.

18.2.1 Simulation under Highway Driving Conditions

Figure 18.3 shows the optimal control sequences (engine power and motor power) obtained by the dynamic programming algorithm in the FTP75 highway driving cycle for the 1200-kg passenger car with 3 kW electrical drive and 25 kW electrical drive. Figure 18.3a shows that

TABLE 18.1

Major Parameters of Full-Size-Engine HEV

Vehicle mass	1200 kg
Rolling resistance coefficient	0.01
Aerodynamic drag coefficient	0.30
Front area	$2.00\,m^2$
Four-gear Transmission	
Gear ratio	
1st gear	2.73
2nd gear	1.82
3rd gear	1.30
4th gear	1.00
Final gear ratio	3.29

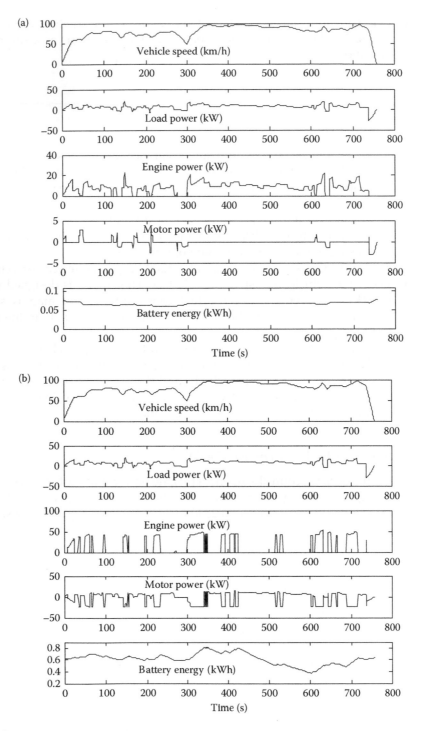

FIGURE 18.3
Simulation in FTP 75 highway drive cycle: vehicle speed, load power, optimal engine power, optimal motor power, and battery energy: (a) 3 kW electrical drive and (b) 25 kW electrical drive.

TABLE 18.2

Weights of Electrical Drives

Electrical Drive Power (kW)	Weight (kg)
40	130.0
35	114.5
30	99.0
25	82.5
20	66.5
15	49.8
10	33.3
5	16.8
3	10.6

the 3-kW hybridized electrical drive cannot generate enough power to assist in propulsion. Therefore, most of the load power is supplied directly by the full-size engine. The minimal fuel consumption achieved is 6.85 L/100 km (34.33 mpg).

When the rated electrical drive power is increased to 25 kW, the minimal fuel consumption (maximum mpg) of this full-size-engine HEV reaches a value of 5.50 L/100 km (42.80 mpg). It is clear from Figure 18.3b that with the help of the 25-kW electrical drive, the full-size-engine HEV has more flexibility to take advantage of the two power sources, engine and electrical drive. This leads to a significant improvement in the fuel economy.

The optimal engine operating points, overlapping with the engine fuel consumption map, for a 1200-kg conventional passenger car (with a full-size engine as its sole power source), the same car hybridized with a 25-kW electrical drive and a 35-kW electrical drive, are shown in Figure 18.4. In Figure 18.4, the different pattern of the operating points indicates different energy management modes, which lead to the minimum fuel consumption for the overall driving cycle. The circles show that the engine-traction-alone mode is used. The squares mean that the engine is producing more power than the load demand, and the excess power charges the battery.

As no electrical traction can be used in the conventional car, in Figure 18.4a, the engine always supplies the exact power that the load demands. This makes the engine operate far from its high-efficiency area: the inner circles of the fuel consumption map. Therefore, the fuel consumption of this conventional vehicle has the highest value: 7.02 L/100 km (lowest value of miles per gallon: 33.49).

As the hybridization ratio increases, Figure 18.4b shows that for the passenger car with a 25-kW electrical drive, many of the engine operating points move to the engine's high-efficiency zone, compared to Figure 18.4a. The squares in Figure 18.4b indicate that the improvement in vehicle efficiency is achieved by charging the battery from the engine. Thus, the minimal fuel consumption for this passenger car is 5.50 L/100 km (maximum mpg of 42.80).

After further increasing the rated power of the electrical drive to 35 kW, as shown in Figure 18.4c, most of the engine operating points lie in its high-efficiency zone.

If we keep on increasing the hybridization ratio, as there is almost no room to improve the engine fuel efficiency, shown in Figure 18.4c, the additional electrical drive power can only increase the weight of the vehicle, which negatively affects the fuel economy.

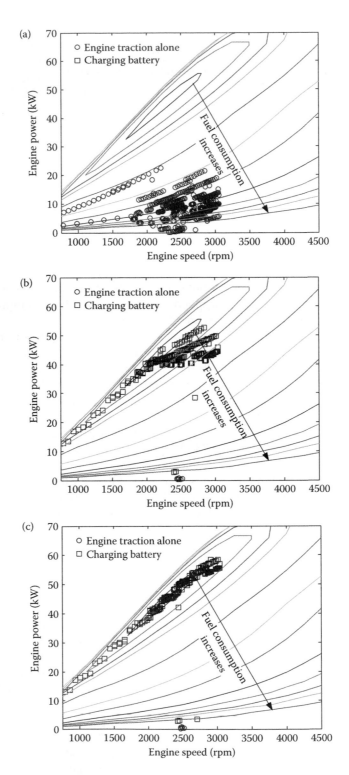

FIGURE 18.4
Optimal engine operating points in FTP75 highway drive cycle: (a) conventional car, (b) car with 25-kW hybridized
electrical drive, and (c) car with 35-kW hybridized electrical drive.

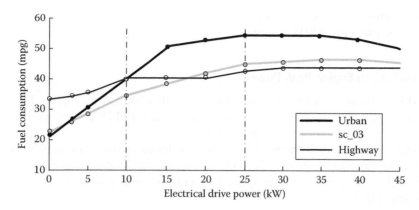

FIGURE 18.5
Effects of hybridization on 1200-kg passenger car with 80-kW engine.

18.2.2 Optimal Hybridization of Electrical Drive Power

To analyze the fuel economy sensitivity to hybridization ratios on a highway drive cycle, based on the simulation results obtained previously, the hybridized electrical drive power versus maximum mpg are shown in Figure 18.5. The same simulations were also made on FTP75 urban and sc_03 drive cycles. Their hybridized electrical drive power versus maximum mpg curves are also shown in Figure 18.5. The mpg, where the motor rated power equals 0 kW, represents a conventional vehicle.

For all three typical drive cycles in Figure 18.5, the maximum mpg starts to increase as the rated power of the hybridized electrical drive increases from 0 kW. As the hybridization ratio increases, the electrical drive power versus maximum mpg curve begins to bend. This is because the fuel economy improvement is competing with the increasing energy demand brought on by the additional weight of the electrical drive. As mentioned in Section 18.2.1, the maximum mpg decreases with the electrical drive power up to a certain value. The bending curves suggest a variable sensitivity to the hybridization ratio.

In Figure 18.5, for all three drive cycles, the maximum mpg has high sensitivity to increasing electrical drive power from 0 to 10 kW. After the electrical drive power reaches 25 kW, this sensitivity diminishes. This means the hybridization starts to lose its benefit, while the weight and cost of the vehicle still increase.

Based on the previous discussion, the range from 10 to 25 kW is determined to be the optimal hybridization electrical drive power range for this 1200-kg passenger car with an 80-kW engine.

18.3 10–25 kW Electrical Drive Packages

In Section 18.2, the 10–25 kW optimal hybridization power range is determined based on simulation results on the 1200-kg passenger car with a 80-kW IC engine. In this section, the same analysis is done on passenger cars with larger IC engine peak power and vehicle mass from 1200 to 1800 kg, which represent most of the passenger cars sold in the North

American automobile market, from the compact car, for example, Toyota Corolla, to the full-size car, for example, BMW 5 series.

18.3.1 Sensitivity to Engine Peak Power

The simulation is based on a 1200-kg passenger car with a 100-kW IC engine. The hybridized electrical drive power versus maximum mpg curves for FTP75 urban, sc_03, and FTP75 highway drive cycles are shown in Figure 18.6.

Comparing Figure 18.6 with Figure 18.5, the sensitivity curves with a 100-kW IC engine are similar to those with an 80-kW IC engine, except for the mpg values. Here the 10–25 kW electrical drive power range still spans the high fuel economy sensitivity range of this full-size-engine HEV.

18.3.2 Sensitivity to Vehicle Mass

To test the sensitivity of optimal hybridization to vehicle mass, the weight of the test vehicle was increased from 1200 to 1500 kg. This is the weight of a midsize passenger car, such as the Ford Fusion. The simulation results for three typical drive cycles are shown in Figure 18.7.

In Figure 18.7, the 10–25 kW hybridization power range is still valid for the 1500-kg passenger car. The same conclusions can also be drawn from Figure 18.8, with an 1800-kg passenger car. It can be seen from Figures 18.7 and 18.8 that the maximum mpg values for each hybridized electrical drive power are smaller than the value in Figure 18.5 with a 1200-kg passenger car. This is to be expected as the increased vehicle weight demands larger traction power, which needs to be satisfied by the power train.

18.3.3 10–25 kW Electrical Drive Power Window

In this section, the optimal hybridization electrical power range, 10–25 kW, for passenger cars is analyzed based on the characteristics of the IC engine and the load for typical drive cycles.

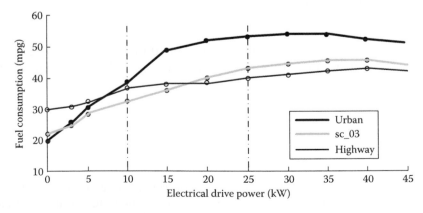

FIGURE 18.6
Effects of hybridization on 1200-kg passenger car with 100-kW engine.

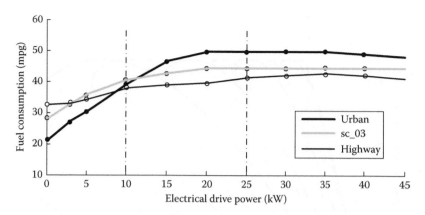

FIGURE 18.7
Effects of hybridization on 1500-kg passenger car with 80-kW engine.

For a conventional passenger car, there are two effective ways to improve the overall fuel economy with a parallel electrical drive. One is to improve the low vehicle speed fuel efficiency. The other is to recover the regenerative braking energy and charge the battery for future traction use.

Figure 18.9 shows the fuel consumption map for a typical 80-kW IC engine in various gear ratios to the driving wheels. The low-speed region of the engine is its least efficient area. This is also true of IC engines with greater peak power. An effective way to improve the fuel economy of the vehicle is to turn off the engine in this low-speed region and use the electric motor alone to supply the load power.

The load points (on the vehicle speed vs. load power plane) from an FTP75 urban drive cycle for a 1200-kg vehicle model are shown in Figure 18.10, along with the performance characteristics of both a 10-kW electric motor and a 25-kW electric motor. The 25-kW electric motor has the capacity to supply all the traction load power in the low-speed region. As the mass of the HEV increases to 1500 and 1800 kg, the traction load powers, in the FTP75 urban drive cycle, are increased as shown in Figures 18.11 and 18.12, respectively. Nevertheless, the 25-kW motor can still supply enough power to meet most of the traction

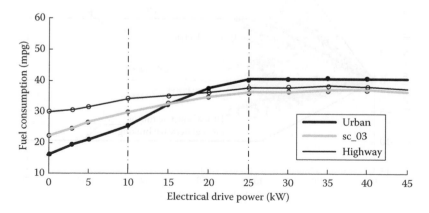

FIGURE 18.8
Effects of hybridization on 1800-kg passenger car with 80-kW engine.

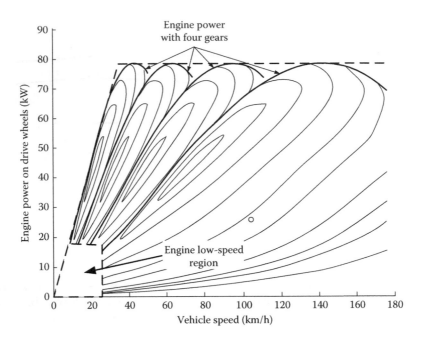

FIGURE 18.9
Fuel consumption map for 80-kW IC engine under each gear on drive wheels.

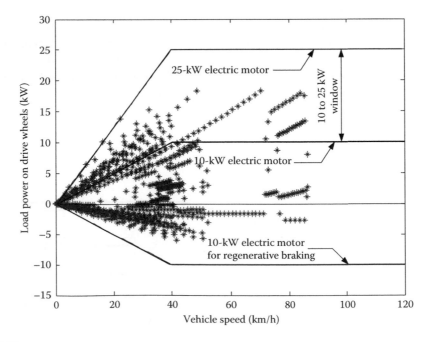

FIGURE 18.10
Load power vs. vehicle speed in FTP75 urban drive cycle for 1200-kg passenger car.

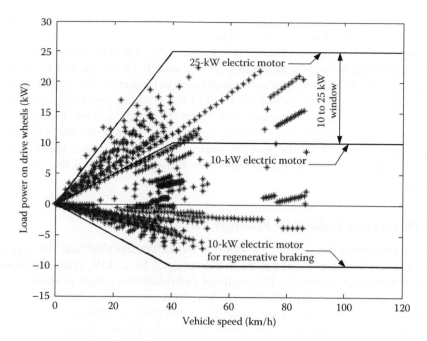

FIGURE 18.11
Load power vs. vehicle speed in FTP75 urban drive cycle for 1500-kg passenger car.

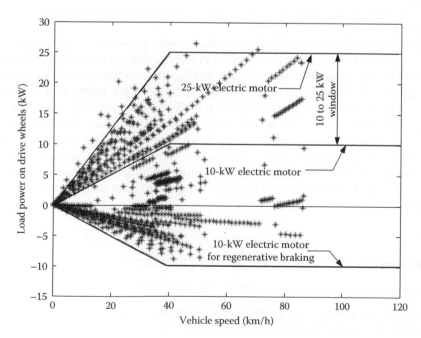

FIGURE 18.12
Load power vs. vehicle speed in FTP75 urban drive cycle for 1800-kg passenger car.

load demand in the low-speed region. This makes 25 kW the upper value of the optimal power range of electrical drive hybridization.

Considering the regenerative braking energy recovery, for passenger cars with masses from 1200 to 1800 kg, as shown in Figures 18.10 through 18.12, the 10-kW electric motor has enough capacity to recover this amount of energy. However, the 10-kW motor cannot supply enough tractive power to replace most of the engine operating points in the low-speed region to make the fuel consumption reach its lowest values. However, by effectively recovering the regenerative braking energy, the 10-kW motor still significantly improves the fuel economy of the vehicle. Further, the 10-kW motor still has the advantage of low weight and cost, which makes it the minimum value of the optimal hybridization range.

18.3.4 Electrical Drive Package for Passenger Cars

Based on the preceding discussion, the optimal hybridization electrical drive power for a full-size-engine HEV is determined to be in the range of 10–25 kW. This was based on fuel consumption sensitivity analysis. This optimal hybridization range is shown to be valid for most passenger cars sold in the North American automotive market, from compact to full-size passenger cars. This conclusion indicates that, to achieve better performance and significant improvement in fuel economy with a short cost payback period, most passenger cars can be hybridized with *the same electrical drive package* with its rated power being in the aforementioned optimal hybridization range. Alternatively, the vehicle buyers themselves can choose several electrical drive packages, from 10 to 25 kW, as options. The comparison with the commercially available passenger cars in the next section shows that a full-size engine with a 10-kW hybridized electrical drive has an advantage in cost payback period. However, with a 25-kW electrical drive, the vehicle enjoys both high fuel economy and significant improvement in acceleration performance, especially from stand still.

In the first three quarters of 2012, the number of new passenger cars sold in the United States was 5.6 million.[7] Assuming that the 25-kW electrical drive is selected as the most common motor-battery package for automobile hybridization, for the best fuel saving and acceleration performance, even with 20% of new passenger cars using this full-size-engine hybridization technology, the production of the 25-kW motor-battery package will reach more than 1 million each year. This number will be much higher based on this year's number of vehicles sold. Based on this, either "one package fits all" or "a few packages" will significantly increase the production volume of specific types of electric motors, batteries, and power electronic devices. This large-volume production will, in turn, significantly decrease the cost of these electrical drives, further increasing the market penetration of the full-size-engine HEV.

18.4 Comparison with Commercially Available Passenger Cars

To demonstrate the advantages of the full-size-engine HEV over commercially available passenger cars, comparisons of fuel economy, cost of power plant, and acceleration performance are made between a typical conventional car, the 2011 Toyota Corolla, and the hybrid vehicle 2011 Toyota Prius.

TABLE 18.3

Fuel Consumption for Corolla-Based Hybridization in Typical Drive Cycles

	Urban	Highway	Combined
Corolla	11.83 (19.89)	7.66 (30.71)	9.50 (24.76)
HEV (10 kW)	6.28 (37.46)	6.30 (37.31)	6.29 (37.39)
HEV (25 kW)	4.54 (51.81)	5.89 (39.93)	5.06 (46.46)

Unit: L/100 km (mpg).

18.4.1 Comparison with 2011 Toyota Corolla

The 2011 Toyota Corolla is a conventional passenger car with a 98-kW IC engine. After hybridizing the 98-kW engine with 10-kW and 25-kW electrical drives, which are the minimum and maximum rated powers of the optimal hybridization range, two full-size-engine HEVs will result. The three vehicle models were simulated on both the FTP75 highway and urban drive cycles.

The fuel consumption for three different vehicles (2011 Corolla, 10-kW hybridized HEV, and 25-kW hybridized HEV) in each drive cycle are listed in Table 18.3. To compare the fuel economy of these three vehicles, the combined fuel consumption is used in Table 18.3, which takes 55% from the urban fuel consumption and 45% from the highway.

Clearly, the full-size-engine HEVs show significant improvements in fuel economy with the help of the hybridized electrical drive, as shown in Table 18.3.

For comparing the cost of power plants for these three vehicles, the following assumptions were made: The induction motor drive has a typical price of $110/kW, and the IC engine has a price of $35/kW. The price of Li-ion battery packs is $1000/kWh.[7] Although the price of Li-ion batteries is decreasing each year, the $1000/kWh was used for the worst-case analysis. The cost of power plants for the three vehicles is listed in Table 18.4. For the full-size-engine HEV, the cost of the power plant is higher than that of conventional vehicles. Assuming the fuel price is $1.06/L ($4/gal), the hybridization cost is recovered after driving about 3 years with the 10-kW hybridized HEV, and 5 years with the 25-kW hybridized vehicle, based on the average annual miles per U.S. driver of about 10,000 miles,[8] as shown in Table 18.4.

Comparing these results with the long cost payback period for the conventional HEV and PHEV, the full-size-engine HEV pays for itself in a shorter period of time. Moreover, even within the payback period, the vehicle owner can enjoy better acceleration performance with the help of the hybridized electrical drive. The acceleration time for each vehicle is listed in Table 18.5.

TABLE 18.4

Power Plant Cost and Cost Payback Mileage for Corolla-Based Hybridization

	Cost ($)	Payback	
		km	miles
Corolla	3430	–	–
HEV (10 kW)	4930	44,250	27,500
HEV (25 kW)	7080	77,710	48,300

TABLE 18.5

Comparison of Acceleration Time for Corolla-Based Hybridization

	0–100 km/h	0–16 km/h
Corolla	8.63 s	1.57 s
HEV (10 kW)	7.81 s	1.33 s
HEV (25 kW)	6.92 s	1.09 s

Two types of acceleration time are given by Table 18.5. One is for acceleration from 0 to 100 km/h (60 mph), where the full-size-engine HEVs show better acceleration performance than the conventional car. The other one is for takeoff-from-standstill performance, 0–16 km/h (10 mph), which is more likely for the driver. In this comparison, the 25-kW hybridized HEV shows a 30% decrease in acceleration time.

18.4.2 Comparison with 2011 Toyota Prius Hybrid

To compare with the Prius hybrid, a conventional vehicle model was first made, "Prius conventional." The power plant of the Prius hybrid is replaced with an 80-kW IC engine, which guarantees the "Prius conventional" has the same performance as the Prius hybrid. The 10- and 25-kW electrical drives are parallel coupled with the IC engine of the "Prius conventional."

The fuel consumption for each vehicle in typical driving conditions is shown in Table 18.6. The fuel consumption for the Prius hybrid was scaled to match the fuel efficiency of the IC engine used in this simulation. Based on the comparison results in Table 18.6, the full-size-engine HEVs show even better fuel economy than the Pius hybrid.

Since the fuel consumption for the full-size-engine HEV was obtained by the optimized control in a typical drive cycle, the comparison might be unfair to the Prius hybrid. In the following simulation, a very simple control strategy, called the maximum state of charge (SOC) for parallel HEVs, introduced previously, was used. The philosophy of the maximum SOC control strategy is to use the engine to propel the vehicle as much as possible, and the electric power is used for compensating for the shortage in the engine power, such as in sudden acceleration. Therefore, the electric energy is not fully used. The fuel consumption for the full-size-engine HEVs using the maximum SOC control strategy is listed in Table 18.7.

For the maximum SOC control strategy, where the electric power is only used during sudden accelerations, it is difficult for the full-size-engine HEV to show its advantage, especially

TABLE 18.6

Fuel Consumption for Prius-Based Hybridization on Typical Drive Cycles

	Urban	Highway	Combined
Prius (conventional)	10.21 (23.04)	6.76 (34.81)	8.30 (28.34)
HEV (10 kW)	5.48 (42.89)	5.91 (39.81)	5.80 (40.55)
HEV (25 kW)	4.22 (55.70)	5.48 (42.92)	4.71 (49.90)
Prius (hybrid)	5.40 (43.59)	5.73 (41.03)	5.50 (42.74)

Unit: L/100 km (mpg).

TABLE 18.7

Fuel Consumption for Prius-Based Hybridization Using Maximum SOC Strategy

	Urban	Highway	Combined
Prius (conventional)	10.21 (23.04)	6.76 (34.81)	8.30 (28.34)
HEV (10 kW)	5.93 (39.64)	6.86 (34.30)	6.32 (37.24)
HEV (25 kW)	6.04 (38.92)	6.91 (34.03)	6.41 (36.72)
Prius (hybrid)	5.40 (43.59)	5.73 (41.03)	5.50 (42.74)

TABLE 18.8

Comparison on Power Plant Cost and Acceleration Performance for Prius-Based Hybridization

	Cost ($)	Acceleration Time
Prius (conventional)	2800	9.70 s
HEV (10 kW)	4300	8.76 s
HEV (25 kW)	6450	7.79 s
Prius (hybrid)	9955	9.70 s

under highway driving conditions. However, as Table 18.7 shows, even with this low-efficiency control strategy, the fuel economy of the full-size hybrid HEVs is still close to that of the Prius hybrid. Further, any improvement in the control strategy narrows or even reverses this gap, as the simulation results show in Table 18.6.

Besides the fuel economy, the full-size-engine HEV still has the advantages of low cost and better acceleration performance, as shown in Table 18.8.

For the power plant cost comparison in Table 18.8, the cost of $1000/kWh for the Li-ion battery is used in the full-size-engine HEV, a worst-case analysis, whereas the Prius hybrid is using the cheaper NiMH battery.

Bibliography

1. Energy efficiency and renewable energy. Vehicle technologies program, fact of the week [Online]. Available: http://www1.eere.energy.gov/vehiclesandfuels/index.html.
2. E. Vinot, R. Trigui, B. Jcanneret, J. Scordia, and F. Badin, HEVs comparison and components sizing using dynamic programming, In *Vehicle Power and Propulsion Conference*, pp. 314–321, Arlington, Texas, September 2007.
3. S. Lukic and A. Emadi, Effects of drive train hybridization on fuel economy and dynamic performance of parallel hybrid electric vehicles, *IEEE Trans. Vehicular Technology*, 53(2), March 2004: 385–389.
4. U. Zoelch and D. Schroeder, Dynamic optimization method for design and rating of the components of a hybrid vehicle, *International Journal of Vehicle Design*, 19(1): 1998.
5. M. Kim and H. Peng, Power management and design optimization of fuel cell/battery hybrid vehicles, *Journal of Power Sources*, 165, March 2007: 819–832.

6. New vehicle sales [Online]. Available: http://www.motorintelligence.com/
7. M. Lowe, S. Tokuoka, T. Trigg, and G. Gereffi, *Lithium-ion batteries for electric vehicles: The U.S. value chain*, Center on Globalization Governance & Competitiveness, October 2010.
8. Federal Highway Administration. Average annual miles per driver by age group [Online]. Available: http://www.fhwa.dot.gov.
9. L. Lai and M. Ehsani, Design study of parallel HEV drive train with full size engine, In *IEEE Transportation Electrification Conference*, Dearborn, Michigan, June 2013.

19

Powertrain Optimization

19.1 Powertrain Modeling Techniques

The basic modeling technique for evaluating vehicle powertrain performance is via longitudinal dynamics. This technique involves dividing a given drive cycle into several time steps and calculating the state of the vehicle at the end of each time interval. To achieve the desired longitudinal acceleration levels for a given drive cycle, powertrain components must be sized appropriately to meet the desired performance levels. To allow for reserve power, conventional powertrains are often oversized for the intended use of the vehicle, leading to the operating point deviating from the optimal operating range. A hybrid powertrain optimized for an intended usage could simultaneously improve vehicle performance in terms of reduced emissions.

A class of models called "backward-facing" models are often used in the domain of powertrain size optimization.[1,2] Backward-facing models do not require a driver model, as the vehicle speed trace is obtained directly from the drive cycle. In such models, the speed trace is imposed on the vehicle model to calculate the angular velocity and torque at the wheels. Subsequently, the angular velocity and torque of the internal combustion engine (ICE), in the case of a conventional vehicle, is determined "backwards" from the wheels through each drivetrain component via efficiency models or maps. The efficiency maps are obtained from steady-state testing of real components; hence, this is why backward-facing models are also called "quasi-static" models. Because the transient characteristics are ignored in backward-facing models, they only need large time steps, resulting in quicker simulation times. These attributes have enabled backward-facing models to be used extensively in the area of powertrain component size optimization.[3]

However, because of their quasi-static nature, backward-facing models give very limited information about measurable quantities in a vehicle such as throttle and brake position. As a result, backward-facing models are less meaningful for implementation in hardware-in-the-loop (HIL) test systems.

In contrast, a dynamic system contains differential equations that describe the state of a system, and it includes elements such as inertia and inductance. As a result, dynamic systems are used in the so-called "forward-facing" modeling approach. Such models deal with quantities that are measurable in real drivetrains and with the correct causality. Forward-facing models also feature a driver model, which is typically modeled as a proportional-integral (PI) controller.

19.1.1 Forward-Facing Vehicle Model

Using a conventional vehicle (CV) as an example, the driver model provides torque demand in the form of desired ICE torque and brake torque to match the speed trace from a drive cycle. The topology of a representative forward-facing model is shown in Figure 19.1.

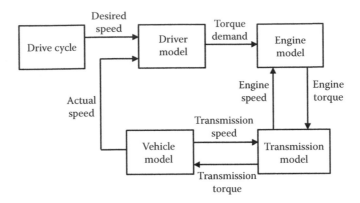

FIGURE 19.1
Forward-facing (dynamic) vehicle model. (G. Mohan, A toolbox for multiobjective optimisation of low carbon powertrain topologies, PhD thesis, Cranfield University, 2016.)

A basic driver model typically uses one or more PI controllers to meet the torque demand, with reference to the desired speed trace. The torque produced by the ICE propagates through the transmission and final drive ratios before ending up as torque applied at the wheels. This is then exerted on the vehicle mass via force on the tire contact patch. The vehicle speed that results from the applied force is propagated back through the drivetrain and returns to the ICE as angular velocity at the crankshaft. Brake torque is applied directly at the wheels.

Unlike backward-facing models, the speed trace is not imposed on the vehicle, and, therefore, there will inevitably be a small margin of error between the actual vehicle speed and the speed trace. It is the role of the driver model to minimize this margin of error. This is similar to the role of a real-world test driver carrying out an emissions test for vehicle type approval.

A forward-facing model provides insight into the vehicle model drivability, and it captures the limits of the physical system. It also facilitates control systems development and implementation on HIL systems. However, with the presence of multiple state equations, the vehicle speed (and subsequently drivetrain angular velocity) is computed via multiple state integration, resulting in the need to use sufficiently small time steps in the simulation. This results in longer simulation times when compared to backward-facing models.

Furthermore, resizing the powertrain components in the model alters the dynamics of the system such that the driver model needs to be retuned to maximize the performance in such parameters as fuel economy and longitudinal acceleration.

19.1.2 Backward-Facing Vehicle Model

The ability of the vehicle model to meet the demands of the drive cycle is the principal assumption of a backward-facing model. Based on the speed trace from the drive cycle, the vehicle acceleration and resistive forces are calculated to determine the resultant tractive force at the tire contact patch. It is then converted into wheel torque and propagated back to the ICE via the drivetrain, along with the angular velocity. Therefore, in backward-facing models, the power information is monodirectional (effort and flow are in the same direction), as seen in Figure 19.2.

With both speed and torque imposed on the powertrain components, a backward-facing model can also be treated as noncausal (i.e., physically nonrealizable). This is in

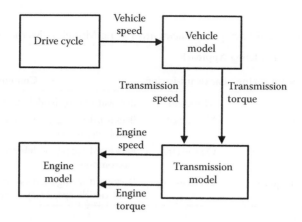

FIGURE 19.2

Backward-facing (quasi-static) vehicle model. (G. Mohan, A toolbox for multiobjective optimisation of low carbon powertrain topologies, PhD thesis, Cranfield University, 2016.)

contrast to forward-facing models where the transfer of power information is bidirectional, that is, the direction of effort (torque) is opposite to the direction of flow (speed). Additionally, a driver model is not needed in backward-facing models.

Backward-facing models rely on efficiency maps that are usually created based on torque and speed data and produced during steady-state real-world testing. This results in the calculation being relatively simpler than forward-facing models (essentially lookup tables instead of state equations) and can therefore be implemented over relatively larger time steps. However, the use of steady-state maps also limits the ability to accurately capture the transient performance of the powertrain.

19.1.3 Comparison of Forward-Facing and Backward-Facing Models

The backward-facing modeling approach is well documented for the study of the fuel consumption and emissions of vehicle powertrains. Nevertheless, the computational costs remain relatively low. The main drawback of this method is that the physical causality is not respected, and the speed profile of the drive cycle has to be known a priori. Therefore, this method is not able to handle feedback control problems or to correctly deal with state events.

In contrast, the forward-facing approach is desirable for hardware development and detailed control simulation but at the expense of the relatively higher computational cost. A pure forward approach is, therefore, often less suitable for preliminary powertrain design. Table 19.1 summarizes the key differences between the forward-facing and backward-facing modeling approaches.

19.2 Defining Performance Criteria

Performance criteria provide a systematic and objective approach to comparing different hybrid powertrain topologies. It enables powertrain topologies to be compared based on

TABLE 19.1

Comparison between Forward-Facing and Backward-Facing Modeling Methods

Criteria	Modeling Approach		Comment
	Forward-Facing	Backward-Facing	
Simulation time step	~1 ms	~1 second	Backward-facing models have shorter simulation times
Physical causality	Causal	Noncausal	Backward-facing models are not suitable for HIL implementation
Model type	Dynamic	Quasi-static	Backward-facing models are map based and quicker to compute
Driver model	Required	Unnecessary	Backward-facing models assume ideal driver model, and vehicle speed trajectory is known a priori

metrics that are defined by vehicle class and usage profile. This enables powertrain topologies to be optimized for a given application.

The following criteria are often used for powertrain design optimization:

- Fuel economy and emissions
- Powertrain mass
- Estimated powertrain costs
- Longitudinal acceleration performance

Fuel economy and emissions are a major driver for pushing alternative powertrain technologies. This is compounded by the cost of fuel and increased concern of global warming caused by harmful emissions. In tandem with alternative powertrains, lowering the overall vehicle mass also aids in reducing emissions. There is a trade-off between lowering emissions and reducing powertrain costs; for example, electric vehicles (EVs), which have lower tailpipe emissions when compared to CVs, are more expensive to manufacture. These performance criteria form the basis of the cost functions and the constraint definitions used in powertrain optimization.

19.2.1 Tank-to-Wheel Emissions

Tank-to-wheel emissions are a measure of comparing only the tailpipe emissions of a vehicle. This criterion was largely driven by the goal set by the European Automobile Manufacturers Association (ACEA) with the European Commission (EC) in 1998. This goal called for manufacturers to produce more fuel-efficient and lower-emission vehicles. They voluntarily agreed to limit the fleet-specific CO_2 emissions produced by new passenger vehicles to 140 g CO_2/km by 2008.

Additionally, European Union (EU) CO_2 targets are predicted to drive a dramatic shift in the types of powertrain produced over the next decade. In the short term, a new European fleet average target for less than 130 g/km of CO_2 emission has been set for all new vehicles produced after 2015, as per the ACEA agreement. This is a further 7% reduction from the 2008 levels. Hence, it can be assumed that tank-to-wheel emission performance is a growing concern and, therefore, of high importance for powertrain technologies comparison. Additionally, if the fleet average CO_2 emissions of a manufacturer exceeds this limit, a penalty is imposed on the excess emissions for each car registered. This penalty amounts to a premium of €5 for the first g/km that is exceeded, €15 for the second g/km, €25 for the third g/km, and €95 for each subsequent g/km thereafter. From 2019, the cost increases to a flat rate of €95 for every g/km exceeded.

There are also additional incentives given to manufacturers to produce vehicles with extremely low emissions (below 50 g/km). Each low-emitting car was counted as 3.5 vehicles in 2012 and 2013, 2.5 in 2014, 1.5 vehicles in 2015, and then one vehicle from 2016 to 2019. This approach will help manufacturers further reduce the average emissions of their new car fleets.[5]

19.2.2 Well-to-wheel Emissions

The tank-to-wheel analysis is a subset of the well-to-wheel analysis, which is used to determine the energy consumption and greenhouse gas emission of a system. The system is defined as every stage involved from fuel production (the well) to its end use in a vehicle (the wheel). Well-to-wheel studies in general form the basis for assessing the impacts of future fuel and powertrain options, particularly in terms of energy usage and greenhouse gas emissions. To assess the well-to-wheel CO_2 emissions of various powertrain topologies, it is necessary to consider CO_2 emissions associated with production of the fuel or source of energy (well-to-tank).

One example of well-to-wheel CO_2 emissions for various energy sources is summarized in Table 19.2.

In the table, each energy source is paired with its corresponding powertrain type, such as CV, EV, and fuel cell EV (FCEV). Hydrogen CO_2 emissions are estimated to be 76.9 g CO_2 MJ^{-1}, based upon a value of 11 kg CO_2 kgH_2^{-1} for steam reforming natural gas and a calorific value of 143 MJ kgH_2^{-1}.[6] Electricity CO_2 emissions are assumed to be 150 g CO_2 MJ^{-1} based upon the 2008 UK average electricity emissions of 540 g CO_2 kWh^{-1}, which included 5.5% of electricity generation from renewable sources (in 2011, this figure was increased to 594 g CO_2 kWh^{-1}, according the Department for Environment, Food and Rural Affairs (DEFRA), a public UK body). Well-to-tank conversion factor for petrol is 14.10 g CO_2/MJ.

The vehicle type used in the example shown in Table 19.2 is assumed to be a "medium vehicle" as defined by the National Travel Survey (NTS).[7] After completing a drive cycle, the amount of electrical energy consumed by an EV or PHEV is determined by replenishing the charge in the battery back to its initial state from the electric grid. Subsequently, the amount of well-to-wheel CO_2 emitted is then calculated by converting this consumed electrical energy into gram-CO_2 using the data published by DEFRA. For a PHEV, its well-to-wheel CO_2 output combines emissions from both its electrical and fossil fuel energy sources.

TABLE 19.2

Example of Well-to-Wheel CO_2 Emissions for Each Fuel Type

Powertrain Type	CV	FCEV	EV
Energy Source	**Petrol**	**Hydrogen**	**Electricity**
Well-to-tank emissions/gCO$_2$ MJ^{-1}	14.1	76.9	150
Tank-to-wheel emissions/gCO$_2$ MJ^{-1}	77.6	–	–
Well-to-wheel emissions/gCO$_2$ MJ^{-1}	91.7	76.9	150
Given fuel consumption/MJ mile^{-1}	2.93	1.46	0.73
Well-to-wheel emissions/gCO$_2$ mile^{-1}	267	112	110
Well-to-wheel emissions/gCO$_2$ km^{-1}	167	70	68

Source: Adapted from G. J. Offer et al. *Energy Policy*, 39, 2011: 1939.

The purpose of this example is to clarify distinctions and significance of the well-to-tank, tank-to-wheel, and well-to-wheel CO_2 emissions.

It ought to be mentioned that this estimate does not include the emissions from construction and decommissioning of the infrastructure that is used to create and process the fuel; the emissions that result from commissioning and decommissioning of the electrical power plant, transmission lines, and charging station; or the manufacture and end-of-life disposal of the powertrain components in a vehicle.

19.3 Powertrain Simulation Methods

To evaluate the performance of the powertrain architectures, modeling and simulation tools are indispensable. This is particularly true as prototyping and testing each design combination is cumbersome, expensive, and time consuming. New hybrid powertrain configurations and controllers are also continuously being developed, so the ability to simulate a powertrain before prototyping is important.

Simulating a vehicle powertrain requires dedicated simulation software. Considerable research has been carried out by Argonne Labs and the National Renewable Energy Laboratory (NREL) in the area of powertrain simulation and optimization. Advanced powertrain researches from these two institutes have resulted in the creation of two simulation tools, respectively PSAT (PNGV System Analysis Toolkit) and ADVISOR (advanced vehicle simulator).

Both tools are frequently cited in the literature in connection with system-level powertrain simulation and optimization. Other powertrain simulation tools such as AVL Boost and GT-Suite are also available; however, they have been noted to be more suited for simulation of detailed ICE attributes at the expense of greater computational time.

There are fundamental differences in the approach used by PSAT and ADVISOR; the former uses forward-facing models, while the latter uses a hybrid approach. In ADVISOR, the models are primarily backward-facing, with forward-facing methods only active when component performance limits are encountered; when they are not, ADVISOR operates strictly as a backward-facing model. Another proponent of the backward-facing modeling method is the QSS toolbox, developed at ETH Zurich by a team led by Lino Guzzella.[8] Unlike ADVISOR, the QSS toolbox is backward-facing only.

Table 19.3 shows a list of reviewed publications that have used these simulation tools for powertrain size optimization. This table summarizes the simulation tools employed and the types of powertrains that were analyzed. Each simulation tool has a modular approach to powertrain modeling and, therefore, provides the flexibility of simulating a wide variety of topologies.

The development of PSAT was backed by the U.S. government for the Partnership for a New Generation of Vehicles (PNGV) initiative. This initiative included a comprehensive forward-facing HEV simulation environment developed by a consortium of three U.S. automotive manufacturers: Ford, GM, and Daimler-Chrysler. One fundamental strength of PSAT is the fact that it features modular implementation of powertrain components within a powertrain architecture. This provides the flexibility to scale the powertrain components as well as replace the models with different model blocks (such as proprietary blocks) if the need arises. This was made possible by strong reference to "power bonds" as seen in bond graph modeling techniques.

TABLE 19.3

Review of Utilization of Powertrain Simulation Software in Literature

	References								
	W. Gao and S. K. Porandla[11]	J. P. Ribau et al.[12]	A. F. Burke and M. Miller[13]	D. Karbowski et al.[14]	X. Wu et al.[15]	L. Guzzella and A. Amstutz[16]	G. Delagrammatikas and D. Assanis[17]	A. Rousseau et al.[18]	T. Hofman et al.[19]
ADVISOR	*	*	*		*		*		*
PSAT				*				*	
QSS toolbox						*			*
Single source						*	*		
Series hybrid		*	*			*		*	*
Parallel hybrid	*		*	*	*				*
Compound hybrid			*						*
CV						*	*		
HEV/PHEV	*		*	*	*	*		*	*
HEV–Fuel cell		*						*	

In comparison, ADVISOR is a hybrid vehicle simulator that incorporates both forward-facing and backward-facing methods. ADVISOR compares the required values (backward-facing results) with achievable values (forward-facing results). Nevertheless, this approach requires the definition of two models for each powertrain component, leading to larger programming overheads for introducing new components. Finally, the QSS toolbox is a fully backward-facing model and can utilize a relatively larger time step, generally on the order of 1 s.

Similar to PSAT, both ADVISOR and the QSS toolbox also follow a modular approach. Users can alter both the model inside the block as well as the MATLAB m-files associated with the block to suit their modeling needs. For example, the user may need a more precise model for the electric motor subsystem. A different model can replace the existing model if the inputs and the outputs are the same. On the other hand, the user may leave the model intact and only change the MATLAB m-file associated with the block diagram. This is akin to choosing a different manufacturer of the same powertrain component. Therefore, all three software packages provide modeling flexibility for the user.

In addition to being modular, the powertrain components within ADVISOR are scalable. This was achieved by including routines that allow variation of component size through scaling of maps. ADVISOR also follows an open-source model and thus receives support from the industry and academia to validate and improve the model database. The key similarity in all three simulation tools is that they use MATLAB and Simulink as their underpinnings to run the simulations. However, though modular, the powertrain architecture in all three simulations tools was fixed during simulation. As a result, trying to compare the results from optimizing different powertrain architectures, such as a pure EV with a PHEV, would require running two sets of optimizations separately. This is because the structure of the powertrain architecture is fixed during the optimization run. For example, ADVISOR requires reconfiguration when comparing series and parallel hybrid, as shown by Same et al.[9] for optimizing a formula student vehicle.

19.4 Modular Powertrain Structure

A modular powertrain structure (MPS) was developed in[10] to facilitate the switching of powertrain topologies and scalable powertrain components during optimization runtime. The high-level structure of this framework is shown in Figure 19.3.

19.4.1 Framework of Proposed Toolbox

All powertrain architectures have at least one type of energy storage and energy converter. Using the powertrain components discussed earlier, energy storages and energy converters can be organized within the MPS based on the arrangement shown in Figure 19.3. The optimizer, interfaced with the MPS, selects and sizes the powertrain components. The MPS also holds information on the various configurations of the powertrain architectures and topologies.

19.4.2 Modular Powertrain Structure

The key feature of the MPS is the layout of the powertrain component placeholders, which can be seen in the high-level block diagram in Figure 19.4. This feature allows permutations

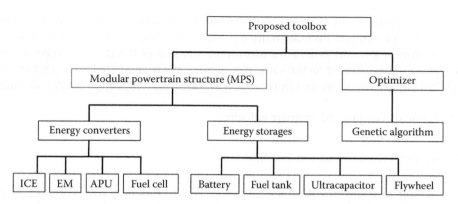

FIGURE 19.3
Structure framework within proposed toolbox.

of different energy converters and energy storages. This was made possible by making a clear distinction between energy storage, energy converter, and power transformer.

Several powertrain topologies can be created based on the available powertrain components that were discussed earlier. These topologies are shown in Table 19.4. The last row of this table shows that each powertrain topology is assigned a "powertrain variant," which is an integer used by the optimizer to select the appropriate powertrain during the optimization runtime.

However, it should be mentioned that the layout of the MPS allows the implementation of many more types of powertrain components (and thus creating a larger selection of

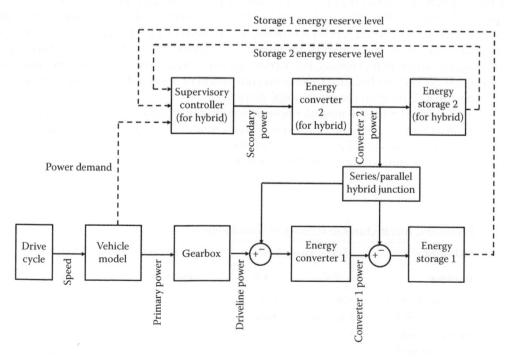

FIGURE 19.4
High-level block diagram for layout of powertrain component placeholders in MPS.

powertrain topologies), as long as the power "bond" between each powertrain component is observed. The power bond is essentially the communication line between the energy storage, energy converter, and power transformer. Each power bond is modeled in terms of effort and flow. This is based on the same distinctions as are used for bond graph modeling.

The following are examples of efforts and flows for their respective energy domains:

- Mechanical: Torque and angular velocity,
- Electrical: Voltage and current,
- Hydraulic: Pressure and volume flow rate.

The switching mechanism within the framework is facilitated by the "variant subsystem" feature of Simulink. Each energy storage and energy converter subsystem acts as a placeholder that contains a library of components. When the optimizer assigns a powertrain variant to the MPS, the corresponding energy converters and energy storages are populated. Depending on the powertrain configuration selected by the optimizer, the variant subsystem selects the correct energy converter and energy storage, as defined by the corresponding powertrain variant shown in Table 19.4.

For example, when the optimizer selects "Variant 1" as the desired powertrain topology, the corresponding energy converter and energy storage are the ICE and fuel tank. Similarly, if the optimizer selects "Variant 2" as the desired powertrain topology, the corresponding energy converter and energy storage are EM and battery. This is shown in Figures 19.5 and 19.6, respectively. In both these nonhybrid topologies, only the primary energy converter and storage are activated; the secondary energy converter and storage are not applicable and subsequently grayed out. The placeholders for the switching components are outlined in red, and the differences in the components between Variants 1 and 2 can be seen here.

Variants 3 and 4 are the series and parallel hybrid architectures, shown in Figures 19.7 and 19.8, respectively. In Variant 3, both sets of energy converters and storages are enabled. The switchable component placeholders are once again highlighted in red to aid clarity when comparing the different topologies. For Variant 4, however, the secondary energy converter is once again grayed out because there is no conversion of energy domain between the flywheel and the ICE (both in the mechanical domain).

Another key enabler for the modular powertrain structure is the power split junction box, which can be seen located between the primary and secondary energy converters and storages. The purpose of this junction box is to regulate the power flow between these two systems, depending on the prevailing hybrid powertrain type. For example, in a series hybrid

TABLE 19.4

Powertrain Architectures to Investigate Model Framework

Architecture	Single Power Source		Series Hybrid	Parallel Hybrid
Topology	CV	EV	PHEV	MHV
Energy converter 1	ICE	EM	EM	ICE
Energy storage 1	Fuel tank	Battery	Battery	Fuel tank
Energy converter 2	–	–	APU	–
Energy storage 2	–	–	Fuel tank	Flywheel
Powertrain variant	1	2	3	4

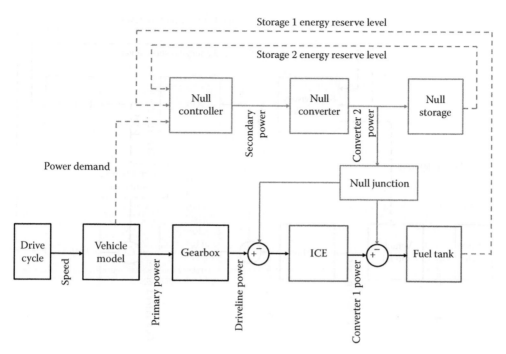

FIGURE 19.5
MPS layout switched to a CV (Variant 1).

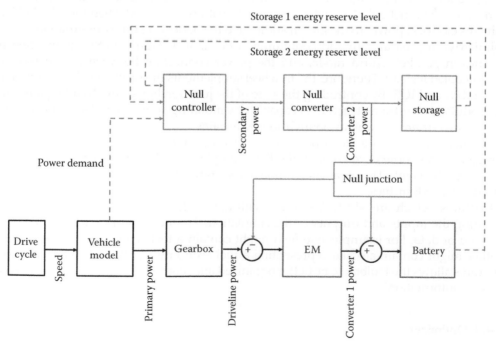

FIGURE 19.6
MPS layout switched to an EV (Variant 2).

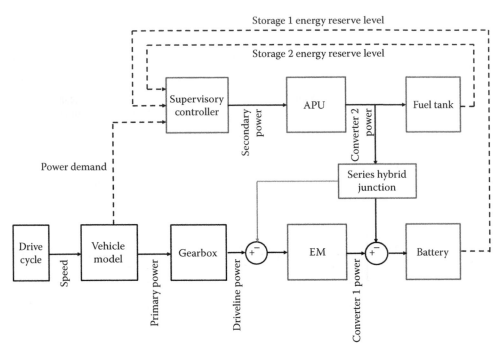

FIGURE 19.7
MPS layout switched to a series PHEV (Variant 3).

configuration (Variant 3, Figure 19.7), the junction box switches in series mode, and the power connection between the electric motor (EM) and the battery is enabled. Therefore, the APU supplements electrical power to the EM, in addition to the battery, in this configuration.

Similarly, in a parallel hybrid configuration (Variant 4, Figure 19.8), the junction box switches in parallel hybrid mode, and the power connection between the gearbox and ICE is enabled instead. Therefore, the flywheel supplements the mechanical power between the gearbox and ICE. By controlling the state of the junction box or by disabling it entirely, the modular powertrain structure is capable of simulating series, parallel, and compound hybrid powertrains, as well as nonhybrid powertrains.

The modular structure described so far ensures that the energy domains between storage and converters are compatible. For example, the EM only connects to a battery and not to a fuel tank. The parameterization of each powertrain component is stored in individual MATLAB m-files.

Structures, which are MATLAB arrays with data fields, are used to store information regarding the inputs and outputs of the simulation. The fields of a structure can contain any kind of data. For example, one field might contain a text string representing a name, another might contain a scalar representing a fuel economy result, and so on. The use of structures allows the toolbox to be better organized and, consequently, provides convenient access to information.

19.4.3 Optimizer

Discontinuities and nonlinarites are present in this optimization problem because of switching between different powertrain topologies. For instance, when considering an

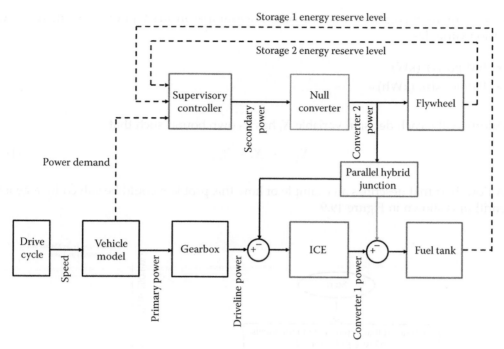

FIGURE 19.8
MPS layout switched to a parallel MHV (Variant 4).

optimal size between a conventional powertrain and an EV powertrain with regard to minimization of well-to-wheel CO_2 for a given drive cycle, there are two possible solutions or minima. A gradient-based method is less suitable for handling such problems. Evolutionary computing-based optimization techniques such as genetic algorithms, swarming theory, or simulated annealing are commonly used to solve such complex and discontinuous engineering problems.

19.5 Optimization Problem

The optimization routine can be represented in a standard form. Given a set of decision variables X and a cost function $\phi(X)$, the optimizer aims to find X^* that minimizes $\phi(X)$ within bounds of G, where G represents a set of design constraints. This can be represented as

$$\min_{X_i, G} \phi(X) \tag{19.1}$$

where X_i contains all the decision. Regardless of which optimization algorithm is used, the decision variables can be collected in vector form as

$$X_i \equiv \begin{bmatrix} X_1 \\ X_2 \\ \vdots \\ X_n \end{bmatrix}, \tag{19.2}$$

where "i" represents each decision variable. For example, in an EV powertrain, there are two decision variables:

- EM power (kW)
- Battery size (kWh).

Additionally, each decision variable X_i has its own bound such that

$$X_{i_{lower}} < X_i < X_{i_{upper}}. \tag{19.3}$$

A flowchart that describes an example of how this problem could be solved by a genetic algorithm is shown in Figure 19.9.

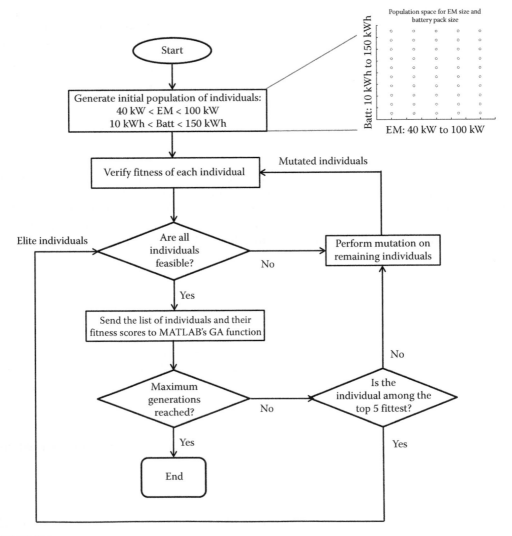

FIGURE 19.9
Flowchart of implementation of a genetic algorithm for powertrain sizing optimization.

19.5.1 Extending Optimizer to Support Multiple Powertrain Topologies

To extend the optimizer to support more than one type of powertrain topology, the representation of the decision variables has to be more generic. Therefore, instead of specifying, for instance, "EM power" and "battery size," the decision variables are now "energy converter size" and "energy storage size," respectively. This allows the inclusion of powertrain topologies that do not contain an EM or a battery, such as a CV powertrain.

Additionally, a third decision variable needs to be added to the individual: the "powertrain variant." Hence, the new set of decision variables is as follows:

- Energy converter size (normalized into an integer scale from 1 to 100),
- Energy storage size (normalized into an integer scale from 1 to 100),
- Powertrain variant (discrete selection of 1 to 4, as shown in Table 19.4).

Consequently, each individual X now contains three variables, which can be denoted by

$$X = [\text{energy converter size, energy storage size, powertrain variant}]. \quad (19.4)$$

By normalizing the energy converter and energy storage sizes into a discrete scale from 1 to 100, the optimizer is able to handle different types of powertrain topologies during a single optimization routine, while preserving the scaling limits of the associated powertrain components. An example of the normalization of the powertrain components are shown in Table 19.5.

The value of the decision variable "powertrain variant" determines the topology of the powertrain during the optimization runtime. This interfaces with the MPS, which switches to the appropriate powertrain topology according to this value. Hence, by adding the powertrain variant as a decision variable within the individual (in addition to the components sizes), both the powertrain topology and powertrain components can be optimized simultaneously to minimize a given cost function. This addresses the problem of simultaneously selecting a powertrain topology and optimize its component sizes for a given cost function.

19.5.2 Multiobjective Optimization

Unlike single-objective optimization (i.e., single cost function), it is possible to have more than one optimal solution in a multiobjective optimization routine. This is true if the solution to the objectives is in conflict. Therefore, improving one objective (i.e., minimizing one of the cost functions) may deteriorate another. A balance in trade-off solutions is achieved when a solution cannot improve any objective without deteriorating one or more of the other

TABLE 19.5

Normalization of the Scalable Powertrain Component Sizes for the Optimizer

Powertrain Component	Dimension	Minimum Size (Scaled to 1)	Maximum Size (Scaled to 100)
ICE	Displacement	0.5 L	3 L
EM	Power	40 kW	100 kW
Battery	Capacity	10 kWh	150 kWh
APU	Power	10 kW	100 kW
Fuel Tank	Capacity	10 L	100 L

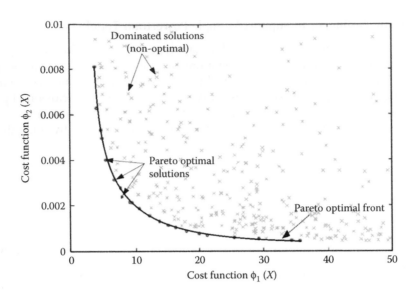

FIGURE 19.10
Pareto solution example.

objectives. These solutions are called Pareto optimal. When plotting these solutions, a Pareto optimal curve is created, as shown in Figure 19.10. Identifying the solutions lying on or near the Pareto-optimal front makes it possible to, for example, find the trade-offs between cost and CO_2 emissions.

19.6 Case Studies: Optimization of Powertrain Topology and Component Sizing

Table 19.6 lists the combination of powertrain architectures that will be considered in these case studies, along with the associated energy converters and energy storages.

The use of different drive cycles gives an opportunity to understand the variation in topology selection and component sizing resulting from differing power and energy demands from each drive cycle. For example, the NEDC, being a modal cycle, has lower power

TABLE 19.6

Powertrain Topologies Investigated

Architecture	Single Power Source		Series Hybrid	Parallel Hybrid
Topology	CV	EV	PHEV	MHV
Energy converter 1	ICE	EM	EM	ICE
Energy storage 1	Fuel tank	Battery	Battery	Fuel tank
Energy converter 2	–	–	APU	–
Energy storage 2	–	–	Fuel tank	Flywheel
Powertrain variant	1	2	3	4

TABLE 19.7

Normalization of Scalable Powertrain Component Sizes for Optimizer

Powertrain Component	Dimension	Minimum Size (Scaled to 1)	Maximum Size (Scaled to 100)
ICE	Displacement	0.5 L	3 L
EM	Power	10 kW	100 kW
Battery	Capacity	8 kWh	80 kWh
APU	Power	10 kW	100 kW
Fuel Tank	Volume	10 L	100 L
Flywheel	Energy	100 kJ	600 kJ

demands than the ARTEMIS cycle and, therefore, results in comparatively smaller powertrain component sizes (for a given topology). Here, only the ARTEMIS cycle is used as a case study. Given the vehicle parameters and drive cycle, the toolbox evaluates the powertrain topologies and creates Pareto fronts to show the trade-offs between cost functions. This could provide insights to manufacturers and practitioners into the influence of powertrain design targets, such as emissions and powertrain costs.

The normalizations of the powertrain components are shown in Table 19.7.

19.6.1 Case Study 1: Tank-to-Wheel versus Well-to-Wheel CO_2

The definitions of the cost functions that are evaluated in this case study are shown in Table 19.8. The powertrain optimization was carried out for each increment of the ARTEMIS cycle, and the results were recorded at the end of each increment. The scope of this procedure is to simultaneously optimize the powertrain sizing and architecture selection while meeting the power and energy demands of the drive cycle.

Three types of optimization routines were carried out in this case study:

- Single-objective optimization for well-to-wheel CO_2 emissions,
- Single-objective optimization for tank-to-wheel CO_2 emissions,
- Multiobjective optimization combining both the aforementioned cost functions.

TABLE 19.8

Decision Variables (X) and Constraints (G)

Term	Definition	Units
$\phi_1(X)$	Tank-to-wheel CO_2	kg
$\phi_2(X)$	Well-to-wheel CO_2	kg
X_1	$1 \leq$ Energy converter $1 \leq 100$	–
X_2	$1 \leq$ Energy storage $1 \leq 100$	–
X_3	$1 \leq$ Energy converter $2 \leq 100$	–
X_4	$1 \leq$ Energy storage $2 \leq 100$	–
X_5	$1 \leq$ Powertrain variant ≤ 4	–
G_1	$1000 \leq$ total vehicle mass ≤ 1600	kg
G_2	Drive cycle speed constraints	m/s

For multiobjective optimization, the cost function ϕ_{multi} is formulated as a weighted sum of the two individual costs as

$$\phi_{multi} = \beta\phi_1 + (1 - \beta)\phi_2, \tag{19.5}$$

where parameter β is used to weigh the two costs for a given

$$\beta \in [0, 1]. \tag{19.6}$$

19.6.1.1 Lowest Well-to-Wheel CO_2 Emissions

Figure 19.11 shows the results for optimizing the powertrains for the lowest well-to-wheel CO_2 emissions. The optimization was carried out for each increment of the ARTEMIS drive cycle. Based on Figure 19.11, an EV powertrain was selected by the optimizer for the first two increments of the drive cycle (autonomy range of up to 150 km).

This was because the EV powertrain achieved the lowest well-to-wheel CO_2/km when compared to the CV, PHEV, and MHV powertrains. The transition from EV to PHEV then occurred when the vehicle mass limit of 1600 kg was reached or exceeded. The EV powertrain could no longer support a battery large enough to cover the necessary range within this mass limit. This is observed in Figure 19.12, which shows the total vehicle mass of the corresponding powertrain and the range for which it was sized.

To cater to an autonomy range of over 150 km in the ARTEMIS cycle, a vehicle with an EV powertrain would exceed the mass limit of 1600 kg. Additionally, the rate of increase in vehicle mass as a function of the autonomy range is steeper for the EV when compared to the other powertrains. This is because the energy density per unit mass of batteries, as discussed previously, is smaller than that of fossil fuels by two orders of magnitude.

Another transition occurred between the fourth and fifth increments of the drive cycle, where the MHV is selected instead of the PHEV. In this scenario, it was more efficient to use the lighter mechanical hybrid for the longer travel range instead of the heavier

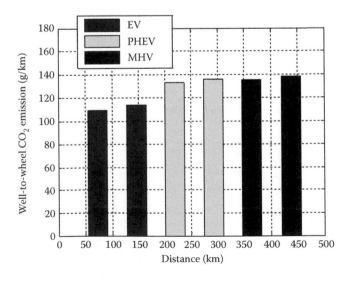

FIGURE 19.11
Well-to-wheel emission.

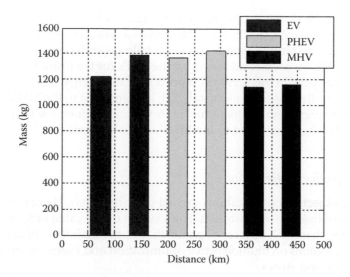

FIGURE 19.12
Total vehicle mass for the respective architecture selected by the optimizer.

plug-in hybrid. Since the well-to-wheel CO_2 encapsulates the total energy used by the vehicle to cover the drive cycle, the heavier PHEV would have emitted higher well-to-wheel CO_2 than the comparatively lighter MHV.

19.6.1.2 Lowest Tank-to-Wheel CO₂ Emission

Figure 19.13 shows the results for optimizing the powertrains for lowest tank-to-wheel CO_2 emissions instead.

Comparison of Figures 19.13 and 19.11 reveals a couple of similarities and differences with regard to the selection of powertrain architecture between the two cost functions. Although

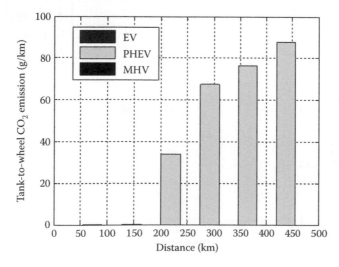

FIGURE 19.13
Tank-to-wheel emission.

not visible in Figure 19.13, the optimizer has once again selected the EV powertrain for the first two increments of the ARTEMIS cycle. This was not reflected in Figure 19.13 because the EV powertrain emits no tank-to-wheel CO_2 (zero tailpipe emissions).

In contrast, the PHEV was selected instead of the MHV for the fifth and sixth increments of the ARTEMIS drive cycle. In this scenario, a portion of the energy used to propel the PHEV was sourced "externally" from the electrical grid, as compared to the MHV, which only has a single source of energy from the fossil fuel in its tank. Therefore, the PHEV emitted lower tank-to-wheel CO_2 when compared to the MHV in this scenario.

19.6.1.3 Multiobjective Optimization

Unlike the single-objective optimizations carried out previously, there is no single optimal solution for a multiobjective optimization. For a sweep of the value of β, as shown in Equations 19.5 and 19.6, there will be a set of points that fit the definition of an "optimum" solution. Further discussions on multiobjective optimization and Pareto fronts can be found in the work by Marler and Arora.[20]

In this investigation, the multiobjective optimization produced Pareto fronts for well-to-wheel CO_2 versus tank-to-wheel CO_2. To illustrate the workings of the optimizer, the Pareto fronts from a single increment and from quadruple increments of the ARTEMIS drive cycle are shown in Figures 19.14 and 19.15, respectively. In each of these figures, five particular Pareto-optimal points were shown for clarity. These five points correspond to

$$\beta = [0, \ 0.25, \ 0.5, \ 0.75, \ 1]. \tag{19.7}$$

As a test, based on Figure 19.14, there was no Pareto front formed after a multi-objective optimization was carried out on a single increment of the ARTEMIS cycle. This was because the resultant tank-to-wheel CO_2 of the selected powertrain (EV) was always zero, regardless

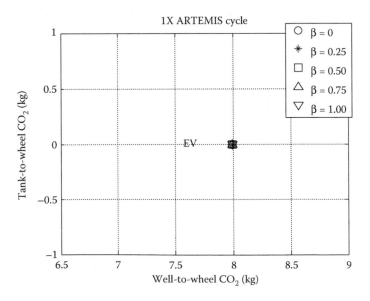

FIGURE 19.14
Pareto front for single increment of ARTEMIS drive cycle.

of the ratio between the two cost functions. This was also reflected in Figure 19.13, where the EV powertrain emitted zero tailpipe emissions.

As a result, the simulation has simply "reverted" to a single-objective optimization because the tank-to-wheel emission was always equal to zero, and, therefore, the optimizer only optimized the powertrain for well-to-wheel CO_2, thereby producing the same results over the entire sweep of β. The optimized EV powertrain produced the same well-to-wheel CO_2 output regardless of the value of β.

In contrast, Figure 19.15 shows the Pareto front for four increments of the ARTEMIS cycle (around 400 km of autonomy range). As discussed previously, the EV powertrain was not selected for higher increments of the ARTEMIS cycle because the vehicle would exceed the mass limit of 1600 kg. The remaining powertrain options (CV, PHEV, and MHV) produce both tank-to-wheel and well-to-wheel CO_2 emissions and, therefore, generated a Pareto front across different values of β, as observed in Figure 19.15.

Figure 19.16 combines the Pareto fronts from all six increments of the ARTEMIS cycle. The numbers in circles denote the Pareto front for each specific drive cycle increment. As observed from the earlier single-objective optimizations, the EV powertrain was selected for the first two increments of the ARTEMIS cycle regardless of the cost function because it produced the lowest tank-to-wheel and well-to-wheel emissions. This is also reflected in Figure 19.16.

For increments 3–6 of the ARTEMIS cycle, the powertrain selection is dominated by the PHEV. However, as the ratio of the optimization favors well-to-wheel CO_2 (i.e., as β approaches zero), the MHV is selected instead of the PHEV. This relates back to the transition seen in Figure 19.11. It is also noteworthy that the ratio between the selection of the MHV over the PHEV rises as the autonomy range in which the vehicle is optimized for (i.e., drive cycle increments) increases. The shaded regions in between each drive cycle increment are the interpolation of the powertrain topology selected in each Pareto front.

The use of this methodology lends itself naturally to identifying the transitions between powertrain topologies. Practitioners intending to use it can generate such comparisons by

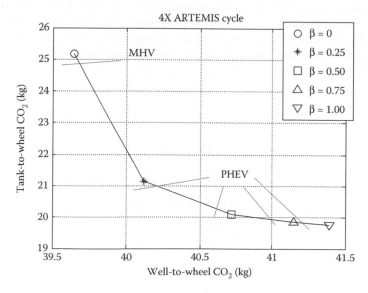

FIGURE 19.15
Pareto front for quadruple increments of ARTEMIS drive cycle.

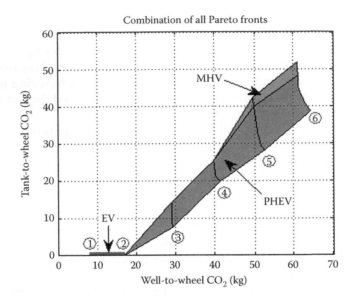

FIGURE 19.16
Sweep of Pareto fronts and powertrain selection for multiple increments of ARTEMIS cycle (numbers in circles denote cycle increment).

defining their own cost function(s), along with the relevant vehicle parameters, powertrain components, and drive cycle.

19.6.2 Case Study 2: Powertrain Cost versus Well-to-Wheel CO_2

In this case study, the trade-offs between powertrain cost and well-to-wheel CO_2 are investigated using the same powertrain topologies shown in Table 19.6.

The optimization parameters for this case study are shown in Table 19.9. Based on these parameters, the multiobjective optimization produced Pareto fronts, as shown in Figure 19.17. The results shown in Figure 19.17 are for a single increment of the ARTEMIS cycle.

Among the powertrain topologies considered in this case study, the EV arguably has the highest powertrain cost relative to its autonomy range. Therefore, it returned the lowest

TABLE 19.9

Decision Variables (X) and Constraints (G)

Term	Definition	Units
$\phi_1(X)$	Estimated powertrain cost	USD
$\phi_2(X)$	Well-to-wheel CO_2	kg
X_1	$1 \leq$ Energy converter $1 \leq 100$	–
X_2	$1 \leq$ Energy storage $1 \leq 100$	–
X_3	$1 \leq$ Energy converter $2 \leq 100$	–
X_4	$1 \leq$ Energy storage $2 \leq 100$	–
X_5	$1 \leq$ Powertrain variant ≤ 4	–
G_1	$1000 \leq$ Total vehicle mass ≤ 1600	kg
G_2	Drive cycle speed constraints	m/s

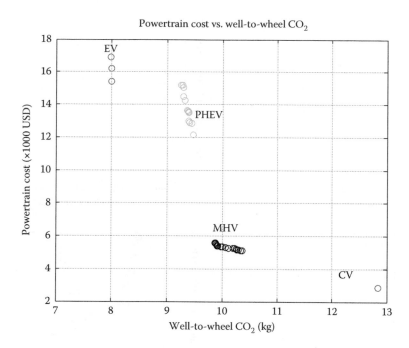

FIGURE 19.17
Pareto front for single increment of ARTEMIS drive cycle.

well-to-wheel CO_2 emissions but at the expense of higher powertrain costs. Conversely, the CV produced the highest emissions but with the lowest powertrain costs. It is also observed that the CV does not have a Pareto front. Based on the Willans ICE model used in this example, a more expensive ICE results in a larger displacement and hence produces more emissions. Therefore, there is no trade-off as such when optimizing the CV for those two cost functions, resulting in only one solution point instead of a Pareto front.

The Pareto fronts for each powertrain in Figure 19.17 are shown in Figure 19.18.

An EV powertrain with a smaller battery (and thus lower powertrain cost) produces more well-to-wheel CO_2 for a single increment of the ARTEMIS because of the higher power losses within the battery (from higher current rates) compared to a larger battery. In contrast, an EV with a larger battery (and thus more expensive), produces lower well-to-wheel CO_2 emissions for a single increment of the ARTEMIS cycle, which supports the Pareto front seen in Figure 19.18b.

Additionally, it is observed that there is a bigger spread of solution points on the Pareto fronts for the hybrid powertrains (both the PHEV and MHV). This is inherent from the larger solution space offered by the hybrid powertrains due to the greater permutation of component size combinations gained from having more energy converters and storages when compared to the single-source powertrains (CV and EV).

By combining several increments of the ARTEMIS cycle, a three-dimensional Pareto plot can be created, as shown in Figure 19.19. It is observed that in the first two increments of the cycle, all four powertrain topologies are present on the Pareto front. However, the PHEV powertrain was eliminated starting with the third increment of the cycle. The reason for this was that, to achieve a higher autonomy range, the PHEV must have either a larger battery, where it approaches the cost of the EV powertrain, or a larger fuel tank, where it

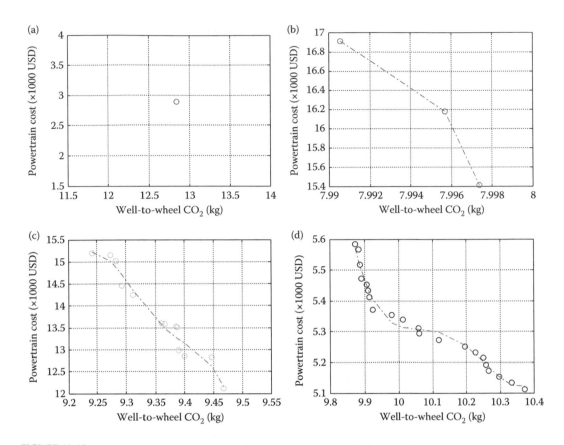

FIGURE 19.18
Pareto fronts for individual powertrain topologies: (a) CV; (b) EV; (c) PHEV; and (d) MHV.

approaches the emissions of the MHV (because of its comparatively higher mass coupled with lower efficiency when running in charge-sustaining mode).

Hence, the region on the Pareto front that was once populated by the PHEV is now over-lapped by the EV and the MHV, thus eliminating the need for a PHEV altogether. Although this may seem counterintuitive, it should be stressed that the emission metric used in this case study is well-to-wheel CO_2; it is envisaged that the results would favor a PHEV over an MHV if tank-to-wheel CO_2 was used instead, such as indicated in Figure 19.13.

After four increments of the cycle, the EV powertrain also gets eliminated because its pos-sible solution points would exceed the vehicle mass constraints of 1600 kg in order to achieve the autonomy range on a single charge. This leaves the MHV and CV powertrains on the Pareto front for the fifth and sixth increments of the cycle.

The creation of a three-dimensional Pareto plot such as in Figure 19.19 is a natural output of the proposed toolbox. If the same results were replicated using the existing optimization methods seen in the literature, it is envisaged that it would take significantly longer because each powertrain topology would have to be optimized separately to create the individual Pareto fronts. Additionally, some of the optimization routines would be computationally wasteful because the particular powertrain topology may no longer be cost-effective for the levels of CO_2 produced, such as the PHEV in this case study, which was eliminated from the third increment of the ARTEMIS cycle.

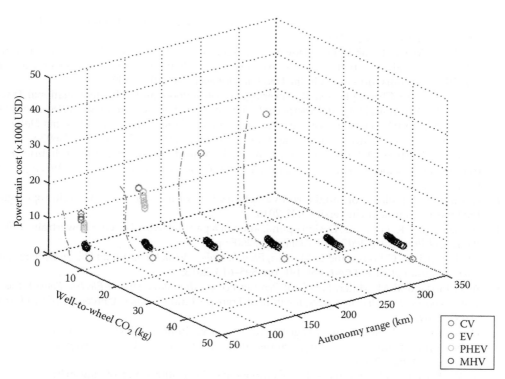

FIGURE 19.19

Combination of Pareto fronts for all increments of ARTEMIS cycle.

Additionally, with the increase in the autonomy range, there is a divergence observed among powertrain architectures, particularly between the EV and MHV. The cost of the EV is highly sensitive to its range, with an increment of approximately $171 per kilometer of autonomy. However, the cost increment of the MHV and CV remain relatively constant, as the cost of extra fossil fuel to cover additional autonomy is insignificant. The reason for the EV's high sensitivity is due to the battery costs. For the PHEV, a mixture of this effect can be seen when comparing its spread of powertrain cost between the first and second increment of the ARTEMIS cycle.

Bibliography

1. G. Rizzoni, L. Guzzella, and B. M. Baumann, Unified modeling of hybrid electric vehicle drivetrains, *Mechatronics, IEEE/ASME Transactions on*, 4(3), 1999: 246–257.
2. K. B. Wipke, M. R. Cuddy, and S. D. Burch, ADVISOR 2.1: A user-friendly advanced power-train simulation using a combined backward/forward approach, *Vehicular Technology, IEEE Transactions on*, 48(6), 1999: 1751–1761.
3. D. W. Gao, C. Mi, and A. Emadi, Modeling and simulation of electric and hybrid vehicles, *Proceedings of the IEEE*, 95(4), 2007: 729–745.
4. G. Mohan, A toolbox for multi-objective optimisation of low carbon powertrain topologies, PhD thesis, Cranfield University, 2016.
5. Regulation, E. 443/2009 of the European Parliament and of the Council of 23 April 2009 setting emission performance standards for new passenger cars as part of the Community's integrated

approach to reduce CO2 emissions from light-duty vehicles. Official Register of the European Union, eur-lex.europa.eu/Notice.do.

6. G. J. Offer, M. Contestabile, D. A. Howey, R. Clague, and N. P. Brandon, Techno-economic and behavioural analysis of battery electric, hydrogen fuel cell and hybrid vehicles in a future sustainable road transport system in the UK, *Energy Policy*, 39, 2011: 1939.

7. T. Anderson, O. Christophersen, K. Pickering, H. Southwood, and S. Tipping, National Travel Survey 2008 Technical Report, *Department for Transport, National Centre for Social Research*, 2009.

8. L. Guzzella and A. Sciarretta, *Vehicle Propulsion Systems: Introduction to Modeling and Optimization*, Springer, 2005. https://link.springer.com/content/pdf/10.1007/978-3-540-74692-8.pdf

9. A. Same, A. Stipe, D. Grossman, and J. W. Park, A study on optimization of hybrid drive train using Advanced Vehicle Simulator (ADVISOR), *Journal of Power Sources*, 195(19), 2010: 6954–6963.

10. G. Mohan, A toolbox for multi-objective optimization of low carbon powertrain topologies, PhD thesis, Cranfield University, UK, 2016.

11. W. Gao and S. K. Porandla, Design optimization of a parallel hybrid electric powertrain, *Vehicle Power and Propulsion, 2005 IEEE Conference*, 2005: p. 6.

12. J. P. Ribau, J. M. C. Sousa, and C. M. Silva, *Plug-In Hybrid Vehicle Powertrain Design Optimization: Energy Consumption and Cost*, vol. 191, pp. 595–613, 2013.

13. A. F. Burke and M. Miller, *Simulated Performance of Alternative Hybrid-Electric Powertrains in Vehicles on Various Driving Cycles*, Institute of Transportation Studies, University of California, Davis, 2009.

14. D. Karbowski, A. Rousseau, S. Pagerit, and P. Sharer, Plug-in vehicle control strategy: from global optimization to real time application, *22nd Electric Vehicle Symposium, EVS22*, Yokohama, Japan, 2006.

15. X. Wu, B. Cao, X. Li, J. Xu, and X. Ren, Component sizing optimization of plug-in hybrid electric vehicles, *Applied Energy*, 88(3), 2011: 799–804.

16. L. Guzzella, and A. Amstutz, CAE tools for quasi-static modeling and optimization of hybrid powertrains, *Vehicular Technology, IEEE Transactions on*, 48(6), 1999: 1762–1769.

17. G. Delagrammatikas and D. Assanis, Development of a neural network model of an advanced, turbocharged diesel engine for use in vehicle-level optimization studies, *Proceedings of the Institution of Mechanical Engineers, Part D: Journal of Automobile Engineering*, 218(5), 2004: 521–533.

18. A. Rousseau, S. Pagerit, and D. Gao, Plug-in hybrid electric vehicle control strategy parameter optimization, *Argonne National Laboratory*, 2007.

19. T. Hofman, M. Steinbuch, R. van Druten, and A. Serrarens, Parametric modeling of components for selection and specification of hybrid vehicle drivetrains, *WEVA J*, 1(1), 2007: 215–224.

20. R. T. Marler and J. S. Arora, Survey of multi-objective optimization methods for engineering, *Structural and Multidisciplinary Optimization*, 26(6), 2004: 369–395. https://link.springer.com/article/10.1007/s00158-003-0368-6.

20

User Guide for Multiobjective Optimization Toolbox

20.1 About the Software

With the modeling techniques described in the previous chapter and plenty of power train components described in this book, the problem of finding an optimal power train topology emerges. The software described below is a MATLAB®/Simulink®-based engine, which solves the problem of finding the optimal, in terms of a predefined cost function, power train topology, and components, for a given drive cycle. It employs a genetic algorithm (GA) that generates populations of diverse power trains and evaluates them using a backward-facing vehicle model as described in the previous chapter. Software (©2018 by Ganesh Mohan, Francis Assadian, Marcin Stryszowski, and Stefano Longo) designed to accompany the material in this bookwill be hosted on the book's CRC Press website: www.crcpress.com/9781498761772.

20.2 Software Structure

The software is organized into modules, structured in a cascade of loops as presented in Figure 20.1 below.

20.2.1 Input Sheet

The input sheet is a file called runGrid.m. This file gathers the input information necessary to run the simulation. A user can select:

- The type of drive cycle, choosing among steady-state cruising, New European Driving Cycle (NEDC), Artemis, a "Cranfield cycle" (created with real data collected in the UK).
- The desired cost function, that is, choosing whether the solution should be optimized in terms of well-to-wheel or tank-to-wheel CO_2 emission, mass, financial cost, or mass of the primary power source (PPS).
- The mass and cost limit. It is important that if a vehicle is unable to follow the velocity profile (i.e., to achieve satisfactory accelerations caused by, for example, high mass or insufficiently powerful engine), it will be ruled out as infeasible.
- The type of power train to be considered. Possible choices are the conventional internal combustion (IC) engine, mechanical hybrid and configurations of an electric vehicle (EV), and plug-in hybrid electric vehicle (PHEV).

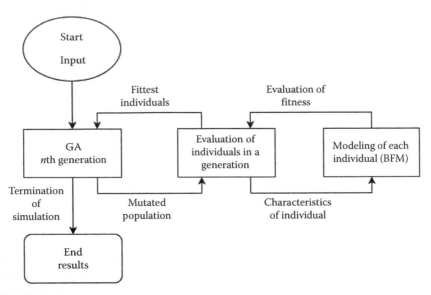

FIGURE 20.1
Structure of optimization software.

20.2.2 Genetic Algorithm

The GA is based on MATLAB's existing toolbox, but it was fine-tuned to improve the convergence rate and stability. In each generation, the algorithm creates a population of feasible solutions, evaluates their fitness, selects the "elite" ones, and uses them to serve as parents for the mutated population in the next generation. See the flowchart in Figure 20.1.

20.2.3 Fitness Evaluation Algorithm

Each solution in the current generation is simulated for its fitness (cost) to be calculated. For example, when minimum well-to-wheel CO_2 emission is the desired criterion, the individuals receive a score based on the calculated (simulated) emissions. Then, the top five individuals are moved to the next generation unchanged to accelerate the calculation, while the remaining undergo mutation.

20.2.4 Simulation of Vehicle Configurations

To model the various configurations of topologies and sizes of power trains, the modular power train structure (MPS), described in the previous chapter, is used. If a Simulink module block is not used within a given solution, a null function is used instead. A backward-facing model is employed to simulate the conditions encountered during the drive cycle and to calculate the performance of the vehicle configuration. The simulation is quasi-static with a 1-s time interval during which steady state is assumed.

20.2.5 Component Models Available

For the simulations, the placeholders in the MPS are replaced by the appropriate models. The models used are a trade-off between speed of calculation and precision of results. Models must be scalable to allow their sizes to be adjusted throughout the optimization routine. The available models are as follows.

- Energy converters:
  - *Conventional engine*: This is modeled by a set of linear equations, with efficiency depending on piston pressure and mean piston velocity, taking torque and speed as input and returning fuel mass flux as output. The engine is scalable between 0.5 and 3.0 L. When used in a hybrid architecture as an *auxiliary power unit* (APU), it is scaled between 10 and 100 kW.
  - *The electric motor* model was built on measurements on a Smart ED vehicle and is scalable between 40 and 100 kW.
- Energy storage:
  - *A fuel tank* stores the fuel. Its model integrates the fuel mass flux.
  - *Battery*: This is a simple electric energy model with voltage and resistance components. It is scaled between 10 and 150 kWh by changing the number of cell strings.

20.2.6 Running a Simulation

Once MATLAB is launched, the user should open the runGrid.m file, where the simulation inputs, parameters, and goals of the simulation can be defined.

20.2.6.1 Definition of Drive Cycle

The first option is the selection of the drive cycle. It is possible to choose from the NEDC standard drive cycle, the Cranfield cycle, and the Artemis cycle. If the "blank" option is selected, the vehicle is modeled as idling at 0 mph (used, for example, for EV charging scenarios). The last option is the steady state, where the number selected represents the velocity of the vehicle cruising, in meters per second.

The drive cycle is chosen by defining the value of the variable dcType. In the example that follows, the Cranfield drive cycle is selected.

```
%% Drivecycle to Use
% 0 - Blank
% 1 - NEDC
% 2 - Cranfield
% 3 - Artemis
% 4 - Constant Speed where type is speed value

dcType = 2;
dcIncrement = 1;% Keeping trac of the results by incrementing them
dcLowerLimit = 1;% Lowest number of multiple drivecycle runs
dcUpperLimit = 1;% Highest number of multiple drivecycle runs
```

For more advanced simulations—which allow for larger distances, necessary, for example, to assess battery wear—there is a possibility to choose a variable distance, where the user can select lower and upper limits for drive cycle repetitions and the number of cycles in each increment.

20.2.6.2 Selection of Cost Function

The cost (or objective) function defines the goal of the optimization. The available cost functions are as follows:

- Well-to-wheel CO_2. This considers all the emissions used to propel the vehicle.

- Tank-to-wheel CO_2. This calculates the CO_2 emitted at the tailpipe of the vehicle (in case of EVs, it is zero).
- The mass of the vehicle might also be considered as an optimization goal.

For multiobjective cost functions, which are used to find Pareto optimal solutions, a weighting factor β needs to be chosen. Typical multiobjective problems are:

- WtW CO_2 and mass,
- WtW CO_2 and cost,
- Mass of battery and mass of ultracapacitor

```
%% Cost Function
% 1 - Well-to-Wheel CO2
% 2 - Tank-to-Wheel CO2
% 3 - Mass
% 4 - WtW CO2 and Mass (multi-objective)
% 5 - WtW CO2 and Cost (multi-objective)
% 6 - Mass of Battery vs Mass of UCap (multi-objective)

costFunction = 1;
```

20.2.6.3 Power Train Type Selection

Finally, the user can choose the power train topologies that are to be considered during the optimization procedure. A conventional power train is a simple IC engine with a fuel tank as energy storage. EV is a simple electric vehicle with an electric motor and a battery as energy storage. The last two choices allow one to consider the state of health (SOH) of the battery in the calculation. The battery SOH is modeled as a scaled sum of the absolute value of the battery current.

The user can select the topologies to be considered by adding the corresponding number to the array *powertrainSelection*. It is important to note that numbers should be in ascending order.

```
%% Powertrain Selection
% 1 - Conventional Powertrain
% 2 - EV
% 3 - Series PHEV
% 4 - Fuel cell + Ultracap
% 5 - Parallel Flywheel + CVT
% 6 - Battery + Ultracap
% 7 - EV (with SOH)
% 8 - Battery (with SOH) + Ultracap

powertrainSelection = [1 2 3 4]; %Please ensure selection array is in
ascending order!
```

20.2.6.4 Advanced Settings

An advanced user who wishes to modify the simulation to finer detail can do so in another m.file provided within the toolbox. What follows is a brief description of the main files available.

runModel.m—combines fuel with oxygen. Within this file, the user can define the initial variables of the vehicle, such as initial battery SOC and SOH, wheel radius, aerodynamic coefficient, electric bus voltage, and transmission ratios. Further changes to component models are possible within the Simulink module.

createGA.m—Within this file, the user can adjust the genetic algorithm termination limits, such as the time limit, stall limit, and parameters within the mutation function.

runDCRange.m—Within this file, the user can adjust the general GA settings, such as the upper limits of generations and number of elite individuals in each generation.

20.2.7 Running the Simulation

Once the simulation is started by running the file runGrid.m, the user can track the process in the MATLAB console, where information after each generation iteration is updated.

The information provided is in a table that contains the number of generations, cumulative count of computed simulations (f-count), best and mean fitness values, and numbers of stalled generations in a row. Stalled generations are symptomatic of a slow convergence rate, meaning that a convergence might have already been achieved.

```
Loading Genetic Algorithm...
Creating Initial Population...
Running Genetic Algorithm...
Checking license for GADS_Toolbox...

                          Best        Mean         Stall
Generation    f-count     f(x)        f(x)       Generations
    1          324        5.928       2903           0
    2          648        5.924       1501           0
   ...         ...         ...         ...           ...
   15         4860        5.914       6.247          7
   16         5184        5.914       6.251          8
Optimization terminated: average change in the fitness value less than
options.TolFun
```

A sample window with the progress of the GA procedure is provided in Figure 20.2, where we can see nine plots that allow us to evaluate the performance of the GA. Starting from the top left of the figure, the plots show:

Top row:

- The first plot presents the mean and best (minimal) fitness value in each generation.
- The second plot shows the elite individuals of the most recent generation.
- The third plot shows the diversity.

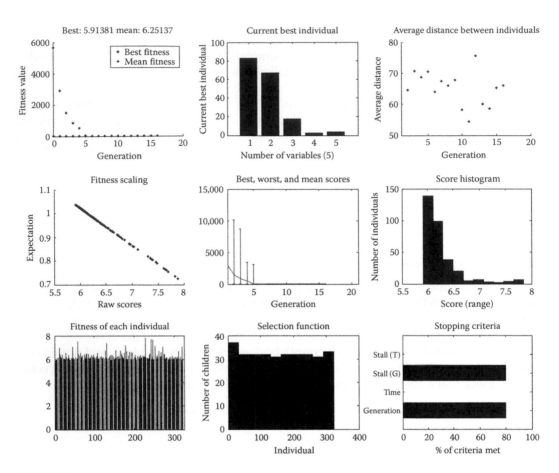

FIGURE 20.2
Charts showing status of running simulation.

Middle row:

- The fourth plot shows the ratio of the expected fitness obtained as a function of score.
- The fifth plot shows the mean, best, and worst scores of individuals in each generation.
- The sixth plot is a histogram of scores achieved by all individuals.

Bottom row:

- The seventh plot shows the fitness values of each individual in the generation.
- The eight plot shows the number of children for each individual.
- The ninth plot shows the simulation termination conditions. A simulation can be stopped once a sufficient number of subsequent stalled generations occurs or once the upper limit of time elapsed of generations is reached (Figure 20.2).

20.2.8 Results

Once the optimization routine has terminated, the results are stored in the variable *resultArray*. The user can access the general results by exploring that variable. The screenshot that follows shows the case for a single pass of the Cranfield cycle to optimize for WtW CO_2 emission choosing between conventional IC engine, EV, PHEV, and FC + UC.

```
resultArray =

          cycleNumber: 1
                value: 5.9138
              wtw_CO2: 5.9138
              ttw_CO2: 5.9138
             conScale: 83
             stoScale: 67
            conScale2: 18
            stoScale2: 3
              vehMass: 1.4843e+03
              vehCost: 1.0426e+04
    powertrainVariant: 4
              in_simu: [1×1 struct]
```

It is important to note that the size of energy storages and converters are scaled from 0 to 100. The user can convert the normalized values back to physical quantities with the data from Table 19.5 in the previous chapter.

From the preceding screenshot the results can be read as can read as follows:

- CO_2 emitted: 5.91 kg
- Power train variant: 4, which is a fuel cell with an ultracapacitor
- 1st energy converter scale: 83, which corresponds to electric motor power rating at 83 kW ($90 * 0.83 + 10$)
- 1st energy storage scale: 67, which corresponds to an electric energy storage of 104 kWh ($140 * 0.67 + 10$)
- 2nd energy converters: 18, which corresponds to a fuel cell power rating of 26 kW ($90 * 0.18 + 10$)
- 2nd energy storage: 3, which corresponds to fuel tank storing 12.7 L ($90 * 0.03 + 10$)
- Vehicle mass: 1484 kg
- Vehicle cost: $10,430.

The *in_simu* array shows further details of the results, with the most important being found in *in_sumu.in* (where all the power train component parameters are stored). Furthermore, if the user, for instance, wants to see the parameters of the electric motor, these can be found in *resultArray.in_simu.in.em*. Other parameters can be explored in a fill-in suggestion using the TAB button.

```
resultArray.in_simu.in.em =

         s: [1×100 double]
         t: [1×150 double]
```

```
    eff: [150×100 double]
    pow: 83000
  peakT: [1×100 double]
  peakS: 1.0023e+03
   mass: 120
   cost: 2.2261e+03
```

In the electric motor case, we can see the confirmation of an earlier estimate, that is, that the motor has a power of 83 kW, is estimated to weigh 120 kg, and is expected to cost $2200.

20.3 Capabilities and Limitations of Software

The provided toolbox allows the user to find optimal topologies and component sizes for given vehicle applications. It utilizes a GA to generate topologies, which are then evaluated with a backward-facing model. Hence, the simulation time step is large (1 s), and transient behavior cannot be captured. Another limitation of the software is the simplicity of the models used. Such models can be replaced by the user with other, more complex ones, but they need to be scalable for the optimization to work.

Appendix: Technical Overview of Toyota Prius

More and more hybrid vehicle products are being introduced to the market. Among them, Toyota Prius was the pioneer and has the largest number of units on the road. To give the reader a case study of a commercially successful hybrid vehicle, the Toyota Prius technology is described in this appendix. This appendix reviews the important technical features of this product, including the architecture, control, and component characteristics. The main resource for this material is autoshop101.com (http://www.autoshop101.com/forms/Hybrid01.pdf). However, the diagrams have been redrawn. We gratefully acknowledge the use of material available in autoshop101.com.

A.1 Vehicle Performance

Prius is a Latin word meaning "to go before." When the Prius was first released, it was selected as the world's best-engineered passenger car for 2002. The car was chosen because it is the first hybrid vehicle that seats four to five people plus their luggage, and it is one of the most economical and environmentally friendly vehicles available. In 2004, the second-generation Prius won the prestigious Motor Trend Car of the Year award and Best-Engineered Vehicle of 2004.

Both the Toyota Hybrid System (THS) power train in the original Prius and the Toyota Hybrid System II (THSII) power train in the second-generation Prius provide impressive electric power steering (EPS) fuel economy numbers and extremely clean emissions, as shown in Table A.1.

A.2 Overview of Prius Hybrid Power Train and Control Systems

The hybrid power train of Toyota Prius uses the series–parallel architecture as discussed in Chapters 6 and 10. Figure A.1 shows an overview of the hybrid power train and control systems. The hybrid system components include the following:

- Hybrid transaxle, consisting of motor/generator 1 (MG1), motor/generator 2 (MG2), and a planetary gear unit (refer to Figure A.3 for more details)
- 1NZ-FXE engine
- Inverter assembly containing an inverter, a booster converter, a DC–DC converter, and an AC inverter
- Hybrid vehicle electronic control unit (HV ECU), which gathers information from the sensors and sends calculated results to the engine control module (ECM), inverter assembly, battery ECU, and skid control ECU to control the hybrid system
- Shift position sensor

TABLE A.1

EPA Fuel Economy and Emissions

THS (2002–2003 Prius)		THS-II (2004 and Later)	
City	52 mpg	City	60 mpg
Highway	45 mpg	Highway	51 mpg
SULEV		AT-PZEV	

Note: SULEV standards are about 75% more stringent than ULEV and nearly 90% than LEV for smog forming exhaust gases. Super ultra-low emission vehicles (SULEVs) emit less than a single pound of hydrocarbons (HCs) during 100,000 miles of driving (about the same as spilling a half quarter of gasoline). Advanced technology partial zero emission vehicles (AT-PZEVs) use advanced technology capable of producing zero emissions during at least part of the vehicle's drive cycle.

- Accelerator pedal position sensor, which converts accelerator angle into an electrical signal
- Skid control ECU that controls regenerative braking
- ECM
- High-voltage (HV) battery
- Battery ECU, which monitors the charging condition of the HV battery and controls cooling fan operation
- Service plug, which shuts off system
- System main relay (SMR) that connects and disconnects the HV power circuit
- Auxiliary battery, which stores 12-V DC for vehicle's control systems

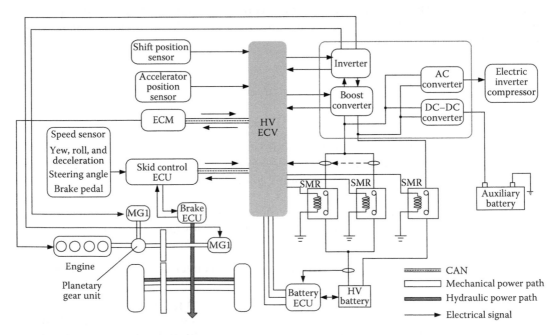

FIGURE A.1
Overview of Prius power train and control systems.

A.3 Major Components

A.3.1 Engine

The 1NZ-FXE engine is a 1.5-L inline four-cylinder gasoline engine with variable valve timing intelligence (VVTi) and electric throttle control system with intelligence (ETCS-i). In the 2004 and later models, there is a special coolant heat storage system that recovers hot coolant from the engine and stores it in an insulated tank where it stays hot for up to three days. Later, an electric pump precirculates the hot coolant through the engine to reduce HC emission normally associated with a cold start.

Table A.2 shows the specifications of the 1NZ-FEX engine.

TABLE A.2

Specifications of 1NZ-FEX Engine

Model	2004 Prius	2003 Prius
Engine type	1NZ-FXE	←
No. of cycles and arrangement	Four-cylinder, in-line	←
Value mechanism	Sixteen-value DOHC, chain drive (with VVTi)	←
Combustion chamber	Pentroof type	←
Manifolds	Cross-flow	←
Fuel system	SFI	←
Displacement (cm^3) (cu. in.)	1497 (91.3)	←
Bore × Stroke (mm) (in.)	75 × 84.7 (2.95 × 3.33)	←
Compression ratio	13.0:1	←
Max. output (SAE-Net)	57 kW at 5000 rpm	52 kW at 4500 rpm
	(76 hp at 5000 rpm)	(70 hp at 4500 rpm)
Max. torque (SAE-Net)	111 Nm at 4200 rpm	←
	(82 ft.lbf at 4200 rpm)	
Value Timing		
Intake		
Open	18° to −15° BTDC	18° to −25° BTDC
Close	72° to −105° ABDC	18° to −15° ABDC
Exhaust		
Open	34° BBDC	←
Close	34° ATDC	←
Firing order	1-3-4-2	←
Research octane number	91 or higher	←
Octane rating	87 or higher	←
Engine service mass[a] (kg) (lb) (References)	86.1 (198.8)	86.6 (190.9)
Oil grade	API SJ, SL, EC, or ILSAC	API SH, SJ, EC or ILSAC
Tail emission regulation	SULEV	←
Evaporative emission regulation	AT-PZEV, ORVR	LEV-II, ORVR

[a] Weight shows the figure with the oil and the engine coolant fully filled.

A.3.2 Hybrid Transaxle

Referring to Figure A.2, the hybrid transaxle contains

- MG1 that generates electric power
- MG2 that drives the vehicle
- A planetary gear unit that provides continuously variable gear ratios and serves as a power splitting device
- A reduction unit consisting of a silent chain, counter gears, and final gears
- A standard two-pinion differential

Table A.3 shows the main parameters of the hybrid transaxle. Table A.4 shows the specifications of MG1 and MG2.

A.3.3 HV Battery

The HV batteries are Ni–MH. Six 1.2-V cells connected together in series constitute a battery module that has a voltage of 7.2 V.

In the 2001–2003 Prius, 38 battery modules are divided into 2 holders and connected in series and have a rated voltage of 273.6 V.

In the 2004 and later models, 28 battery modules are connected for a rated voltage of 201.6 V. The cells are connected in two places to reduce the internal resistance of the battery.

Table A.5 shows the HV battery information.

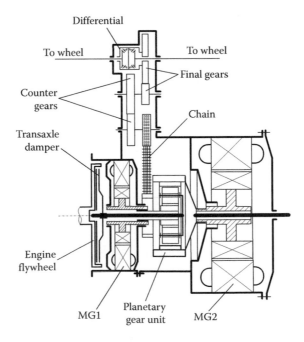

FIGURE A.2
Schematic illustration of hybrid transaxle.

TABLE A.3

Main Parameters of Transaxle

	2004 Model	2003 Model
Transaxle type	P112	P111
Planetary Gear		
Number of ring gear teeth	78	←
Number of pinion gear teeth	23	←
Number of sun gear teeth	30	←
Differential gear ratio		
Chain		
Number of links	72	74
Drive sprocket	36	39
Drive sprocket	35	36
Counter Gear		
Drive gear	30	←
Driven gear	35	←
Final Gear		
Drive gear	26	←
Driven gear	75	←
Fluid capacity		
Liters (U.S. qts, lmp qts)	3.8 (4.0, 3.3)	4.6 (4.9, 4.0)
Fluid type	ATF WS or equivalent	ATF type T-IV or equivalent

TABLE A.4

Specification of MG1 and MG2

MG1 Specification	2004 Model	2003 Model
Item		
Type	Permanent magnet motor	
Function	Generate engine starter	
Maximum voltage (V)	AC 500	AC 273.6
Cooling system	Water-cooled	
MG2 specification	2004 model	2003 model
Item		
Type	Permanent magnet motor	
Function	Generate engine starter	
Maximum voltage (V)	AC 500	AC 273.6
Maximum output kW(PS)/rpm	50 (68)/1200–1540	33 (45)/1040–5600
Maximum torque Nm (kgf · m)/rpm	400 (40.8)/0–1200	350 (35.7)/0–400
Cooling system	Water cooled	

TABLE A.5

HV Battery Information

HV Battery Pack	2004 and Later	2001–2003
Battery pack voltage	206.6 V	273.6 V
Number of Ni–MH battery modules in pack	28	38
Number of cells	168	228
Ni–MH battery module voltage	7.2	←

The battery ECU provides the following functions:

- It estimates the charging/discharging amperage and the output charge and discharge requests to the HV ECU so that the state of charge (SOC) can be constantly maintained at a medium level.
- It estimates the amount of heat generated during charging and discharging and adjusts the cooling fan to maintain HV battery temperature.
- It monitors the temperature and voltage of the battery, and if a malfunction is detected, it can restrict or stop charging and discharging to protect the HV battery.

The SOC of the battery is controlled by the HV battery ECU. The target SOC is 60%. When the SOC drops below the target range, the battery ECU informs the HV ECU and then signals the engine ECM to increase its power to charge the HV battery. The normal low to high SOC deviation is 20% as shown in Figure A.3.

The HV battery is air cooled. The battery ECU detects battery temperature via three temperature sensors in the HV battery and one intake air temperature sensor. Based on their readings, the battery ECU adjusts the duty cycle of the cooling fan to maintain the temperature of the HV battery within the specified range.

Three switched reluctance motor (SRMs) are used to connect or disconnect power to the HV circuit based on commands from the HV ECU. Two SRMs are placed on the positive side, and one is placed on the negative side, as shown in Figure A.4.

When the circuit is energized, SMR_1 and SMR_3 are turned on. The resistor in line with SRM1 protects the circuit from excessive initial current (called inrush current). Next, SRM2 is turned on and SRM1 is turned off.

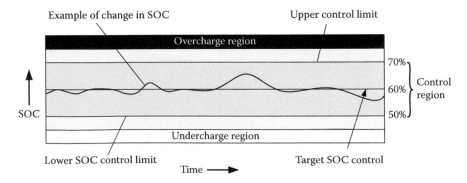

FIGURE A.3
Battery SOC control region.

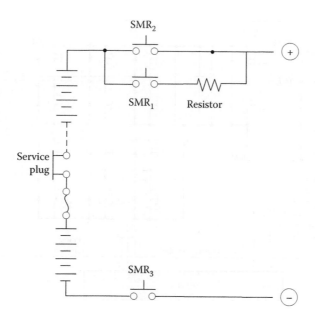

FIGURE A.4
SMRs and service plug.

When de-energized, SMR_2 and SMR_3 are turned off in that order, and the HV ECU verifies that the corresponding relays have been properly turned off.

A service plug is placed between the two battery holders. When the service plug is removed, the HV circuit is shut off. The service plug assembly also contains a safety interlock reed switch. Lifting the clip on the service plug opens the reed switch, shutting off the SMR. There is also a main fuse for the HV circuit within the service plug assembly.

Toyota Prius uses an absorbed glass mat 12-V maintenance-free auxiliary battery. This 12-V battery powers the vehicle's electrical system similar to a conventional vehicle.

A.3.4 Inverter Assembly

The inverter assembly includes an inverter, a booster converter, a DC/DC converter, and an AC converter, as shown in Figures A.1 and A.5.

A.3.4.1 Booster Converter (2004 and Later)

The booster converter boosts the nominal voltages of 206.1 V DC that is the output of the HV battery to a maximum voltage of 500 V DC. To boost the voltage, the converter uses a boost integrated power module with a built-in insulated-gate bipolar transistor (IGBT) for switching control and a reactor to store the energy, as shown in Figure A.5.

When MG1 or MG2 acts as a generator, the AC inverter, generated by either motor, is converted to DC. Then the booster converter drops the voltage to 201.6 V DC to charge the HV battery.

A.3.4.2 Inverter

The inverter changes the HV DC from the HV battery into three-phase AC for MG1 and MG2, as shown in Figure A.5. The HV ECU controls the activation of the power transistors.

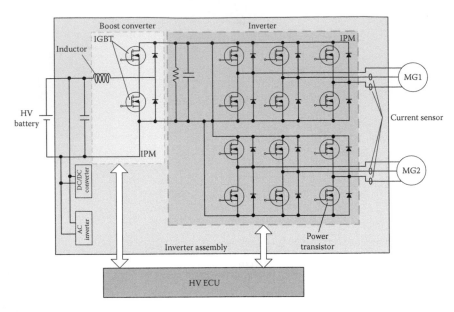

FIGURE A.5
Inverter assembly.

In addition, the inverter transmits information that is needed to control the current, such as the output amperage or voltage, to the HV ECU.

The inverter, MG1, and MG2 are cooled by a dedicated radiator and coolant system that is separated from the engine coolant system. The HV ECU controls the electric water pump for this system.

A.3.4.3 DC/DC Converter

A DC/DC converter is used to transform the HV into 12 V to recharge the 12-V auxiliary battery. The structure of the DC/DC converter is shown in Figure A.6. In the 2001–2003

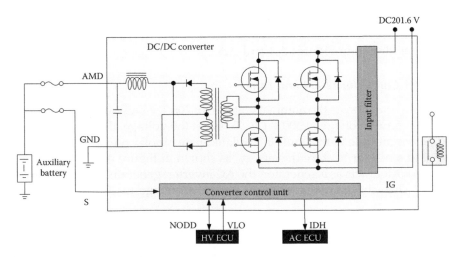

FIGURE A.6
DC/DC converter.

models, it transforms 273.6 V DC to 12 V DC. In the 2004 and later models, it transforms 201.6 V DC to 12 V DC.

A.3.4.4 AC Inverter

The inverter assembly in the 2004 and later models includes a separate inverter for the air conditioning system that changes the HV battery's nominal voltage of 201.6 V DC into 206.6 V AC to power the air conditioner system's electric motor, as shown in Figure A.7.

A.3.5 Brake System

The hybrid vehicle brake system includes both standard hydraulic brakes and a regenerative braking system that uses the vehicle's kinetic energy to recharge the battery. As soon as the accelerator pedal is released, the HV ECU initiates the regenerative braking. MG2 is turned by the wheels and used as a generator to recharge the battery. During this phase of braking, the hydraulic brakes are not used. When more rapid deceleration is required, the hydraulic brakes are activated to provide additional stopping power. To increase energy efficiency, the system uses the regenerative brakes whenever possible. Selecting "B" on the shift lever maximizes regenerative efficiency and is useful for controlling downhill speeds. The overall structure of the hybrid brake system is shown in Figure A.8.

A.3.5.1 Regenerative Brake Cooperative Control

Regenerative brake cooperative control balances the brake force of the regenerative and the hydraulic brakes to minimize the amount of kinetic energy lost to heat and friction. It recovers the energy by converting it into electrical energy.

A.3.5.2 Electronic Brake Distribution Control (2004 and Later Models)

In the 2004 and later models, brake force distribution is performed under electrical control of the skid control ECU. The skid control ECU precisely controls the braking force in accordance with the vehicle's driving conditions.

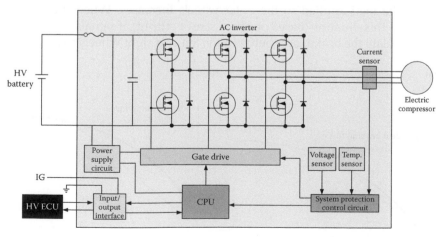

FIGURE A.7
AC inverter.

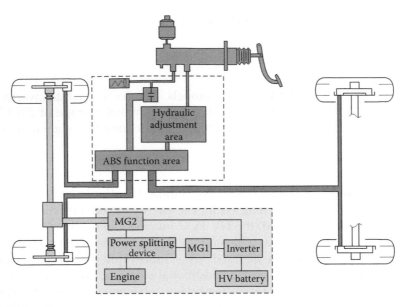

FIGURE A.8
The Hybrid brake system.

1. *Braking force distribution, front and rear (2004 and later models)*: Generally, when the brakes are applied, a vehicle's weight shifts forward, reducing the load on the rear wheels. When the skid control ECU senses this condition (based on the speed sensor output), it signals to the brake actuator to regulate the rear braking force so that the vehicle remains under control during the stop. The amount of brake force applied to the rear wheels varies based on the amount of deceleration. The amount of brake force that is applied to the real wheels also varies based on road conditions. Figure A.9a, b shows the braking force on the front and rear wheels without and with load on the rear wheels.

2. *Brake force distribution, left and right (2004 and later models)*: When the brakes are applied while a vehicle is cornering, the load applied to the inner wheels decreases, whereas the load applied to the outer wheels increases. When the skid control ECU senses this situation (based on speed sensor output), it signals to the brake actuator to regulate the brake force between the left and right wheels to prevent a skid.

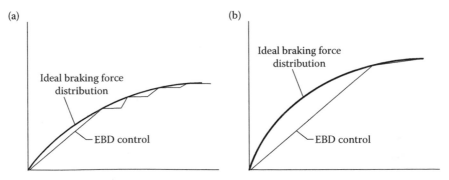

FIGURE A.9
Braking force on front and rear wheels: (a) without load on rear wheels and (b) with load on rear wheels.

A.3.5.3 Brake Assist System (2004 and Later Models)

In emergencies, drivers often panic and do not apply sufficiently fast pressure to the brake pedal, so in the 2004 and later models, a brake assist system (as shown in Figure A.10) is used to interpret a quick push of the brake pedal as emergency braking and supplements braking power accordingly.

To determine the need for an emergency stop, the skid control ECU looks at the speed and the amount of brake pedal application based on signals from the master cylinder pressure sensors and the brake pedal stroke sensor. If the skid control ECU determines that the driver is attempting an emergency stop, it signals to the brake actuator to increase the hydraulic pressure.

A key feature of the brake assist system is that the timing and degree of braking assistance are designed to ensure that the driver does not discern anything unusual about the braking operation. As soon as the driver eases up on the brake pedal, the system reduces the amount of assistance it provides.

A.3.6 Electric Power Steering

A 12-V motor powers the EPS system so that steering feel is not affected when the engine shuts off. The EPS ECU uses torque sensor output along with information from the skid control ECU about vehicle speed and torque assist demand to determine the direction and force of the power assist. It then actuates the DC motor accordingly. The structure of the EPS system is shown in Figure A.11.

The EPS ECU uses signals from the torque sensor to interpret the driver's steering intention. It combines this information with data from other sensors regarding the current vehicle condition to determine the amount of steering assist that is required. It can then control the current to the DC motor that provides steering assist.

When the steering wheel is turned, torque is transmitted to the pinion, causing the input shaft to rotate. The torsion bar that links the input shaft and the pinion twists until the torque and reaction force equalize. The torque sensor detects the twist of the torsion bar and

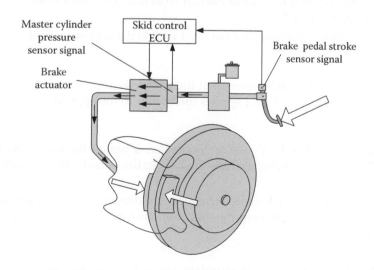

FIGURE A.10
Brake assist.

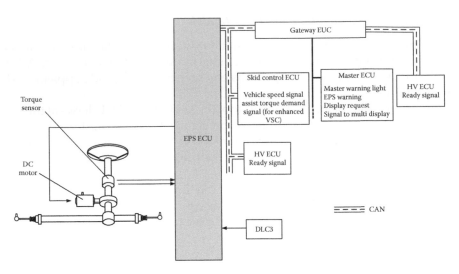

FIGURE A.11
The EPS system.

generates an electrical signal that is proportional to the amount of torque applied to the torsion bar. The EPS ECU uses that signal to calculate the amount of power assist that the DC motor should provide.

A.3.7 Enhanced Vehicle Stability Control (VSC) System (2004 and Later Prius)

The enhanced vehicle stability control (VSC) system in the 2004 and later models helps maintain stability when the vehicle's tires exceed their lateral grip. The system helps control the vehicle by adjusting the motive force and the brakes at each wheel when

- The front wheels lose traction, but the rear wheels do not (front wheel skid tendency known as "understeer,"
- The rear wheels lose traction, but the front wheels do not (rear wheel skid tendency, or "oversteer").

When the skid control ECU determines that the vehicle is in understeer or oversteer condition, it decreases engine output and applies the brakes to the appropriate wheels individually to control the vehicle.

- When the skid control ECU senses understeer, it brakes the front and rear inside wheels. This slows the vehicle, shifts the load to the outside front wheel, and limits front wheel skid.
- When the skid control ECU senses oversteer conditions, it brakes the front and rear outside wheels. This restrains the skid and moves the vehicle back toward its intended path.

Enhanced VSC also provides the appropriate amount of steering assist based on driving conditions by coordinating EPC and VSC control, as shown in Figure A.12.

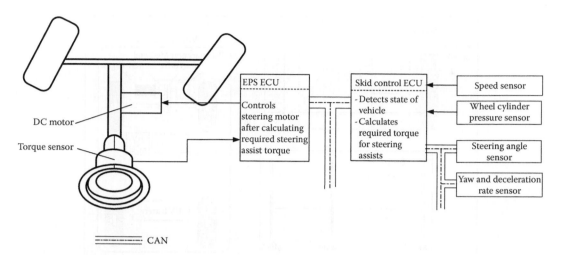

FIGURE A.12
Cooperative control with EPS.

A.4 Hybrid System Control Modes

The Toyota Prius hybrid system uses a series–parallel hybrid configuration that has many operation modes, as discussed in previous chapters. Prius uses the following control strategies:

1. When starting off and traveling at low speeds, MG2 provides the primary motive force. The engine may start immediately if the HV battery SOC is low. As the speed increases above 24–32 km/h (15–20 mph), the engine will start.

2. When driving under normal conditions, the engine's power is divided into two paths: a portion drives the wheels and a portion drives MG1 to produce electricity. The HV ECU controls the energy distribution ratio for maximum efficiency.

3. During full acceleration, the power generated by the engine and MG1 is supplemented by power from the HV battery. Engine torque combined with MG2 torque delivers the power required to accelerate the vehicle.

4. During deceleration or braking, the wheels drive MG2. MG2 acts as a generator for regenerative energy recovery. The recovered energy from braking is stored in the HV battery pack.

The operation modes of the engine, MG1, and MG2 are depicted in response to different driving conditions as follows.

Stopped: If the HV battery is fully charged and the vehicle is not moving, the engine may stop. The engine starts up automatically if the HV battery needs charging. Also, if MAX AC is selected in the 2001–2003 models, the engine runs continuously due to the engine-driven compressor. The 2004 and later models use an electrically driven compressor. Figure A.13 shows the depicted operation modes of the engine, MG1, and MG2.

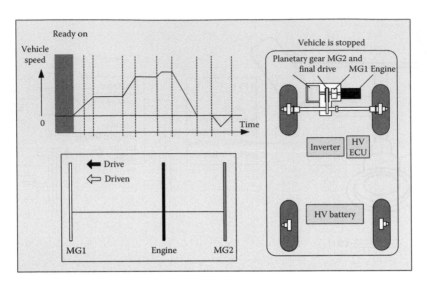

FIGURE A.13
Operation with stopped vehicle.

Starting out: When starting out under light load and light throttle, only MG2 turns to provide power. The engine does not run, and the vehicle runs on electric power only. MG1 rotates backward and just idles. It does not generate electricity, as illustrated in Figure A.14.

Engine starting: As the speed increases above 24–32 km/h (15–20 mph), the engine starts. The engine is started by MG1. The operations of the engine, MG1, and MG2 are shown in Figure A.15.

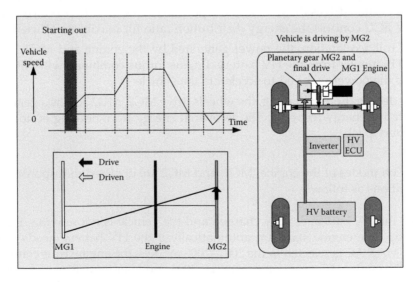

FIGURE A.14
Operation mode while starting out.

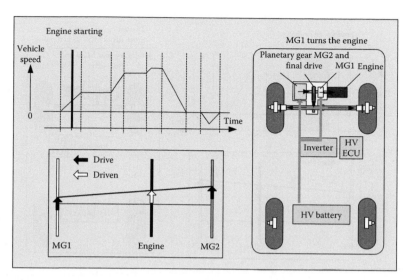

FIGURE A.15
Operation mode of engine starting.

Light acceleration with engine: In this mode, the engine delivers its power to the drive wheels and MG1, which is generating. MG2 may assist the engine in propulsion if required, depending on the engine power and the requested driving power. In this mode, the energy generated by MG1 may be equal to the energy delivered to MG2. The drivetrain operates as an continuously variable transmission (CVT). The operations of the engine, MG1, and MG2 are shown in Figure A.16.

Low-speed cruising: This mode is similar to the mode of light acceleration with the engine as shown in Figure A.17.

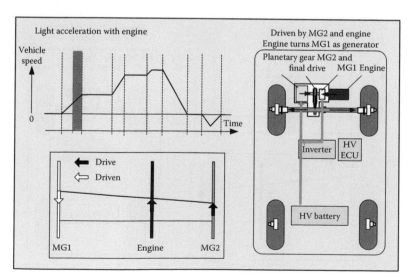

FIGURE A.16
Operation mode of light acceleration with engine.

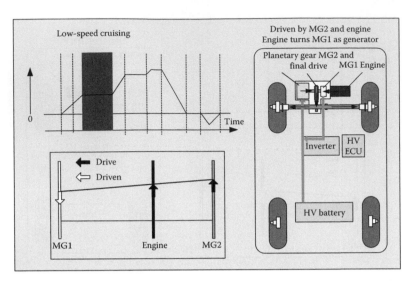

FIGURE A.17
Operation mode of low-speed cruising.

Full acceleration: In this mode, the engine delivers its power to the wheels and to MG1, which is in the generating mode. MG2 adds its power to the engine power and is delivered to the wheels, as shown in Figure A.18. The power drawn by MG2 from the HV battery power is greater than the power generated by MG1. Thus, the HV battery pack contributes energy to the drivetrain, and its SOC drops.

High-speed cruising: In this mode, the shaft of MG1 is fixed to the vehicle stationary frame, and the drivetrain is operated in pure torque-coupling mode. Both the engine and MG2 propel the vehicle, as shown in Figure A.19.

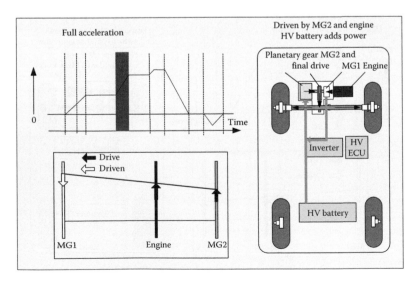

FIGURE A.18
Operation mode in full acceleration.

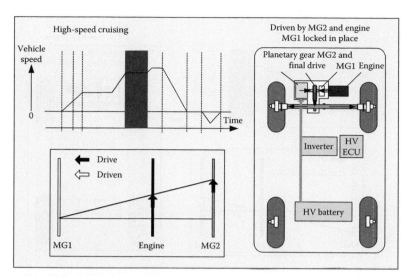

FIGURE A.19
Operation mode at high-speed cruising.

Driving at maximum speed: In this mode, both MG1 and MG2 receive power from the HV battery pack and deliver their mechanical power to the drivetrain. In this case, MG1 turns in the opposite direction, as shown in Figure A.20.

Deceleration or braking: When the vehicle is decelerating or braking, the engine is shut down. MG2 becomes a generator and is turned by the drive wheels and generates electricity to recharge the HV battery pack. The operation is shown in Figure A.21.

Reverse: When the vehicle moves in reverse direction, MG2 turns in the reverse direction as an electric motor. The engine is shut down. MG1 turns in the forward direction and just idles, as shown in Figure A.22.

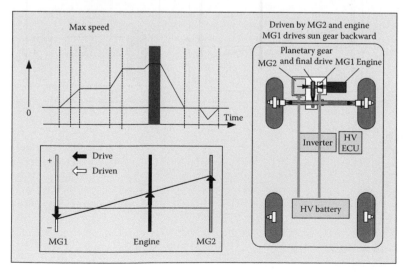

FIGURE A.20
Operation mode while driving at maximum speed.

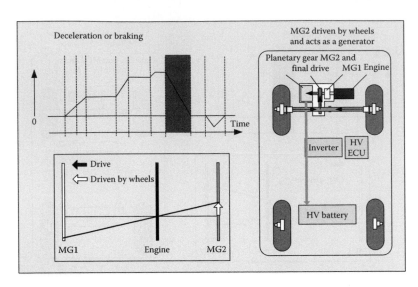

FIGURE A.21
Operation mode in deceleration or braking.

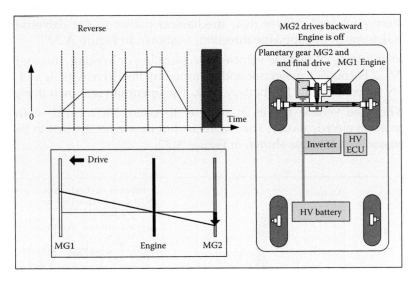

FIGURE A.22
Reverse operation.

Index

A

AB$_2$ alloys, 354
AB$_5$ alloys, 354
ABS, *see* Antilock brake system
Acceleration, 103–104
    performance, 31–34
    performance of vehicle, 102
    performance verification, 247
    time, 33
Accelerator pedal position sensor, 508
AC/DC converter, 437
ACEA, *see* European Automobile Manufacturers Association
Acid rain, 2, 7
AC inverter, 515
Acoustic noise in SRM, 218–220
Activated carbons, 358
Activation losses, 401
Active hybrid energy storage with battery and ultracapacitor, 370–371
Actuation system, 77
Adhesive coefficient, 40
Advanced flywheel, 363
Advanced lead–acid batteries, 353
Advanced vehicle simulator (ADVISOR), 478, 480
AER, *see* All electric range
AER-focused control strategy, 307; *see also* Blended control strategy
    energy consumption, 310
    energy consumption by drive wheels *vs.* driving distance, 309
    engine operating points overlapping its fuel consumption map, 311
    engine power *vs.* traveling distance in FTP75 urban drive cycle, 310
    fuel and electric energy consumption *vs.* number of FTP75, 312
    motor power *vs.* traveling distance in FTP75 urban driving cycle, 311
    powers of motor and energy storage, 309
    SOC and remaining energy in energy storage *vs.* traveling distance, 311
    traction power, 307
    vehicle parameters used in power computation, 308
    vehicle speed and traction power, 308
Aerodynamic drag, 21, 41, 431

AFCs, *see* Alkaline fuel cells
Air–fuel mixture, 63, 65–66
Air gap, 222
Air pollution, 1
    carbon monoxide, 2
    nitrogen oxides, 2
    pollutants, 2–3
    unburned HCs, 2
Alkaline fuel cells (AFCs), 406–409
Alkaline metal hydrides, 416
All electric range (AER), 306
Alnico, 189
Alternative fuels, 69
    compressed natural gas and natural gas engine, 70–72
    engines, 69–72
    enhanced hydrogen, 72
    ethanol and ethanol engine, 69–70
AM methods, *see* Amplitude modulation methods
Ammonia
    AFCs, 419
    as hydrogen carrier, 418
Amplitude modulation methods (AM methods), 212, 213
AMT, *see* Automated manual transmission
Angular velocity, 56, 265
    drivetrain, 474
    electrical, 155
    of flywheel, 363
    of ICE, 473
    mechanical, 156
    negative, 127, 287
    positive, 127
    of rotating stator mmf, 155
    of rotor, 156, 191, 340
ANNs, *see* Artificial neural networks
Antilock brake system (ABS), 42, 45
APU, *see* Auxiliary power unit
Arithmetic method, self-tuning with, 215
    optimization in presence of parameter variations, 216
    optimization with balanced inductance profiles, 216
Armature, 142
    circuit, 143–144, 151
    connection, 150, 151
    current, 142–146, 151, 152, 168, 169, 193